Designing Complex Products with Systems Engineering Processes and Techniques

Completely revised including six new chapters, this new edition presents a more comprehensive knowledge of issues facing developers of complex products and process management. It includes more tools for implementing a Systems Engineering approach to minimize the risks of delays and cost overruns and helps create the right product for its customers.

Designing Complex Products with Systems Engineering Processes and Techniques, Second Edition highlights how to increase customer satisfaction, quality, safety, and usability to meet program timings and budgets using a Systems Engineering approach. It provides decision-making considerations and models for creating sustainable product design and describes many techniques and tools used in product development and the product life-cycle orientation. The book also offers techniques used in Design for Manufacturing, Design for Assembly, and product evaluation methods for verification and validation testing. Many new examples, case studies, six new chapters, and updated program and data charts held on our website are offered.

The book targets practicing engineers, engineering management personnel, product designers, product planners, product and program managers in all industrialized and developing countries. In addition, the book is also useful to undergraduate, graduate students, and faculty in engineering, product design, and product project and program management.

Designing Complex Products with Systems Engineering Processes and Techniques

Second Edition

Vivek D. Bhise

CRC Press
Taylor & Francis Group
Boca Raton London New York

CRC Press is an imprint of the
Taylor & Francis Group, an **informa** business

Second edition published 2023
by CRC Press
6000 Broken Sound Parkway NW, Suite 300, Boca Raton, FL 33487-2742

and by CRC Press
4 Park Square, Milton Park, Abingdon, Oxon, OX14 4RN

CRC Press is an imprint of Taylor & Francis Group, LLC

© 2023 Vivek D. Bhise

First edition published by CRC Press 2013

Reasonable efforts have been made to publish reliable data and information, but the author and publisher cannot assume responsibility for the validity of all materials or the consequences of their use. The authors and publishers have attempted to trace the copyright holders of all material reproduced in this publication and apologize to copyright holders if permission to publish in this form has not been obtained. If any copyright material has not been acknowledged please write and let us know so we may rectify in any future reprint.

Except as permitted under U.S. Copyright Law, no part of this book may be reprinted, reproduced, transmitted, or utilized in any form by any electronic, mechanical, or other means, now known or hereafter invented, including photocopying, microfilming, and recording, or in any information storage or retrieval system, without written permission from the publishers.

For permission to photocopy or use material electronically from this work, access www.copyright.com or contact the Copyright Clearance Center, Inc. (CCC), 222 Rosewood Drive, Danvers, MA 01923, 978-750-8400. For works that are not available on CCC please contact mpkbookspermissions@tandf.co.uk

Trademark notice: Product or corporate names may be trademarks or registered trademarks and are used only for identification and explanation without intent to infringe.

ISBN: 978-1-032-20369-0 (hbk)
ISBN: 978-1-032-20371-3 (pbk)
ISBN: 978-1-003-26335-7 (ebk)

DOI: 10.1201/9781003263357

Typeset in Times
by SPi Technologies India Pvt Ltd (Straive)

Support Material: https://www.routledge.com/9781032203690

Contents

Preface for the First Edition xxiii
Preface for This Second Edition xxv
Website Materials xxvii
Acknowledgments xxix
Author xxxi

Part I Systems Engineering Concepts, Issues, and Methods in Product Design

Chapter 1 Introduction to Products, Processes, and Product Development 3

    Introduction and Objectives 3
    Understanding Products, Customers, Processes, and Systems 4
        What Is a Product? 4
        Who Is the Customer? 5
        What Are Customer Needs? 6
        What Is a Process? 6
    Designing a Complex Product 9
        Definition of a System 9
        Systems, Subsystems, and Components 10
        Systems Work with Other Systems 11
        Product Families and Component Sharing 11
        Product Development 12
            Processes in Product Development 12
            Flow Diagram of Product Development 14
            Managing the Complex Product 15
        Life Cycle Stages of a Product 17
            Program Phases, Reviews, and Milestones 17
    Concluding Remarks 20
    References 21

Chapter 2 Systems Engineering and Other Disciplines in Product Design 23

    Introduction 23
    Systems Engineering Fundamentals 23
        What Is Systems Engineering? 23
        Managing a Complex Product 26
    Systems Engineering Processes in Product Development 27
        Systems Engineering Process 27
        Five Loops in the Systems Engineering Process 28

	Major Tasks in the Systems Engineering Process 30
	Requirements Analysis 30
	Functional Analysis and Allocation 31
	Design Synthesis 31
	Verification 32
	Validation 32
	Verification versus Validation 33
	Subsystems and Components Development 33
	Example of Cascading a Requirement from the Product Level to a Component Level 34
	Iterative Nature of the Loops within the Systems Engineering Process 35
	Incremental and Iterative Development Approach 35
	Systems Engineering "V" Model 38
	NASA Description of the Systems Engineering Process 40
	Managing the Systems Engineering Process 42
	Relationship between Systems Engineering and Program Management 42
	Role of Systems Engineers 43
	Integrating Engineering Specialties into the Systems Engineering Process 44
	Role of Computer-Assisted Technologies in Product Design 45
	CAD and CAE 45
	Model-Based Systems Engineering 45
	Importance of Systems Engineering 46
	Advantages and Disadvantages of the Systems Engineering Process 47
	Some Challenges in Complex Product Development 47
	Concluding Remarks 48
	References 49
Chapter 3	Decision-Making and Risks in Product Programs 51
	Introduction 51
	Problem-Solving Approaches 52
	Decision-Making 53
	Alternatives, Outcomes, Payoffs, and Risks 53
	Maximum Expected Value Principle 53
	Other Principles 55
	Techniques Used in Decision-Making 58
	Analytical Hierarchical Method 58
	Weighted Total Score for Concept Selection 62
	Informational Needs in Decision-Making 63
	Decision-Making in Product Design 64
	Key Decisions in Product Life Cycle 64

Trade-Offs during Design Stages 64
Risks in Product Development and Product Uses 65
    Definition of Risk and Types of Risks in Product
    Development 68
    Types of Risks during Product Uses 69
Risk Analysis 69
    Risk Matrix 70
    Risk Priority Number and Nomographs 71
    Problems in Risk Measurements 72
Importance of Early Decisions during Product Development 73
Concluding Remarks 73
References 74

Chapter 4   Product Attributes, Requirements, and Allocation of
Functions 75

Introduction 75
Attributes and Requirements 75
    What Is an Attribute? 75
    Importance of Attributes 76
    What Is a Requirement? 76
    Attribute Requirements 77
    Why "Specify" Requirements? 77
    How Are Requirements Developed? 77
    Characteristics of a Good Requirement 78
Types of Requirements 79
    Customer Requirements 79
    Functional Requirements 79
    Performance Requirements 80
    Interface Requirements 80
    Reliability Requirements 80
    Environmental Requirements 80
    Human Factors Requirements 81
    Safety Requirements 81
    Security Requirements 81
    Designed-to-Conform versus Manufactured-to-Conform
    Requirements 81
    Where Are Requirements Stored? 82
Requirements Allocation and Analysis 82
    Requirements Allocation 82
    Requirements Analysis 82
Attributes Development 83
    Cascading Attribute Requirements to Lower Levels 84
    Dividing the Product into Manageable Levels 85
Relating Attribute Structure to Systems 86

	An Example: Attributes, System Decomposition, and	
	Requirements for Vehicle Exterior Lighting System	87
	Attributes	88
	Systems and Subsystems	88
	Relationship between System Components and	
	Requirements	89
	Requirements of Exterior Lighting System	89
	Verification Tests	89
	An Example: Cascading of Vehicle Level Sub-attribute	
	Requirements into Powertrain Subsystem Requirements	91
	An Example: Attributes, Requirements, and Trade-Offs in	
	Suspension Systems of a Sports Car	91
	Attributes	91
	Requirements	91
	Trade-Offs	92
	Factors Affecting Requirements	93
	Role of Standards in Setting Requirements	94
	Types of Standards	94
	Advantages of Standards	95
	Disadvantages of Standards	96
	Problems with Standards	96
	Standards Development Process	97
	Concluding Remarks	98
	References	98
Chapter 5	Understanding and Managing Interfaces	99
	Introduction	99
	Interface Definition, Types, and Requirements	99
	What Is an Interface?	99
	Types of Interfaces	100
	Interface Requirements	102
	Visualizing Interfaces	103
	Interface Diagram	104
	Interface Matrix and N-Squared Diagram	104
	Examples of Interface Diagrams and Interface Matrices	106
	Laptop Computer Interfaces	106
	Automotive Fuel System Interfaces	110
	Illustration of Use of Information Contained in Interface	
	Matrix	113
	Clustering and Sequencing of Matrix Data	115
	Teamwork in Interface Management	118
	Establishment of Interface Control	119
	Concluding Remarks	119
	References	120

Contents ix

Chapter 6 Detailed Engineering Design during Product Development 121

    Introduction 121
    Engineering Design 121
    Six Product Examples.. 123
    Illustration of Wind Turbine Design Using Systems
    Engineering "V" Model... 129
        Left Side of the "V"—Design and Engineering 129
        Right Side of the "V"—Verification, Manufacturing, and
        Assembly 131
        Right Side of the Diagram—Operation and Disposal 131
    Activities in Engineering Design... 132
    Concluding Remarks 135
    References 135

Chapter 7 Product Evaluation, Verification, and Validation............................. 137

    Objectives and Introduction... 137
    Why Evaluate, Verify, and Validate? 137
        Testing, Verification, and Validation 137
        Distinctions between Product Verification and Product
        Validation 138
    Overview on Evaluation Issues.. 138
        Types of Evaluations... 140
    Evaluation Methods: An Overview... 141
    Methods of Data Collection and Analysis 143
        Observation Methods.. 143
        Communication Methods.. 143
        Experimentation Methods... 144
    Objective Measures and Data Analysis Methods 144
    Subjective Methods and Data Analysis 145
        Rating on a Scale 145
        Analysis of 10-Point Ratings Data 150
        Paired Comparison-Based Methods.. 151
    Evaluations during Product Development 152
        Verification Plan and Tests... 153
        Validation Plan and Tests... 154
    Concluding Remarks 158
    References 158

Chapter 8 Program Planning and Management 159

    Introduction 159
    Program versus Project Management 159
        Program Management Functions.. 160
        Development of Detailed Project Plans.................................... 160

Project Management 161
   Steps in Project Planning 162
Tools Used in Project Planning 162
   Gantt Chart 162
   Critical Path Method 162
   Program (or Project) Evaluation and Review Technique 164
   Work Breakdown Structure 165
   Project Management Software 166
   Other Tools 166
Systems Engineering Management Plan 166
   Contents of SEMP 167
   Checklist for Critical Information 170
   Role of Systems Engineers 171
   Value of Systems Engineering Management Plan 171
Complexity in Program Management 171
   Time Management 172
   Cost Management 172
   Challenges in Program Management 172
Concluding Remarks 174
References 174

Chapter 9   Costs and Benefits Considerations and Models 175

Introduction 175
Types of Costs 175
   Nonrecurring and Recurring Costs 175
      Nonrecurring Costs 176
      Recurring Costs 176
   Revenues Buildup over Time as the Product Is Sold 176
   Make versus Buy Decisions 177
   Fixed versus Variable Costs 179
   Quality Costs 180
   Manufacturing Costs 180
   Safety Costs 181
   Product Termination Costs 182
   Total Life Cycle Costs 182
Effect of Time on Costs 182
Benefits Estimation 183
Project Financial Plan 183
   An Example: Automotive Product Program Cash Flow 183
   Effect of Interest and/or Inflation 188
Product Pricing Approaches 188
   Traditional Costs-Plus Approach 188
   Market Price-Minus Profit Approach 189
   Software Applications 190
   Trade-Offs and Risks 190

Concluding Remarks 190
References 190

Part II Quality, Human Factors, Safety, and Sustainability Approaches

Chapter 10 Quality Management and Six-Sigma Initiatives 193

Introduction 193
Definition of Quality... 193
Key Concepts in Quality Management... 194
    Quality Gurus and Their Findings 194
Product Quality Measurements 195
Customer Satisfaction and the Kano Model of Quality................... 196
Quality Initiatives 198
    Total Quality Management 198
ISO 9000.. 199
    Malcolm Baldridge Award Criteria...200
    Six-Sigma Methodologies 200
Overview of Tools Used in Quality Management 201
Concluding Remarks 202
References 202

Chapter 11 Human Factors Engineering in Product Design 205

Introduction 205
Human Factors Engineering...206
    What Is It?...206
    Human Factors Engineering Approach.......................................206
    Human Factors Research Studies 209
Human Factors Engineer's Responsibilities in Designing
Complex Products...210
Importance of Human Factors Engineering...211
    Characteristics of Ergonomically Designed Products 211
    Why Apply Human Factors Engineering?...................................211
    Human Factors Engineering Is Not Commonsense...................212
A Brief Overview of Human Characteristics and Capabilities........212
    Physical Capabilities..212
    Information Processing Capabilities...213
    Other Factors Affecting Human Capabilities.............................214
    Percentile Values 214
Human Errors 214
    Definition of an Error..215
    Types of Human Errors..215

Human Interface 216
User Performance Measurements 217
    Types and Categories of User Performance Measures 218
    Characteristics of Effective Performance Measures 218
Human Factors Methods: An Overview 220
Considerations in the Applications of Human Factors
Guidelines 221
Concluding Remarks 221
References 224

Chapter 12 Safety Engineering in Product Design 225

Introduction 225
Background: Safety Engineering 225
    Definition of Safety Engineering 225
    Safety Problems 226
    Importance and Need of Safety Engineering 227
    3Es of Safety Engineering and Countermeasures 228
    Methods Used in Safety Engineering 228
    Historic Background 229
Definition of an Accident 230
Accident Causation Theories 231
Safety Performance Measures 234
    Why Measure Safety Performance? 234
    Currently Used Accident Measures 234
    Accident-Based Incident Rates 235
    Advantages and Disadvantages of Current
    Accident-Based Measures 236
    Non-accident Measures 236
Safety Analysis Methodologies 237
    Two Possibilities: Accident versus Hazard 237
    Accident Analysis Methods 237
    Hazard Analysis Methods 238
Product Safety and Liability 239
    Terms and Principles Used in Product Litigations 239
    Product Defects 240
    Warnings 240
Safety Costs 241
Security Considerations in Product Design 242
Concluding Remarks 242
References 242

Chapter 13 Design for Sustainability 245

Introduction 245
    What Is Sustainability? 245
    What Is a Sustainable Product? 246

　　　　Life Cycle Consideration 246
　　Tools/Methods Used for Sustainability Analyses 247
　　　　Design for Environment 247
　　　　Design for Disassembly 249
　　　　Goal of DFD 249
　　DFD and DFA Guidelines 250
　　　　Use of Fasteners 251
　　Recycling and Material Recovery 251
　　　　Selection and Use of Materials 252
　　　　Product Design Guidelines for Recycling 252
　　　　Design for Active Disassembly 253
　　Concluding Remarks 253
　　References 253

Part III Tools Used in Product Development, Quality, Human Factors, and Safety Engineering

Chapter 14　Methods and Toolbox 257

　　Introduction 257
　　Overview of Methods 257
　　Classification of Methods 258
　　　　Observation Methods 258
　　　　Communication Methods 261
　　　　Experimentation Methods 261
　　　　Data Presentation Methods 262
　　Methods in Product Development, Quality, Human Factors,
　　　Safety, and Program Management 262
　　Integration of Tools in Applications 263
　　Concluding Remarks 264
　　References 265

Chapter 15　Product Development Tools 267

　　Introduction 267
　　Benchmarking and Breakthrough 267
　　　　Benchmarking 268
　　　　Breakthrough 276
　　　　Differences between Benchmarking and Breakthrough 277
　　Pugh Diagram 277
　　　　An Example of Pugh Diagram Application 278
　　Quality Function Deployment 280
　　　　An Example of the Quality Function Deployment Chart 285
　　　　Cascading Quality Function Deployments 287

Advantages and Disadvantages of Quality Function
Deployment 287
Failure Modes and Effects Analysis 289
An Example of a Failure Modes and Effects Analysis 290
Failure Modes and Effects and Criticality Analysis 291
Other Product Development Tools 294
Business Plan 294
Program Status Chart 295
Standards 296
Model-Based Systems Engineering 296
Computer-Aided Design Tools 299
Prototyping and Simulation 300
Physical Mock-Ups 300
Technology Assessment Tools 301
Concluding Remarks 301
References 301

Chapter 16 Design for Manufacturing and Assembly 303

Introduction 303
Design, Functioning, Manufacturing, and Assembly 303
Principles of DFMA 304
Materials, Manufacturing and Assembly Considerations: An
Example of IC Engine Piston 305
Manufacturing and Assembly Considerations 306
Manufacturing and Assmbly Costs 307
Assembly Engineer's Recommendations to Component
Designers for Assembly Cost Reductions 307
Methods to Estimate Assembly Time 310
Methods-Time Measurement 310
Boothroyd et al. Assembly Evaluation Methods 311
Boothroyd et al. Manual Assembly Evaluation Method 311
Other Boothroyd et al. Assembly Evaluation Methods 312
An Example of Applications of MTM-1 vs. Boothroyd's
Manual Assembly Time Estimating Methods 312
Similarities between MTM-1 and Boothroyd et al. Manual
Assembly Methods 314
Dissimilarities between MTM-1 and Boothroyd et al.
Methods 316
Boothroyd et al. Methods for Estimating Assembly Times for
High-Speed Automatic Assembly and Robotic Assembly 317
High-Speed Automatic Assembly 317
Robotic Assembly 320
Concluding Remarks 320
References 320

Contents   xv

Chapter 17   Traditional and New Quality Tools 321

     Introduction 321
     Traditional Quality Tools 321
        Pareto Chart 321
           Purpose..321
           Description..321
           Example: Pareto Chart of Customer Complaints....................322
        Cause-and-Effect Diagram 323
           Purpose..323
           Description..323
           Example: C-E Diagram for Misaimed Headlamps 324
           Cause-and-Effect Process Diagram 325
        Check Sheet 326
           Purpose..326
           Description..326
           Example: Checklist for Door Trim Defects 327
           Example: Check Sheet for Defects in Painted Car Body........328
        Histogram 328
           Purpose..328
           Description..328
           Example: Histogram of Resistance of an Electrical
           Component..329
        Scatter Diagram 329
           Purpose..329
           Description..330
           Example: Scatterplot of Sitting Height versus Standing
           Height of 30 Human Operators...330
        Stratification..330
           Purpose..330
           Description..331
           Example: Stratification of Anthropometric Data by
           Gender...331
        Control Charts...331
           Purpose..331
           Description..332
        Some Examples of Control Charts 333
           Variables Control Charts 333
           Attributes Control Charts 336
     New Quality Tools..338
        Relations Diagram 338
           Purpose..338
           Description..338
           Example: Understanding Causation of Headlamp
           Misaim 339
        Affinity Diagram...341
           Purpose..341

Description 341
Example: Grouping Causes of Headlamp Misaim 341
Systematic Diagram 341
Purpose 341
Description 341
Example: Alternatives to Reduce Product Development
Time 343
Matrix Diagram 345
Purpose 345
Description 345
Example: Relationship between Vehicle Parameters and
Vehicle Performance 345
Matrix Data Analysis 346
Purpose 346
Description 347
Examples of Matrix Data Analysis 347
Process Decision Program Chart 347
Purpose 347
Description 347
Example: PDPC for Reducing Problems in a Product
Development Process 348
Arrow Diagrams 348
Purpose 348
Description 348
Examples 349
Experiment Design 349
An Example: Experiment to Select a Display with the
Highest Luminance 350
Multivariate Experiment Designs 352
Taguchi's Three-Step Product Design Approach 353
Taguchi's Product Robustness and Quadratic Costs 353
Taguchi Experiments 354
Concluding Remarks 355
References 356

Chapter 18 Human Factors Engineering Tools 357

Introduction 357
Databases on Human Characteristics and Capabilities 357
Anthropometric and Biomechanical Human Models 358
Human Factors Checklists and Scorecards 358
Checklist 359
An Example: A Checklist for Evaluation of an
Automotive Control 359
Scorecard 359
An Example: Ergonomic Scorecard for Automotive
Interior Evaluation 359

Task Analysis 362
  An Example: Task Analysis for Opening a Liftgate and
  Removing a Jack 365
Human Performance Evaluation Models..365
Laboratory, Simulator, and Field Studies 366
Human Performance Measurement Methods 367
  Range of Human Performance Measures 367
  Types and Categories of Human Performance Measures 368
  Examples of Behavioral Human Performance Measures 369
  Methods to Measure Human Operator Workload......................370
    Operator Performance Measurements....................................370
    Physiological Measurements 370
    Subjective Assessments..371
    Secondary Task Performance Measurement 373
Product Psychophysics 374
Concluding Remarks 375
References 375

Chapter 19 Safety Engineering Tools...379

Introduction 379
Hazard Identification and Risk Reduction Tools 379
  Hazard Analysis 379
  General Hazard Analysis 379
  Detailed Hazard Analysis 380
  Methods Safety Analysis 380
  Checklists to Uncover Hazards 381
  Risk Analysis 382
Systems Safety Analysis Tools 382
  Failure Modes and Effects Analysis 382
  Fault Tree Analysis 384
    Purpose...384
    Description...384
Accident Data Analysis Tools...390
  Purpose of Accident Data Collection 390
  Flow of Accident Data Collection 391
  Accident Data Reporting Thresholds...391
  Accident Investigations 391
  Accident Data Sources and Users...392
Safety Performance Monitoring, Evaluation, and Control 393
  Interview and Observational Techniques for Non-Accident
  Measurement of Safety Performance...393
  Critical Incident Technique 393
  Behavioral Sampling..394
  Control Charts 396
  Before versus after Studies 396
  Cost–Benefit Analysis...396

Reliability Analyses 396
  Definitions of Reliability and Maintainability 396
  Reliability of a Series System 397
  Reliability of a Parallel System 399
  Reliability of Hybrid Systems 400
  Designing for Reliability 400
  Approaches for Reliability Improvements 403
  A Reliability Engineer's Tasks 404
Concluding Remarks 405
References 405

Chapter 20 Cost–Benefit Analysis 407

Introduction 407
  Cost–Benefit Analysis: What Is It? 407
  Why Use Cost–Benefit Analysis? 407
  Steps Involved in Cost–Benefit Analysis 408
Some Examples of Problems for Application of Cost–Benefit
Analysis 408
Cost–Benefit Analysis of Residential Solar Panels:
An Example 410
  Problem 410
    Cost–Benefit Analysis and Calculations 412
      Installed Costs 412
      Operation and Maintenance Cost 415
      Insurance 415
      Present Value of Cost 415
      Avoided Electric Utility Cost 415
      SREC, Net Metering, and Tax Credit Revenue 416
      Net Present Value 417
      Conclusions of the Cost–Benefit Analyses 417
Exercising Cost–Benefit Model for Sensitivity Analysis 418
Risks and Uncertainties in Cost–Benefit Analysis 418
  Uncertainties 419
  Controversial Aspects 419
Concluding Remarks 420
References 420

Chapter 21 Life Cycle Analyses 423

Introduction 423
What Is Product Life Cycle? 424
Life Cycle Analysis 424
  Objectives of LCA 424
  LCA Impact Categories 426

Carbon Footprint 427
Four Phases of Life Cycle Assessment 427
Life Cycle Cost Analysis 429
Objectives of LCCA 429
Some Examples of LCA and LCCA Applications 429
Examples of LCA 429
Emissions from Automotive Products 429
Examples of LCCA 430
Cost–Benefit Analysis of Photovoltaic
Solar Panels 430
Levelized Cost of Technologies 430
Concluding Remarks 432
References 433

Part IV Applications, Case Studies, and Integration

Chapter 22 Applications of Systems Engineering Tools: A Case Study on an Automotive Powertrain System 437

Introduction 437
Automotive Powertrain Project 437
Project Objectives 437
Project Steps 437
Systems, Subsystems, and Sub-Subsystems 438
Engine Sub-Subsystem 438
Transmission Sub-Subsystems 439
Drivetrain Sub-Subsystems 439
Fasteners 439
Decomposition Tree for the Powertrain System 440
Interfaces 440
Cascading Vehicle Attribute Requirements to Powertrain
Requirements 444
Trade-Offs in Powertrain Development 444
Concluding Remarks 450
Reference 450

Chapter 23 Case Studies and Integration 451

Introduction 451
Case Study 1: Motorcycle Systems 451
Objectives 451
Project Description 451
Motorcycle Attributes to Systems Relationships 453

Case Study 2: Benchmarking and Evaluation of Steering Wheels ... 456
  Objectives ... 456
  Project Description ... 456
  Benchmarking Study ... 456
  Evaluation in a Driving Simulator ... 458
Case Study 3: Pugh Analysis of an Automotive Concept ... 459
  Objective ... 459
  Problem: New Product Concept ... 459
  Analysis of the Problem ... 461
Case Study 4: Cyclone Grinder Development ... 462
  Objective ... 462
  Project Description ... 463
  Customer Requirements for the Grinder ... 463
  Functional Requirements for the Grinder ... 463
  Systems and Components of the Grinder ... 464
  Grinder Development Project Schedule ... 464
  Key Concepts for Successful Cyclone Grinder Design ... 464
  Risk Management ... 467
  Key Observations ... 468
Case Study 5: Smart Car Design and Production ... 468
  Objectives ... 468
  Project Introduction ... 468
  Smart Car's Customer Needs ... 469
  Benchmarking of the Smart Car ... 471
  Key Product Design Development Issues ... 472
  Key Business and Supply Chain Issues ... 472
Case Study 6: Problems during Boeing 777 Development ... 472
  Objective ... 472
  Project Description and Uncovered Problems ... 473
Case Study 7: Boeing 787 Dreamliner Design and Production ... 474
  Objective ... 474
  Project and Product Description ... 474
  Production Issues ... 476
Case Study 8: Flexible Assembly Line for Laptop Computers ... 477
  Objectives ... 477
  Background ... 477
  Assembly Line Configuration ... 478
Case Study 9: Specifications for an Electric Car ... 480
  Objective ... 480
  Project Background ... 480
  Application of the Matrix Data Analysis ... 480
Concluding Remarks ... 483
References ... 483

Contents

Chapter 24 Case Studies in Cost–Benefit Analysis 485

Introduction 485
Case Study 1: Cost–Benefit Analysis of Automotive Product
Development Programs 485
Case Study 2: JEDI Model Applications for Comparison
of Costs and Economic Benefits of Wind Turbine Power
Plant with Natural Gas Power Plant 493
Case Study 3: Evaluation of Five Electric Power Generation
Alternatives 496
    Methodology 498
Case Study 4: NHTSA/EPA Cost–Benefit Analysis: Increases
in Vehicle Costs, Fuel Savings, and Avoided Pollution 498
    Increase in Vehicle Price vs. Fuel Savings 499
Case Study 5: Manufacturing and Assembly Line: Robotic
Assembly of an Automotive Differential Gear Carrier 501
    Baseline Manual Assembly 503
    Robotic Assembly Method 504
        Fixture 505
        Bolt Feeding 506
        Shim and Bearing Feeding 506
        Grippers 506
    Row and Column Codes for Components 507
    Assumptions Regarding Parts and Assembly Considerations 508
        Cost–Benefit Analysis 511
Life Cycle Cost Analysis 512
Concluding Remarks 512
References 512

Chapter 25 Challenges and Future Issues in Systems Engineering 513

Introduction 513
Challenges in Systems Engineering 513
Need for Tools in Complex Product Development 514
    Tools to Manage Multifunctional and Multiple
    Requirements 515
Coordination of Global Design Teams 515
Commonality 515
Modularity 516
CAD and CAE Integration 516
Ergonomic Needs in Designing Products 516
Future Technological Challenges 517
Bright Future for Systems Engineers 517
    Characteristics of a Good Systems Engineer 517
Teaching Systems Engineering 518
    Objectives of the Projects 519

Project Work 519
Brief Descriptions of the Projects 520
Concluding Remarks 522
References 522

Appendix 1 Product Development Case Studies 523

Appendix 2 Benchmarking, Quality Function Deployment, and Design Specifications 525

Appendix 3 Vehicle Systems Analyses: Requirements, Interfaces, Trade-Offs, and Verification 527

Appendix 4 Business Plan and Systems Engineering Management Plan for the Proposed Vehicle 529

Appendix 5 Conceptual Design of the Proposed Vehicle 531

Appendix 6 Vehicle Assembly Process Plan 533

Appendix 7 Term Project: Final Report 535

Appendix 8 Calculations of Centerline and Control Limits for Control Charts 537

Index 543

Preface for the First Edition

The objective of this book is to present systems engineering processes and techniques that can be used to design future products. The book is especially aimed at the design phases of complex products. The complexity in a product can be attributed to an increase in the number of parts, number of systems needed to accomplish product functions, number of external systems affecting the product, types of technologies associated with the systems, number of interfaces among the systems, number of variables associated with the systems and their interfaces, number and types of users and uses and variations in the operating environments, and number of disciplines or specialized fields needed to analyze, design, and evaluate various components and systems.

With the increasing feature content demanded by more sophisticated customers, the products, such as computers, cell phones, home appliances, automobiles, and airplanes, are becoming more complex. It is important to design the "right products the first time" and avoid costly redesigns, modifications, and delays. Further, the increased global competition is demanding that product design cycles be shortened. The programs to design and market such more complex products also face increased risks of cost overruns, substantial delays in their completion, and failures in meeting the customer expectations.

The answers to improved product designs and reducing costs and timings lie within how the accumulated knowledge in various fields, advances in technologies, and design trends in the markets have been applied during the life cycle stages of the product. The power of "The Systems Approach" or "Systems Thinking" has been recognized as the key in designing complex systems. The systems approach has become more powerful due to advances in techniques, implementation processes, and growing number of applications and case studies.

Over the past forty-plus years, as I taught many industrial engineering courses and worked on complex automotive design challenges and other related product issues and product programs, it has become increasingly clear to me that many of the concepts and methods used in various specialized areas such as quality engineering, human factors engineering, safety engineering, project and program planning, product design, product planning, production planning, and manufacturing systems, are all focused on satisfying customers and can be studied under the umbrella of "Systems Engineering." Further, the boundaries between different specialized areas have blurred due to the need to consider systems aspects of the problems. However, the basic issues and understanding into key considerations underlying the products and processes have not changed. Some key considerations are:

- Products (or systems) should be created to satisfy the needs of their customers.
- Processes are where the "work" gets done.
- Systems (or products) are used to create and run the processes.
- Complex systems and products need Systems Engineering knowledge.

- Flexible framework involving systems engineering processes, techniques, "lean" thinking, and continuous improvement approach are needed for efficient problem solving.
- "Big Picture" involving the integrated view of systems is needed to conceptualize and understand problems.
- Understanding into the "connectedness" of systems ("systems within systems" or "systems of systems") is essential to producing long-lasting solutions.
- Decision-making can be greatly facilitated by applications of many tools used in quality, human factors, safety, product development, and systems engineering areas.

This book, thus, is the result of my curiosity to organize concepts, principles, processes and tools available from various industrial engineering and related engineering and management activities and present them in a form to facilitate decision-making throughout the life cycles of complex product programs. The book is organized into four parts. Part I covers systems engineering concepts, issues, and basics of decision-making, program management, risks, and costs in product development. Part II covers useful concepts, principles, and approaches in quality, human factors, and safety engineering. Part III describes important tools and methods used in the above fields during the product programs, and finally, Part IV presents case studies to illustrate the usefulness of simple but effective techniques to help understand the connectedness and integrated nature of the systems engineering processes. It also discusses future challenges and issues related to advancing the systems engineering field. The book is intended to serve as a reference book for professionals in the industry involved in product design, development, engineering, and program management; and it can be used as a textbook for courses in systems engineering, product design, and automotive systems engineering.

Writing this book was a fun ride and a lot of work. It raised a number of questions and presented many challenges in implementing and teaching systems engineering. Many of the examples used in this book were extracted from projects created and used in many of the graduate courses that I offered at the University of Michigan-Dearborn.

Preface for This Second Edition

In this new edition, I have added six new chapters and expanded coverage in several existing chapters. I have also reorganized some of the topics, added more material (e.g., Model-Based Systems Engineering) in several existing chapters, and eliminated some material that I felt is no longer relevant. While this new edition maintains, the four parts contained in the original book, new chapters are added to provide a more complete and up-to-date coverage of issues and techniques needed to develop new complex products with added emphasis on comfort, convenience, safety, economy, sustainability, and efficiencies in manufacturing and assembly.

The new chapters cover the following topics: (1) Detailed Engineering Design during Product Development, (2) Design for Sustainability, (3) Design for Manufacture and Design for Assembly, (4) Cost–Benefit Analysis as a Decision-Making Tool, (5) Life Cycle Analyses as a Product Planning and Design Tool, and (6) Case Studies in Cost–Benefit Analysis. The case studies included in the first edition have been retained to provide a broader range of products and their unique development issues. I have also included new case studies (Chapter 24) and a number of examples, especially from the energy field, to elaborate on current topics such as electric energy generation, fuel efficiency, and pollution reduction in several chapters.

The new revised edition, thus, is intended to provide practitioners and students a more comprehensive knowledge on issues facing the developers of complex products and management of processes involved in their development. It provides more tools for implementing the systems engineering to minimize the risks of delays and cost overruns, and more importantly, helps create the right product for its customers.

More specifically, this new edition provides the following:

1. Deeper understanding into what is involved during engineering design phase of a product
2. Better understanding into what is involved in achieving a sustainable product and environmentally conscious company and brand image
3. Usefulness of the concepts of design of disassembly and recycling of materials
4. Ability to conduct cost-benefit analyses in decision-making during product development
5. Reasons for performing life cycle analyses and life cycle costing analyses to make better decisions
6. Knowledge to deal with multi-criteria problems and create a product that is economical to produce, environmentally friendly, user friendly, and safe.

7. Ability to evaluate product designs to ensure high efficiency in manufacturing and assembly
8. Many techniques used in product development and their applications through additional examples and case studies
9. Understanding into how model-based systems engineering can reduce time and improve accuracy in systems engineering implementation
10. Comprehensive set of tools to create the right product, the first time, for its customers

It is my hope that the book will motivate you to develop and implement new tools to improve development and production of future products that will be well liked by its customers.

Vivek D. Bhise
June 30, 2022

Website Materials

The following files are in the Download section of this book's web page on the CRC Press website (https://www.routledge.com/9781032203690).

- A. Computer programs and models
    1. Automotive Program Chart with Cost Model
    2. Program Cost Flow by Month
    3. Program Cost Flow by Quarters
    4. Anthropometric Dimensions of Seven Populations
- B. Slides for Chapters 1–25

Acknowledgments

This book is a culmination of my education, experience, and interactions with many individuals from the automotive industry, academia, and government agencies. While it is impossible for me to thank all the individuals who influenced my career and thinking, I must acknowledge the contributions of the following individuals.

My greatest thanks go to Bob Himes, Eulie Brayboy, Lyman Forbes, and Dave Turner from the Ford Motor Company who provided me with many opportunities to observe, study, analyze, and learn many processes and understand issues involved in designing complex automotive products. My assignments exposed me to many problems in areas such as analyzing and proposing human factors and safety standards, developing vehicle systems specifications, evaluating vehicle lighting, field of view, and display systems, developing new concept vehicles in the advanced design studios, implementing package and ergonomics attribute within the attribute engineering framework of the systems engineering, reviewing and benchmarking many design concepts and production vehicles in the U.S., Europe, and Japan, conducting market research studies, conducting six-sigma projects to improve vehicle quality, experimenting and talking with lots of users and customers of different automotive products from sports cars to heavy trucks.

The University of Michigan-Dearborn campus provided me with unique opportunities to integrate my industrial experience in developing and teaching various courses. Our Automotive Systems Engineering, Engineering Management, Industrial Engineering and Manufacturing Systems Engineering, and Energy Engineering Programs allowed me to interact with hundreds of students who in turn implemented many of the techniques taught in our graduate programs in solving problems within many of the automotive OEMs and supplier companies. I would like to thank Profs. Munna Kachhal and Armen Zakarian for giving me opportunities to develop and teach many courses in the Industrial and Manufacturing Systems Engineering, and Dr. Roger Schulze, Director of the Institute for Advanced Vehicle Systems got me interested in working on a number of multidisciplinary programs in Automotive Design. Together, we created a number of design projects by forming teams of our engineering students with students from the College for Creative Studies in Detroit, Michigan. I would like to also thank Prof. Pankaj Mallick, who directed our Automotive Systems Engineering and Multidisciplinary Programs for working together on many projects and fascinating discussions on how to teach systems engineering to our graduate students. My special thanks also go to James Dowd from Collins and Aikman and the Advanced Cockpit Enablers (ACE) team members for sponsoring a number of research projects on various automotive interior components and creation of a driving simulator to evaluate a number of advanced concepts in vehicle interiors.

Over the past forty-plus years, I was also fortunate to meet and discuss many automotive design issues with members of many committees of the Society of Automotive Engineers, Inc., the Motor Vehicle Manufacturers Association, the Transportation Research Board and the Human Factors and Ergonomics Society.

I would like to also thank Cindy Carelli from the CRC Press/Taylor & Francis Group—for her encouragement in preparing the proposal for this book, and her production group for turning the manuscript into this book.

Finally, I want to thank my wife, Rekha, for her constant encouragement and her patience while I spent many hours working on my computers in writing the manuscript and creating the figures included in this book.

Vivek D. Bhise
Ann Arbor, Michigan
June 16, 2022

Author

Vivek D. Bhise is currently a LEO Lecturer/Visiting Professor and Professor in post-retirement of Industrial and Manufacturing Systems Engineering at the University of Michigan-Dearborn. He received his B.Tech. in Mechanical Engineering (1965) from the Indian Institute of Technology, Bombay, India, M.S. in Industrial Engineering (1966) from the University of California, Berkeley, California, and Ph.D. in Industrial and Systems Engineering (1971) from the Ohio State University, Columbus, Ohio.

During 1973–2001, he held a number of management and research positions at the Ford Motor Company in Dearborn, Michigan. He was the manager of Consumer Ergonomics Strategy and Technology within the Corporate Quality Office, and the manager of the Human Factors Engineering and Ergonomics in the Corporate Design of the Ford Motor Company where he was responsible for the ergonomics attribute in the design of car and truck products.

Dr. Bhise has taught graduate courses in Automotive Systems Engineering, Vehicle Ergonomics, Vehicle Package Engineering, Automotive Assembly Systems, Energy Evaluation, Risk Analysis and Optimization, Human Factors Engineering, Total Quality Management and Six Sigma, Product Design and Evaluation, Safety Engineering and Computer-Aided Process Design and Manufacturing and Methods Engineering over the past 42 years (1980–2001 as an Adjunct Professor, 2001–2009 as a Professor, and 2009–present as a LEO Lecturer/Visiting Professor) at the University of Michigan-Dearborn. His publications include over 100 technical papers in the design and evaluation of automotive interiors, vehicle lighting systems, field of view from vehicles, and modeling of human performance in different driver/user tasks.

He is also the author of the following books:

Decision-Making in Energy Systems. ISBN 978-0-367-62015-8. Boca Raton, FL: CRC Press, 2022.
Automotive Product Development: A Systems Engineering Implementation. ISBN: 978-1-4987-0681-0. Boca Raton, FL: CRC Press, 2017.
Designing Complex Products with Systems Engineering Processes and Techniques. ISBN: 978-1-4665-0703-6. Boca Raton, FL: CRC Press, 2014.
Ergonomics in the Automotive Design Process. ISBN: 978-1-4398-4210-2. Boca Raton, FL: CRC Press. 2011. (Also translated in Chinese language and published by China Machine Press in China, 2016).

He received the Human Factors and Ergonomics Society's A. R. Lauer Award for Outstanding Contributions to the Understanding of Driver Behavior in 1987. He has served on a number of committees of the Society of Automotive Engineers, Inc., Vehicle Manufacturers Association, Human Factors and Ergonomics Society, and Transportation Research Board of the National Academies. He is a member of the Human Factors and Ergonomics Society, the Society of Automotive Engineers, Inc., and the Alpha Pi Mu.

Part I

Systems Engineering Concepts, Issues, and Methods in Product Design

1 Introduction to Products, Processes, and Product Development

INTRODUCTION AND OBJECTIVES

Designing a complex product containing many systems and components is a very challenging undertaking. It not only requires a design team involving many disciplines but also requires complex coordination of many requirements, understanding customer priorities and trade-offs between product characteristics, and managing of a number of tasks to ensure that the product program remains within the planned schedule and budget. And finally, the resultant product must meet the expectations and needs of its customers. Systems Engineering (SE) has emerged as the key discipline to help manage the complex processes throughout the life cycle of the product. Failures in implementing, integrating, and executing the SE process in the technical and management activities of the product program can result in costly program overruns and delays. The failures may also produce a product that will not completely satisfy its customers. The competition among different producers in the global product development area is fierce and some major problems during the product development can affect the financial health and survivability of the corporations.

There is no magic solution or process to create any complex product. Even creation of a simple product such as a screwdriver with minimally two components (namely, steel shank and a plastic handle) will require considerations of many issues to define its dimensions related to shape and size, surface characteristics, materials for the handle and the shank, manufacturing processes, and so on. There is no unique process to create a screwdriver as it will depend on its characteristics such as the type and size of the screwhead to turn, type of materials for the handle and the shank, method of joining the handle to the shank, hardness of the shank, and so on. The complexity increases enormously when the complex products with many components and many functions such as computers, automobiles, airplanes, spaceships, and ships are to be designed and produced in manufacturing and assembly plants involving many different materials, many manufacturing processes, and hundreds of workstations, tools, machines, and people.

Because of the need to develop very complex products for the Air Force, Navy, Military and Space Explorations, the U.S. Department of Defense (DoD) and the National Aeronautics and Space Administration (NASA) applied their massive research capabilities and advanced the field of SE to ensure that the complex products and systems could be developed with coordination of many of their suppliers and contractors. In spite of the SE implementations, there are many examples of the

DOI: 10.1201/9781003263357-2

DoD programs that incurred huge cost overruns, substantial delays, or even cancellations (Witus and Ellis, 2011).

Over the past 40-plus years, many applications of SE and other related fields such as Total Quality Management, Human Factors Engineering, and Safety Engineering along with many approaches such as Baldrige Quality Award Criteria, ISO 9000 Standards for Quality Management Systems, and Six Sigma Methodologies (including Design for Six Sigma) have demonstrated the usefulness of many principles, methods, and procedures to improve customer satisfaction.

The objective of this book is to provide an understanding of the well-accepted processes, techniques, and their applications in designing complex products.

UNDERSTANDING PRODUCTS, CUSTOMERS, PROCESSES, AND SYSTEMS

What Is a Product?

A product can be defined as hardware, software, or an integrated combination of the hardware and software that people use to perform one or more functions. Hard products are usually created by assembling a number of physical components. The simplest product can have one component, for example, a sewing needle, a wrench, or a bolt. A hammer is another simple product that typically has two components, namely, a hammerhead and a handle. A more complex product such as a refrigerator has over a hundred components.

For convenience in developing the complex products, they can be decomposed into systems based on their functions. A refrigerator thus can have the following systems: cabinet system (body with compartments, doors and seals, handles, insulation, and wheels), cooling system (refrigerant, compressor, tubing, heat exchanger, and expansion valve), electrical system (wiring, electric motor, light bulbs, and switches), and a control system (thermostat, temperature sensors, and actuators). An automobile is another complex product that typically has about 8,000–10,000 components and contains many systems such as body system, powertrain system, chassis system, electrical system, fuel system, climate control system, lighting system, and so on.

A software product will have a collection of lines of code (instructions to a computer). Complex software programs can also be divided into various modules, subprograms, and subroutines to perform certain functions and computational tasks, and to manage their development, debugging, and operation. A software product can be stored in an internal memory system in a computer or in a hard product such as a compact disc or a flash memory stick.

There are also other types of products such as food products, chemicals, energy-related products (e.g., gasoline and power generators), and artwork (e.g., paintings and decorative items) that are complex.

It is important to realize that a product is generally designed to perform at least one or more functions. By performing its functions, the product extends the capabilities of people to perform certain tasks. For example, a sewing needle helps with stitching and making clothes. A wrench allows tightening a nut. A refrigerator allows storing foods to extend their shelf-life and make frozen foods. An automobile allows traveling to farther distances. Thus, any experience that the people get while using a product should be "satisfying" to its users; that is, the product should meet the needs of its customers and the customers should like and enjoy using it.

Some products are purely "decorative," that is, they serve to provide certain sensory information and evoke some emotions, pleasures, or responses within their receivers. Some examples of such products are pictures, art pieces, sculptures, music/songs, perfumes, stuffed toys, and so forth. However, they do not help or serve in performing any physical function. Such products are not considered in this book. However, the sensory perception provided by functional products by their visual, auditory, tactile, or even olfactory characteristics is important in creating the overall perceptions related to pleasing or non-pleasing impressions, belongingness to a certain class (e.g., "economy vs. luxury" or made with "genuine vs. fake [or man-made]" materials), or "brand image" that need to be considered during designing of the products. The sensory perceptions generated by the products can serve a useful role in improving customer satisfaction. Some examples of such sensory perceptions are the softness of a seat or materials used for its seating surfaces, the sound of a car engine, the tactile feel of keyboard switches, the colors in a visual display, the smell of a leather seat, and so forth.

Thus, how a product is designed, that is, the product design, is very important as it directly affects its usability, customer satisfaction, its acceptance, and success in the marketplace.

WHO IS THE CUSTOMER?

Each product must be designed to meet the needs of its customers. Customers are people who purchase and use the products. The customers are generally the end-users in a chain of supplier–receiver (customer) transactions. Everyone in the chain who receives a component from a supplier can be considered as a customer. Thus, a component supplier can provide a part to a customer who assembles the part in his system. The system is then supplied to the next customer who assembles it in his product and sells it to the end-user. Thus, in a long chain of supplier–customer relations such as a tier-3 supplier providing a component to the tier-2 supplier who in turn is supplying a subassembly to the tier-1 supplier who finally supplies an assembled system to an automobile manufacturer are all examples of internal customers.

The end-user is the final customer who purchases the product. It should be noted that in some cases the purchaser may be different than the actual user of the product. For example, a truck product is purchased by a fleet owner (or a company) who in turn provides the truck to an employee (the driver) who becomes the end-user. Furthermore, other persons who work on a product for purposes such as maintenance, repair, or service (e.g., repairman, programmer, or inspector) are also customers. Thus, a product can involve many suppliers and customers and one must consider the needs of all the customers (within the chain and the end users) while designing the product. Some organizations define the customers in the chain as the "internal customers" to separate them from the end users who are considered as the "external customers" (i.e., outside the organization). Some organizations use the word "stakeholders" to define the owners, customers, vendors/suppliers, and every person directly or indirectly related to the product that have a stake in the product (i.e., affected by the product). The main objective of the product designers should be to carefully define their customers/stakeholders and ensure that their requirements and concerns are satisfied.

What Are Customer Needs?

Customers purchase products for their use. When different customers use a given product, they may have different needs depending on when, where, and how they use the product. For example, an automotive product depending on its body style (e.g., car, truck, SUV, or van) will be used for different purposes in different environments (e.g., day, night, weather conditions, or geographic regions) under different situations (e.g., going to work vs. going on a vacation with the family, or delivering a package). Thus, to understand the customer needs, that is, what do they want in a product, different methods can be used. Some of the currently used methods are customer observations, customer interviews, mail surveys, market research surveys, customer clinics, feedback from the customers on their current and previously owned or used products, feedback from the dealers, sales agents, distributors, insurers, maintenance personnel, and so on.

There are also other internal customers such as company employees who work in the various activities (e.g., product planning, design, engineering, purchasing, manufacturing, assembly, and maintenance) in the organizations that produce and service the products, suppliers who build and provide components or systems that are installed in the product, and persons who invest their money in the organizations (e.g., investors and shareholders). The needs of these customers should also be considered as inputs in the product design process.

The needs of all internal and external customers should be obtained before the beginning or during the very early phases of the product design process. The customer needs are generally categorized to determine different attributes that the product must have to satisfy its customers. The product attributes are also prioritized by considerations such as their importance to the customers, frequency of different types of uses, and affordability (or costs related to delivering each attribute). The product designers must study customer needs and understand the product attributes and use the information to design the product. Applications of many methods and processes used to translate the customer needs into product attributes, attribute requirements, functions, and functional requirements are covered in Chapters 4, 14, and 15.

What Is a Process?

A process is where the "work" gets done. A process generally consists of a series of steps or operations that are performed on a part (or workpiece) or with a product (or equipment) by one of more operators such a human operator, a robot, a computer, or a machine (or equipment). The operators may also use one or more tools (e.g., a hand tool, power tool, or software application) in performing any of the tasks. The process can be also defined by following a component (a part, an assembly, a transaction tracking/initiating or recording device, e.g., a bar-code label, a radio frequency identification device [RFID]), a tracking paper (e.g., a Kanban card in an assembly plant), or a person (e.g., who moves from a station to other stations) through a series of steps in performing a function or a task.

To create (i.e., to design and produce) a product, one or more processes are required. For example, the process to produce a component such as a forged aluminum window for a commercial aircraft will involve the following steps: (1) cut slug (a piece of raw material) from an aluminum bar stock, (2) transport the slug to an oven, (3) place the slug on a conveyor passing through the oven, (4) pick up the hot slug and place it in a preform die, (5) forge the slug into a preformed part in a forging press, (6) place the part on another conveyor leading to another oven, (7) pick up the part and forge it in the next die, (8) transport the part to a sawing machine, (9) pick up the part and saw the flash (excess material around the forged part), (10) transport the part to a grinding machine, and (11) grind the edges at the parting lines.

There are many processes to perform different functions in an organization. For example, some processes in a manufacturing organization can include process planning process, production process, product review and approval process, raw material acquisition (purchasing) process, tools ordering process, operator training process, tools and machine setting process, tool inspection process, machine programming process, machining process, part inspection process, parts transportation process, heat treating process, painting process, assembly process, testing process, inspection process, packing process, shipping process, processes improvement process, and defect reduction process.

Since a process is where the work gets done, it is very important to design every process such that it is efficient (i.e., it reduces time, errors, waste, toxic emissions, and expenses) and provides consistent product characteristics or functions needed to satisfy the customers. The process engineers and industrial engineers, who design and maintain the process, study a number of possible alternate processes or reengineer existing processes to achieve the best possible process performance. During the process design, one should ask a number of questions such as: Is the process really needed? Can the process be eliminated or combined with other existing processes? Can a new technology be used to achieve a "breakthrough" and improve the performance of the process substantially? Doses the process reduce carbon-footprint (carbon emissions) of the product?

Figure 1.1 presents a flow diagram of basic activities in producing a manufactured product. It is important to note that the diagram shows the "customers" in the middle top part. It illustrates the key consideration that the process must begin with the customers to determine their needs and the process must end with the feedback from the customers on their use of the product (past or existing) to ensure that the product is created to satisfy the customers. The customer needs are used by the product planning team to set requirements on the product, the requirements are given to the product design and development team, the detailed product design is used to plan production processes, the produced components are assembled to create the product, and finally, the product is shipped, that is, distributed for sale, and subsequently purchased and used by the customers. After the customers use the product, their feedback should be used to improve the product (or its next version). The top of the diagram in Figure 1.1 begins with the words "market segment" (a particular type of

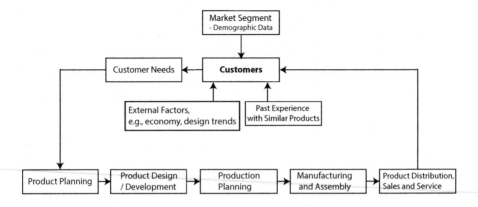

FIGURE 1.1 Flow diagram of basic activities in producing a manufactured product.

product in a specified market) that includes the population and characteristics of its end-users/customers. The market segment is generally determined by marketing researchers in the organization along with other members of the product design and development team.

After the detailed design of a product, the prototype parts are made and verified (i.e., tested to ensure that they meet their requirements), the parts are assembled into products, and the assembled products are validated (i.e., tested to prove that the product works and meets the customer needs). Other concurrent product-related activities include designing of the production processes, tools, fixtures, and machines and installation of machines and materials transportation systems in the manufacturing plants. The manufacturing plants with the equipment, people, and processes are used to produce the components of the product, to assemble the components, and to test the product.

The ISO 9000 documents define the entire process of product creation as the "product realization" process (ISO, 2012). The product realization process consists of the following six processes: (1) planning of product realization, (2) customer-related processes (involving determination of product requirements, review of product requirements, and customer communications), (3) design and development, (4) purchasing, (5) production and service provision, and (6) control of monitoring and measurement devices.

The process of producing the product (after it is designed) is generally called the "production process." It is very important to realize that the product design (i.e., how it is designed—its configuration, dimensions, components, materials, recyclability, surface treatment, assembly method, and so forth) determines its production process. The costs of production equipment and production plants are generally much larger than the product design costs. Therefore, it is very important to design the product "right" so that the product can be manufactured and assembled easily and with the least amount of costs and variability in its characteristics that affect its customers. (Note: Reduced variability will produce a consistent product [process output] from its manufacturing process.)

It is also important to realize that processes are also involved in the product uses (i.e., how a customer uses a product). Some examples of the products and their use processes are a drill-driver (product) is used to tighten screws (i.e., the process of tightening the screws), welding machines are used to join parts (e.g., the joining process involving operation of the welding electrodes and their movements incorporated in the welding machine), and cars are used by the drivers (e.g., the processes involved in entering the car, adjusting the seat, starting the car, changing gears, and driving to a destination).

DESIGNING A COMPLEX PRODUCT

A product can be considered to be "complex" when it contains many systems and many components, and thus, it would require designing many components and many interrelationships (interconnections or interfaces) between many components and systems. Further, many steps and iterations are involved in many processes to design and produce the product. Some examples of steps involved in product creation are developing requirements, understanding many design issues, analyzing trade-offs between requirements and product characteristics, checking if the design meets all the requirements, evaluating if the product will be acceptable to the customers, conducting design reviews, performing evaluations tests, designing equipment, manufacturing and assembly plants, building plants, checking all the production processes, producing parts, assembling the products, and so on. Each step in a process also takes time, resources (people, their skill levels, machines/equipment, environment [e.g., plant lighting and climate control]), and capital. Thus, designing and producing a complex product requires substantial quantities of resources and coordinated actions of many people and equipment. Any problems encountered in creating the product generally can cause long delays to find and correct the errors and expensive cost overruns.

This book deals with the understanding and designing complex products that are developed and managed by people (i.e., designers, engineers, product planners, project/program managers, and others—usually working in many teams) to perform certain specified functions.

DEFINITION OF A SYSTEM

A system consists of a set of elements (e.g., components) that work together to perform one or more functions. The components of a system generally consist of people, hardware (e.g., parts, tools, machines, computers, and facilities), or software (i.e., instructions, codes, and databases). The systems also require operating procedures (or methods) and organization policies (e.g., documents with goals, requirements, and rules) and they work under a specified range of environmental and situational conditions.

Some definitions of systems are provided as follows:

1. A system is a set of functional elements organized to satisfy specified objectives. These include hardware, software, people, facilities, and data.

2. A system is a set of interrelated components working together toward some common objective(s) or purpose(s) (Blanchard and Fabrcky, 2011).
3. A system is a set of different elements so connected or related as to perform a unique function not performable by the elements alone (Rechtin, 1991).
4. A system is a set of objects with relationships between the objects and between their attributes (Hall, 1962).

The set of components have the following properties (Blanchard and Fabrycky, 2011):

1. Each component has an effect on the whole system.
2. Each component depends on other components.
3. The components cannot be divided into independent subsystems.

The work performed by a system can involve tasks (or work elements) such as generating motion (e.g., movement of brake pads in a vehicle braking system), processing of materials (e.g., machining), assembling components (e.g., transporting, positioning, aligning and attaching parts with fasteners), processing information (e.g., reading instructions, storing data, and making a decision), printing papers, and so on. These tasks generally require other tasks (performed by other systems) such as supplying (1) raw materials (e.g., metals in the form of bar stock, sheet metal, ingots, and plastic grains), (2) standardized components or parts (e.g., fasteners, wires, tubes, hoses, papers, data storage strips or disks, paints, inks, and cooling fluids), and (3) energy (e.g., electrical, pneumatic, or hydraulic power).

Thus, any system is also a "part of" or "works with" other systems. A system (or product) is generally embedded in a super-system (e.g., an automotive product is a part of the highway transportation system, or a power plant is a part of the national electric distribution network or the "power grid").

Systems, Subsystems, and Components

Any large system or a complex product can be further divided (or decomposed) into a number of systems and the systems can be divided into subsystems. And each subsystem can be further decomposed into a sub-subsystem, and so on—in a hierarchical manner into lower-level systems. Each subsystem or sub-subsystem may consist of one or more components. Thus, a large system or a complex product can be divided into a hierarchical structure involving systems, subsystems, sub-subsystems, and components. For example, a complex product such as a car will have a number of systems such as body system, powertrain system, fuel system, steering system, electrical system, braking system, lighting system, climate control system, and so on. Each of the vehicle systems can be further divided into subsystems. For example, the vehicle lighting system will have a head-lighting system, tail-lighting system, signaling system, light-switching system, and wiring system. Furthermore, any subsystem can be further divided into sub-subsystems or components. For example, a headlamp will have components such as a lens, a reflector, a light source, a light source retaining clip or ring, and aiming screws.

Introduction to Products, Processes, and Product Development

SYSTEMS WORK WITH OTHER SYSTEMS

Systems generally function within other systems and therefore the systems are connected with each other. Systems must function with other systems to perform the functions of their parent system (or product). Thus, systems would have interfaces (or connections) with other systems. The interfaces can physically connect two or more systems together and/or transmit (or communicate) the necessary flow of materials (e.g., oil, air, gases, and coolant), power, and/or data. The physical products are generally formed by physical connections (e.g., with fasteners, welds, or connectors) of a number of systems, subsystems, or components together and they generally share a common space (i.e., packaged together inside a three-dimensional envelope bounded by exterior surfaces).

It is therefore important to realize that if any new system is to be designed, it must generally work with other existing systems. For example, a new powertrain for a vehicle can be incorporated into the same previously designed vehicle body. Furthermore, many different types of automotive products (i.e., cars, trucks, buses, and vans) must operate in the existing highway transportation infrastructure involving the system of roadways, traffic control systems, fuel stations, service facilities, parking structures, and so on. Similarly, a computer product must work with the existing Internet, wireless devices, universal serial bus (USB) connections, disk drives, hard drives, and so forth.

PRODUCT FAMILIES AND COMPONENT SHARING

One important characteristic of a complex product is that it generally shares a large number of components, subsystems, and even systems with other similar products (or product models). For example, tires, seats, engines, suspensions, braking systems, electrical systems, lighting sources, or even vehicle platforms in some of the automotive products can be shared across many models of the same automotive manufacturer and its partnering companies (similarly in the aircraft and the computer industry). Thus, a complex product generally belongs to one or more families of similar products.

The sharing of components and systems allows reducing costs in all categories (e.g., product design and development, manufacturing, operation, repair, and maintenance) during the life cycle of the product. Another advantage of components-sharing is that it leads to higher quality levels because components do not have to be engineered more than once, thus eliminating the chance of flaws. The complex products also use many standardized (or common) components and systems. The standardized components benefit from further reductions in costs as they can be used in many products by many product manufacturers and across many industries. The cost reductions are achieved by increasing production volumes of the shared components and systems and spreading the fixed costs over a larger number of production units (see Chapter 9 for fixed vs. variable costs).

The sharing of the components and systems also can have some disadvantages as it could prevent or slow down the introduction of some new technologies and thus reduce opportunities for rapid innovations and improvements in future complex products. The shared or existing components are also called the "carryover" components. The decision to use many carryover components also restricts creation of "totally new" or "breakthrough" designs or configurations of future products (see Chapter 15 for the breakthrough approach).

Product Development

The majority of product development programs do not involve designing a product from "scratch" (a totally new product) or a product of the type that did not exist before. The process of designing a product is therefore typically called the "product development process" in most industries as compared to the "product design process." (But the terms "product development" and "product design" can also be interchanged or used in the same context in many industries.) After the product is designed, the process of producing (or manufacturing and assembling) the product is generally called the "production process."

Processes in Product Development

It is important to realize that work is generally performed by using one or more processes. A process usually involves equipment (one or more workstations with tools, machines, robots, or computers) and human operators that are arranged in a sequence of steps (operations or tasks). Designing a product also is performed by using a process (defined earlier as the product development process). The product development process, depending on the complexity of the product, can involve many processes within and outside the organization (e.g., supplier facilities) responsible for developing the product. Product development processes vary due to differences in the products (i.e., their characteristics, functions, features, and demand volume), type of product development program (e.g., refreshing an existing product or designing a totally new product), and the design organization.

A generic process of product creation and use involves the entire product life cycle that generally includes the following phases:

1. Preconcept or preprogram (preprogram planning)
2. Product concept exploration (alternative concepts development, advanced engineering)
3. Product definition and risk reduction (feasible design, preliminary design, and risk analysis)
4. Engineering design (detailed engineering design including testing)
5. Manufacturing development (processes, tooling and plant development)
6. Production (manufacturing and assembly)
7. Product distribution, sales, marketing, and operational support
8. Product updating or discontinuation (including recycling and disposal)

The first five of the above-described phases can be defined as the product development process and the fifth and sixth phases can be considered as the production process. It should be noted that the fifth phase of manufacturing development can be considered to be the transition from the product design to manufacturing. It is very

Introduction to Products, Processes, and Product Development

important to include the product manufacturing considerations very early during the product design (i.e., during phases 1–4 by implementing simultaneous [concurrent] engineering) to ensure that the transition in the fifth phase (involving designing of manufacturing processes and the creation of required tools and equipment in the manufacturing plants) occurs seamlessly without changes in the product design in the later phases to meet the production needs.

The work performed in each of the above phases is performed by undertaking specialized processes. For example, the preconcept phase can involve a process of understanding the customer, corporate needs, and regulatory requirements to decide on the type and characteristics of the new product and prepare a plan for the subsequent activities.

Ulrich and Eppinger (2016) described the generic product development process with the following phases:

1. Planning
2. Concept development
3. System-level design
4. Detail design
5. Testing and refinement
6. Production ramp-up

It should be noted that Ulrich and Eppinger (2016) in their last phase of production ramp-up consider processes involved in getting ready for the production (i.e., design of manufacturing and assembly processes, tools, and facilities building and prove-out of the production processes) and then gradually increasing the production rate to the desired level.

An automotive product development and subsequent life cycle processes typically include the following major phases:

1. Market research—Determining customer needs
2. Product planning—Defining vehicle attributes and preparing a business plan
3. Design—Creating product concepts
4. Select suppliers and their integration in the product design teams
5. Market research—Selecting a design theme
6. Refining the selected design and preparing a product verification plan
7. Engineering the vehicle
8. Detailed engineering and testing
9. Production (manufacturing and assembly) processes and tooling design
10. Build verification prototypes
11. Build/modify assembly plant
12. Build, verify, and conduct validation tests with production prototypes
13. Production—Build vehicles for sale
14. Test, verify, and perform quality audits
15. Dealership training and vehicle distribution
16. Marketing promotions—Product brochures, advertising, and dealer training
17. Sales and service

18. Customer feedback—Repairs, warranty costs, and customer surveys
19. Planning model updates or product retirement

The product development phases will include the first 10 of the above-described phases. The production process will overlap with the product development process primarily during the production processes and tooling design phases (phase 9). The production process will extend over from phases 9 to 14.

Flow Diagram of Product Development

Product development process generally begins with the customers in understanding their needs and ends up with the customers providing their feedback after using the product. Figure 1.2 shows the major phases in the product development process

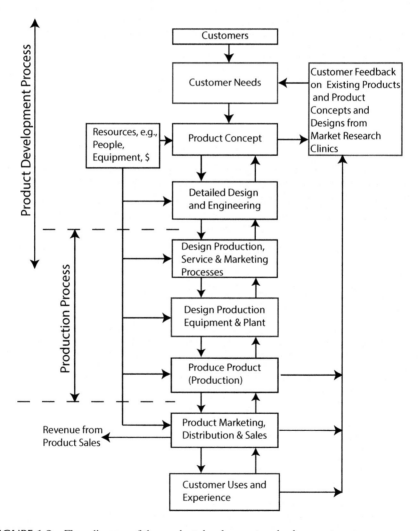

FIGURE 1.2 Flow diagram of the product development and subsequent processes.

along with the production, marketing, sales, and product usage phases. Based on an understanding of the customer needs, a design team consisting of members from different disciplines (e.g., industrial design, product architects, engineers, manufacturing personnel, product planners, and market researchers) generally develops one or more product concepts. The product concepts are iteratively improved, and customer feedback is obtained to determine if a leading concept could be selected for the detailed design and engineering work (see Chapter 6). Based on the selected product design, manufacturing capabilities, and marketing plans, the production processes and suppliers are selected, and the production equipment and plants are designed for their manufacturing and assembly. The produced products are shipped to the distributors and retailers for sale. As the purchased products are used by the customers, feedback from the customer experience (i.e., data on field operating performance, customer likes/dislikes, product repair, and maintenance) is continuously collected and provided for improving existing products and/or designing future products.

Managing the Complex Product

A complex product with a number of components can be organized and divided by its systems, subsystems, and components for the management of design, engineering, manufacturing, assembly, and testing purposes. The systems are generally defined by specialized engineering disciplines such as body engineering, chassis engineering, structural engineering, aerodynamics, electrical and electronics engineering, propulsion engineering (or powertrain engineering), climate control engineering, and so on.

For example, a refrigerator would have the following systems: (1) cold food storage system (above freezing temperatures), (2) freezer storage (below freezing temperatures), (3) cooling system (compressor, motor, heat exchangers, and piping), (4) electrical power system, and (5) temperature control system.

Each system in a product can be further subdivided into subsystems. For example, a braking system in an automotive product can have the following subsystems: (1) brake pedal system, (2) master cylinder system, (3) brake fluid lines and hoses, (4) in-wheel brake systems (drum or disc brake system), and (5) parking brake system. Each subsystem can be further divided into its components.

The division of a complex product into its systems, subsystems, and components allows management of work involved in designing each system to ensure that the product meets its various functional requirements. The design of connections or interfaces between various systems, their subsystems, and components needs considerable understanding of the type of interfaces such as physical connection (e.g., alignment between two adjoining faces and joining techniques), flow of fluids, data, or energy between interfacing components, providing certain functional capabilities (e.g., movements and transmission of forces between adjoining parts). Chapter 5 provides more information on interfaces and interface development.

Figure 1.3 provides a more detailed product development process showing how the engineering work is performed by incorporating the product decomposition considerations. This concept is also incorporated in the SE "V" model presented in Figure 2.4 and discussed in Chapter 2.

To facilitate the design of a complex product, the customer requirements on the product level are cascaded down to systems, subsystems, and components, and their

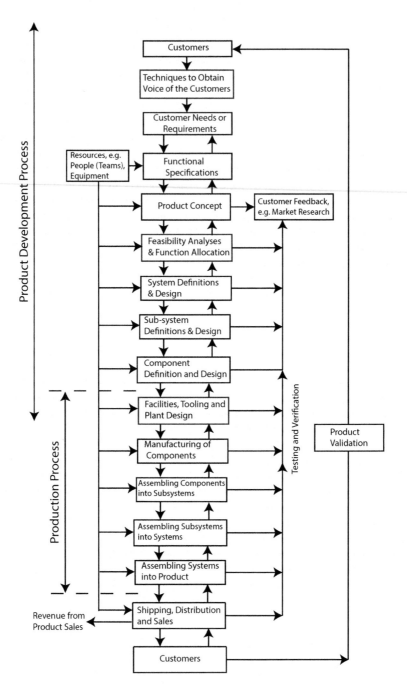

FIGURE 1.3 A more detailed product development and subsequent processes with product decomposition.

functions are established (see Chapter 4 for more details). The flow diagram in Figure 1.3 also shows feedback loops (upward) between each step and its preceding steps. Thus, the entire process is iterative until a feasible and satisfactory balanced design (involving trade-offs between a number of competing design considerations such as performance, costs, timings, and program risks) is achieved. This process is further discussed in Chapter 2.

Managing a complex product with many systems and interactions between different systems also requires a complex management and organizational structure. The entire program of developing a complex product would be divided into many separate projects each involving several disciplines and integration of activities and outputs to ensure that the product program meets its overall goal. Chapter 8 covers the planning and management issues related to the complex product programs. It is very important that the team members from different project teams communicate on interfacing issues to ensure that the whole product functions well and meets all the product-level requirements.

LIFE CYCLE STAGES OF A PRODUCT

Program Phases, Reviews, and Milestones

The activities performed during a product program can be divided into a series of different phases. A program is generally mapped in a timing chart along with timings of a number of program reviews (by various subject matter experts and the management) and milestones defining completion dates of certain key activities. The descriptions and timings of the activities, phases, reviews, and milestones in a product program will vary and depend on the product, its characteristics, and the organization undertaking the program.

The life cycle of a product program typically involves the following phases:

1. *Concept development*: Customer needs are understood; existing products are benchmarked; new technologies and design trends are investigated; various product concepts and configurations are explored; and viable product concepts are proposed.
2. *Concept selection*: Various proposed product concepts are analyzed; market research clinics are held to determine customer likes and dislikes and preferences to proposed concepts; and a leading concept is selected and modified (if necessary) for the detailed design and engineering.
3. *Detailed design and engineering*: System requirements are refined and cascaded down to lower levels; suppliers are selected; engineering analyses are performed to determine feasible configurations; systems are designed and refined; detailed drawings or CAD models are made; production and assembly issues are analyzed; early samples of entities (systems, subsystems, and components) are tested to ensure that they meet requirements; overall product functions and performance are validated to assure that the product will meet its customer needs.

4. *Manufacturing development*: Manufacturing processes are developed; plant and production equipment are designed, installed, and tested.
5. *Production*: Operation of the plant and production equipment is verified; the people are trained to perform all production and assembly tasks; produced products are checked for compliance; and products are shipped to distributors or retailers.
6. *Utilization and support*: Products are sold and used by the customers; the service and maintenance functions are managed; customer feedback data are gathered and provided to various management groups (production, engineering, financial, marketing, etc.) within the organization; the product may be refined, and new models/version introduced to continue production and sales.
7. *Product discontinuation/retirement*: Production is halted; the existing products withdrawn from service at the end of their life; the retired products and plant equipment may be recycled for reuse or retrieval of raw materials.

Figure 1.4 presents a timing chart of a product program showing typical phases and activities in a product program.

The NASA's *Systems Engineering Handbook* (NASA, 2007) describes the life cycle phases of a product program as follows:

1. Pre-phase A: Concept studies (i.e., identify feasible alternatives)
2. Phase A: Concept and technology development (i.e., define the project and identify and initiate necessary technological needs development)
3. Phase B: Preliminary design and technology completion (i.e., establish a preliminary design and develop necessary technology)
4. Phase C: Final design and fabrication (i.e., complete the system design and build/code the components
5. Phase D: System assembly, integration and test, and launch (i.e., integrate components and verify the system, prepare for operations, and launch)
6. Phase E: Operations and sustainment (i.e., operate and maintain the system)
7. Phase F: Closeout (i.e., disposal of systems and analysis of data)

Figure 1.5 presents a program time chart with the preceding phases. The NASA's *Systems Engineering Handbook* (NASA, 2007) also describes key decision points or milestones at the end of each phase and the various reviews during the above-mentioned phases.

The milestones shown in Figure 1.5 with the triangles represent different reviews. The acronyms used for the reviews are defined below:

CDR = Critical Design Review
CERR = Critical Events Readiness Review
DR = Decommissioning Review
FRR = Flight Readiness Review

Introduction to Products, Processes, and Product Development 19

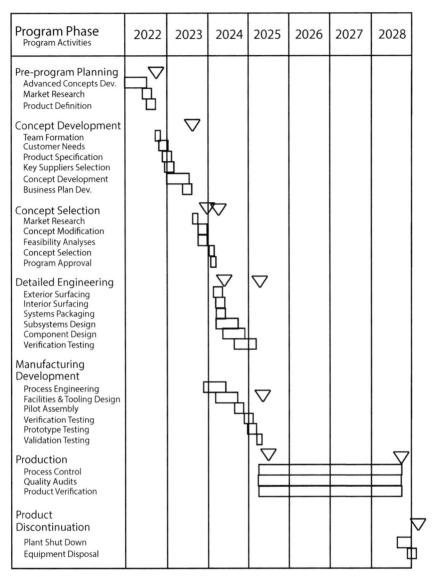

FIGURE 1.4 Timing chart of a product program.

KDP = Key Decision Point
MCR = Mission Concept Review
MDR = Mission Definition Review
ORR = Operational Readiness Review
PDR = Preliminary Design Review
PFAR = Postflight Assessment Review

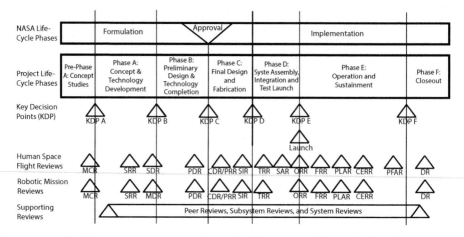

FIGURE 1.5 NASA project life cycle. (Redrawn from NASA, *Systems Engineering Handbook*. Report SP-2007-6105, Rev 1, 2007.)

PIR = Program Implementation Review
PLAR = Postlaunch Assessment Review
PRR = Production Readiness Review
P/SDR = Program/System Definition Review
P/SRR = Program/System Requirements Review
PSR = Program Status Review
SAR = System Acceptance Review
SDR = System Definition Review
SIR = System Integration Review
SRR = System Requirements Review
TRR = Test Readiness Review

CONCLUDING REMARKS

This chapter provided background information in terms of definitions, key issues, and considerations in understanding customers, customer needs, products, complex products, systems, system decomposition, processes, product development process, production process, and product life cycle phases. The purpose of products is to allow the customers to perform certain functions that cannot be performed by the individuals alone. The products must be designed to satisfy the customers in meeting their needs. The next chapters in Part I of this book are prepared to progressively develop the SE approach and management considerations in developing the products. Part II provides background on quality engineering, human factors engineering, and safety engineering, which are key disciplines in the SE implementation. Part III provides details and illustrations of various techniques used during the product programs, and Part IV provides case studies on various product development issues, applications of the techniques, and future challenges.

REFERENCES

Blanchard, B. S. and W. J. Fabrycky. 2011. *Systems Engineering and Analysis*. Fifth Edition. Upper Saddle River, NJ: Prentice Hall PTR.

Hall, A. D. 1962. *A Methodology for Systems Engineering*. New York: D. Van Nostrand Company, Inc.

International Organization for Standardization (ISO). 2012. *ISO 9000 Standards*. http://www.iso.org/iso/iso_catalogue/management_and_leadership_standards/quality_management.htm (accessed February 24, 2012).

National Aeronautics and Space Administration (NASA). 2007. *Systems Engineering Handbook*. Report SP-2007-6105, Rev 1, http://www.acq.osd.mil/se/docs/NASA-SP-2007-6105-Rev-1-Final-31Dec2007.pdf (accessed April 30, 2013).

Rechtin, E. 1991. *Systems Architecting, Creating and Building Complex Systems*. Englewood Cliffs, NJ: Prentice Hall.

Ulrich, K. T. and S. D. Eppinger. 2016. *Product Design and Development*. Sixth Edition. New York: Irwin McGraw-Hill.

Witus, G. and R. D. Ellis. 2011. Current Challenges in DoD Systems Engineering. In *Systems Engineering Tools and Methods* (Eds.) Kmarani, A. K. and M. Azimi. Boca Raton, FL: CRC Press, pp. 19–33.

2 Systems Engineering and Other Disciplines in Product Design

INTRODUCTION

The emphasis in this book is on products that are designed, made, and used by people. Thus, these products must be developed to satisfy certain needs of people. Some products are very complex, and Systems Engineering (SE) can help greatly during their life cycle phases. The complexity in a product can be directly related to the number of systems contained in the product; number of subsystems in each of the systems; number of components in each of the subsystems; and number of interfaces between different components, subsystems, and systems. The development process of complex products can be made more effective through a systematic implementation of the SE processes and application of a number of its techniques.

The objective of this chapter is to provide a deeper understanding of the SE, its approach, processes, issues, techniques, advantages, and disadvantages. In this chapter, we will develop the basic understanding into the SE and the product development processes by defining many terms and considerations used in the implementation of the processes. We would also study the other disciplines that are closely associated to the SE such as Quality Engineering, Human Factors Engineering, and Safety Engineering.

SYSTEMS ENGINEERING FUNDAMENTALS

What Is Systems Engineering?

Systems Engineering is a multidisciplinary engineering decision-making process involved in designing and using systems and products. The SE activities involve both the technical and management activities from an early design stage to the end of the life cycle of a product, or a system (i.e., when the product is removed from service and disposed off). The SE begins with understanding customers and their needs and the development of an acceptable concept of the product (or system). The SE is also a multidisciplinary approach, that is, it involves people from many different disciplines to work together and consider many design and operational issues and trade-offs between product characteristics simultaneously to enable the realization of a successful product or a system. SE focuses on defining customer needs and required functionality early in the development cycle, documenting requirements, and proceeding with the design synthesis and system (product) validation while considering the complete problem (INCOSE, 2006).

The objective of the SE is to see that the product (or the system) is designed, built, and operated so that it accomplishes its purpose of satisfying its customers in the most cost-effective way possible by considering performance, safety, costs, schedule, and risks.

The basic characteristics of the SE approach are as follows:

1. *Multidisciplinary*: The SE is an activity that knows no disciplinary bounds. It involves a collection of disciplines throughout the design and development process. It involves professionals from different disciplines working together (simultaneously and co-located under one roof) constantly communicating and helping each other on all aspects of the product. The types of disciplines to be included depend on the type and characteristics of the product and the scope of the product program.

    For example, the SE application for an automotive product will require disciplines such as engineering (e.g., mechanical, electrical, computer and information science, chemical, manufacturing, industrial, human factors, quality, environmental, and safety engineering), sciences (e.g., physics, chemistry, and life sciences), industrial designers (who define the sensory form of the product, i.e., the look, feel, sound, and smell of interior and exterior of the product, such as styling and appearance of surfaces of the product, touch feel of the surface and material characteristics, sounds of operating equipment, and smell of materials), market researchers (who define the product needs, its market segment, customers, market price, and sales volumes), management (e.g., program and project management personnel involving product planners, accountants, controllers, and managers), plant personnel involved in its manufacturing and assembly, insurers, distributors, and dealers.

    It is important to get inputs from all the disciplines that affect or are affected by the characteristics and uses of the product at the early stages of product development. This ensures that their needs, concerns, and trade-offs between different multidisciplinary issues are considered and resolved early, and costly changes or redesigns are avoided.

2. *Customer focused*: The SE places continuous focus on the customers, that is, the product design should not deviate from the needs of the customers. The customers should be involved in defining the product needs and in subsequent evaluations to ensure that the product being designed will meet their needs.

3. *Product-level requirements first*: The SE places concentrated effort on the initial definition of the requirements at the overall or the whole product level. For example, at the product level, the requirements on an automotive product will be based on all basic major attributes (that are derived from the needs of its external and internal customers) of the vehicles such as safety, fuel economy, drivability (ability to maneuver, accelerate, decelerate, and cornering or turning), seating comfort, thermal comfort, styling, and costs.

    It is important to realize that the customer buys the "integrated (whole)" product for his/her use and not as a mere collection of independent components that form the product. Thus, the requirements on the systems, subsystems, and components of the product should be derived only after the product-level requirements are clearly understood and defined.

4. *Product life cycle considerations*: The SE includes considerations of the entire life cycle of the product being designed—throughout all stages from "Concept Development to Disposal of the Product" (from lust to dust). Thus, it is the application of all relevant scientific and engineering disciplines in all the phases of the product such as designing, manufacturing, assembly, testing and evaluation, uses under all possible operating conditions, service and maintenance, and disposal or retirement from service—which the product encounters throughout its life cycle.
5. *Top-down orientation*: The SE considers "top-down" approach that views the product (or the entire system) as a whole and then sequentially breaks down (or decomposes) the product into its lower levels such as systems, subsystems, and components.
6. *Technical and management*: The SE is both a technical and management process. It involves making all the technical decisions related to the product life cycle as well as the management of all the tasks to be completed in a timely manner to implement the SE process and apply the necessary techniques.
7. *Technical process*: The technical process of the SE is an analytic effort necessary to transform the operational needs of the customers into a design of the product (or a system) with proper size, capacity, and configuration. It creates a documentation of the product requirements and drives the entire technical effort to evolve and verify an integrated and life cycle balanced set of solutions involving the users and the product in its usage situations.
8. *Management process*: The management process of the SE involves assessing costs and risks, integrating the engineering specialties and design groups, maintaining configuration control, and continuously auditing the effort to ensure that the cost, schedule, and technical performance objectives are satisfied to meet the original operational need of the product and the product program.
9. *Product-specific orientation*: The SE implementation (i.e., its details such as steps, methods, procedures, team structure, tasks, and responsibilities) depends on the product being produced (i.e., its characteristics) and the company producing it (i.e., different companies generally have somewhat different processes, timings, organizational responsibilities, and brand-specific requirements).

The *NASA Systems Engineering Handbook* (NASA, 2007) describes the SE as both the art and science of developing an operable system (or a product) capable of meeting requirements within often-opposed constraints. It further describes the SE as a holistic, integrative discipline, wherein the contributions of structural engineers, electrical engineers, mechanism designers, power engineers, human factors engineers, and many more disciplines are evaluated and balanced, one against another, to produce a coherent whole that is not dominated by the perspective of a single discipline (NASA, 2007).

MANAGING A COMPLEX PRODUCT

The need for SE arose with the increase in complexity of the products. (Note: Designing very simple products can be accomplished with a very small number of people.) The increased product complexity in turn increases the number of interactions (i.e., relationships) between many components and also increases the challenges in designing for high levels of reliability. The complexity is not only due to the engineering aspects of the systems but also due to the organization and management of people from multiple disciplines, data, and making a multitude of decisions.

Therefore, a complex product can be divided (or decomposed) into a number of manageable entities. This decomposition is an important task in the management of the product development process. The product can be decomposed into many systems, systems into subsystems, and subsystems into components. Some products can be divided into many hierarchical levels, that is, systems can be divided into many levels of subsystems such as subsystem, sub-subsystem, and sub-sub-subsystem. The number of levels of divisions depends on many factors such as the past design experience and problems encountered in developing similar products, ability of the design team to deal with many design issues simultaneously, level of the design and engineering details that need to be analyzed and evaluated, stringency in meeting requirements, program schedule, and so forth.

Figure 2.1 shows an illustration of a product decomposition tree. This figure presents a tree diagram (an upside-down tree) showing a top-down progressive

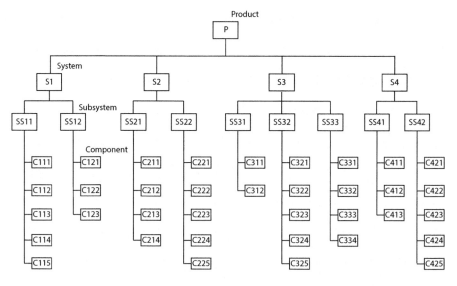

FIGURE 2.1 Illustration of top-down decomposition of a product into systems, subsystems, and components.

TABLE 2.1
Functions, Systems, and Components of a Laptop Computer

Function	System	Components
Hold all parts together	Chassis system	Top cover, hinges, magnesium chassis, palm rest, expansion cover, bottom cover
Display visual information	Display system	14-inch TFT display, protective glass, magnesium bezel
Present audio data	Audio system	Speakers, audio cable, stereo audio system, display connector
Input data	Input system	Keyboard, touchpad and track point, microphone
Process data	Electronic processing system	Printed circuit board (system board with a dual-core processor)
Store data	Memory system	DVD Drive, 64 GB solid-state drive, memory chips
Provide electrical power	Power system	AC to DC converter, replaceable battery, connectors, cables, internal battery
Communicate with the Internet	Wireless communication system	Wireless local network, wireless wide network
Protect components from overheat	Cooling system	Cooling fan, fan motor, heat sink, and shield

decomposition of a product (P) into its systems (S1 to S4), each system into its subsystems (e.g., SS11, SS12, SS21, ..., SS42), and each subsystem into its components (C111, C112, ..., C425).

Table 2.1 provides an example of systems and components of a laptop computer. The first column of the table provides a brief description of the functions of each system. The second and third columns provide the system name and its components, respectively.

It should be noted that each system exists to serve at least one or more functions required for the product to meet its requirements. It is also important to understand and keep track of the functions of each system because a design team involved in designing each system must make sure that the system performs its functions.

SYSTEMS ENGINEERING PROCESSES IN PRODUCT DEVELOPMENT

SYSTEMS ENGINEERING PROCESS

The SE Process during the product development includes the following basic tasks:

1. Define the objectives of the product (or product program)
2. Establish product performance requirements (requirement analysis)
3. Establish the functionality of the product (functional analysis)
4. Develop alternate design concepts for the product (architectural synthesis)
5. Select a baseline product design (selection of a balanced product design)
6. Verify that the baseline product design meets requirements (verification)

7. Validate that the baseline product design satisfies its users (validation)
8. Iterate the above process through lower levels (cascading product requirements to lower [decomposed] levels through allocation of functions and design synthesis).

It is important to realize that the product development typically involves many activities and tasks. The tasks can be organized and described differently depending on the objectives or viewpoints of the describer (e.g., a product design engineer vs. a systems engineer). A product design engineer may describe the product development and manufacturing process shown in Chapter 1, Figure 1.2, with the following major phases: (1) understanding the customer needs and product planning, (2) product concept development, (3) detailed design and engineering, (4) production planning process (includes production equipment and facilities design process), (5) production (manufacturing and assembly), (6) product distribution and sales, and (7) collection of customer feedback, whereas a systems engineer will describe the product development process differently as shown in Figure 2.2.

Figure 2.2 shows the following major SE tasks in the product development process: (1) defining the customer needs, (2) conducting requirements analysis (defining the requirements on the product), (3) conducting functional analysis (i.e., determining what functions need to be performed by the product, its systems, subsystems, and components) and allocating functions (i.e., assigning the functions to each system, subsystem, and component of the product), (4) conducting design synthesis (i.e., integrating, evaluating, and refining the whole product configuration and its architecture), and (5) formulating a balanced product design by considering trade-offs between requirements and different issues related to performance, safety, quality, costs, and timing schedules.

These tasks are performed iteratively (in the sense that the process is applied again in each iteration) as more design issues and product details are considered at a greater depth during each successive iteration. The overall product design with its systems is evaluated and redesigned (reconfigured, adjusted, or refined) in each iteration until an acceptable product is realized. A multifunctional co-located design team with constant communication and design reviews between team members and simultaneous (concurrent) engineering approach in performing the above activities is important for the successful completion of the product design activities.

FIVE LOOPS IN THE SYSTEMS ENGINEERING PROCESS

It is important to realize that the SE process shown in Figure 2.2 contains the following five types of loops: (1) requirements loop, (2) design loop, (3) control loop, (4) verification loop, and (5) validation loop. The loops are iterated to modify the product architecture and configuration until a balanced product design (i.e., satisfactorily meeting all requirements with the necessary trade-offs decisions between various design considerations) is achieved. The five types of loops are described as follows.

1. *Requirements loop*: This loop helps in refining the definition of requirements as they are used in analyzing the functions required by the requirements by

Systems Engineering and Other Disciplines in Product Design

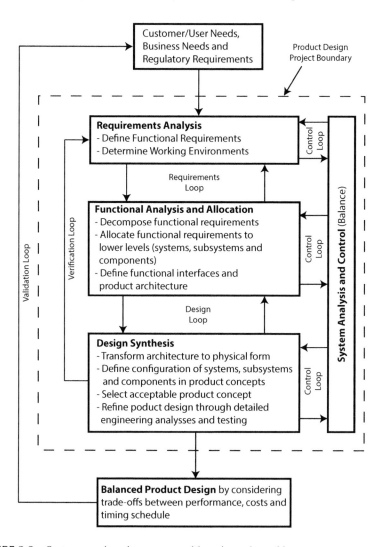

FIGURE 2.2 Systems engineering process with major tasks and loops.

allocating the functions to systems, subsystems, and components at various levels.

2. *Design loop*: This loop involves iterative applications of the functional analysis and allocation results to design the product such that the entire product with interfaces between various systems, subsystems, and components can perform to meet all its requirements.

3. *Control loops*: These loops make sure that right issues are considered and analyzed at the right time and right decisions are made to control the three basic tasks (requirements analysis, functional analysis and allocation, and design synthesis) shown in the diagram as "Systems analysis and control" (see the vertical block on the right side of Figure 2.2). Thus, the control

loops ensure that communications (e.g., questioning and refining design details in various design review meetings) occur with all personnel involved in the entire product development program. The communications and design reviews help attain a balance between various design issues by considering trade-offs between different product characteristics. The control loops thus facilitate timely communication of all the required tasks according to the Systems Engineering Management Plan (SEMP; see Chapter 8) and thus they help in meeting the budgetary and timing requirements of the product program.

4. *Verification loop*: This loop involves conducting tests on the designed product, its systems, subsystems, and components to ensure (i.e., to verify) that all requirements are met at every level. The testing can be accomplished by using computer applications (e.g., simulations), laboratory (or bench tests), or field tests depending on the availability of test facilities and the state of the hardware and/or the software to be tested. The verification process is iterative until the accepted design meets all the applicable requirements.
5. *Validation loop*: This loop involves tests and evaluations that are conducted to ensure that the product will meet all its stated customer needs, that is, the product will be acceptable by the customers. The validation tests are generally performed at the product level. However, system- or subsystem-level tests using customers are also useful in validating designs at the lower levels. The iterations of the validation loops and any changes resulting from the failures or deficiencies in the validation testing are generally expensive. If the product is designed right with the proper analyses in all the preceding tasks and their iterations, then the validation tests can be performed with a minimal number of iterations.

Major Tasks in the Systems Engineering Process

The major tasks shown in the three middle blocks and verification and validation loops of Figure 2.2 are described in the following section with more details.

Requirements Analysis

Requirements analysis is critical to the success of a product. The requirements should be documented, actionable, measurable, testable, traceable, related to identified customer and business needs (including meeting government rules and requirements), and defined to a level of detail sufficient to account for all the product attributes (i.e., product characteristics desired by the customers). This analysis either verifies that the existing requirements are appropriate or develops new requirements that are more appropriate for the mission/operation of the product.

The analysis includes the following: (1) development of measures suitable for ranking alternative designs in a consistent and an objective manner and (2) evaluations of the impact of environmental factors and operational characteristics on the performance of the product and minimum acceptable functional requirements. These measures and evaluations should also consider the impact of the design on costs and schedule. Each requirement should be periodically examined for validity, consistency, desirability, and attainability (see Chapter 4 for more information on requirements and Chapter 13 on sustainability/environment-related factors).

Functional Analysis and Allocation

The purpose of the functional analysis is to determine functions of each system, subsystem, or component of the product. The functions are determined from the performance requirements. For a given system, subsystem, or a component to meet its specified performance requirement, the systems engineer must progressively identify and analyze system functions and subfunctions in order to identify alternatives to meet the system requirements. The function identification task is performed in conjunction with function allocation and design synthesis activities that generally involve making assumptions regarding possible design configurations of the product and its systems. All specified modes of operation and situations (e.g., normal, unusual, or emergency) are considered in this analysis, and functions and subfunctions of all systems, subsystems, and components are identified. The process of allocation of function involves the following: (1) assignment of a requirement to a function, (2) assignment of a system element to a requirement, and (3) division of a requirement among entities and assignment of each component to its higher level entity (e.g., subsystem or system). For example, a requirement on the total weight of the system can be considered by allocating the target weight to each component (or percentage of the requirement satisfied or supported by each component).

During the allocation process, the SE activity allocates performance and design requirements to each system function and subfunction. These derived requirements are stated in sufficient detail to permit allocation of one or more functions to entities such as hardware, software, procedural data, or personnel. The SE also identifies any special personnel skills or peculiar design requirements that must be considered to perform the allocated functions. Allocation activities are performed in conjunction with the functional analysis and synthesis activities. Traceability of the allocated system requirements should be maintained to every function allocated to each entity.

The functions assigned to a system, subsystem, and component will be dependent on the type of technologies selected for the product. For example, for an automotive product, selection of technology for its power train system will affect the functions of many other systems, their subsystems, and components. If a gasoline engine is selected, then the fuel system and the entities (systems, subsystems, and components) of the engine will be very different than if an electric power train is selected. Thus, during the early iterations of the functional analysis, many different technologies and the level of readiness of the technologies should be considered to determine how various different requirements can be met.

Design Synthesis

During the synthesis, the SE together with the representatives of the hardware, software, and other appropriate engineering specialties develops a product architectural design (i.e., overall configuration of the product) that is sufficient to meet the performance and design requirements that are allocated in the detailed design. Design of the product architecture occurs simultaneously with the allocation of requirements and analysis of the product and system functions.

The design is generally documented by using block diagrams, flow diagrams, and/or package engineering drawings (or computer-aided design [CAD] models). These diagrams will (1) portray the arrangement of items that make up the baseline design, (2) identify each element along with techniques for its test, support, and operation, (3) identify the internal and external interfaces, (4) permit traceability to source requirements through each decomposition level, and (5) provide procedures for comprehensive change control. Several interface and flow diagramming techniques that can be used in the design synthesis are covered in Chapter 5 and Part III of this book. Various block diagramming techniques used for system breakdown and functional analyses are also described by Blanchard and Fabrycky (2011).

The documentation of work performed in this phase becomes the primary source of data for developing, updating, and completing the product, systems, and subsystems specifications, interface control documentation, specification trees, and test requirements. Interface control requirements and drawings should be established, coordinated, and maintained. Changes to these documents are maintained and disseminated to all appropriate participating engineering groups. (Note: Interface diagrams and interface matrices are covered in Chapter 5.)

During the final configuration and requirements definition, the SE uses the specifications as a mechanism to transfer information from the systems requirements analysis to system architecture design and system design tasks. Joint sign-offs of specifications by the specification authors, the designers, and the design engineers pertaining to the SE and the design engineering disciplines ensure understanding and buy-in into the overall product design. The specifications should ensure that the requirements are testable and are stated at the appropriate specification level.

Specialty engineering functions (e.g., human factors engineering, safety engineering, quality engineering, and others; also called attribute engineering functions [see Chapter 4]) participate in the SE process in all phases. They are responsible for reliability, maintainability, testability, manufacturing capability, quality management, human factors, safety, and design to meet their cost targets. Specialty engineering shall be involved in the issuing of design requirements and the monitoring of the progress of the design and performance analysis to ensure that the design requirements are met, and the trade-offs made between various product attributes are acceptable.

Verification

During requirements verification, the SE and test engineering conduct tests to verify that the completed product and its systems meet all the requirements contained in the requirements specifications. Tests conducted to verify requirements are performed using hardware and software (or acceptable computer simulations of both hardware and software) configured to the final design.

Validation

The entire product is tested under all types of situations expected during the operating life of the product. The testing should involve a representative sample of customers to ensure validity. In addition, industry experts (i.e., subject matter experts) who are very knowledgeable (and even more critical than the customers) about the product

Systems Engineering and Other Disciplines in Product Design

attribute issues should perform the validation evaluations. The outcomes of the tests are used to determine whether the product will be acceptable by its customers.

Verification Versus Validation

The difference between verification and validation can be better understood by studying the locations of the verification and validation blocks in the flow diagram presented in Figure 2.3. It shows the validation loop on the right side and the verification loop on the left side of the diagram. The validation is generally performed at the product level (see the right-side loop beginning from the "Balanced product" in Figure 2.3). The verification loop generally evaluates systems, subsystems, or components that are lower-level functions (as compared with the product level evaluated for validation purposes). Chapter 7 provides more information on verification and validation procedures and methods.

SUBSYSTEMS AND COMPONENTS DEVELOPMENT

Cascading requirements from a higher level to a lower level (e.g., product to systems, systems to subsystems, and subsystems to components) require considerable knowledge about possible configurations, technologies, and trade-offs between product characteristics in developing a basic architecture of the products and its systems. The term "cascading requirements" means transfer or allocation of a higher-level requirement to a lower-level entity. This means that a lower-level entity must be designed to meet a function allocated to a higher-level entity. The cascading of requirements

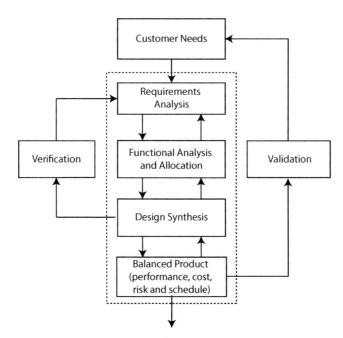

FIGURE 2.3 Verification and validation processes in product design.

from a higher to a lower level involves the following: (1) careful use of functional analysis and allocation to ensure that no requirements are dropped or misallocated (i.e., the requirement is allocated to an entity as compared with another entity that can perform the required functions better or when designed in a different configuration), (2) specifying subsystems with minimal interfaces with other systems and subsystems to satisfy the requirement with lower costs and reduced variability, and (3) specifying each subsystem requirement with clearly defined targets, test procedures, evaluation measures, and criteria for acceptable performance.

Design reviews with multifunctional team of experts are generally very useful to evaluate alternate configurations with different allocations of functions needed to fulfill the cascading of the product (or higher)-level requirements. The reviews should be conducted openly with a clear intention of working together to improve the customer acceptability of the product. The following issues should be discussed during the design reviews: (1) results of functional analysis and function allocation issues, (2) interfaces between systems and subsystems, (3) trade-offs to be considered between functions and features of different systems and subsystems, (4) effects on costs, reliability, and major product attributes (e.g., safety or weight), and (5) feasibility analysis results, design challenges, and problems. The feasibility analyses should consider issues such as (1) manufacturing feasibility and meeting or exceeding the targets specified in all applicable requirements, (2) feasibility of meeting time schedules and budgets, and (3) technical and personnel support, and development challenges.

EXAMPLE OF CASCADING A REQUIREMENT FROM THE PRODUCT LEVEL TO A COMPONENT LEVEL

Let us assume that an automotive climate control engineer is asked to design a climate control for a passenger car. The requirement of the climate control system would be that the occupants of the vehicle should be comfortable inside the passenger compartment by controlling the temperature to a customer-preferred level inside the compartment within 20°–29°C (68°–85°F) when the outside air temperature is anywhere between −7°C and 49°C (−20°F in the winter to 120°F in the summer). The subsystems of the climate control system are as follows: (1) hot air exchanger, (2) cold air exchanger, (3) air blower with a variable speed motor, (4) duct system, (5) hot coolant fluid (water) delivery from the engine cooling system, (6) compressed refrigerant from the air-conditioning pump, (7) hoses and tubes connecting the heat exchangers of the refrigerant and hot coolant fluid, (8) thermostat, (9) air blower speed controller, and (10) electrical system. Other vehicle systems that affect the occupant thermal comfort are vehicle body (window glass areas and body insulation) and seating subsystem (i.e., insulation of seat materials).

The climate control system requirements can be cascaded down to: (1) the heating system involving hot air exchanger, air blower, blower speed controller, thermostat, and hot coolant supply system, (2) the cold air supply system involving the cold air exchanger, air blower, blower speed controller, thermostat, and the compressed refrigerant supply system, (3) windows, glass, and vehicle body that conducts heat from outside air temperature and radiant heat from the sun load through the glass areas, and (4) heat transfer characteristics of the seat upholstery.

The cascaded requirements are shown in Table 2.2. The vehicle (product)-level requirement is stated in the second column from the left. The cascaded system-, subsystem-, and component-level requirements are stated in the fourth, sixth, and eighth columns from the left, respectively.

The requirements and the requirements-setting process are further discussed in Chapter 4.

ITERATIVE NATURE OF THE LOOPS WITHIN THE SYSTEMS ENGINEERING PROCESS

The SE process is based on an iterative, top-down, hierarchical decomposition of the product-level requirements to the component-level requirements. The decomposition process is supported by studies and analyses performed by specialists from different disciplines that analyze the basis for significant decisions and the options considered. The iterative, top-down, hierarchical decomposition methodology includes the parallel activities of the functional analysis, allocation, and synthesis. The iterative process begins with product-level decomposition and proceeds through the major system level, the functional subsystem level, and the hardware/software configurations of entities at the lower levels. As each level is developed, the activities of functional analysis, allocation, and synthesis should be completed before proceeding to the next lower level.

When carryover components are used, bottom-up iterative processes also need to be implemented to ensure that the subsystems and systems are designed to meet the needs to interface the carryover components. However, the bottom-up iterations can reduce design flexibilities in designing the upper-level systems. The constraints placed by the carryover components can reduce the performance and efficiencies of the upper-level systems and the product.

INCREMENTAL AND ITERATIVE DEVELOPMENT APPROACH

In some cases, the product development process is started with incomplete information before the customer needs are fully known or understood. This could be because of time pressures or other urgent demands (e.g., political, security, competition) to begin the product development process quickly. As more and more information on the customer needs and competitive products is available, the customer requirements are refined, and the product being developed is modified. The product development process under such situations can become very chaotic, volatile, and inefficient. However, providing flexibility in responding to new information is also important. INCOSE (2006) refers to such process as "evolutionary development" where speed, responsiveness, flexibility, and adaptability are important. Despite the downside of unstable, chaotic project, this approach can allow learning through continuous changes and investigations of a number of possible product features and product architectures. And thus, it avoids the loss of large investments and may generate short-term or localized solution optimizations.

TABLE 2.2
An Illustration of Requirements Cascade from Product Level to Component Level

Product Level	Vehicle-Level Requirement	System Level	System-Level Requirement	Subsystem Level	Subsystem-Level Requirement	Components	Component-Level Requirement
Product = Passenger car	Comfortable thermal environment inside the passenger compartment	Climate control system	Provide hot or cold air according to thermostat and blower speed in accordance with outside heat load	Heating system	Provide hot air according to thermostat and blower speed in accordance with outside heat load	Hot air exchanger	Allow hot water to circulate
						Air blower	Blow air at the set speed
						Hot water supply	Send hot water to hot air exchanger
						Thermostat	Control hot water inflow
						Air blower speed controller	Control blower speed to speed setting
						Heater air ducts	Pass hot air through floor vents
						Hot water hoses and pipes	Maintain hot water flow
				Cooling system	Provide cold air according to thermostat and blower speed in accordance with outside heat load	Cold air exchanger	Allow compressed refrigerant to pass through expansion valve in the cold air exchanger
						Air blower	Blow air at the set speed
						Compressed refrigerant supply	Send refrigerant to the expansion valve

Systems Engineering and Other Disciplines in Product Design

System	Function	Component	Sub-function	Sub-component	Sub-sub-function
				Thermostat	Control refrigerant flow
				Air blower speed controller	Control blower speed to speed setting
				A/C air ducts	Pass cold air through air registers in the instrument panel
				Refrigerant hoses and pipes	Maintain refrigerant flow
Vehicle Body	Protect occupants from exterior heat	Body-in-white	Hold positions of climate control, seats, window openings and window glass.	Window openings	Reduce incident radiant heat load
				Window glass	
			Secure insulation materials in body cavities	Insulation	Reduce conductive heat load
		Seat	Position occupants in the vehicle interior space	Seat upholstery material	Reduce heat conductance of upholstery material
Electrical system	Provide electrical energy to the blower and thermostat	Control power flow through climate controller unit	Control circuits to the blower, thermostat, and hot water and refrigerant fluid flow		

SYSTEMS ENGINEERING "V" MODEL

All the above-discussed steps and considerations can be described along with a horizontal time axis in a model. The model is presented in Figure 2.4, and it is known as the SE "V" model (refer to Blanchard and Fabrycky [2011] for more information). The model is described in the context of development of a new automotive product. The model shows the basic phases of the entire product program on a horizontal time axis which represents time (t) in months before Job#1. In the automotive industry, "Job#1" is defined as the event when the first production product (vehicle) is shipped out of the assembly plant. The product program generally begins many months prior to the Job#1. The beginning time of the program depends on the complexity of the program (i.e., the changes in the new product as compared with the out-going product) and the management's approval to begin the product development process.

In the early stages prior to the official start of a product program, an advanced product planning activity (usually the advanced product planning department or a special product planning team) determines the product characteristics and preliminary architecture (e.g., vehicle type, size, and type of power train), performance characteristics, the intended market (i.e., countries where the product will be sold), and a list of reference products (used for benchmarking) that the new product may replace or compete with. A small group of engineers and designers from the advanced design group are selected and asked to generate a few early product concepts to understand design and engineering challenges. A business plan including the projected sales volumes, the planned life of the product, the program timing plan, facilities and tooling plan, manpower plan, and financial plan (including estimates of costs, capital needed, revenue stream, and projected profits) is developed and presented to the senior management along with all other product programs planned by the company (to illustrate how the proposed program will fit in the overall corporate product plan). The product program typically begins officially after the approval of the business

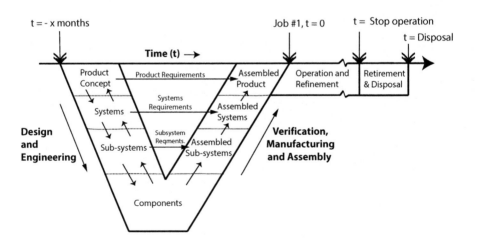

FIGURE 2.4 Systems engineering "V" model showing life cycle activities from the concept design to the disposal of a complex product.

Systems Engineering and Other Disciplines in Product Design

plan by the company management. This program approval event is considered to occur at x-months prior to Job#1 in Figure 2.4. (Note: The business plan is described in Chapter 15.)

At minus x-months, the chief product program manager is selected, and each functional group (such as design, body engineering, chassis engineering, power train engineering, electrical engineering, aerodynamics engineering, packaging and ergonomics/human factors engineering, and manufacturing engineering) within the product development and other related activities are asked to provide personnel to support the product development work. The personnel are grouped into teams and the teams are organized to design and engineer various systems and subsystems of the product.

Figure 2.4 shows that the first major phase after the team formation is to create an overall product concept. During this phase, the designers (industrial designers) and the package engineers work with different teams to create the product concept that involves creating (1) early drawings or CAD models of the proposed product, (2) computer-generated 3D life-like images or videos of the product (fully rendered with color, shading, reflections, and textural effects), and (3) physical mock-ups (foam core, clay, wooden, or fiberglass bucks to represent the exterior and interior surfaces of the product). The images and/or models of the proposed product are shown to prospective customers in market research clinics and to the management. Their feedback is used to further refine the product concept.

As the product concept is being developed, each engineering team decides on how each of the systems can be configured to fit within the product space (defined by the product concept) and how the various systems can be interfaced with other systems to work together to meet all the functional, ergonomic, quality, safety, and other requirements of the product. This phase is shown as the "Systems" phase in Figure 2.4. If problems about any of the systems are discovered during this phase, the designs are sent back to refine or modify the product concept. (The feedback is represented as the up arrow from the systems to product concept shown in Figure 2.4.)

As the systems are being designed, the next phases involve a more detailed design, that is, design of subsystems of each system and components within each of the subsystems. These subsequent phases, straddled in time to the right, are shown as "Subsystems" and "Components." The above phases from the left half of the "V" represent the time and activities involved in "Design and Engineering." The up arrows in the left half of the "V" indicate the iterative nature of the SE loops covered earlier (see Figures 2.2 and 2.3).

The right half of the "V," moving from the bottom to the top, involves testing, assembly, and verification where the components are produced and tested to ensure that they meet their functional characteristics and requirements (developed during the left half of the "V"). The components are assembled to form subsystems that are tested to ensure that they meet their functional requirements. Similarly, the subsystems are assembled into systems and tested; and finally, the systems are assembled to create the whole product. At each phase, the corresponding assemblies are tested to ensure that they meet the requirements considered during their respective design phases (i.e., the assemblies are verified). These requirements are shown as the horizontal arrows between the left and the right sides of the "V" in Figure 2.4. It should be noted that down arrows between various assembly steps in the right half of "V" are not shown in Figure 2.4. The down arrows will indicate failures in the verification steps. When failures occur, the information is transmitted to the respective design team to incorporate design changes to avoid repetition of such failures.

The engineers and technical experts assigned to various teams in the product program work throughout all the above phases and continuously evaluate the product design to verify that the product users can be accommodated, and they will be able to use the product under all foreseeable usage situations. Early production vehicles developed just before the Job#1 are usually used for additional whole vehicle evaluations for the product validation purposes (see Chapter 7 and Figure 7.5 for additional information on verification and validation testing).

After Job#1, the product is transported to the dealerships and sold to the customers. The model in Figure 2.4 also shows a time period (on the right side of the V) called "Operation and refinement." During this time period, the product is being purchased, used, and maintained by the customers and serviced by dealers (or other repair shops). The product may also be refined (revised with minor changes during the existing model cycle or updated as refreshed or new models every few years) with some changes during its operational time. As the product becomes old and outdated, the product is pulled from the market. This marks the end of the life cycle of the product. At that point, the assembly plant and its equipment are recycled or retooled for the next product (as a next model year product), a totally new product is produced in the plant, or the plant is closed. As the products reach the end of their useful life, the products are sent to the scrapyards where many of the components may be disassembled. The disassembled components are either recycled for extraction of the materials or sent to the junkyards.

The website of this book contains an Excel spreadsheet illustration of a "V" model of an automotive product development process. It also shows a Gantt chart with various activities and monthly cost estimates of the activities (see Chapter 9, Figure 9.1 and Table 9.2).

NASA DESCRIPTION OF THE SYSTEMS ENGINEERING PROCESS

The *NASA Systems Engineering Handbook* describes the SE to include 17 processes that are grouped into the following three sets of processes: (1) system design processes, (2) product realization processes, and (3) technical management processes (see Figure 2.5). The descriptions of each set of three processes as provided in the *NASA Systems Engineering Handbook* (2007) are provided below.

System design processes: These NASA-defined system design processes (see the left side of Figure 2.5) for a complex product design problem are really the part of the "Design and Engineering" activities that take place on the left side of the SE "V" model shown in Figure 2.4. (These processes are numbered as 1 to 4 in Figure 2.5.) The four-system design processes define the stakeholder (customer) expectations, generate technical requirements, logically decompose the problem and the product, and convert the technical requirements into a technical solution (i.e., a product concept) that will satisfy the stakeholder's expectations. These processes are applied

Systems Engineering and Other Disciplines in Product Design 41

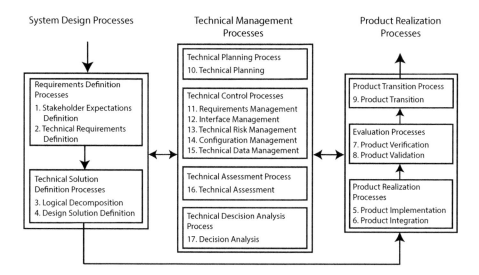

FIGURE 2.5 Systems engineering process described in the NASA handbook. (Redrawn from National Aeronautics and Space Administration [NASA], *NASA Systems Engineering Handbook*, Report no. NASA/SP-2007-6105 Rev1, NASA Headquarters, Washington, DC, 2007.)

iteratively to each level of the product, that is product to systems, systems to subsystems, and subsystems to component level until the lowest level components are defined to the point where they can be built, bought, or reused (from carryover components). The outputs of the system design processes are fully defined technical requirements on all levels of the product and the design solution defining the product.

Product realization processes: The product realization processes are shown on the right side of Figure 2.5. These processes are numbered as 6 to 9 in Figure 2.5. These processes are applied to each operation/mission of the product in the system structure starting from the lowest level product (component level) and working up to higher-level integrated entities. These processes are used to create the design solution for each entity (e.g., by assembling components into subsystems, subsystems into systems, and systems into product—referred here as the product implementation or product integration processes [numbered 5 and 6]). The processes numbered 7, 8, and 9 are for verification, validation, and transitioning up to the next hierarchical level of the product. The realized product design solutions thus meet the stakeholder expectations. (Note that these processes take place on the right side of the SE "V" model shown in Figure 2.4.)

Technical management processes: The technical management processes (shown in the middle section of Figure 2.5 and numbered as 10 to 17) are used to establish and evolve technical plans for the project; to manage communication across interfaces; to assess progress against the plans and requirements for the systems, products, or services; to control technical execution of the project to completion; and to aid in the decision-making process.

The above SE processes are used both iteratively and recursively. It should be noted that the horizontal two-way arrows between the three sections of Figure 2.5 indicate the iterative and recursive nature of the entire process. The "iterative" is the "application of a process to the entities at different levels with the same product or set of products to correct a discovered discrepancy or other variation from requirements," whereas "recursive" is defined as adding value to the product (or system) "by the repeated application of processes to design next lower level system or to realize next upper level of the products within the product (or system) structure." This also applies to repeating application of the same processes to the system structure in the next life cycle phase to mature the system definition and satisfy phase success criteria. The technical processes are applied recursively and iteratively to break down the initializing concepts of the product (or system) to a level of detail concrete enough that the technical team can implement a product from the information. Then, the processes are applied recursively and iteratively to integrate the smallest product entity into greater and larger systems until the whole of the product (or system) has been assembled, verified, validated, and transitioned.

The above-described system design processes, product realization processes, and the technical management processes can thus be considered as the SE implementation in the traditional product development process. The performance and effectiveness of the product development process can thus be managed by the technical and management processes within the SE process. This issue is covered in the next section.

MANAGING THE SYSTEMS ENGINEERING PROCESS

The Systems Engineering Management Plan (SEMP) is a documentation created to plan the SE activities in terms of what needs to be done (e.g., types of analyses, methods, and procedures), when the activities need to be done (e.g., sequence of activities in the process), and who performs the activities (i.e., certain specified engineering departments or disciplines).

The SEMP is basically a planning document, and it should be prepared by the SE activity after the concept design phase. The SEMP is covered in detail in Chapter 8.

Relationship between Systems Engineering and Program Management

The SE emphasizes technical issues by analyzing, measuring, and risk reduction, whereas the program (or project) management deals with the management issues. The main aim of the systems engineer is to create a successful product whereas that of the program manager is to successfully complete the program on time and within the cost constraints. Both the disciplines, the SE and the program management, overlap in terms of understanding the importance of the various tasks that need to be completed in the process of designing and producing the product. Figure 2.6 presents a Venn diagram showing responsibilities of the SE and the project (or program) management in a product program. The SE covers the technical tasks and issues, and the project (or program) management covers management of timings of planned tasks, resources, schedules, documentation, and data. The tasks that are common to both

Systems Engineering and Other Disciplines in Product Design

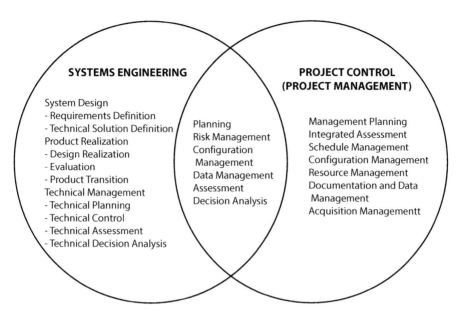

FIGURE 2.6 Relationship between systems engineering and project management. (Redrawn from National Aeronautics and Space Administration [NASA], *NASA Systems Engineering Handbook*, Report no. NASA/SP-2007-6105 Rev1, NASA Headquarters, Washington, DC, 2007.)

are shown in the overlapping parts of the Venn diagram in Figure 2.6. These overlapping areas contain common or shared data needed by both disciplines in areas such as overall product planning (with schedule and budget), product configuration, internal and external risks, decisions and analyses that affect the program, data, and so forth.

The systems engineer, as compared with the program management, has to concentrate more on the early stages of the conceptual and preliminary design to ensure that technical aspects such as the requirements are properly considered, and systems are being designed by proper consideration of trade-offs and allocation of functions within the packaging space of the product. These early steps have a major influence on ensuring that the right product is being designed.

Program management is the function of planning, overseeing, and directing the numerous activities required to achieve the requirements, goals, and objectives of the customer and other stakeholders within specified cost, quality, and schedule constraints. Chapter 8 provides more information on program management functions and also covers issues that distinguish the program management from the project management.

ROLE OF SYSTEMS ENGINEERS

The systems engineers will usually play the key role in leading the development of the product and/or system architecture, defining and allocating requirements, evaluating design trade-offs, balancing technical risks between systems, defining and assessing interfaces, and providing oversight on verification and validation activities. The systems engineers will usually have the prime responsibility in developing many of the project documents, including the SEMP, requirements/specification documents, verification and validation documents, certification packages, and other technical documentation (NASA, 2007).

The SE is about trade-offs and compromises, about generalists rather than specialists. The SE is about looking at the "big picture" and not only ensuring that they get the design right (meets requirements) but that they get the right design. Thus, a system engineer needs to perform the following tasks:

1. Understand customer and program needs
2. Obtain required data
3. Develop SEMP
4. Communicate the SEMP to program teams
5. Provide recommendations to program teams on SE tasks
6. Assist teams in conducting necessary trade-off analyses
7. Continuously communicate with program teams to perform the above tasks.

Chapter 25 provides additional information on the characteristics of a "good" systems engineer and challenges facing the SE profession.

INTEGRATING ENGINEERING SPECIALTIES INTO THE SYSTEMS ENGINEERING PROCESS

As part of the technical effort, specialty engineers (who have specialized in an engineering discipline, e.g., mechanical engineering) in cooperation with the SE and subsystem designers often perform tasks that are common across disciplines. Foremost, they apply specialized analytical techniques to create information needed by the project/program manager and the systems engineer. They also help define and write system requirements in their areas of expertise, and they review data packages, engineering change requests (ECRs), test results, and documentation for major project reviews. The project manager and/or systems engineer needs to ensure that the information and products so generated add value to the project commensurate with their cost.

The specialty engineering technical effort should be well integrated into the project. The roles and responsibilities of the specialty engineering disciplines should be summarized in the SEMP. The specialty engineering disciplines to be included in a product program will vary depending on the product needs; however, the following disciplines are generally included:

1. Quality Engineering
2. Mechanical Engineering
3. Electrical/Electronics/Computer Engineering
4. Industrial Engineering/Operations Research
5. Human Factors Engineering
6. Safety Engineering
7. Reliability Engineering
8. Manufacturing, Processing, Assembly, and Plant Engineering.

Systems Engineering and Other Disciplines in Product Design 45

In addition to the above engineering disciplines, the product design team generally includes members from the following disciplines:

1. Styling/Industrial Design
2. Environmental/Sustainability Engineering
3. Energy systems engineering
4. Marketing, Sales, and Dealerships
5. Finance, Product Planning, and Program/Project Management
6. Legal Support on Regulations, Patents, and Liabilities.

ROLE OF COMPUTER-ASSISTED TECHNOLOGIES IN PRODUCT DESIGN

CAD AND CAE

The way we use computers in the life cycle of a product has changed dramatically over the past 15–30 years. The computer-assisted technologies form the backbone of product visualization, configuration, and engineering evaluations. The CAD systems are used to create three-dimensional solid models of parts, assemblies, and movements of parts. The CAD helps enormously in visualization of the product and in communicating product configuration, features, and details. The CAD also helps in early phases of packaging systems and components in the product space. Many different product configurations with different allocations of spaces for different systems, subsystems, and components can be created in a short time to enable visualization of different possible arrangements (or layouts). The CAD models also help in communicating and comparing product information (e.g., drawings, dimensions, tolerances for assembly, fits, clearances, or interferences between parts) between different product design, engineering, and testing activities of the equipment producers and their suppliers. The CAD models can also be used to plan production processes, program production machines, and management of supply chains (e.g., ordering materials and parts). Chapter 15 provides additional information on these CAD tools.

Computer-assisted engineering (CAE) applications can be used to analyze structural, thermal, fluid flow, and many functional aspects of the products. Prototype parts and systems can be created for evaluations. Many computer simulations can be conducted to evaluate different product uses over a large number of use cycles under different usage conditions before costly field testing is undertaken.

The computer-assisted/integrated technologies can save time, increase speed and efficiency, reduce costly errors (e.g., avoiding misreading of dimensional values), and rework in accomplishing many tasks during the life cycle of complex products.

MODEL-BASED SYSTEMS ENGINEERING

Model-based systems engineering (MBSE) is now rapidly gaining interest among the SE and product design community. It is a systems engineering methodology that focuses on creating and using predeveloped models as the primary means of information exchange between engineers, rather than on document-based information exchange. It helps to support system requirements, design, analysis, verification and validation activities from the beginning in the conceptual design phase and continuing throughout development and later life cycle phases. The MBSE methodology is used in many industries such as aerospace, defense, rail, and automotive.

MBSE is the formalized application of modeling to support system or product development activities involving system requirements, design, analysis, verification, validation, and all life cycle phases of the system (or product). It is, thus, an implementation of the systems engineering process that focuses on the use of a software application (i.e., a model) as the primary means of information exchange, rather than on document-based information exchange. It helps in developing and managing requirements and interfaces between systems with complex functionality and capabilities.

Model-based approach as compared to the traditional document-based approach is considered superior because of the ability of the software to obtain inputs from many team members and to maintain detailed database on relationships and traceability created by the software applications. The model-based approach is also found to reduce the time taken to develop product/system specifications and it is also more accurate and error-free. Additional information on MBSE is provided in Chapter 15.

IMPORTANCE OF SYSTEMS ENGINEERING

The SE with the assistance from the other engineering disciplines establishes the baseline product (or system) design, allocates system requirements, establishes measures of effectiveness for ranking alternative designs, and integrates the design among the design disciplines. The SE is responsible for verifying that the developed product (or system) meets all the requirements defined in the product specifications. The SE also provides the analyses to ensure that all the requirements will be met. Thus, the products developed by applications of the SE principles, processes, and techniques will benefit from the following:

1. Right products will be developed because the SE will make sure that (a) the customer needs are obtained and translated into requirements, (b) the requirements are used by multidisciplinary teams for product development, (c) best product configurations are selected because of the iterative and recursive applications, (d) all product entities are verified to assure compliance with their requirements, and finally, (e) the product is validated. Thus, the customers will like the products and will be very satisfied.
2. Product development time can be reduced by avoiding costly delays.
3. Costly redesign and rework problems will be reduced.
4. Product will remain in the market for a longer time.

ADVANTAGES AND DISADVANTAGES OF THE SYSTEMS ENGINEERING PROCESS

The major advantages of the implementation of the SE process in the development of a complex product program are as follows:

1. It will help in reducing costs and time overruns.
2. It will help in creating products that the users want (i.e., ensures customer's satisfaction).

The disadvantages of incorporation of SE functions in a product development program are as follows:

1. It adds people to the payroll and thus increases costs to the program.
2. It creates an additional documentation burden by creating the SEMP.
3. It creates more work for the team members in communicating with the SE activities.

SOME CHALLENGES IN COMPLEX PRODUCT DEVELOPMENT

Implementation of the SE process in a major product program faces many challenges. There are many basic needs in successful implementation of the SE. Some important considerations in the SE implementation include the following:

1. Commitment of everyone in the organization from the top management to the team members to understand and follow the SE process
2. Availability of resources to recruit experienced systems engineers with formal training in the SE process and its techniques
3. Organization culture to work in teams and focus on satisfying customer needs
4. Commitment in systematically following the iterative and recursive process from requirements analysis to development of a balanced product concept
5. Availability of detailed product engineering capability with test facilities and resources to perform the verification and validation tests
6. Top-notch program management and communication capability
7. Availability of the state-of-art computer-assisted design, analysis, and engineering capabilities.

Some challenges and issues in the SE implementation in the product development programs are provided below.

1. Reducing the time required in implementation, integration, and execution of the SE activities. (This in turn can reduce program costs.)
2. Reducing chances of long delays in reworking and cost overruns or even cancellation of programs. Some reasons for such problems are as follows:

a. Customer needs may change with time due to technological, political, economic factors—many of which cannot be predicted well during the early stages of the program.
  b. Technology readiness may be overestimated (sufficient time not allocated to develop new systems).
  c. Impossible or overly ambitious requirements are given to the product development teams, for example, during the development of the F-111 aircraft, the Air Force and the Navy provided very different requirements for the aircraft (Kamrani and Azimi, 2011).
  d. Changes in priorities during the product program, for example, new information during the Desert Storm operation required the Department of Defense to change priorities of the basic needs for a new airplane (Kamrani and Azimi, 2011).
  e. Late inclusion of new customers with different requirements (e.g., shifting the market strategy of selling a vehicle designed for one market to another such as from the U.S. market to the European or Asian market or vice versa) can have a major impact on the success of the product.
  f. Unexpected results or surprises in the product evaluations (tests) requiring a lot of reanalysis and redesign of the product.
  g. Personnel with the required expertise and training not available for the program. (New team members must be trained in the SE processes and techniques.)
3. Initial estimates of program time and budget may be overly optimistic or unrealistic. (This would apply more to new products with no historic data on similar products.)
4. Difficulties in resolving trade-offs between different product characteristics or product attributes. For example, the systems engineer is faced with decisions such as:
  a. To reduce cost at a constant level of risk, performance must be reduced.
  b. To reduce risk at a constant cost, performance must be reduced.
  c. To reduce cost at a constant performance level, higher risks must be accepted.
  d. To reduce risk at a constant performance level, higher costs must be accepted.

Additional challenges in the implementation of the SE and future needs and issues are presented in Chapter 25.

CONCLUDING REMARKS

The SE process is a powerful process in supporting complex decision-making tasks involved in multidisciplinary considerations. The process is very useful in designing complex products and systems. It is however difficult to master as there is no unique implementation procedure. In fact, it is a very flexible process in the sense that it can be applied in different level of details depending on the resources available and the implementer's knowledge of the process and its techniques. The implementation of the SE process can vary between organizations and also within different products of the same organization. Not all programs require a very detailed and formal SE implementation. Smaller product programs cannot generally afford to have a separate SE department or systems engineers dedicated to the program.

The SE process is especially useful to design very large and complex products (or systems) that have hundreds or thousands of components. Beginning the process with the customer needs and performing requirements analysis iteratively and recursively provides a tremendous advantage in defining what needs to be done. Conducting functional analysis and design synthesis iteratively helps in creating and evaluating a number of possible alternate configurations. The processes of verification and validation ensure that the product meets its requirements and the product itself is the right product for the customers.

REFERENCES

Blanchard, B. S. and W. J. Fabrycky. 2011. *Systems Engineering and Analysis*. Upper Saddle River, NJ: Prentice Hall PTR.

International Council on Systems Engineering (INCOSE). 2006. *Systems Engineering Handbook—A Guide for System Life Cycle Processes and Activities*. Report no. INCOSE-TP-2003-003-03, Version 3.

Kamrani, A. K. and M. Azimi (Eds.). 2011. *Systems Engineering Tools and Methods*. Boca Raton, FL: CRC Press.

National Aeronautics and Space Administration (NASA). 2007. *NASA Systems Engineering Handbook*. Report no. NASA/SP-2007-6105 Rev1. NASA Headquarters, Washington, DC. 20546. http://ntrs.nasa.gov/archive/nasa/casi.ntrs.nasa.gov/20080008301_2008008500.pdf (accessed October 15, 2012).

3 Decision-Making and Risks in Product Programs

INTRODUCTION

Decisions are made throughout the life cycle of every product program. Decisions are made whenever alternatives exist and the most desired alternative needs to be selected. The selected alternative should result in reducing risks and increasing benefits. Many different criteria can be used in selecting an alternative. The early decisions are related to the type of product to be designed (e.g., an automotive manufacturer needs to decide on the type of vehicle to be designed), requirements on the product, and its characteristics (e.g., how far should an electric vehicle travel on a fully charged battery). Later, the decisions are related to the number of systems in the product, their functions, and how the systems should be configured and packaged within the product space.

Decisions are also made during each phase and at each milestone of a product development program where the management decides whether to proceed to the next phase, make changes to the product being designed or even to scrap the program. Early decisions have a major impact on the overall costs and timings of the program—because the later decisions depend on the design-specific parameters and their values selected in the earlier phases of the program. For example, powertrain type, size, its location in the vehicle space (e.g., front-wheel drive or rear-wheel drive), and the technologies to be implemented in a new automotive product will affect decisions related to the design of its systems (e.g., fuel system, cooling system, and space available to package suspensions).

After the product is introduced in the market, the customer feedback is received. The manufacturer needs to decide on whether to make any changes and what to change if negative feedback is received. The reasons for the customer dissatisfaction need to be understood and decisions need to be made on whether to recall the product or how and when to fix any defects in the product. Furthermore, after the product is marketed for an extended period of time, decisions need to be made on what product characteristics should be revised, how to revise, and when to revise.

All the above-described decisions involve risks. For example, adding more features (or capabilities) than what the customers need, and over-designing will waste resources. Conversely, failures in incorporating any customer-desired major changes and under-designing the product will result in loss of sales or even degrade the reputation of the product and its brand in the marketplace.

This chapter covers various decision-making approaches and models and also provides an understanding into issues related to risks and methods to analyze the risks.

DOI: 10.1201/9781003263357-4

PROBLEM-SOLVING APPROACHES

Many different approaches are used to solve problems in various scientific and engineering fields. The approaches are generally similar and there are some variations in their applications. The scientific method is an efficient method of doing research (learning by doing) (Konz, 1990). Table 3.1 compares the steps in the scientific method with that in the engineering design method and the six-sigma quality engineering approaches. DMAIC and IDOV are the acronyms of the two six-sigma approaches. The acronym, DAMES, was suggested by Konz (1990) to remember the steps in the engineering design method. (Note: The acronyms can be found in third, fifth, and seventh columns of Table 3.1 by reading vertically down. Also see Chapters 10 and 14 for more information on the six-sigma methods.) All the four approaches presented in Table 3.1 are generally used in an iterative manner until a satisfactory solution is found. Each step of any of these problem-solving approaches requires decision-making. Decisions are made during each step to decide on what to do, how to do, when to do, what alternative to choose, and so forth.

Early steps in all problem-solving methods involve data collection and analysis of the available data to understand the problem and to develop possible alternatives. The alternatives are evaluated by conducting experiments or exercising available models (which are generally verified by earlier research). The results of these experiments and model applications are used to make the decisions to select acceptable solutions.

TABLE 3.1
Problem-Solving Approaches

Step No.	Scientific Method		Engineering Design Method		Six-Sigma Improvement Method		Design for Six-Sigma (DFSS)
1	State the problem	D	*D*efine Problem	D	*D*efine Problem	I	*I*dentify Requirements
2	Construct a hypothesis or model	A	*A*nalyze—Determine relevant variables	M	*M*easure—Gather data on issues and variables	D	*D*esign—Characterize it by invention and innovation
3	Apply analysis to the model. Use the model to predict what can happen	M	*M*ake—Search for Possible Solutions	A	*A*nalyze—Analyze gathered information	O	*O*ptimize the Design—Conduct analyses and/or experiments
4	Design and conduct an experiment under real situations	E	*E*valuate possible solutions to determine the best solution	I	*I*mprove—Develop improvements	V	*V*erify the Design—Conduct evaluations/tests to verify and validate the design
5	Compare model predictions with experimental results	S	*S*pecify Solution	C	*C*ontrol—Establish process controls		

DECISION-MAKING

ALTERNATIVES, OUTCOMES, PAYOFFS, AND RISKS

Systems Engineering (SE) involves decision-making such as what needs to be done, when, how, and how much, taking into account the trade-offs between possible design considerations. In a decision-making situation, the decision maker (e.g., engineer, designer, or program manager) is faced with the task of deciding on an acceptable alternative among several possible alternatives. The decision maker also needs to consider possible future outcomes (i.e., what will happen in the future), and the costs or benefits (called the payoffs) associated with each combination of an alternative and an outcome. Further, each possible outcome may or may not occur in the future. There are many different decision models to determine a desired or an acceptable alternative (Blanchard and Fabrycky, 2011). A few of the models are described here.

Let us assume the following:

A_i = ith alternative, where $i = 1, 2, \ldots, m$
O_j = jth outcome, where $j = 1, 2, \ldots, n$
P_j = the probability that jth outcome will occur, where $j = 1, 2, \ldots, n$
E_{ij} = evaluation measure (payoff—positive for benefit [profit] and negative for cost [loss]) associated with ith alternative and jth outcome.

The decision evaluation matrix associated with the above problem is presented in Table 3.2. Many principles can be used to select a desired alternative. The principles are described here.

MAXIMUM EXPECTED VALUE PRINCIPLE

One commonly used principle to select an alternative is based on the maximum expected value. The expected value of $A_i = \{E_i\}$ can be computed as: $\sum_j [P_j \times E_{ij}]$.

TABLE 3.2
Decision Evaluation Matrix

	Probabilities of Outcomes					
	P_1	P_2	P_3	·	·	P_n
	Outcomes					
Alternative	O_1	O_2	O_3	·	·	O_n
A_1	E_{11}	E_{12}	E_{13}	·	·	E_{1n}
A_2	E_{21}	E_{22}	E_{23}	·	·	E_{2n}
A_3	E_{31}	E_{32}	E_{33}	·	·	E_{3n}
·	·	·	·	·	·	·
A_m	E_{m1}	E_{m2}	E_{m3}			E_{mn}

Thus, under this principle, the decision maker will select the alternative with the maximum expected value, which is defined as: max $\{E_i\}$ for $i = 1, 2, \ldots, m$.

The selection of the alternative and application of the above principle are illustrated in the following example.

Let us assume that an automotive manufacturer wants to select a powertrain for its new small vehicle. The manufacturer is considering the following five alternatives:

A_1 = Design a new small car using the current gasoline powertrain
A_2 = Do not design a new small car—continue with the present model
A_3 = Design a new small car with an electric powertrain
A_4 = Design a new small car with a diesel powertrain
A_5 = Design a small car with all three (gasoline, diesel, and electric) powertrain options.

Six possible outcomes assumed by the manufacturer are as follows:

O_1 = Economy does not change and the battery technology does not improve
O_2 = Economy improves by 5% and the battery technology does not improve
O_3 = Economy degrades by 5% and the battery technology does not improve
O_4 = Economy does not change and the battery technology improves by 50%
O_5 = Economy improves by 5% and the battery technology improves by 50%
O_6 = Economy degrades by 5% and the battery technology improves by 50%.

The evaluation measures (benefits or costs in dollars) associated with the combinations of the above five alternatives and the six outcomes are provided in Table 3.3. The table also provides probabilities for each of the outcomes assumed by the manufacturer.

TABLE 3.3
Data for Powertrain Selection Decision Problem

	Probability of Outcome					
	0.15	0.3	0.15	0.15	0.2	0.05
	Outcomes					
Alternative	O_1	O_2	O_3	O_4	O_5	O_6
A_1	$100,000	$120,000	$50,000	$80,000	$200,000	$100,000
A_2	−$200,000	$150,000	−$300,000	−$100,000	$100,000	$50,000
A_3	$50,000	$75,000	−$100,000	$100,000	$50,000	$150,000
A_4	$150,000	$50,000	$100,000	$75,000	$25,000	$25,000
A_5	$200,000	$100,000	$75,000	$100,000	$100,000	$75,000

Decision-Making and Risks in Product Programs

The following computation illustrates the computation of the expected value of A_1:

$$\begin{aligned}\text{Expected value of } A_1 = \{E_1\} &= (0.15 \times 100{,}000) + (0.3 \times 120{,}000)\\&+ (0.15 \times 50{,}000) + (0.15 \times 80{,}000)\\&+ (0.2 \times 200{,}000) + (0.05 \times 100{,}000)\\&= \$115{,}500\end{aligned}$$

The expected values of A_2, A_3, A_4, and A_5 are −$22,500, $47,500, $70,000, and $110,000, respectively. Thus, the alternative A_1 has the maximum expected value of $115,500 among the five alternatives and it will be selected under the maximum expected value principle (see the column labeled as "Expected Value Principle" in Table 3.4).

OTHER PRINCIPLES

Six additional principles that can be used to select an alternative are described below.

1. *Aspiration level*: The principle of aspiration level is based on the assumption that the decision maker needs to meet certain aspiration (or desired) level such as a minimum acceptable profit level or a maximum amount of tolerable loss. If we assume that the decision maker in the above example (Table 3.3) wants to make at least $200,000 profit, then he would consider alternatives A_1 and A_5 (because these two alternatives include the outcomes with a payoff of $200,000). On the other hand, if he does not want to incur any loss, he would not consider alternatives A_2 and A_3 (as these two alternatives can incur a loss in at least one outcome).
2. *Most probable future*: The decision maker may decide based on the most likely outcome (which has the highest probability of occurrence). In our above example (Table 3.3), the outcome O_2 has the highest probability (0.3) of occurrence. Under this situation (outcome O_2), selection of alternative A_2 will ensure the maximum profit of $150,000.
3. *Laplace principle*: The Laplace principle assumes that the decision maker does not have any information on the probability of occurrences of any of the outcomes, and thus, he assumes that all the outcomes are equally likely. In our above example, under this principle, all the occurrence probabilities will be equal to 1/6. Thus, the decision maker can simply take the average value of all E_{ij} for each alternative (i.e., over each i) and select the alternative with the maximum profit. In our above example, under this principle, the decision maker would select alternative A_1 or A_5 with the maximum average profit of $108,333 (see the column labeled as "Laplace Principle" in Table 3.4).
4. *Maximin principle*: This principle is based on the "extremely pessimistic view" of the decision maker (i.e., the nature will do its worst under every alternative). Therefore, the decision maker will select the alternative that maximizes the value of the proceed (profit) among the minimum values of all alternatives (i.e., the decision maker will reduce his loss by selecting

TABLE 3.4
Alternatives Selected by Five Principles

	Probability of Outcome									
	0.15	0.3	0.15	0.15	0.2	0.05				
	Outcomes						Expected Value Principle	Laplace Principle (Average Value)	Maximin Principle (Min Values)	Maximax Principle	Hurwicz Principle (with $\alpha = 0.5$)
Alternatives	O_1	O_2	O_3	O_4	O_5	O_6					
A_1	$100,000	$120,000	$50,000	$80,000	$200,000	$100,000	<u>$115,500</u>	<u>$108,333</u>	$50,000	<u>$200,000</u>	<u>$125,000</u>
A_2	-$200,000	$150,000	-$300,000	-$100,000	$100,000	$50,000	-$22,500	-$50,000	-$300,000	$150,000	-$75,000
A_3	$50,000	$75,000	-$100,000	$100,000	$50,000	$150,000	$47,500	$54,167	-$100,000	$150,000	$25,000
A_4	$150,000	$50,000	$100,000	$75,000	$25,000	$25,000	$70,000	$70,833	$25,000	$150,000	$87,500
A_5	$200,000	$100,000	$75,000	$100,000	$100,000	$75,000	$110,000	<u>$108,333</u>	<u>$75,000</u>	<u>$200,000</u>	<u>$137,500</u>

Note: The selected alternatives are shown by underlining the most desired payoff under each of the five principles shown in the last five columns of this table.

Decision-Making and Risks in Product Programs

the alternative with the least loss [or select the alternative with the highest profit among the minimum values]). The profit (P_i) in ith alternative can be defined as follows:

$$P_i = \max_i \left\{ \min_j E_{ij} \right\}$$

Laplace Principle	All outcomes are equally likely
Maximin Rule	Extremely pessimistic view (nature will do its worse). Take max of mins
Maximax Rule	Extremely optimistic view. Take max of max
Hurwicz Rule	A compromise between optimism and pessimism

Table 3.4 shows that, under this principle, the decision maker will select alternative A_5, which has the highest value ($75,000) among the lowest possible values of the evaluation measure among all the alternatives (see the column labeled as the "Maximin Principle" in Table 3.4).

5. *Maximax principle*: This principle is based on the "extremely optimistic view" (think about the best possible) of the decision maker. The decision maker will select the alternative that maximizes the maximum values in each alternative, that is, to take the maximum of the maximum values in each alternative. The profit (P_i) in ith alternative can be defined as follows:

$$P_i = \max_i \left\{ \max_j E_{ij} \right\}$$

Table 3.4 shows that under this principle, the decision maker will select alternative A_1 or A_5, which has the highest value ($200,000) among the highest possible values of the evaluation measure among all the alternatives (see the column labeled as the "Maximax Principle" in Table 3.4).

6. *Hurwicz principle*: This principle is based on a compromise between optimism (Maximax principle) and pessimism (Maximin principle). The profit (P_i) in ith alternative is computed based on the selection value of index of optimism (α) as follows:

$$P_i = \alpha \left[\max_i \left(\max_j E_{ij} \right) \right] + (1-\alpha) \left[\max_i \left(\min_j E_{ij} \right) \right]$$

where α = index of optimism
And α can vary as follows: $0 \leq \alpha \leq 1$
Note: $\alpha = 1$ indicates that the decision maker is extremely optimistic
$\alpha = 0$ indicates that the decision maker is extremely pessimistic.

The value of P_i should be computed for each alternative using the above formula and the alternative with the maximum value of P_i should be selected.

The last column of Table 3.4 illustrates that for $\alpha = 0.5$, alternative A_5 will be selected because it has the highest value ($137,500) in the last column when the values were computed using the above expression for P_i.

TECHNIQUES USED IN DECISION-MAKING

Decision makers need the information to help decide on all the basic parameters (e.g., variables covered in the earlier section, such as the number of alternatives, possible outcomes, probabilities of the outcomes, costs or benefits associated with each alternative in each outcome) of the problem. Without the availability of reliable information, the decisions made by the decision maker may not be very useful or could even be very misleading. Further, care must be taken in selecting the decision maker to ensure that he or she is not biased and does not have any misconceptions or preconceived notions related to the product concepts, technologies considered in the concepts, customer expectations, and so forth.

Many techniques are available to gather information and display the information in a format that will help the decision maker to understand the problem. An overview of the tools is provided in Chapter 14, and the chapters in Part III present more details on the descriptions and formats of the tools. The formats of many of the tools are visual in nature and thus they promote a better understanding of the magnitudes and relationships between variables or events. Some examples of the tools and their applications are: (1) Fish diagram is used commonly to illustrate the causes of a problem, (2) Pareto chart helps in identifying the top few issues that contribute to most of the problems, (3) the quality function deployment (QFD) chart provides a list of engineering specifications and their relative importance ratings, (4) Pugh chart helps in selecting a product concept by considering multiple attributes, and (5) a risk analysis helps assess the level of risk in a given situation.

The following section describes the procedure for application of a technique called the analytical hierarchical method. The technique is used to extract judgments of experts in decision-making situations.

ANALYTICAL HIERARCHICAL METHOD

The analytical hierarchical method is a simple technique to determine the relative importance of different alternatives. It is based on subjective judgments made by one or more decision makers. Each decision maker is assumed to be an expert in the problem area and is free from any biases. The decision maker's task is further simplified by paired comparisons of alternatives. For example, if there are n possible alternatives, then there will be $n(n-1)/2$ number of possible pairs of alternatives. The decision maker is given each pair separately and is asked to select the better of the two alternatives and assign a relative importance rating (weight) to the selected alternative based on a preselected criterion (or a product attribute). The important ratings are then used to compute relative weights of each of the alternatives. The alternative with the highest weight is selected as the most preferable alternative. The method is described by Satty (1980) and Bhise (2012). The following example illustrates the procedure.

In the analytical hierarchical method, the products (or alternatives) are compared in pairs. The better product in each pair is also rated in terms of the strength of the attribute (used for evaluation) it possesses in relation to the strength of the same attribute in the other product in the pair. The strength of the attribute is expressed using a ratio scale. The scale (or the weight) value of 1 is used to denote equal strength of the attribute in both the products in the pair. And the scale value of 9 is used to indicate the extreme or absolute strength of the attribute in the better product. And the product with the weaker strength is assigned the inverse of the scale value of the better product. The following example will illustrate this rating procedure.

Let us assume that there are two products, W and S, in a pair; and the attribute to compare the products is "ease of use." The scale values assigned to the products using the ratio scale would be as follows:

1. If product W is "extremely or absolutely easy" to use as compared to product S, then, the weight of W preferred over S will be 9, and the weight of S preferred over W will be 1/9.
2. If product W is "very easy" to use as compared to product S, then, the weight of W preferred over S will be 7, and the weight of S preferred over W will be 1/7.
3. If product W is "easy" to use as compared to product S, then, the weight of W preferred over S will be 5, and the weight of S preferred over W will be 1/5.
4. If product W is "moderately easy" to use as compared to product S, then, the weight of W preferred over S will be 3, and the weight of S preferred over W will be 1/3.
5. If product W is "equally easy" to use as compared to product S, then, the weight of W preferred over S will be 1, and the weight of S preferred over W will be also 1.

Satty (1980) described the nine-point scale with the following adjectives to indicate the level of preference (or importance) for comparing the two items in each pair.

1. **= Equal preference**
2. = Weak preference
3. **= Moderate preference**
4. = Moderate plus preference
5. **= Strong preference**
6. = Strong plus preference
7. **= Very strong or demonstrated preference**
8. = Very, very strong preference
9. **= Extreme or absolute preference**.

From the viewpoint of making the scales more understandable and easier to apply, usually only the odd-numbered scale values (shown in bold case above) are described and presented to the subjects (decision makers). To allow the subjects to decide on the weight, the author found that the scale presented in Figure 3.1 works very well. Here, the subject will be asked to put an "X" mark on the scale on the left side if product W is preferable over product S. The higher numbers on the scale indicate higher preference. If both products are equally preferred, then the subject will be asked to place the "X" mark at the mid-point of scale with value equal to 1. If product S is preferred over product W, then the subject will use the right side of the scale.

Let us assume that we have to compare six products, namely, W, S, Q, M, P, and L, by using the analytical hierarchical method. A subject will be asked to compare the products in pairs. The 15 possible pairs of the six products will be presented to the subject in a random order. (Note: for $n = 6$, $n(n - 1)/2 = 15$.) The subject will be given a preselected attribute (e.g., ease of use) and asked to provide strength of preference ratings for the preferred product in each of the 15 pairs by using scales such as the one presented in Figure 3.1. The data obtained from the 15 pairs will then be converted into a matrix of paired comparison responses as shown in Table 3.5. Each cell of the matrix indicates the ratio of preference weight of the product in the row over the product in the column. Thus, the ratio 9/1 in the first row and the second column indicates that the product in the row (W) was "absolutely preferred" (i.e., considered extremely or absolutely easy = rating weight of 9) over the product in the column (S).

To compute the relative weights of preference of the products, the fractional values in Table 3.5 are first converted into decimal numbers as shown in the left-side

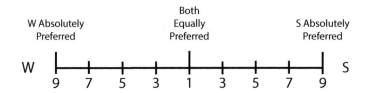

FIGURE 3.1 Scale used to indicate the strength of the preference when comparing two products (W and S).

TABLE 3.5
Matrix of Paired Comparison Responses for One Evaluator

	W	S	Q	M	P	L
W	1	9/1	1/1	5/1	1/3	1/1
S	1/9	1	1/2	5/1	1/1	3/1
Q	1/1	2/1	1	3/1	5/1	3/1
M	1/5	1/5	1/3	1	1/1	1/3
P	3/1	1/1	1/5	1/1	1	1/3
L	1/1	1/3	1/3	3/1	3/1	1

Note: The value in a cell indicates the preference ratio for comparing the product in a row with the product in a column corresponding to the cell.

matrix in Table 3.6. All the six values in each row are then multiplied and entered in the column labeled as "Row Product" in Table 3.6. The geometric mean of each row product is computed. It should be noted that the geometric mean of the product of n numbers is the $(1/n)$th root of the product (e.g., 1/6th root of 90.00 is 2.1169. Note: $2.1169^6 = 90$). All the six geometric means in the column labeled as "Geometric Mean" are then summed. The sum, as shown in Table 3.6, is 6.9813. Each of the geometric means is then divided by their sum (6.9813) to obtain the normalized weight of the products (see the last column of Table 3.6). It should be noted that due to the normalization, the sum of the normalized weights over all the products is 1.0.

The normalized weights (called the normalized preference values) are plotted in Figure 3.2. The figure, thus, shows that the most preferred product (based on the ease

TABLE 3.6
Computation of Normalized Weights of the Products

	W	S	Q	M	P	L	Row Product	Geometric Mean	Normalized Weight
W	1.00	9.00	1.00	5.00	0.33	1.00	15.0000	1.5704	0.2249
S	0.11	1.00	0.50	5.00	1.00	3.00	0.8333	0.9701	0.1390
Q	1.00	2.00	1.00	3.00	5.00	3.00	90.0000	2.1169	0.3032
M	0.20	0.20	0.33	1.00	1.00	0.33	0.0044	0.4055	0.0581
P	3.00	1.00	0.20	1.00	1.00	1.00	0.6000	0.9184	0.1315
L	1.00	0.33	0.33	3.00	3.00	1.00	1.0000	1.0000	0.1432
							Sum-->	6.9813	1.000

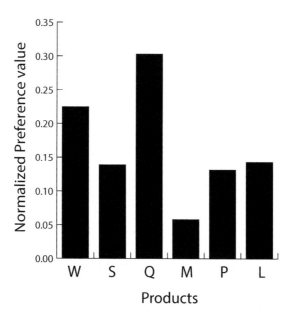

FIGURE 3.2 Normalized preference values of the six products.

of use) was Q (with its normalized weight of 0.3032) and the least preferred product was M (with its normalized weight of 0.0581).

The above example was based on data obtained from one subject. If more subjects are available, then the normalized weights for each subject can be obtained by using the above procedure and then average weight of each product can be obtained by averaging (i.e., by using the geometric mean) over the normalized weights of all the subjects for each product.

WEIGHTED TOTAL SCORE FOR CONCEPT SELECTION

During the product development process, the decision makers (e.g., usually top management) are faced with the decision to select a concept and proceed with its detailed design and engineering work. The selection is complicated because the product concepts need to be evaluated by considering many attributes of the product. The attributes are generally developed from the customer needs obtained from extensive interactions with the customers (e.g., by conducting market surveys or from customer feedback). The customers can also be asked to provide important ratings (or weights) of each of the attributes. For example, in the QFD, the customers are asked to provide important ratings for each of the attributes (or customer needs) using a 10-point scale, where 1 = least important and 10 = very important (see Chapter 15 for details on the QFD). The weights can also be developed by using the analytical hierarchical method covered in the previous section. The customers (or the design team members) can also be asked to rate each product concept on each attribute. All the above information can then be used to determine total weighted score of each product concept. The product concepts can be compared based on the total weighted score and the concept with the highest total score can be selected.

The computation of the total weighted score (T_j) of the jth product concept is described by the following mathematical expression.

$$T_j = \sum_{i=1}^{n} w_i R_{ij}$$

where T_j = total weighted score of jth product concept by considering all the n attributes. Note: $i = 1, 2, \ldots, n$
w_i = weight of ith product attribute
R_{ij} = ratings of concept C_j on ith attribute

Table 3.7 provides an example of the above weighting scheme. Each product concept is defined as C_j, where $j = 1$ to 4; and each product attribute is defined as A_i, where $i = 1$ to 5. The ratings (R_{ij} s) are provided using a 10-point scale, where 1 = poor and 10 = excellent. The attribute weights (w_1 to w_5) were obtained by using a five-point scale, where 1 = not important and 5 = very important. The total weight score ($T_1 = 119$) of concept C_1 was the highest and concept C_3 with $T_3 = 92$ was the lowest. Thus, product concept C_1 can be selected for further development, or the rating data

TABLE 3.7
Illustration of Total Weighted Score of Product Concepts Based on Attribute Weights and Ratings of Product Concepts by Attributes

Attribute	Attribute Weight (w_i)	Product Concepts			
		c_1	c_2	c_3	c_4
A_1	5	10	8	5	7
A_2	3	5	8	9	4
A_3	5	7	4	5	7
A_4	1	5	8	3	6
A_5	2	7	9	6	8
		T_1	T_2	T_3	T_4
Total weighted score		119	110	92	104

can be used to come up with further modifications of the product concepts. New ratings can be obtained after the modifications to iterate the above procedure until an acceptable product concept is achieved.

The above method is used in the QFD (to compute absolute importance scores of functional specifications), and it can be considered as a modified scoring method for the Pugh diagram. Both the QFD and Pugh diagram are described in Chapter 15.

INFORMATIONAL NEEDS IN DECISION-MAKING

The key to making good decisions is to have sufficient information and good understanding of issues related to the alternatives, outcomes, trade-offs, and payoffs associated with the decision situation. Therefore, it is important to select a decision maker carefully and make sure that the person is familiar with the product and its uses. In some situations, customers who have used similar products are asked to provide their ratings on each product (or alternative) used in the evaluation. On the other hand, experts who are very familiar with the product and have extensive knowledge about the product can be very discriminating (much more than even the most familiar customers) and can provide unbiased evaluations.

In addition, the experts can obtain additional information through other methods such as: (1) benchmarking other products, (2) literature surveys, (3) exercising available models (e.g., models to predict the performance of products under different situations) and using the information obtained from the model results, and (4) conducting experiments (see Chapter 7 on more methods and issues in product evaluations. Also see Chapter 17 for information on experiment design).

Exercising available models under various "what if" scenarios (i.e., conducting sensitivity analyses) can provide more insights into the variability (or robustness) in the performance of the product and thus can prepare the decision maker to make more informed decisions. Design reviews with different groups, disciplines, and experts can also generate information on the strengths and weaknesses of the product (or product concept) being reviewed.

DECISION-MAKING IN PRODUCT DESIGN

KEY DECISIONS IN PRODUCT LIFE CYCLE

Some of the key decisions made in a product program typically involve the following:

1. *Program kick-off*: Top management of the organization decides that a new product (or revisions to an existing product) should be developed and a project should be kicked off to plan the product development and budget planning activities.
2. *Program confirmation*: Based on the additional information obtained from the design team's presentation of the created product concepts, market research results, trends in the new technologies and design, and the competitors' capabilities, the top management confirms the decision to select a product concept. Additional decisions are also made to allocate budget and dates for the product introductions in the selected markets.
3. *Product concept freeze*: Management decides that the selected product concept is sufficiently developed (i.e., all design and engineering managers feel confident that the product could be produced [i.e., it is feasible] within the planned budget and schedule). Thus, the concept will be frozen (i.e., no major changes will be made) and succeeding program activities will be continued.
4. *Engineering sign-off*: All key managers of engineering activities sign a document stating that the product "as designed" will meet all applicable requirements with a high probability (e.g., 90%).
5. *Production release*: All product testing (verification and validation tests) is completed, and product is determined to be ready for the market. Product is released for production; that is, the factories begin production of units for sale.
6. *Periodic reviews*: Periodic (monthly, quarterly, or annual) reviews of the product sales, customer satisfaction, and comparison of data with the competitors' products are conducted to determine whether any changes in the product volume or product characteristics are needed.
7. *Product discontinuation and replacement*: Based on the market data and the customer feedback, the management decides to terminate the production of the product on a certain date and requests marketing department to plan for future product(s) or model(s) for its replacement.

TRADE-OFFS DURING DESIGN STAGES

Teams involved in designing any product need to make a number of decisions involving trade-offs between a number of conflicting design considerations (e.g., product characteristics and attributes). Some examples of trade-off considerations in designing passenger cars are described below.

1. *Space for vehicle systems versus space for occupants*: The space within the vehicle is occupied by various vehicle systems and their components, and

Decision-Making and Risks in Product Programs 65

space is used to accommodate occupants in the passenger compartment and other items in the trunk (or cargo areas). In order to provide more space for the occupants (interior passenger space), the space occupied by vehicle systems (e.g., vehicle body structure sections, engine, chassis and suspension components, fuel tank) needs to be reduced. Thus, a vehicle designer can make trade-offs by designing more compact vehicle systems to allow more space for the occupants. This trade-off is commonly referred to in the auto industry as "Machine Minimum and Man Maximum" (i.e., minimizing the space for mechanical components and maximizing the space for the occupants).

2. *Fuel economy versus performance*: A vehicle with a high acceleration capability (commonly referred to as the time required for accelerating from 0 to 60 mph [called the 0-to-60 time in seconds]) requires higher engine power, which in turn reduces fuel economy (measured in miles per gallon of gasoline consumed).

3. *Vehicle weight versus performance*: This trade-off is commonly referred to by considering the horsepower-to-weight ratio. An increase in vehicle weight reduces the acceleration capability of the vehicle with the same engine power.

4. *Ride comfort versus handling*: Better (more comfortable) riding cars require softer suspensions that reduce the handling (maneuvering) capability of the vehicle.

5. *Lightweight materials versus cost*: Lightweight materials (e.g., aluminum, magnesium, high-strength steels, and carbon-fiber materials) can reduce vehicle weight. However, these lightweight materials are several times more costly than the commonly used steel sheet (mild steel) material.

6. *High-raked windshield versus costs*: Windshields with higher rake angle (more sloping windshields; see Figure 3.3) can reduce aerodynamic drag, increase fuel economy, and provide a sleeker more-aerodynamic appearance than conventionally styled vehicles (with more up-right windshield). The high-raked windshields are longer in length (see Figure 3.3 where $L_1 > L_0$) than more conventional low-raked windshields. The longer length (L_1) also requires thicker glass, longer wipers, more powerful wiper motor, higher capacity windshield defroster, and higher-capacity air-conditioning (due to higher heat/sun load). The thicker glass also reduces light transmission of the windshield, which in turn reduces driver visibility. The thicker glass also increases vehicle weight, which in turn can reduce fuel economy. The higher raked windshields thus can increase vehicle costs.

RISKS IN PRODUCT DEVELOPMENT AND PRODUCT USES

The product programs involve many risks. All important decisions in business and life involve some level of risk. A risk is considered to be present when an undesired event (which generally incurs substantial loss) is probable (i.e., likely to occur with some level of probability). Risks are possible anytime during or after the product development process. If the decision maker takes too little risk by over-designing

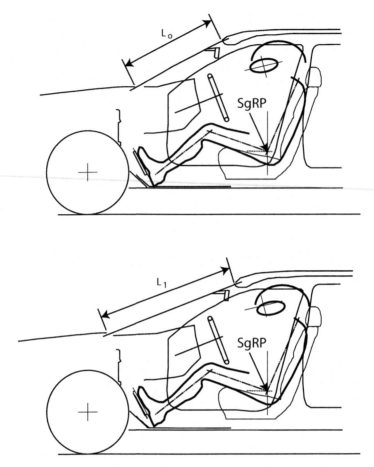

FIGURE 3.3 Comparison of conventional versus high-raked windshield.

Note: The upper picture shows a vehicle with a conventional windshield rake angle. The lower picture shows a vehicle with a higher-rake-angle windshield.

(or using too high a safety factor), the product will be more costly and the extra cost most likely will be wasted. On the other hand, if the decision maker takes too much risk by under-designing (e.g., product will have insufficient strength, or use cheap low-quality materials and/or components), then the program will be too costly due to high costs from product failures. The product failures can cause accidents that can incur additional costs due to occurrences of (1) injuries, (2) property damages, (3) loss of income, (4) interruptions or delays in work situations, (5) product liability cases, and so forth.

Figure 3.4 presents a flow diagram showing relationships between various causes of product failures (i.e., the product does not work as expected by the customer) and the risk evaluation process. The top part of Figure 3.4 shows a cause-and-effect process chart (Fish diagram) for an undesired head event defined as "Product does not work as expected" (the cause-and-effect process chart is described in Chapter 17; see

Decision-Making and Risks in Product Programs

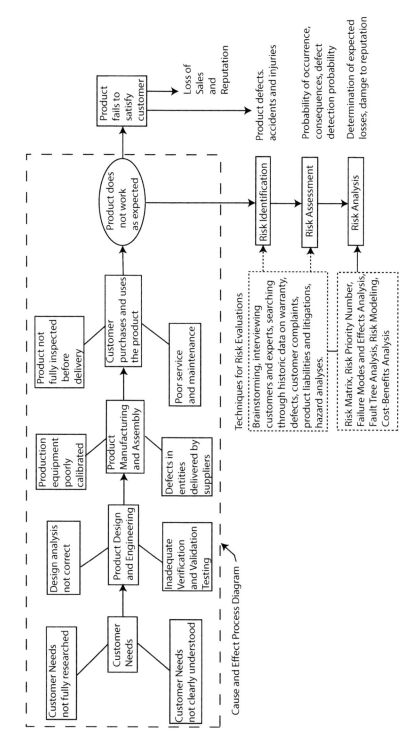

FIGURE 3.4 Flow diagram of product failures and risk analysis.

Figure 17.4). The risks due to product defects can be analyzed by using techniques in the following three sequential steps: (1) risk identification, (2) risk assessment, and (3) risk analysis. The techniques used in the risk evaluation process are covered in the following section.

DEFINITION OF RISK AND TYPES OF RISKS IN PRODUCT DEVELOPMENT

A risk is generally associated with an occurrence of an undesired event such as a financial loss and/or an injury resulting from a product-related failure. The risk can be measured in terms of the magnitude of the consequence due to the occurrence of an undesired event. The consequence due to a risk can be measured by costs associated with customer dissatisfaction, loss due to product defects or resulting accidents, loss due to interruption of work, loss of revenue, loss of reputation, and so forth.

The risk is generally assessed by consideration of the following variables: (1) probability of occurrence of the undesired event, (2) the consequence (or severity) of the undesired event (e.g., amount of loss or severity of injuries), (3) probability of detection of the undesired event before or when it occurs, and (4) preparedness of the risk fighting unit (e.g., fire department, emergency response units, and police) that can attempt to contain the severity of the loss or injury.

The risks during the product development process can be categorized as follows:

1. *Technical risk*: This type of risk occurs due to one or more technical problems with the design of the product. For example, a design flaw discovered during testing of an early production component. Such a problem may prevent the product to achieve the required technical capability or performance. To eliminate the technical problem or the flaw, additional analyses, engineering changes and/or technology changes, and testing may be needed. These additional tasks usually result in an increase in the costs and delays in the schedule. Adoption of new technologies before adequate developmental work often leads to serious delays (e.g., problems in developing carbon fiber components for airplanes [see Case Study #7 in Chapter 23], manufacturing problems with lightweight materials to improve fuel consumption in automotive products).
2. *Cost risk*: This risk is associated with the cost overruns due to technical problems and resulting delays in the schedule. The risk also may be due to underbudgeting caused by assuming optimistic estimates or underestimation of required tasks, time, and costs (e.g., not providing sufficient allowance for rework).
3. *Schedule risk*: This risk is related to not being able to meet the schedule due to delays from a number of possible reasons (e.g., parts not delivered by suppliers on time, late changes made in the design due to failures uncovered in testing, or planned schedule may be too optimistic) (see Case Studies #6 and 7 in Chapter 23).
4. *Programmatic risk*: This risk is associated with the product development program (e.g., being over budget, delayed, modified, or even canceled due to a number of reasons). Since most of the complex products have many

components that are made and supplied by various suppliers, selection of suppliers with unproven or low technical capabilities often leads to program delays, lower quality, and cost overruns.

These above four categories of risks are generally interrelated; that is, a risk in any one of the categories also affects the risks associated in other categories. The risks also cause backward cascading effects of the problems in the work completed in the early phases but discovered in the later phases. These problems affect the progress in the succeeding phases due to factors such as redesign, rework, retests, delays, and cost overruns.

TYPES OF RISKS DURING PRODUCT USES

The risks after the product is introduced in the market and used by end-users can be categorized as follows:

1. *Loss of user confidence in the product*: The end-users may be afraid to use a product because of a defect in the product. The defect may be caused due to a design or manufacturing defect or some "hidden-danger" that can cause an undesired event (e.g., a sudden loss of control, a fire or an explosion, an accident, and an exposure to toxic substances).
2. *Loss in future sales*: The likelihood of an undesired event can cause a loss in the reputation of the producer and thus can affect future sales.
3. *Excessive repair or recall costs*: The producer will need to fix the product problem by repairing under warranty or by initiating a product recall.
4. *Product litigation costs*: The costs of defending the product in product liability cases and costs related to settlements before the court trials or payments of penalties, fines, and so forth.

RISK ANALYSIS

A risk analysis can be defined as a decision-making exercise conducted to determine the next course of action after a potential undesired event has been identified and the magnitude of the consequence of the undesired event has been estimated. The phase of identification of an undesired event can be called the "risk identification" phase, and the phase of estimation of the magnitude of the consequence due to the undesired event can be called the "risk assessment" phase.

Some commonly used methods for risk identification, risk assessment, and risk analysis are given here.

1. *Risk identification methods*: Brainstorming, interviewing experts, hazard analysis (see Chapter 19), failure modes and effects analysis (FMEA) (see Chapter 15), and use of checklists and historic data (e.g., past records of product defects, warranty problems, and customer complaints)
2. *Risk assessment methods*: Estimation of probability (or frequency) of occurrence of undesired events, magnitude of the consequence (or severity) of the

undesired event and probability of detection of the undesired event by using brainstorming, interviewing experts, safety analysis (e.g., fault tree analysis [see Chapter 19] and FMEA [see Chapter 15]), and historic data (e.g., costs of past product failures)
3. *Risk analysis methods*: Risk matrix, risk priority number (RPN), nomographs, existing design and performance standards, and specialized risk models

(Floyd et al., 2006)

Risk Matrix

The risk matrix involves simply creating a matrix with combinations of relevant variables associated with the degree of risk due to the undesired outcomes. A risk matrix is a simple graphical tool. It provides a process for combining (1) the probability of an occurrence of an undesired event (usually an estimate) and (2) the consequence if the undesired event occurred (usually cost estimates in dollars).

The risk (in dollars) can be computed as follows:

$$\text{Risk}(\$) = [\text{Probability of occurrence}] \times [\text{Consequence of the undesired event}(\$)]$$

Figure 3.5 shows a plot of the above relationship of probability of occurrence and consequence (loss in dollars) to the risk (expected loss in dollars) due to an undesired

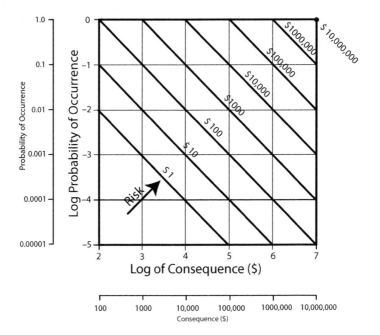

FIGURE 3.5 Relationship of risk to probability of occurrence and consequence of occurrence of an undesired event.

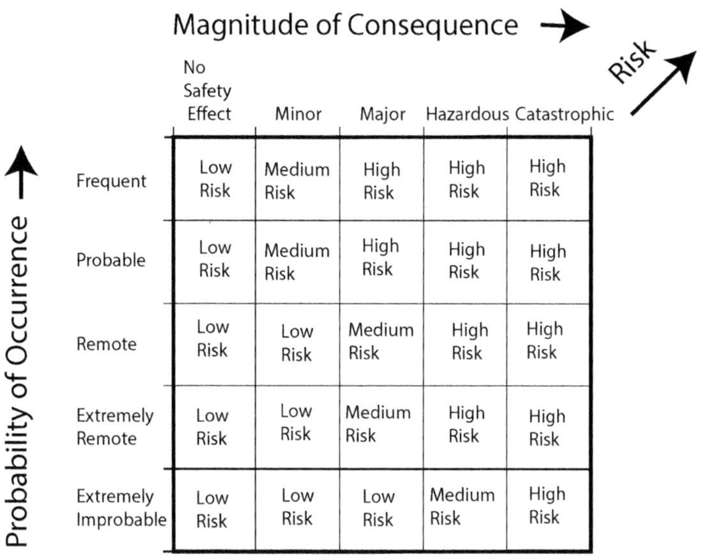

Decision-Making and Risks in Product Programs 71

		No Safety Effect	Minor	Major	Hazardous	Catastrophic
	Frequent	Low Risk	Medium Risk	High Risk	High Risk	High Risk
	Probable	Low Risk	Medium Risk	High Risk	High Risk	High Risk
	Remote	Low Risk	Low Risk	Medium Risk	High Risk	High Risk
	Extremely Remote	Low Risk	Low Risk	Medium Risk	High Risk	High Risk
	Extremely Improbable	Low Risk	Low Risk	Low Risk	Medium Risk	High Risk

Magnitude of Consequence →

Probability of Occurrence ↑

Risk ↗

FIGURE 3.6 An example of a risk matrix. (Redrawn from Federal Highway Administration [2007].)

event. The plot is made by using log scales (logarithm to the base of 10) for both the axes. Thus, the expected loss, which is the result of the multiplication of the values on its X and Y axes, is represented by slanting lines of risk on the logarithmic axes. The magnitude of the risk is the same on any given risk line. The risk lines for $1 to $1,000,000 are shown in Figure 3.5.

A simplified form of the above relationship between the probability of occurrence and the magnitude of the consequence can be presented in a matrix format. Figure 3.6 presents an example of a risk matrix. The cells of the matrix represent different risk levels increasing from low risk to high risk from the lower left corner of the matrix to the top right corner of the matrix. The risk matrix thus allows for a quick assessment of the risk level after the occurrence probability, and the magnitude of the consequence due to an undesired event are estimated.

RISK PRIORITY NUMBER AND NOMOGRAPHS

RPN is another method used to assess the level of risk. It is based on the multiplication of three ratings, namely: (1) severity, (2) occurrence, and (3) detection. This method is used in the FMEA, which is presented in Chapter 15. Examples of rating scales used for severity, occurrence, and detection are presented in Tables 15.4, 15.5, and 15.6, respectively. Different definitions of the rating scales are generally used in different companies, industries, and government agencies.

Nomographs are also used as an alternate method for estimating the RPN. An example of a nomograph is shown in Figure 19.1.

Other methods such as modeling and simulations are used to facilitate decision-making. Exercising models under different assumptions (conducting sensitivity analysis) can provide a good understanding of the underlying variables and their effects on risks and subsequent decisions. Analyses under a range of possibilities with different levels of optimistic and pessimistic assumptions are also useful to estimate the limits of risks. Historic data and judgments of experts can also play a major role in the decision-making.

PROBLEMS IN RISK MEASUREMENTS

Assessing the risks to users/customers involves identifying the hazards, assessing the potential consequences, and the occurrence probability of such consequences. Identification of hazards is particularly difficult when both the potential customers and product uses are difficult to predict. Products involving new technologies are also difficult to evaluate because very little failure data are generally available. It is especially hard to predict the risks during the early stages of product development when the product concept is also not fully developed.

The problems in risk measurements occur due to many reasons. Most problems occur due to (1) lack of data on different types of hazards and risks, (2) subjectivity involved in identification and quantification of the data, and (3) differences in assumptions made during the design phases about how the customers or users will use the product versus the actual uses of the product. The risk assessment models used in this area are therefore not precise. But they can be used as guides along with the recommendations of multiple experts and discussions between the decision makers and the experts.

Subjective assessments of the three component areas (occurrence, severity, and detection) are also difficult and subject to a number of questions such as: Who would collect the data and conduct evaluations? Should the evaluations be conducted by experts, product safety advisory boards, teams, or individuals involved in the design process? Furthermore, the level of understanding and awareness of risks varies considerably between different evaluators. Cost is also another problem in collecting failure-related data as the product tests are generally costly and funds are usually limited to undertake costly data collection studies.

There are trade-offs in the application of risk assessment methods between the consistency in the data, the level of details related to the outcomes, and the time and resources (particularly human and financial) required for the analyses. Apparently, simple methodologies may contain implicit weightings that may not be appropriate for every product being assessed. Judgments may be intuitive, based on implicit assumptions, especially in relation to the boundaries between categories (or ratings). Taken together, these factors can result in a high degree of subjectivity in risk assessment, although the subjectivity can be reduced by the extent of guidance provided to the assessors in applying the various scales and ratings. In general, the potential for inconsistencies in the results will be directly related to the amount of subjectivity involved in the risk measurement process.

IMPORTANCE OF EARLY DECISIONS DURING PRODUCT DEVELOPMENT

"Designing right the first time" is very important as reworking any product design in later phases is always very time-consuming and costly. Early in the product development, key decisions are generally made on what technologies to use and how the product should be configured. Any changes to these early assumptions made during the later stages of product development can increase the costs substantially. Because such changes may require throwing away much of the early design work (and even some hardware development work) and redoing all the analyses again with a different set of assumptions and requirements.

Involvement of specialists from all key technical areas is a very important aspect of the Systems Engineering process as it ensures that all possible technologies and design configurations are considered as possible alternatives before converging on one or a few alternatives. The subsequent decisions are dependent on the selected technologies and design configurations. For example, during the development of Boeing 777, the management decided on a two-engine airplane as compared with the four-engine approach used in the past long-distance commercial aircraft. The two-engine approach required substantially more work in designing bigger engines, improving reliability of the engines and performing additional flight tests to prove to the Federal Aviation Administration that the two-engine Boeing 777 aircraft was safe or safer than the four-engine aircraft in the long trans-oceanic flights (PBS, 1995). Another example in a new material-related technology is as follows. Early during the program planning, Boeing decided to produce the Boeing 787 Dreamliner using the carbon-fiber materials as compared to using aluminum for its structural components. The development of large parts (e.g., airplane wings, tail, and fuselage) with the carbon-fiber materials involved a number of developmental challenges related to understanding and implementing the carbon-fiber technology (see Case Study #7 in Chapter 23).

CONCLUDING REMARKS

This chapter covered many basic models and issues in decision-making. Decision-making in the real world involves consideration of many issues (both internal and external to the organization), many variables and their effects, likelihoods of outcomes, and associated costs that cannot be well quantified due to reasons such as missing facts, uncertainties in the readiness of new technologies, unknown future developments, and changes in global economy. Many models involving varied levels of complexity using many independent variables can be created to analyze the effects of many risk-related variables. The models can be exercised under different assumptions (conducting sensitivity analysis) to get a good understanding of underlying variables and their effect on the decisions. However, a good decision maker will also inject some subjectivity based on his/her intuition or judgment to make the final decisions. The decisions are never final and can be revisited after new and more reliable information is available.

REFERENCES

Bhise, V. D. 2012. *Ergonomics in the Automotive Design Process.* Boca Raton, FL: CRC Press.

Blanchard, B. S. and W. J. Fabrycky. 2011. *Systems Engineering and Analysis.* Fifth Edition. Upper Saddle River, NJ: Prentice Hall PTR.

Federal Highway Administration. 2007. Website: http://international.fhwa.dot.gov/riskassess/images/riskmatrixhh.cfm (Accessed July 19, 2012).

Floyd, P., Nwaogu, T. A., Salado, R. and C. George. 2006. Establishing a Comparative Inventory of Approaches and Methods Used by Enforcement Authorities for the Assessment of the Safety of Consumer Products Covered by Directive 2001/95/EC on General Product Safety and Identification of Best Practices. Final Report dated February 2006 prepared for DG SANCO, European Commission by Risk & Policy Analysts Limited, Farthing Green House, 1 Beccles Road, London, Norfolk, NR14 6LT, UK.

Konz, S. 1990. *Work Design-Occupational Ergonomics.* Third Edition. Worthington, OH: Publishing Horizons.

Public Broadcasting Service (PBS). 1995. *21st Century Jet—The Building of the 777.* Producers: Karl Sabbagh, David Davis and Peggy Case. PBS Home Video (5 hours). Produced by Skyscraper Products for KCTS Seattle and Channel 4 London.

Satty, T. L. 1980. *The Analytic Hierarchy Process.* New York: McGraw-Hill.

4 Product Attributes, Requirements, and Allocation of Functions

INTRODUCTION

The Systems Engineering (SE) process begins with the understanding of customer needs. The customer needs are translated into product attributes (i.e., the characteristics that the product must have) that are used and managed during the product development process to ensure that the product possesses the attributes to satisfy its customers. The attribute management process involves development of requirements for the attributes, cascading the attribute requirements from the product level to its lower levels (i.e., entities such as systems, subsystems, and components), allocating functions to all entities in the product, and developing test procedures to verify that each entity meets its requirements. This chapter describes the development of product attributes, requirements, and allocation functions to the product entities.

ATTRIBUTES AND REQUIREMENTS

What Is an Attribute?

Attribute is a characteristic of a product that it must have to sell well. It is assumed that the customers buy and use products based on their total impact, which can be broken down into a number of attributes. A product can have many attributes. The product attributes must be derived from the needs of its customers. All the product attributes, taken together, should cover all the needs of the customers.

For example, the attributes of a laptop computer are (1) physical size, (2) weight, (3) display size, (4) ergonomics (i.e., ease in using the keyboard, touchpad, audio, display, and disk drive), (5) processor capabilities (e.g., capacity and speed to process data), (6) data storage capacity, (7) battery capacity (e.g., hours of operation between recharges), (8) input/output ports, (9) wireless connectivity, (10) aesthetics (i.e., styling/appearance), (11) durability, and (12) life cycle costs (i.e., costs incurred by the customer during the life cycle of the computer). Thus, the aforementioned list of attributes should cover all the customer needs that the laptop design team should consider during the entire life cycle of the laptop computers.

Another product example will help in understanding the breadth of the areas covered by the product attributes. The attributes of an automotive product can be described as follows: (1) aesthetics/styling, (2) occupant package and ergonomics, (3) affordability (i.e., acquisition, operating, and maintenance costs), (4) performance and fuel economy, (5) interior comfort (e.g., noise, vibrations, and climate

DOI: 10.1201/9781003263357-5

control), (6) ride and handling (i.e., vehicle dynamics considerations related to how the vehicle feels during the driving maneuvers), (7) safety and security, (8) pollution (i.e., emissions of harmful materials generated by the vehicle during its operation), (9) information and entertainment (i.e., providing needed information and entertainment to the vehicle occupants), and (10) customer life cycle experience (i.e., the overall experience of the customer during the vehicle usage in the customer's life stages or changes).

IMPORTANCE OF ATTRIBUTES

Instead of determining product requirements directly from the customer needs as shown in Figure 2.2 in Chapter 2, some organizations have found that defining the product attributes from the customer needs and managing each of the product attributes (i.e., managing all product requirements that deliver or contribute to a given product attribute) is a better approach—especially for the development of complex products. For a product to possess a certain attribute, its attribute requirements can be cascaded (or assigned or allocated) from the product level down to lower levels in the system hierarchy (i.e., from product to its systems to subsystems and components).

The management of product requirements based on its attributes has the following three major advantages.

1. The attribute requirements help everyone in the product development process to understand the traceability of the requirements to certain product attributes (i.e., any requirement can be traced back to one or more product attributes).
2. People specialized in an attribute can be made responsible to ensure that the product is being designed to meet the attribute requirements (so that the product will possess the attribute).
3. The presence of attribute requirements ensures that all the product attributes are studied (i.e., tracked and evaluated) at every product development phase, and the compliance to the attribute requirements is reviewed at all major milestones in the product program.

Thus, the attributes management approach ensures that the customer needs are not overlooked during the design and subsequent phases of the product development, and thus the customers are satisfied.

WHAT IS A REQUIREMENT?

A requirement defines one or more product characteristics and their accomplishment levels needed to achieve a specific objective (e.g., a function to be performed, performance level to be achieved, and maximum weight and/or size limits on the product) for a given set of conditions. Requirements are developed to meet customer needs, government regulations, and corporate needs (e.g., brand-specific features). The requirements are created to achieve certain attribute characteristics, functions, or performance of the product.

ATTRIBUTE REQUIREMENTS

The requirements specified to achieve the product attributes can be defined as the attribute requirements. To manage the attributes, each attribute can be further divided into sub-attributes, sub-sub-attributes, and so on—at different levels. Thus, the product-level attribute requirements can be cascaded to lower levels during the functional analysis and allocation.

In some organizations (e.g., the automobile industry), the customer requirements are specified in terms of attribute requirements. An attribute manager is assigned to each attribute. And the responsibility of each attribute manager is to ensure that the requirements of his attribute are allocated to a proper set of systems and lower-level entities and evaluated constantly to ensure that the product meets its attribute requirements. The attribute management responsibility is generally assigned to independent core engineering functions that are different than the line engineering activities responsible for designing and developing various systems of the product. For example, a manager assigned to the "comfort and convenience" attribute will have to review the entire design of the vehicle being developed and analyze each vehicle system to ensure that all aspects of comfort and convenience, such as ergonomics of driver interfaces, seat comfort, entry/exit ease, luggage loading convenience, engine service ease, and thermal comfort, are considered.

WHY "SPECIFY" REQUIREMENTS?

Clearly stated, requirements provide information and direction needed to begin the product design process. The information provides (1) clear visibility across different teams (responsible for different systems within the product) into how and why the requirements are allocated and thus helps understand cross-functional interactions between all systems within the product; (2) clear responsibilities to the design teams to meet the requirements; (3) early assurance that all top-level requirements are fully satisfied in the product, with traceability to where they are satisfied; (4) checks to prevent unintentional addition of features and costs (i.e., avoids "gold plating"); (5) checks to avoid unwelcomed surprises in later phases of the product development; (6) quick assessment of the impact of any changes made to the requirements; and (7) procedures for early and thorough verification and validation of the product design in meeting the requirements.

HOW ARE REQUIREMENTS DEVELOPED?

Most requirements are not entirely developed during the early stages of a product program. In fact, developing requirements "from scratch" requires a lot of data gathering, analysis, testing, and evaluations, and thus, requirements development is a very time-consuming and expensive process. Most requirements are adopted from previously developed and proven requirements available from various sources such as (1) standards (e.g., internal company standards; external standards developed by international, national, or local government agencies; professional societies; and trade associations); (2) product/system design specifications/guidelines developed by the manufacturer and its suppliers; (3) test and evaluation procedures and practices within product development organizations; and (4) experiences (past failures and successes), customer feedback, lessons learned, and insights gained from previous programs of similar products. The requirements should also be continuously evaluated to ensure that they are not outdated due to advances in technologies, design trends, new materials, and changes in customer needs and government regulations.

Implementation of new technologies and features to new models of previously developed products (e.g., development of an electric vehicle as compared to the vehicles with the traditional internal combustion engines) will require considerable additional work in understanding issues such as how the product will be used by the customers, customer concerns, problems during operation of the product during its life cycle, development of new technologies to get them ready, and so on. Further trade-offs between different attributes need to be also considered during the requirements development process. The requirements development process thus requires inputs and reviews from experts from different departments/disciplines.

CHARACTERISTICS OF A GOOD REQUIREMENT

Many characteristics are to be considered in determining if a requirement is "good" (i.e., useful, nonconfusing, and implementable). Thus, the considerations in developing a "good" requirement are described as follows:

1. The requirement must state "The product shall" (i.e., shall do, shall perform, shall operate, shall provide, shall weigh, etc.) followed by a description of what must be done.
2. The requirement should be unambiguous, clearly stated, and complete. It must precisely state what must be accomplished, the accomplishment level, and the conditions in which it must apply. It should be worded to minimize confusion and differences in its interpretation between different individuals (especially engineers). To ensure that a requirement is complete, it should provide contextual details such as the situation, environment, operating condition, time durations, urgency/priorities, and characteristics of its users, under which the product is expected to function.
3. The requirement should use consistent terminology to refer to the product and its lower-level entities.
4. The requirement should clearly state its applicability (i.e., when, where, types of system, or hierarchical system level where it is applicable and where it cannot be applied).
5. The requirement should be verifiable by a clearly defined test, test equipment, test procedure, and/or an independent analysis.
6. The requirement should be feasible (i.e., it should be possible to create the system or product without extraordinarily large amount of development time and costs).
7. The requirement should be consistent and traceable with other requirements above and below in the system hierarchy.

8. Each requirement should be independent of other requirements. This characteristic will help in controlling and reducing variability in the product parameters and hence its performance.
9. Each requirement should be concise, that is, it should be stated with minimum information content.

(Note that the considerations 8 and 9 meet the two basic axioms [Axiom 1: Independence Axiom—maintain the independence of functional requirements, and Axiom 2: Information Axiom—minimize the information content in the design] of the axiomatic theory considered in product design [Yang and El-Haik, 2003].)

The *NASA Systems Engineering Handbook* (NASA, 2007) also provides more information in its Appendix C on "How to Write a Good Requirement" and it also includes a requirements validation checklist.

TYPES OF REQUIREMENTS

CUSTOMER REQUIREMENTS

Customer requirements are statements that define the customer (or stakeholder) expectations of the product or system in terms of its mission, objectives, functions, performance, environment, and constraints. The requirements are usually specified by defining target values of measures of the product attributes to produce predefined levels of effectiveness (or ratings when rating scales are used, e.g., must achieve a rating of 7 or above on a specified 10-point scale). These requirements are defined from the validated needs of the targeted customers during the entire life cycle of the product. The customer requirements can be categorized by the attributes desired by the customers.

FUNCTIONAL REQUIREMENTS

Functional requirements define what functions need to be performed to accomplish the product objectives (i.e., its operation or use). Functional requirements are the minimum set of independent requirements that completely characterize the operations (or tasks) that the product should perform including details such as what to perform, when to perform, and how to perform. The functional requirements thus define the necessary tasks, actions, or activities that must be accomplished.

The requirements can be categorized as (1) functional requirements or (2) nonfunctional requirements. The functional requirements specify the necessary task, action, or activity that must be accomplished. The functional characteristics can be further categorized into (1) behavioral or operational requirements related to the use of the product, (2) performance requirements for actions, tasks, or activities to be performed, and (3) architectural (e.g., how the product should be configured, modularized, or packaged into available space). Nonfunctional requirements are not related to the performance of any task, but they specify nonfunctional characteristics such as aesthetic characteristics (e.g., the exterior color of an office cabinet [if the color is not used to meet any code or classification]).

Performance Requirements

Performance requirements define how well a product (or a system) needs to perform its functions. For example, the function of a braking system of an automobile can be specified in terms of a requirement on its performance such as the braking system shall be capable of stopping the vehicle within 36.6 m (120 ft.) from 96.6 kph (60 mph) speed on a flat level dry road at 21°C (70°F) external temperature with the brake pedal force of 36.3 kg (80 lbs.).

The performance requirement defines the extent to which a mission or function must be executed, generally measured in terms of quantity, quality, coverage, timelines, or readiness. Performance requirements are initially defined through requirements analyses and trade-off studies using customer needs, objectives, and/or customer or manufacturer's requirement statements. Performance requirements are defined for each identified customer mission and for each primary function and its subfunction. Performance requirements are also assigned to each system and to its system elements (subsystems and components) through the top-down cascading and allocation processes.

Interface Requirements

The interface requirements specify what needs to be done (or the capabilities) at a given interface between different systems, subsystems, or components for the operation of the product. The requirement on a physical interface will specify the performance or design characteristics of the physical connection or a joint between the two entities in terms such as the strength of the joint, pressures encountered at the joint carrying a specified fluid or gas flow, size and type of fasteners, and electrical characteristics such as industry-specified connector code number, the current carrying capacity, resistance, capacitance, or data flow rate at the joint. The interface requirements thus can specify (1) functional performance; (2) electrical, environmental, human, and physical requirements; and (3) constraints on a common boundary between two or more systems or system elements. Chapter 5 provides more information on the interfaces and their design issues.

Reliability Requirements

Reliability can be defined as the probability that a product, system, subsystem, or component will not fail for a given period of time under specified operating conditions. Reliability is an inherent system design characteristic. Chapter 19 provides more information on the reliability computations and analyses.

Environmental Requirements

These requirements are to control the adverse effects of the environment on people, products, or systems under which the product or a system is designed to work. The environmental concerns that must be addressed include effects of acceleration, vibration, shock, static loads, acoustic, thermal, contamination, corrosion, crew-induced loads, total dose or peak level of radiation/toxins, surface and internal charging, weather/atmospheric conditions and air quality (e.g., emissions of greenhouse gases), magnetic, pressure gradients during operation, microbial growth, radio frequency, and so forth.

HUMAN FACTORS REQUIREMENTS

These are requirements for the product- or lower-level systems, subsystems, or components to ensure that humans as operators or maintainers of the product or systems can perform their allocated functions or tasks with ease and comfort. Human capabilities, characteristics, and limitations must be considered during the design stages in the same way as that of the characteristics of mechanical or electronic components. Chapter 11 provides more information on human factors and human performance issues.

SAFETY REQUIREMENTS

These are requirements for the safe operation of the product or system. Safety is generally defined in terms of freedom from accidents or hazardous situations that can result in adverse health effects, injuries, fatalities to people, or property damage. Chapter 12 provides more information on safety and other related issues.

SECURITY REQUIREMENTS

Many complex products may have security-related requirements to ensure that the product cannot be accessed by any unauthorized persons or by persons considered to be a threat to the product or its systems. The requirements should include denying access as well as incorporation of additional considerations if the security is breached. Maintenance of product operation during and after the security intrusions will require special treatment, especially if it involves lives and/or national-level security concerns.

DESIGNED-TO-CONFORM VERSUS MANUFACTURED-TO-CONFORM REQUIREMENTS

In general, a product must be designed to meet its stated requirements by taking into account possible variations in its characteristics (or parameters) due to manufacturing. When a manufactured product is selected at random from its production lot and tested according to specified test procedures, it must meet its requirements. Thus, such a product can be considered to be "manufactured-to-conform" to its requirements. On the contrary, if the design of a product (i.e., product design as described with its design specifications including design tolerances, e.g., in its drawings or CAD models) meets all its stated requirements, it can be considered to be "designed-to conform" to its requirements. However, the product that is shown to be "designed-to-conform" may not conform to its requirements when it is manufactured because of actual manufacturing variations that may exist due to a number of reasons (e.g., tolerance stack-up not considered or not accurately predicted, production machines not calibrated, tool wear, operator errors, and raw materials out of specifications).

This is an important issue because certification of a product as a "design-to-conform" product is not the same as a "manufactured-to-conform" product. Every manufacturer needs to make sure that its product verification processes consider all possible manufacturing problems (i.e., sources of manufacturing variations) that can affect its compliance to the specified requirements. Chapters 10 and 17 provide additional information on approaches and tools to control quality in production.

WHERE ARE REQUIREMENTS STORED?

Product requirements of all the aforementioned types are generally compiled in (1) company practices and engineering standards, (2) professional organization standards (e.g., Society of Automotive Engineers, Inc. [SAE], American Society of Testing Materials [ASTM]), (3) government standards (e.g., local, state, federal/country level [e.g., Consumer Products Safety Commission, Nuclear Regulatory Commission, Federal Motor Vehicle Safety Standards (FMVSS)], or standards followed by a group of countries, e.g., International Organization for Standards [ISO], Economic Commission for Europe [ECE]). Such predeveloped standards reduce time required to gather information to develop requirements needed to begin the product development process. (More information on standards is provided in a later section entitled "Role of Standards in Setting Requirements" of this chapter.)

REQUIREMENTS ALLOCATION AND ANALYSIS

REQUIREMENTS ALLOCATION

Allocation of a requirement involves (1) assignment of a requirement to a function, (2) assignment of a system element to a requirement, or (3) division of a requirement and assignment of a portion of the requirement to each part or a separate element (e.g., weight divided into parts, distribution of sensors assigned to different entities).

The requirements for a higher level are generally allocated to lower levels. The requirements allocation is conducted at the earliest possible product concept generation level. Functional analysis and allocation of functions at different levels of a product or a system generally succeeds requirements allocation. The requirement setting and requirements allocation process in the product development is iterative in nature due to the consideration of many possible design configurations (or concepts), functional allocation, and continuous refinements to arrive at an acceptable product design (see Chapter 2, Figure 2.2).

REQUIREMENTS ANALYSIS

Requirements analysis involves determination of product (or system)-specific requirements based on the analyses of customer needs, all applicable requirements (available from various sources, e.g., standards), objectives (including considerations of missions, projected utilization environments for people, products, and processes and design constraints), and measures of effectiveness of the requirements (i.e., how the compliance to each requirement is to be measured). It is usually an iterative procedure

Product Attributes, Requirements, and Allocation of Functions 83

based on the understanding of the relationships between the above-mentioned needs, requirements, objectives, and the measures of effectiveness. Requirements analysis assists in refining the product requirements in concert with defining functional and performance requirements for the product's primary life cycle functions. It is a key step in establishing achievable requirements that satisfy the customer and the product developer's needs. Chapter 2 (Figure 2.2) also provides context to the requirements analysis in the SE process.

ATTRIBUTES DEVELOPMENT

During the early stages of the product development, the customer requirements should be met to ensure that the product possesses the attributes or the characteristics to satisfy its customers. The total effect (or impact) of the product on its customers is broken down into a set of attributes for the convenience of management of product characteristics relevant to the customers. The attribute development is a process of carefully managing, allocating, and cascading each attribute (i.e., functions and characteristics needed to satisfy the requirements of each attribute) down to different levels of the product entities.

For example, the attributes of a complex product such as laptop computer can be defined after surveying a large number of customers. The attributes can be stated as follows:

1. Easy-to-use input devices (keyboard, cursor control, CD/DVD reader, and USB ports)
2. Easy-to-view display screen
3. Good sound quality
4. Large enough hard disk data storage (RAM)
5. Sufficient memory capacity for anticipated software applications and file storage
6. Processor capacity and speed
7. Time between battery charging (battery capacity)
8. Safe to use (does not cause injuries during usage)
9. Lasts for a long time (durability).

The above-mentioned attributes must be allocated to the following systems of a laptop computer:

1. Chassis system
2. Display system
3. Audio system
4. Input system
5. Electronic processing system
6. Memory system
7. Power system
8. Wireless communication system
9. Cooling system.

The last attribute, durability, would apply to all the preceding systems. The processor capacity and speed attribute would apply to the electronic processing system. Good sound quality attribute will apply to the audio system. The time between battery recharge attribute will apply to all power-consuming and power-handling systems such as the display system, audio system, electronic processing system, wireless communication system, cooling system, and the power system.

The following section will cover additional issues related to the cascading of the attributes, assignment of the attributes, and cascading the attribute requirements to various lower levels of entities within the product.

CASCADING ATTRIBUTE REQUIREMENTS TO LOWER LEVELS

To ensure that all issues related to any given product attribute are taken into account during the design stages of the product, each attribute is systematically subdivided into lower levels such as sub-attribute, sub-sub-attribute, and sub-sub-sub-attributes. Figure 4.1 illustrates those attributes (A1, A2, A3, and A4) are subdivided into their corresponding sub-attributes such as SA11, SA12, SA21, SA22,..., SA42. The requirements defining each of the sub-attributes are shown in Figure 4.1 as R111, R112,..., R425. Development of an attribute tree helps in progressively dividing the attributes into a manageable number of lower-level attributes so that requirements for each sub-attribute issue can be clearly defined. For each requirement, one or more test procedures (verification tests) along with performance measures and minimum acceptance criteria levels are also specified to ensure that the requirement can be verified.

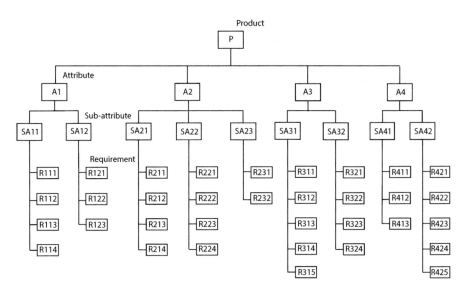

FIGURE 4.1 Attribute tree illustrating subdivision of attributes into lower-level attribute requirements.

Product Attributes, Requirements, and Allocation of Functions

For example, the attribute "Easy to view display screen" of the laptop computer can be broken down into the following sub-attributes: (1) display size, (2) display resolution, (3) display luminance (physical brightness), (4) display color, (5) visibility from large viewing angles, (6) controls for the display brightness and color, (7) display orientation (adjustment) angles, (8) display surface reflectivity, and (9) display surface cleanability.

Some of the requirements for the aforementioned sub-attributes can be specified as follows:

1. The display shall have a minimum size of 33 cm (13 in.) measured diagonally with a length-to-width ratio of 7:4.
2. The display shall have a minimum resolution of 4 pixels/mm.
3. The display shall produce minimum luminance of 600 cd/m^2 (on bright white visual details) and the minimum contrast ratio between white and black areas shall be at least 1000:1.
4. The display shall have rotary controls to adjust luminance, colors, and resolution.
5. The display should be legible to 65-year-old viewers from 70° to the left to 70° to the right of the display axis (normal to the display surface).

DIVIDING THE PRODUCT INTO MANAGEABLE LEVELS

Complex product with many systems and components is divided into a manageable number of systems so that employees with specialized knowledge can be assigned to design teams working on different systems and subsystems. Figure 4.2 shows that product (P) is divided into systems S1, S2, S3, and S4. Each system is further

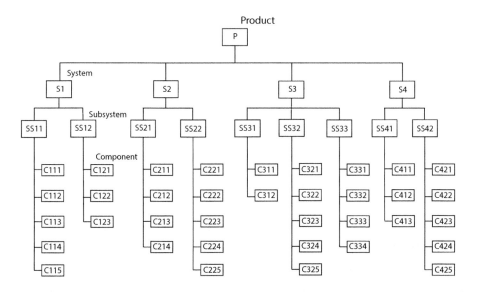

FIGURE 4.2 Decomposition tree of a product into systems, subsystems, and components.

divided into its subsystems. For example, subsystems of system S1 are SS11 and SS12. Furthermore, the components required to accomplish the function of each subsystem are determined. The components of subsystem SS11 are shown in Figure 4.2 as C111, C112, ..., C115. The diagram shown in Figure 4.2 is also referred to as the product decomposition tree.

In our example of the laptop computer, the systems of the laptop are (1) chassis system, (2) display system, (3) audio system, (4) input system, (5) electronic processing system, (6) memory system, (7) power system, (8) wireless communication system, and (9) cooling system. Each of the aforementioned systems will have subsystems. For example, the subsystems of the chassis system are (1) die-cast body, (2) die-cast display frame, and (3) fasteners (screws and clips). The components of the die-cast body subsystems are lower body frame, hinges, upper body frame, and battery service door. The components of the cooling system are fan, electric motor, and heat conducting ribs and vents in the die-cast body.

RELATING ATTRIBUTE STRUCTURE TO SYSTEMS

Relating the attribute structure (i.e., attributes and their requirements) to systems and their lower-level entities requires a lot of thinking and design iterations (see Chapter 2, Figure 2.4). The designer will have to think about how to configure systems within the product to meet various attributes and their requirements. Here, the designer typically thinks about how similar products in the past were configured and begins to develop an overall product package, that is, how would the product be configured in the required amount of space constraints. He would also think about the current and future technologies that can be developed sufficiently to incorporate into the product. He will have to think about the many sub-subsystems and subsystems within the systems that need to be designed and interfaced for the product to perform. Simultaneously, he will have to think about the product attributes and their requirements that must be met. The task of keeping track of all the relationships between many attributes and their sub-attributes and systems and their subsystems becomes very complex. This is where the SE approach is useful. To manage the complexity, implementation of one or more multidisciplinary design teams and concurrent engineering approaches must be considered. The multidisciplinary teams provide experts from various fields needed to consider many issues involved during the creation of the product design. Furthermore, these experts need to constantly communicate and interact with each other to determine and design interfaces between various systems and trade-offs between different considerations to meet various requirements.

The problem of relating an attribute to systems in an automotive product is illustrated by the following two examples.

The first example illustrates how the "safety and security" attribute of the vehicle is related to the vehicle lighting system. The "safety and security" attribute is shown in Figure 4.3 under "product" level on the left-hand side. A sub-attribute of "safety and security" is "safe night driving" (shown on the left side under "safety and security"). The two requirements of "safe night driving" are (1) it should provide adequate "road visibility" to the driver and (2) the driver's vehicle should be visible to other drivers (i.e., "vehicle visibility" is provided). The topmost line of the figure

Product Attributes, Requirements, and Allocation of Functions

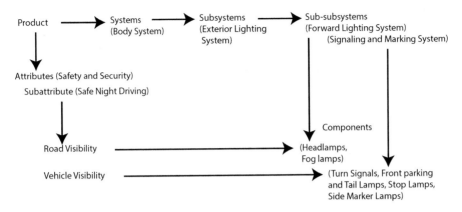

FIGURE 4.3 An example relating automotive safety and security attribute to a vehicle lighting system and its subsystems, sub-subsystems, and components.

shows the product decomposition from product (automobile) to system (body system) to subsystem (exterior lighting system) to its sub-subsystems (forward lighting system and signaling and marking system). Headlamps and fog lamps are components of the forward lighting system, and they provide illumination to make the roadway visible. Similarly, the signaling and marking system with its components such as turn signals, front parking lamps, tail lamps, stop lamps, and side marker lamps make the vehicle visible to other drivers. Thus, the figure illustrates that the functions of the components of the forward lighting system and the signal and marking system are designed to meet two requirements of safe night driving sub-attributes, namely, road visibility and vehicle visibility.

The second example similarly relates the "comfort" attribute of the vehicle to four vehicle sub-subsystems related to seats, armrests, sound absorbing material, instrument panel, air ducts, climate controls, and front and rear suspensions. Figure 4.4 shows that the sub-attributes of "comfort" are seating comfort, quiet interior, thermal comfort, and ride comfort. The automotive systems designed to meet the sub-attributes are body system, climate control system, and chassis system. The subsystems and components of the systems related to the above sub-attributes are shown in Figure 4.4. The vehicle seat components, body cavity sound, and thermal insulation and suspension components are shown (see lower right side of the figure) to be related to four sub-attributes of comfort, namely, seating comfort, quiet interior, thermal comfort, and ride comfort (see lower left side of figure).

AN EXAMPLE: ATTRIBUTES, SYSTEM DECOMPOSITION, AND REQUIREMENTS FOR VEHICLE EXTERIOR LIGHTING SYSTEM

The vehicle lighting system example covered in the preceding section (see Figure 4.3) is illustrated further by reviewing its attributes, subsystems, and requirements. The exterior lighting is a subsystem of the vehicle body system. The exterior lighting system is thus interfaced with the vehicle body system and vehicle electrical system. The attributes of the vehicle lighting system are described as follows.

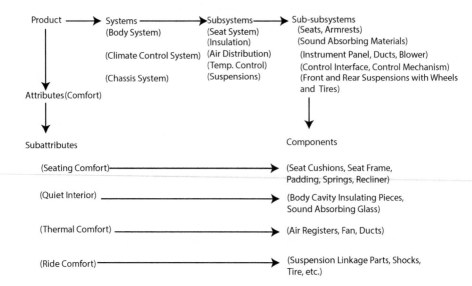

FIGURE 4.4 An example relating automotive comfort attribute to vehicle body system and its subsystems, sub-subsystems, and components.

Attributes

The attributes of the exterior lighting system are (1) road visibility, (2) vehicle visibility, and (3) long life. These three attributes of the exterior lighting system can be traced back to the "safety and security" and "durability (long life)" attributes of the vehicle (as an automobile is the parent product). It should be noted that "road visibility" is a sub-attribute of "safe night driving," which is a sub-attribute of "safety and security" attribute of the vehicle (product) (see Figure 4.3). Similarly, "vehicle visibility" is a sub-attribute of "safe night driving." "Long life" (or durability) is a vehicle attribute.

Systems and Subsystems

The automotive exterior lighting system typically consists of following subsystems:

1. Forward lighting subsystem consisting of the headlamps, parking lamps, front turn signal lamps, fog lamps, and daytime running lamps
2. Rear lighting subsystem consisting of the tail lamps, stop lamps, rear turn signal lamps, reversing lamps (backup lamps), license plate lamps, and rear reflex reflectors
3. Side lamps subsystem consisting of the side marker lamps, side turn signals, and side reflex reflectors
4. User interface subsystem consisting of the headlamp switch, turn signal and high beam switch, hazard switch, brake-pedal switch, and the instrument panel with turn signal display (blinking arrows) and high beam warning lamp

Product Attributes, Requirements, and Allocation of Functions

5. Sensors subsystem (smart headlighting) consisting of an ambient light sensor, steering wheel angle sensor, and body attitude and headlamp aim and headlamp beam switching actuators
6. Electrical power distribution subsystem consisting of the wiring harnesses, relays, and fuses
7. Mechanical support subsystems consisting of the front fascia, rear fascia, headlamp aiming screws, fasteners, clips, and so forth.

Relationship between System Components and Requirements

Table 4.1 shows the relationships between the product, system, subsystem, sub-subsystem, and components of the vehicle lighting system in the left five columns. The last five columns of the table show two vehicle attributes, namely, safety and security and durability. The relationships between vehicle systems, subsystems, sub-subsystems, and components to the attribute, sub-attributes and sub-sub-attributes are shown by "x" marks in the last five columns. The requirements to meet the attribute relationship to systems and their components corresponding to each of the "x" marks can be derived from the following basic requirements for the exterior lighting requirements.

Requirements of Exterior Lighting System

The exterior lighting system requirements are derived from the "safety and security" attribute of the vehicle.

1. All exterior lamps shall meet applicable photometric and installation requirements in the FMVSS 108 (NHTSA, 2010). (Note: Photometric requirements control the light intensity distributions [beam patterns] of the lamps. And installation requirements control the mounting locations of the lamps in the vehicle.)
2. All subsystems shall operate on a nominal voltage of 12.8 V.
3. All lamps shall have a minimum life of 2,000 hours.
4. Switches shall function properly for at least 1,000,000 on/off cycles.
5. Wire harness conductor gage shall be sized to carry the electrical load of all exterior lamps to provide required illumination levels.
6. Lamp assemblies shall withstand impact from 0.25-inch pebbles at 75 mph.
7. Lamp assemblies shall attach to the vehicle body at minimum 3 points.
8. All headlamps and fog lamps shall have horizontal and vertical aiming capability.

The preceding requirements need to be cascaded down to systems, subsystems, and components according to attributes, sub-attributes, and sub-sub-attributes as shown in Table 4.1.

Verification Tests

Each requirement that is indicated by an "x" mark in Table 4.1 needs a test and a corresponding test procedure for its verification. The test and compliance criteria are provided in the FMVSS 108 (NHTSA, 2010). Most of the tests in the FMVSS 108

TABLE 4.1
Requirements Cascade for Automotive Exterior Lighting System

Product	System	Subsystem	Sub-subsystem	Component	Attribute	Sub Attribute	Sub-sub-attribute		Attribute
					Safety and Security	Safe Night Driving	Road Visibility	Vehicle Visibility	Durability
Vehicle									
	Body system								
		Exterior lighting			x				x
			Forward lighting system		x				x
				Headlamps		x	x		x
				Fog lamps			x		x
			Signaling and marking system				x		x
				Parking lamps				x	x
				Tail lamps				x	x
				Stop lamps				x	x
				Turn signal lamps				x	x
				Backup lamps				x	x
				License plate lamp				x	x
				Side marker lamps				x	x
				Reflex reflectors				x	x
		Instrument panel		User interface					x
				Controls/switches					x
				Displays					x
	Electrical system	Lighting system		Wiring					x
				Sensors					x
				Actuators					x
	Chassis system	Steering column		Controls/switches					x
				Displays					x

Note: x = Requirement specified in Federal Motor Vehicle Safety Standard 108 and other referenced SAE standards.

Product Attributes, Requirements, and Allocation of Functions 91

are adopted from various SAE standards on vehicle lighting (SAE, 2009). Thus, the description of the tests and test procedures can be found in the SAE standards referenced in the FMVSS 108.

AN EXAMPLE: CASCADING OF VEHICLE LEVEL SUB-ATTRIBUTE REQUIREMENTS INTO POWERTRAIN SUBSYSTEM REQUIREMENTS

Attribute requirements cascading table is a useful tool to ensure that product level requirements of each sub-attribute are cascaded down to each lower level system and its subsystems. Table 22.2 (in Chapter 22) provides an example of an attribute cascading table for cascading vehicle sub-attribute level requirements into requirements on the powertrain system and its three subsystems (namely, engine subsystem, transmission subsystem, and drivetrain subsystem).

AN EXAMPLE: ATTRIBUTES, REQUIREMENTS, AND TRADE-OFFS IN SUSPENSION SYSTEMS OF A SPORTS CAR

Attributes

One of the attributes of a sports car is that it should have exceptional vehicle handling and braking systems. Thus, the front and rear suspensions (subsystems of the vehicle suspension system) should have the following three sub-attributes: (1) handling performance; (2) noise, vibrations, and harshness (NVH) performance; and (3) braking performance.

Requirements

The important requirements of the sub-attributes can be briefly described as follows:

1. Handling performance
   a. The vehicle shall have sufficient stiffness to prevent body roll of more than 10 degrees and pitch angle of more than 5 degrees.
   b. The vehicle shall have sufficient stiffness for anti-squat (resist compression in the rear suspension) during acceleration (e.g., less than 30 mm rear suspension deflection at 0.5g forward acceleration).
   c. The vehicle shall be able to attain 1g lateral acceleration with body roll angle of less than 10 degrees.
   d. The suspension system shall be adjustable between biasing at 1–10 Hz targeted dampening.
   e. The suspension linkages must be adjustable to maintain precise wheel locations.
   f. The steering system gain should be speed sensitive and have an on-center feel.
2. NVH performance
   a. The suspension system shall have the ability to dampen large-amplitude displacements of more than 40 mm.
   b. The suspension system shall provide the driver with a feeling that the vehicle is well connected to the road (i.e., not feel as if the car is "floating").

c. The suspension linkages and connections shall be robust and not loosen or rattle for over 100,000 miles.
   d. The brakes shall not produce perceptible noise over 60,000 miles of city driving.
3. Braking performance
   a. The braking system shall meet the FMVSS 135 braking requirements.
   b. The brakes shall be fade resistance over at least 60,000 miles.
   c. The suspension and braking systems shall provide the feeling of road-connectedness and be well integrated even during aggressive maneuvers involving vehicle accelerations or decelerations over 0.3 g.
   d. The braking system shall be aesthetically integrated with the wheel package.

The preceding requirements must be cascaded down to develop requirements for subsystems and components of the vehicle handling and braking systems. The requirements cascading task need to be simultaneously considered with the task of designing the configuration of the handling and braking systems. The entire requirement cascade and configuration design process should be iterative and various trade-offs between the primary attributes of the handling performance, NVH performance, braking performance, and costs must be carefully considered.

Trade-Offs

Some important trade-offs to be considered during the suspension design are described as follows:

1. *Performance versus costs*: The suspension performance can be improved by adding springs/dampeners, along with an active suspension system and magneto-rheological shocks. However, the additions will increase costs. Other additional costs are due to (1) multiple stabilizer features, (2) dynamic sensors, (3) control modules to process and communicate information, (4) additional mounting interfaces, and (5) use of expensive lightweight materials to reduce unsprung mass of many aluminum components.
2. *Performance versus increased complexity in other vehicle systems*: Incorporation of additional features will increase the complexity of other vehicle systems and subsystems to interface and process additional data. For example, processing the active handling data may require a body electronic control module to have an additional processing capacity. Additionally, electrical wiring harnesses will be needed to communicate to the sensors. The added complexity will also increase potential failure modes of the complex suspension system and other related systems such as high-performance tires, high-performance transmission, axles, and U-joints.
3. *High-performance versus high-production complexity*: Incorporation of additional features will not only increase the design workload and costs but will also introduce more complexity (due to more parts and interfaces) in manufacturing and assembly and testing and verification tasks.

Product Attributes, Requirements, and Allocation of Functions

The preceding example provides some insights into the large amount of development work that must occur to provide a high-performance suspension system in a sports car.

FACTORS AFFECTING REQUIREMENTS

Organization responsible for developing the product must consider internal and external factors that can affect the requirements for each of the product attributes. The internal factors are factors that are within the organization, and they can be controlled by the company management. Thus, the internal factors affecting the requirements can be considered to be controllable. The external factors are factors that are outside the organization's control that affect or constrain the organizations capabilities in determining requirements.

The internal factors include the following:

1. *Capabilities of the organization*: Resources available to develop and produce the product. These include financial, human (people and their skills and expertise), facilities and equipment, and their capacities to handle the design work and production load.
2. *Existing products*: Availability of systems, subsystems, and components used in existing products that could be reused or modified to include in the future products. Such components or systems are generally referred to as "carryover" components or systems. These components or systems could be "common" or "shared" across many of the existing products (or product platforms).

The external factors include the following:

1. *Government regulations*: All applicable requirements specified in the government regulations must be met to sell the product in the market.
2. *Competitors*: Abilities of the competitors to produce similar products (with product features and pricing) in the same time frame.
3. *Suppliers*: Capabilities of the suppliers of purchased components, subsystems, systems, and services to supplement the organization's capabilities. (Note: The supplier capabilities affect the "make vs. buy" decisions related to acquisition of entities within a product.)
4. *Financial position*: Ability of the consumers and producers to accept financial obligations and market demands affected by the economy, monetary, and fiscal policies of countries where the products could be manufactured and marketed.
5. *Technological developments*: New developments in technologies related to design, operation, and production of the systems and components must be taken into account to ensure that the product will meet customer expectations on the technical, operational, and cost aspects of the product.
6. *Design trends*: Emerging and latest trends in design (e.g., style, fashion, and technologies) related to customer expectations about the product.

ROLE OF STANDARDS IN SETTING REQUIREMENTS

Existing product design standards contain valuable information in designing products. The information is generally organized into the following categories: (1) common terminology and definitions; (2) historical background information related to the development of the standard; (3) rationale considered during the development of the standard; (4) description of the requirements; (5) measurements, test equipment, and test procedures; (6) criteria for acceptable performance; and (7) guidelines on special topics such as labeling, operating instructions, and installation.

A product engineer must be knowledgeable about all the requirements for the product (specified in various applicable standards) and the processes used in designing and producing the product. In the product liability cases, it is assumed that the manufacturer is knowledgeable about his product as any expert in that field (see section on "Product Safety and Liability" in Chapter 12). Thus, the engineer must know all existing standards (published by government agencies, industry associations, company, etc.), innovations, design trends, and competition related to the product and processes involved in its production, verification tests, product distribution, and product uses.

TYPES OF STANDARDS

As the product engineer begins to gather standards available from various sources, it is also important to realize that there are different types of standards. The standards differ depending on the source and their applicability. Furthermore, each standard has some unique advantages, disadvantages, and problems that must be understood before their applications. Some basic information on these issues is presented in the following discussion.

The standards can be categorized as follows:

1. *Design (specifications) versus performance standards*: The design standards specify acceptance criteria based on certain characteristics of the products; for example, a bicycle reflector can be specified based on its size. Thus, the design requirement for a bicycle reflector can be stated as follows: The bicycle shall be equipped with two reflectors, a front and a rear, each with an area of at least 6.45 cm^2 (1 in.^{2}). On the contrary, the performance standard would specify how the product should perform. For example, the performance requirement of the bicycle reflectors can be stated as: the front and rear bicycle reflectors shall be visible to an approaching driver at night from at least 122 m (400 ft.) viewing distance under low beam headlamp illumination. Thus, in a performance standard, the designer has flexibility in selecting many parameters (e.g., shape, size, and reflectance of the reflector) of the product as long as it meets its performance (visibility) criterion.
2. *Mandatory (federal, state, county, and municipal) versus voluntary standards (company, professional society)*: All specified requirements in a mandatory standard must be met. Whereas the requirements specified in a

voluntary standard can be met only if the organization producing the product voluntarily decides to meet the requirements.
3. *Designed-to-conform versus manufactured-to-conform standards*: The difference between design-to-conform versus manufactured-to-conform becomes important if there would be variations in the manufacturing process; the characteristics and performance of the manufactured product will be different than what is specified in the product's documents, drawings, or models created during its design phases.
4. *Horizontal versus vertical standards*: The horizontal standards apply to products designed to meet the same function but are used across many different industries (e.g., requirements for a ladder design may be the same when applied in a factory design vs. in a ship design). The vertical standards are specific to specific industries (or chimney organizations), for example, shipbuilding standards will only apply to products or features incorporated in ships.
5. *Consensus standards*: The consensus standards are developed when the majority of individuals involved in the standards development process agree on the requirements proposed during a balloting process used for the approval of the standard. Product standards developed by many professional societies and organizations (e.g., SAE, American Society of Mechanical Engineers [ASME], ASTM, American National Standards Institute [ANSI], and Underwriters Laboratories [UL]) are consensus standards. Such standards are generally developed under procedures that provide opportunities for discussions on diverse views and that indicate that interested and affected persons have reached substantial agreement on its adoption.
6. *Federal standards*: These standards are developed by a federal government agency created by a federal act (e.g., the Motor Vehicle Safety Act) and published in the Federal Register (called the Code of Federal Regulations). The federal standards are mandatory (i.e., if a federal standard for a product exists, then the product cannot be sold in the United States without compliance to the standard).
7. *Other international standards*: These standards are developed by international organizations, for example, ISO and ECE.

Advantages of Standards

1. Presence of a standard would reduce design time as it would provide information on the rationale for the standard, design issues, design criteria, test procedures, and design guidelines. Some standards also provide checklists for design considerations.
2. Applications of a standard to a set of products (within a brand, an organization, or an industry) can facilitate uniformity and sharing of common, standardized components or designs.
3. Use of standards can promote uniformity by requiring new or unfamiliar designers to design their new products to meet the requirements provided in the standards.

4. The standards can improve the quality of products as their uniform applications would avoid repetitions of previously made design errors. Thus, the standards can alert the designers of potential problems.
5. Use of standardized equipment, product configurations, and operating procedures would reduce the chances of human error in using or operating the products.
6. Properly developed safety standards would result in producing safer products, which can reduce the number of accidents and injuries.
7. Presence of a standard would reduce the likelihood of a violation or wrongful trade-offs between design considerations included in the standard.

Disadvantages of Standards

1. Mandatory compliance to a design standard would reduce design freedom or design flexibility and thus may stifle innovations in future products.
2. Poorly conceived design standards can prohibit development of new products with better (e.g., safer) features.
3. Need for a standard would be questionable if the existing products with widely different design configurations and features are effective in serving customers with different needs.
4. Many consensus standards may not be very stringent. Thus, such standards will not serve a useful purpose in improving product performance and quality.

Problems with Standards

1. The requirements stated in the standard may not be stringent enough (or inadequate) for the product application.
2. The requirements may not be consistently applied to the products used in different industries or different applications.
3. The requirements in different standards on a product may not be compatible or may even be contradictory (e.g., different standards have different criteria that may be impossible to meet simultaneously).
4. Statements of requirements may be difficult to interpret or understand (e.g., test methods and their procedures may be difficult to interpret and hence may be applied differently by different test engineers).
5. A requirement may be incomplete (i.e., it lacks some key details or may not consider all aspects of the entities within the product).
6. Conformance to the standard may not guarantee creation of a successful or a safe product if the standard is incomplete and/or the requirements are too lenient. The standard may be useless (e.g., consensus standards usually are not very stringent).
7. The standards are expensive to develop. The standard development process involving data collection, analysis, and evaluations may be expensive and time-consuming.

Product Attributes, Requirements, and Allocation of Functions

8. The standards need to be reviewed periodically and upgraded to account for new findings, advances in technologies, and design trends.

STANDARDS DEVELOPMENT PROCESS

The standards development process typically involves the following steps:

1. Establish a need for the standard (e.g., through an increased number of accidents or injuries during product uses, a request from a product design or engineering office).
2. Conduct literature reviews to understand issues, processes, considerations, and cause-effect relationships.
3. Gather data to establish or support the needs (e.g., to determine the relationship between product characteristics and accidents, conduct tests to evaluate the performance of similar products).
4. Analyze the data (e.g., conduct hazard analyses or evaluate the products for performance, durability, interchangeability, variations; conduct cost-benefit analyses to assess proposed requirements).
5. Prepare a draft standard.
6. Ask experts to review the draft and propose changes.
7. Conduct additional tests to verify/modify proposed requirements.
8. Ballot the draft of the proposed standard to standards setting committees (on consensus standards).
9. Modify the wording to incorporate acceptable suggestions.
10. Accept the standard and publish or incorporate it in a corporate standards manual.

The standard setting process in the U.S. federal agencies involves a number of steps, as follows:

1. Establish a need for the standard (e.g., increasing number of accidents involving a certain product) or petitions received from organizations or individuals to consider developing a new standard.
2. Petitions are received, reviewed, and a decision is made to either grant or deny the petition.
3. Notices are published in the Federal Register for "Request for Information" or "Advance Notice of Proposed Rule Making" (ANPRM).
4. Information received from responses is reviewed. Additional research or analyses may be conducted and a "Notice of Proposed Rule Making" (NPRM) is published.
5. Responses are reviewed and if the decision is to accept the requirements then a Final Rule (FR) is published with its effective date. If the responses raise more questions, then additional tests are undertaken to gather more data and above steps 3 and 4 are repeated. Sometimes the entire standard setting process can extend over many years.

CONCLUDING REMARKS

The requirements must be developed very carefully and must be based on the customer needs. Developing requirements is one of the major and important steps in designing a new product. The requirements define what the product must achieve. The whole design process should be geared toward meeting the requirements. Therefore, requirements analysis and allocation of functions are important tasks in the SE process. Requirements from the product level must be cascaded down to lower levels. The allocation of functions to all systems, subsystems, and components should be based on the requirements cascade. The balanced design of a product should emerge from iterative and recursive cascading of the requirements and the allocation of functions. Chapter 5 deals with the design of interfaces to ensure that all interfacing entities within the product can perform their functions to meet applicable requirements.

REFERENCES

National Aeronautics and Space Administration (NASA). 2007. *Systems Engineering Handbook*. Report SP-2007-6105, Rev 1. http://www.acq.osd.mil/se/docs/NASA-SP-2007-6105-Rev-1-Final-31Dec2007.pdf (accessed April 30, 2013).

National Highway Transportation Safety Administration. 2010. Federal Motor Vehicle Safety Standards. *Federal Register*, CFR (Code of Federal Regulations, Title 49, Part 571, US Department of Transportation. http://www.nhtsa.gov/cars/rules/import/fmvss/index.html (accessed October 15, 2012).

Society of Automotive Engineer, Inc., 2009. *SAE Handbook*. Warrendale, PA: SAE.

Yang, K. and B. El-Haik. 2003. *Design for Six Sigma*. ISBN 0071412085. New York, NY: McGraw-Hill.

5 Understanding and Managing Interfaces

INTRODUCTION

Interfaces are essential to ensure that various entities (i.e., systems, subsystems, and components) of a complex product can work together to enable the product to possess the attributes derived from the needs of its customers. Interfaces are links or joints between entities. The interfaces can perform many functions. Physical interfaces allow two entities to be attached to each other, whereas other interfaces such as electrical connections allow the transfer of electrical power and signals (for information and control) required for functioning of the product. Since the complex products have many entities and the number of interfaces between the entities are usually very large, the task of managing these interfaces is very complex and time-consuming. Engineers involved in designing the complex products spend a lot of time in meetings with other engineers and designers from all the interfacing entities to understand the requirements, functions, and trade-offs between different attributes related to the interfaces. The interface analysis work is generally conducted during the detailed engineering phase of the product development (see Chapter 6). In this chapter, we will study many aspects of the interface design—in terms of design considerations and management of various tasks (Pimmler and Eppinger, 1994; Sacka, 2008).

INTERFACE DEFINITION, TYPES, AND REQUIREMENTS

WHAT IS AN INTERFACE?

An interface can be defined as a "joint" where two (or more) entities (e.g., systems, subsystems, or components) are linked together to serve certain functions. Thus, the interface affects the design of both the entities and the parameters defining the link (i.e., configuration of connecting elements at the interface). The link or the interface between the two entities must be compatible; that is, the values of the parameters (e.g., dimensions) of the two interfacing entities defining their capabilities must match to accomplish the joint. An interface can involve (1) physical connection (e.g., fastening with bolts), (2) sharing of space (i.e., packaged close to each other), (3) exchange of energy (e.g., transfer of mechanical or electric power, air, or fluid under positive or negative pressure), (4) exchange of material (e.g., oil, coolant, and gases), and/or (5) exchange of data (e.g., digital and/or analog signals).

Knowing the type of interface and its characteristics is important to ensure that the two interfacing entities work with each other to perform certain functions. During the early design phases of the product, as the requirements are allocated and the systems are identified, the interfaces between different entities and their parameters must be identified. As the design progresses further, the parameters that define each interface in terms of its characteristics (e.g., their dimensions, strength of physical attachment forces, amount of current or data flow passing through the interface), and their level of strength must be analyzed and controlled during subsequent detailed design activities. The engineers involved in designing both the interfacing entities must know how the two entities work with each other and how and what the interface must exchange, communicate, or share to get the two entities to work together and perform their intended functions.

It should be realized that since each system in a product performs one or more functions and all systems in a product must work together for the product to function, the interfaces must be carefully designed to ensure that they are compatible with both the interfacing systems.

Types of Interfaces

Interfaces between systems, subsystems, or components of a product and other external systems that affect the operation of the product and their components (e.g., parts, subassemblies, human operators, and software) need to be studied and designed to ensure that the product can function and be used by its customers. Interfaces can be categorized by considering many engineering characteristics and user needs of the product (Lalli et al., 1997). Some commonly considered types of interfaces are described in the following.

1. *Mechanical or physical interface*: This type of interface ensures that any two interfacing components can be physically joined together (e.g., by use of bolts, rivets, threads, couplings, welds, adhesives, and linkages that can allow movements and transmission of forces between entities such as a spring, damper, or frictional element) and have the required strength, transfer capabilities (e.g., for materials, heat, and forces), and durability (ability to work under required number of work cycles involving loads, vibrations, temperatures, etc.).
2. *Fluidic or material transfer interface*: A fluidic or material transfer interface (for transfer of fluids, gases, or powdered or granular materials) can be considered as a different type of interface or it can be considered as a mechanical interface involving pipes, tubes, hoses, ducts, seals, and so on. The fluidic interface will enable flow of fluids, gases, or powdered/granular materials with their characteristics such as flow rates, purity, pressures, temperatures, insulation, sealing, and corrosion resistance.
3. *Packaging interfacing entities*: Physical space is required to package or to accommodate the two interfacing entities. The required space can be determined from (1) the sizes/volumes and shape of spaces (i.e., three-dimensional envelopes) occupied by the two interfacing entities and their interfaces; (2) clearance spaces required around the entities to account for vibrations, movements of parts/linkages, air passages for heat transfer, magnetic linkages, and hand/finger or tool access space for assembly/service/repair; and (3) consideration of minimum and maximum separation distances required for their operations.

4. *Functional interface*: In some cases, depending on certain needs to provide one or more functions, one or more of the preceding types of interfaces may be combined and defined as a functional interface. For example, an automotive suspension system forms a unique functional interface (involving physical links and their relationships with relative movements) between sprung and unsprung masses of the vehicle.
5. *Electrical interface*: An electric interface ensures that two interfacing entities can form an electrical connection/coupling (e.g., with connectors, pins, screws, soldering, and spring-loaded contacts/brushes) that can carry required electrical current or signals, provide necessary insulation protection, transfer data, and may have other characteristics such as resistance, capacitance, electromagnetic fields, and interferences.
6. *Software interface*: A software interface ensures that when data are transferred from an entity (with a software system) to another, the format and transmission characteristics of the coded data are compatible to facilitate the required amount and rate of the data processing and/or transfer.
7. *Magnetic interface*: A magnetic interface generates the required magnetic fields for operation of devices such as solenoids/relays, electric machines (motors and generators), and levitation devices.
8. *Optical interface*: An optical interface (e.g., fiber optics, light paths, light guides, light piping, mirrors or reflecting surfaces, lenses, prisms, and filters) allows transfer of light energy between adjoining entities through luminous or nonluminous (e.g., infrared [IR]) energy transmission, reflection and by shielding, baffling/blocking, or filtering of unwanted radiated energy.
9. *Wireless interface*: This type of interface can communicate signals or data without wires through radio frequency communication, microwave communication (e.g., long-range line-of-sight through highly directional antennas, or short-range communication), and IR short-range communication. The interface applications may involve point-to-point communication, point-to-multipoint communication, broadcasting, cellular networks, and other wireless networks.
10. *Sensor or actuator interface*: A sensor has a unique interface that converts certain sensed energy (e.g., light, motion, touch, distance, or proximity to certain objects, pressure, and temperature) into an electrical output or signal. Whereas, an actuator produces an output (e.g., movement of a control or mechanical links) by converting an input from one type of modality to a different output modality. For example, a steeper motor produces a precise angular movement for each electrical pulse input. Similarly, floats or floating sensor devices can sense fluid levels and convert into electrical signals.
11. *Human interface*: When a human operator is involved in operating, monitoring, controlling, or maintaining a product, the human-machine or human-computer interface (commonly referred as the HMI or HCI, respectively) will include devices such as human accommodating or positioning devices (e.g., chairs, seats, armrests, cockpits, standing platforms, steps, footrests, handles, and access doors), controls (e.g., switches, buttons, touch controls,

stalks, levers, joysticks, pedals, and voice controls), tools (e.g., hand tools and powered tools), and displays (e.g., visual displays, auditory displays, tactile displays, and olfactory displays).

INTERFACE REQUIREMENTS

To design an interface, an engineer must first understand the overall requirements on the product, allocated functions, and characteristics of both the entities attached or linked at the interface. The requirements on the interface should specify the following: (1) the functional performance of both the entities, (2) configuration of the entities, (3) available space to create the interface, (4) environmental conditions for the operation of the product and comfort of the human operators, (5) durability (minimum number of operational cycles the product must function), (6) reliability and safety in performing the required functions, (7) human needs (e.g., viewing and reading needs, hearing needs [sound frequencies and levels], lighting and climate control needs, and product operating needs), (8) electromagnetic interference, and so on. In addition, the requirements should include any other special constraints (e.g., weight requirements, aerodynamic considerations, and operating temperature ranges) that must be met.

Steps involved in the interface requirements development process generally use an iterative approach (with a series of steps and loops as shown in Figure 2.2) unless a previously developed requirements document (or a standard) is available. The series of steps typically involve the following:

1. Gather information to understand how the interfacing entities work, fit into the product, and support the overall functionality, performance, and requirements of the product (e.g., review existing system design documents and standards). Draw an interface diagram (see the section entitled "Interface Diagram"). Meet with the design team members of the interfacing entities (e.g., core engineering functions such as body engineering, powertrain engineering, electrical engineering, and climate control engineering—for an automotive product) and product design teams to understand issues and trade-offs considerations with the product attributes (e.g., packaging space, safety, maintenance, and costs).
2. Document all design considerations such as inputs, outputs, constraints, and trade-offs associated with the interface and its effects on other entities (e.g., develop a cause-and-effect diagram, see Chapter 17; conduct a FEMA, see Chapter 15).
3. Study existing designs of similar interfaces and compare them by benchmarking the competitors' products (see Chapter 15 for benchmarking technique).
4. Study available and new technologies that could be implemented to improve the interface.
5. Create an interface matrix (see the section entitled "Interface Matrix and N-Squared Diagram") to understand the types of interfaces and their characteristics.
6. Create a preliminary set of requirements.

Understanding and Managing Interfaces

7. Translate requirements into design specifications (use of the quality function deployment technique can help in this and the next step, see Chapter 15).
8. Brainstorm possible verification tests (or obtain available test methods from existing standards) that need to be performed to demonstrate compliance to the preliminary requirements.
9. Develop alternate interface concepts/ideas.
10. Review alternate concepts and ideas with subject matter technical experts.
11. Select a leading design by analyzing all other entities that are functionally linked to the entities associated with the interface (develop a Pugh diagram to aid in decision making, see Chapter 15).
12. Modify and refine interface diagram and interface matrix.
13. Iterate the preceding steps until an acceptable interface design is found.

The iterative workload described in the preceding process can be reduced if an internal (company) design guide or standard for the entities being interfaced can be used as a starting document along with the product-level requirements. Experts and other knowledgeable people in the organization can provide information on valuable lessons learned during the development of similar interfaces from the past product programs.

VISUALIZING INTERFACES

An interface between any two entities (which could be systems, subsystems, or components) can be represented by the use of a simple arrow diagram as shown in Figure 5.1.

The arrow indicates a link (or relationship) between the two entities, namely, entity A and entity B. The arrow representing the link can denote any of the following (see Figure 5.1):

1. Output of entity A is an input to entity B.
2. Entity A is mechanically attached to entity B.
3. Entity A is functionally attached to entity B (i.e., function of A is required by B to perform its function).
4. Entity A provides information to entity B.
5. Entity A provides energy to entity B.
6. Entity A transmits or sends signals, data, or material (e.g., fluids, gases) to entity B.

For example, in an automobile, an engine mount (a rubber coupling) between the engine and the chassis is a physical coupling that serves as a physical interface between the engine and the engine mount and another physical interface between

FIGURE 5.1 Interface between two entities.

FIGURE 5.2 Interfaces between the engine and the vehicle chassis.

the engine mount and the chassis frame. The function of the engine mount is to position the engine, physically attach it to the chassis, and also to isolate the vibrations transmitted from the engine to the chassis system (see Figure 5.2). (Note: Letter "P" placed above the arrows in Figure 5.2 indicates "physical" connection.)

INTERFACE DIAGRAM

An interface diagram is a flow (or an arrow) diagram showing how different systems, subsystems, and components of a product are interfaced (i.e., joined or linked). It provides a visual representation of the product or a portion of the product showing where the interfaces occur. It also should show the type of each interface (by use of letter codes, e.g., "P" for a physical connection, "E" for an electrical connection, "M" for material/ fluid transfer, and "I" for information or data transfer, placed next to the arrow).

An interface diagram is a useful tool in understanding how various systems, subsystems, and components are interfaced with each other. The diagram can be created at any level, that is, at the product level showing all the systems of the product, at a system level showing all the subsystems of the system, at a subsystem level showing all components of the subsystem, or at mixed levels showing a system, its subsystems, and also showing other major systems of the product. Two examples of the interface diagram are shown in Figures 5.5 and 5.7 in a later section entitled "Examples of Interface Diagrams and Interface Matrices" of this chapter. (Figure 5.5 presents an interface diagram for a laptop computer and Figure 5.7 presents an interface diagram for an automotive fuel system.)

INTERFACE MATRIX AND N-SQUARED DIAGRAM

An interface matrix and N-squared diagram are two commonly used methods to illustrate existence and types of interfaces between different entities (systems, subsystems, or components). Both are formatted in a matrix-type arrangement with entities described by rows and columns of the matrix. The interface matrix shows names of all the entities as headings for both rows (on the left) and columns (on the top) of the matrix, whereas the N-squared diagram shows names of all the entities in the diagonal cells of the matrix. The N-squared diagram has been used extensively to develop data interfaces, primarily in the software development field. However, it can also be used to develop hardware interfaces. Both methods present descriptions of the interfaces in the cells of the matrix. A cell is defined by the intersection of the row and column represented by the two interfacing entities. The description of the interface is shown by a letter code to indicate the type of the interface. An interface matrix, thus, shows the following: (1) it captures the existence of all interfaces, (2) it shows output-to-input relationships between any two entities (see Figure 5.3), and (3) it presents the type(s) of interfaces between any two entities.

Understanding and Managing Interfaces

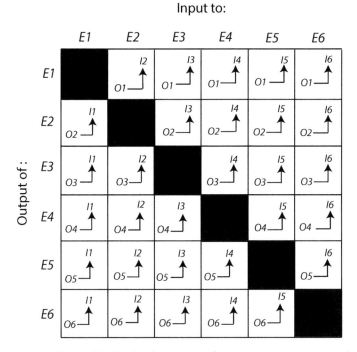

Note: E = Entity (system, subsystem or component)
O = Output
I = Input

FIGURE 5.3 Output-to-input relationships indicated by the cells of the interface matrix.

Figure 5.3 presents the output-to-input relationships between the two entities for each cell of a 6 × 6 interface matrix (except for the cells in the diagonal). The entities are labeled as *E1* to *E6*. The outputs of the entities are labeled as *O1* to *O6* and the inputs are shown as *I1* to *I6*. The arrow shown in each cell indicates that the output of an entity defined by its row is used as an input to an entity defined by the column of the matrix. For example, the cell in the first row and second column shows that *O1* is the output of entity *E1*, and it is interfaced with entity *E2* shown as input *I2* received by entity *E2*. Similarly, Figure 5.4 presents an N-squared diagram illustrating the relationships between the entities *E1* through *E6* shown in the diagonal of the matrix.

The interface matrix (also called as an interaction matrix in some organizations) and the N-squared diagram are useful tools in understanding and displaying interfaces. The contents of the cells typically present coded descriptors of the types of interfaces between the outputting entities and the entities receiving the inputs. The codes typically include P = physical interface, S = spatial-packaging interface, E = electrical interface, M = material flow, I = information or data flow, and 0 (or a blank cell) = no relationship.

The interface diagram and interface matrix are both very useful tools in visualizing relationships and documenting presence of the interfaces (Sacka, 2008). These

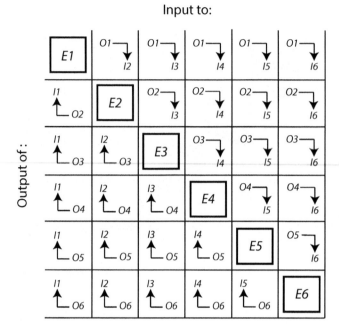

FIGURE 5.4 Output-to-input relationships indicated by the cells of the N-squared diagram.

tools make the design team realize the presence of many interfaces and types of these interfaces in the product. The next step is to understand the connection configuration details and requirements of these interfaces to ensure that the interfacing entities can be designed to work together to perform their allocated functions.

EXAMPLES OF INTERFACE DIAGRAMS AND INTERFACE MATRICES

LAPTOP COMPUTER INTERFACES

Based on the systems, subsystems, and components of a laptop computer presented in Table 2.1, an interface diagram of the systems in the laptop computer is presented in Figure 5.5. The letters defining interface types are placed next to each arrow showing the interfaces between different systems. The chassis system provides space for packaging all other systems, and the size of the chassis system depends on the sizes of other systems. Furthermore, all the systems are physically attached to the chassis systems. Thus, the arrows indicating "output of" other systems to "input to" the chassis system are indicated as of types "P" and "S." The power system provides electrical energy to all other systems except the chassis system. Thus, the arrows from the power system to all other systems (except the chassis system) are identified as of type "E." The cooling system is assumed to provide cooling air (material flow

Understanding and Managing Interfaces

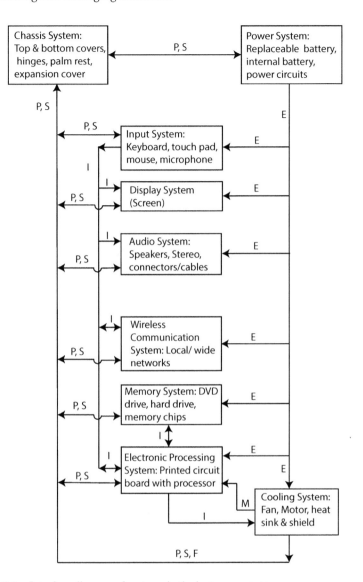

FIGURE 5.5 Interface diagram of systems in the laptop computer.

"M" interface) to cool only the electronic processing system. The electronic processing systems take data (information) outputs (type "I" interface) from the input system (i.e., keyboard and mouse), the wireless system, and the memory system. Whereas, the electronic processing system delivers its outputs (type "I" interface) to the display system, the audio system, the memory system, the wireless system, and the cooling system. The cooling system is also functionally attached to the chassis system to ensure proper airflow (type "M" interface) through inlets and outlets provided through the chassis system.

Figure 5.6 provides the interface matrix for the laptop computer and provides information contained in the interface diagram presented in Figure 5.5. By observing the

		1 Wiring Harness	2 Electronic Control Unit	3 Sensors	4 Air Cleaner	5 Throttle Valve	6 Air Flow Meter	7 Intake Valves	8 Intake Runner	9 Air Intake Chamber	10 Fuel Pressure Regulator	11 Fuel Delivery Pipe	12 Fuel Injectors	13 Fuel Filter	14 Fuel Pump	15 Fuel Filler	16 Fuel Return Pipe	17 Fuel Tank	18 Electrical System	19 Powertrain System	20 Body System	21 Instrument Panel
1	Wiring Harness	■	E																			E
2	Electronic Control Unit	E	■	E																E	P	
3	Sensors		E	■																E		
4	Air Cleaner				■					P										F		
5	Throttle Valve					■	P			P												
6	Air Flow Meter					P	■			P												
7	Intake Valves							■														
8	Intake Runner								■			P								P		
9	Air Intake Chamber				P	P	P			■										P		
10	Fuel Pressure Regulator										■	P										
11	Fuel Delivery Pipe										P	■	P		P							
12	Fuel Injectors								P			P	■									
13	Fuel Filter													■	P							
14	Fuel Pump													P	■	P		P	E			
15	Fuel Filler														P	■		P			P	
16	Fuel Return Pipe																■	P			P	
17	Fuel Tank														P	P	P	■		F	P	
18	Electrical System	E		E											E				■	E	P	E
19	Powertrain System		E	E	F				P	P								F	E	■	P	E
20	Body System		P													P	P	P	P	P	■	P
21	Instrument Panel	E																	E	E	P	■

FIGURE 5.6 Interface matrix for the systems in the laptop computer.

Understanding and Managing Interfaces

		1 Chassis System	2 Display System	3 Audio System	4 Input System	5 Electronic Processing System	6 Memory System	7 Power System	8 Wireless Communi. System	9 Cooling System
1	Chassis System		P, S	P, S	P, S	P, S	P, S	P, S	P, S	P, S
2	Display System	P, S		O	—	—	O	O	O	O
3	Audio System	P, S	O		O	O	O	O	O	O
4	Input System	P, S	—	—		—	—	—	—	O
5	Electronic Processing System	P, S	—	—	—		—	—	—	—
6	Memory System	P, S	O	O	O	—		O	O	O
7	Power System	P, S	E	E	E	E	E		E	E
8	Wireless Communi System	P, S	O	O	O	—	O	O		O
9	Cooling System	P, S, F	O	O	O	M	O	O	O	

FIGURE 5.6 (Continued)

rows of the matrix for the chassis system, electronic processing system, and power system in this matrix, it can be seen that they are interfaced with all systems of the laptop computer. The chassis system contains and packages all other systems in the laptop. The electronic processing system is the heart of all information processing, and the input system controls the information processing activities of the computer. And the power system provides electric energy to all other systems (except the chassis system). The systems with the least interfaces are the wireless and cooling systems. Thus, these systems can be designed independent of most other systems in the laptop.

AUTOMOTIVE FUEL SYSTEM INTERFACES

An automotive fuel system is considered here to include the following three subsystems: (1) fuel delivery system, (2) air induction system, and (3) electronic fuel control system. The components of the three subsystems are presented in Table 5.1.

Figure 5.7 presents an interface diagram of the automotive fuel system. The interface diagram shows interfaces (links) within the fuel system, its subsystems, and other systems in the vehicle. The three subsystems of the fuel system are placed within the three separate boxes shown in the dotted lines in Figure 5.7. Other automotive systems, namely, electrical system and powertrain system, are placed outside the dotted-lined boxes. The fuel delivery system supplies fuel (type M interfaces) to the air induction system, and the air induction system provides fuel monitored

TABLE 5.1
Components in the Three Subsystems of the Automotive Fuel System

Subsystems	Components
Fuel delivery	Fuel filler
	Fuel tank
	Fuel pump
	Fuel filter
	Fuel delivery pipe (fuel rail)
	Pulsation damper
	Fuel injectors
	Fuel pressure regulator
	Fuel return pipe
Air induction	Air cleaner
	Air flow meter
	Throttle valve
	Intake manifold runners
	Intake valves
Electronic control	Electronic control unit (ECU)
	Engine sensors (air flow meter, air temperature, throttle angle, air density, fuel temperature, fuel pressure, oil pressure, coolant temperature, exhaust temperature, crank angle, timing, engine rpm, and speed) Fuel injector assemblies
	Switches, connectors, relays, and wiring

Understanding and Managing Interfaces

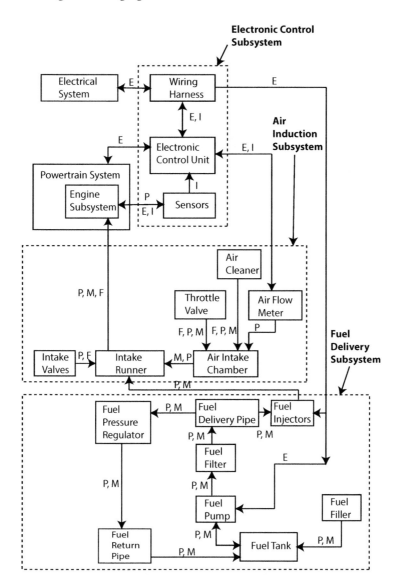

FIGURE 5.7 Interface diagram of an automotive fuel system.

through the electronic control subsystem to the engine subsystem. The electrical system through the wiring systems (type E interfaces) in the electronic monitoring system provides power to the electronic control unit, the airflow meter, and the fuel pump.

Figure 5.8 provides the interface matrix for the automotive fuel system. It includes interfaces between various subsystems of the fuel system and other vehicle systems (i.e., electrical, powertrain, body, and instrument panel systems). The cells where there is no relationship (or interface) are left blank to reduce clutter in the matrix.

	1 Wiring Harness	2 Electronic Control Unit	3 Sensors	4 Air Cleaner	5 Throttle Valve	6 Air Flow Meter	7 Intake Valves	8 Intake Runner	9 Air Intake Chamber	10 Fuel Pressure Regulator	11 Fuel Delivery Pipe	12 Fuel Injectors	13 Fuel Filter	14 Fuel Pump	15 Fuel Filler	16 Fuel Return Pipe	17 Fuel Tank	18 Electrical System	19 Powertrain System	20 Body System	21 Instrument Panel
1 Wiring Harness	■	E																E		P	E
2 Electronic Control Unit	E, I	■	E															E	E, F	P	—
3 Sensors		I	■			E													P, E, I		—
4 Air Cleaner				■	P				P, M, F										F		
5 Throttle Valve					■				P, M, F										F		
6 Air Flow Meter			F			■			P												
7 Intake Valves							■	P, F											P, M, F		
8 Intake Runner								■	P, M												
9 Air Intake Chamber									■		P, M								F		
10 Fuel Pressure Regulator										■	P					P, M					
11 Fuel Delivery Pipe										P, M	■	P, M									
12 Fuel Injectors								P, M			P, M	■									
13 Fuel Filter													■	P							
14 Fuel Pump													P, M	■			P, M				
15 Fuel Filler															■		P			P	
16 Fuel Return Pipe																■	P				
17 Fuel Tank														P, M		P	■	F		P	
18 Electrical System	E	E																■	P	P	E
19 Powertrain System				F														F	■	P	—
20 Body System	P	P																P	P	■	P
21 Instrument Panel	E	—																E	E	P	■

FIGURE 5.8 Interface matrix showing interfaces between components of the automotive fuel subsystems and other vehicle systems.

Understanding and Managing Interfaces 113

The interface diagram and interface matrix are important tools because they provide basic information by identifying all the interfaces and thus help the engineers realize the complex interfacing issues and tasks during the development of any system. For example, the engineer designing the fuel pump realizes that the fuel pump has an electrical interface to receive power for its operation; it has to be physically mounted inside the fuel tank, and the pumped fuel has to pass through the filter to the fuel delivery pipe. Thus, the fuel pump engineer needs to communicate with the electrical systems engineers and other fuel systems engineers to make sure that his fuel pump can physically connect and operate with the interfacing components. In addition, the engineer must understand the bigger picture of how the rest of the fuel system works with the powertrain system, electrical system, body system, and the instrument panel (which displays status of the fuel and other systems) of the vehicle (product level).

ILLUSTRATION OF USE OF INFORMATION CONTAINED IN INTERFACE MATRIX

This section provides an illustration of an interface diagram created to show interfaces between an infotainment system and other systems in an automotive product. An infotainment system consists of the driver information and entertainment system. Let us assume that the interface matrix presented in Figure 5.9 was developed by an automotive instrument panel engineer to understand the tasks of designing the infotainment system and making it functional by interfacing it with the rest of the vehicle systems. The notations of the types of interfaces shown the interface matrix are as follows: F = functional, E = energy transfer, I = information transfer, P = physical connection, M = material transfer, S = spatial packaging space, and 0 = none.

The infotainment system is an integrated driver information and entertainment system, and it is assumed to contain the following subsystems: (1) climate controls, (2) a radio (with AM/FM bands and satellite radio), (3) a CD player, (4) a navigation system, and (5) a vehicle systems monitoring system (which checks and provides the status of different vehicle systems including fuel consumption, trip statistics, and distance to empty). In addition to understanding the customer needs for the infotainment system and engineering requirements of each of the subsystems, the engineer would need to understand the characteristics and requirements of the other interfacing systems.

Using the information provided in the interface matrix given in Figure 5.9, the tasks of the instrumental panel engineer assigned to develop the infotainment system for a passenger car can be analyzed as follows:

1. The body system provides information on when the doors and/or trunk lid are ajar. The body system also provides space for packaging the infotainment system.
2. The powertrain system provides information on vehicle speed, engine speed, engine temperature, and fuel consumption to the infotainment system.

	Body System	Powertrain System	Chassis System	Fuel System	Electrical System	Climate Control System	Safety System	Infotainment System
Body System		P, S	P, S	P, S	P, S,	P, S	P, S	P, S, I
Powertrain System	P, S		E, P, S	0	I, E	I, E, M	I	I
Chassis System	P, S, F	0		0	I	0	I, F	I
Fuel System	P, S	P, S, M	0		I	0	I	I
Electrical System	P, S, E	I, E	I, E	E		E, I	0	I, E
Climate Control System	P, S	I, E, M	0	0	I		0	I
Safety System	P, S	I	P, S, I	I	I	0		I
Infotainment System	P, S	I	I	0	I	I	0	

FIGURE 5.9 Interface matrix showing relationships between the infotainment system and other vehicle systems.

3. The chassis system provides information on the state of brake pads (wear and temperatures) to the infotainment system.
4. The fuel system provides information on the fuel situation (fuel used, fuel consumption, and distance-to-empty) to the infotainment system.
5. The electrical system provides electrical energy and the information on the state of the electrical system (current flow and voltage) to the infotainment system.
6. The climate control system provides information on the exterior temperature and settings of different climate controls to the infotainment system.
7. The vehicle safety system provides information on the status of readiness of the safety subsystems such as braking system temperature, brake fluid level, tire pressure warning system, airbags, and night vision system to the infotainment system.
8. The infotainment system needs space inside the instrument panel (a subsystem of the vehicle body system), and it is packaged in the center stack with a high-mounted screen with touch, voice, and other hand-operated controls.
9. The infotainment system provides information on the driver-selected modes to the powertrain system (e.g., sporty, economy, manual, or auto transmission settings), the climate control system (e.g., selected interior temperature, a/c unit on/off, and air distribution mode), electrical system (e.g., status of driver-selected modes for power demand computations), and the chassis system (e.g., to adjust brake pads for wear).

Understanding and Managing Interfaces

The tasks of the infotainment engineer would be to understand the preceding inputs and outputs and provide the necessary software and hardware systems to create user-friendly controls and displays interface. To obtain the necessary information, the engineer will need to meet and communicate with all engineers from the interfacing systems to ensure that his infotainment system can perform all the preceding communication, control, and status display functions.

CLUSTERING AND SEQUENCING OF MATRIX DATA

The interface diagrams and interface matrices are also very useful in visualizing configuration and links between entities in a system. The usefulness of the techniques can be better understood by the illustration of an example of clustering given in the following discussion.

Let us assume that a mechanical system involving 16 components is being analyzed to determine its configuration and packaging into a self-contained unit. The components are defined as C1, C3, C3, ..., C16. Figure 5.10 presents an interface diagram of the system showing all the mechanical interfaces between the components. The diagram shows that components C4, C10, C11, C12, C13, and C14 are not connected to any other components in the system. The lack of interfaces to these six components suggests that these components are performing functions that are independent of other remaining components. Thus, they must be serving other systems but are merely included in the mechanical system because of some other reasons (e.g., designer needed space to locate these components purely for

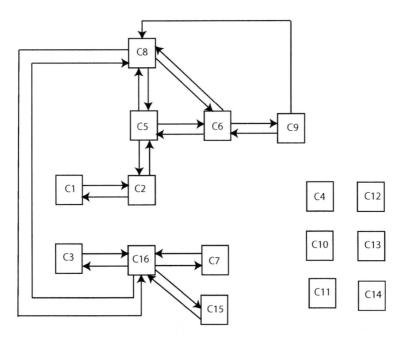

FIGURE 5.10 Interface diagram in the initial configuration.

| Column # | 1 | 2 | 3 | 4 | 5 | 6 | 7 | 8 | 9 | 10 | 11 | 12 | 13 | 14 | 15 | 16 |
Row #	C1	C2	C3	C4	C5	C6	C7	C8	C9	C10	C11	C12	C13	C14	C15	C16
1 C1	■	M														
2 C2	M	■		M												
3 C3			■													M
4 C4				■												
5 C5		M			■	M	M									
6 C6				M		■	M	M								
7 C7							■									M
8 C8					M	M		■								M
9 C9					M	M			■							
10 C10										■						
11 C11											■					
12 C12												■				
13 C13													■			
14 C14														■		
15 C15															■	M
16 C16			M				M	M							M	■

FIGURE 5.11 Interface matrix for the initial configuration.

packaging convenience). The rest of the components are connected and have several connections to other components. The interface matrix of the system is shown in Figure 5.11.

The relationships represented in a matrix diagram can be rearranged using a number of clustering and sequencing methods to help grouping or clustering of components (e.g., defining content of modules in a family of products), systems, or functions and to create product configurations (e.g., sequencing of steps and/or determining product layouts) to reduce or combine interfaces. Design structure matrix (DSM) approach is useful in "product architecturing" (Yassine, 2004; Zakarian, 2008).

Pimmler and Eppinger (1994) showed that using the DSM approach, the components can be rearranged from the initial arrangement in the matrix in Figure 5.11 to the matrix in Figure 5.12 so that they could be packaged into three modules (clusters). The rearranged matrix simply places the rows and columns of the original matrix in a different order selected by a clustering technique. The three modules can be easily seen as three clusters around the diagonal in the rearranged interface matrix shown in Figure 5.12. The three clusters, identified as modules #1, 2, and 3, are also outlined with darker solid and dotted lines around the clusters of cells in the interface matrix in Figure 5.12. Figure 5.13 also shows the clusters identified as modules #1, 2, and 3 in the interface diagram. Component C5 is common or connecting component between modules #1 and 2. Similarly, component C8 is common or connecting component between modules #2 and 3.

Understanding and Managing Interfaces

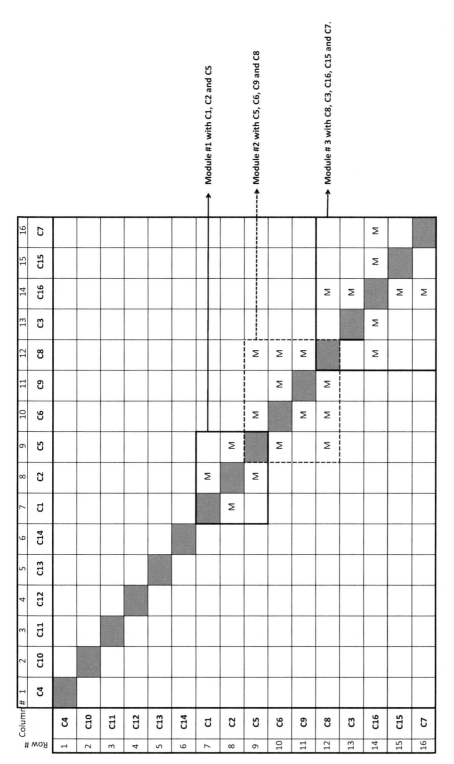

FIGURE 5.12 Interface matrix for the rearranged configuration.

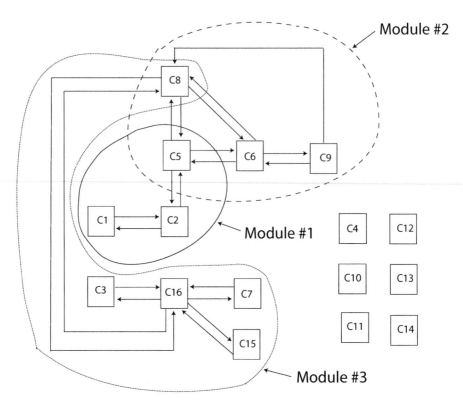

FIGURE 5.13 Interface diagram showing three modules in the rearranged configuration.

The DSM and clustering techniques thus are useful in determining alternate configurations and modularization of components or subsystems. The modularization generally reduces complexity and costs by sharing of common modules across various product models.

TEAMWORK IN INTERFACE MANAGEMENT

Each engineer responsible for delivering a system (with all its subsystems and components) must have design expertise on functioning details, manufacturing processes, and trade-offs between the characteristics of the system and its entities (i.e., its subsystems and their components). The engineer must also have knowledge about how his system fits, connects, and works within a bigger (parent) system involving the complex product and its operating conditions. A thorough understanding of the requirements on the product is necessary to ensure that these higher-level requirements can be cascaded down to the lower levels of systems in a complex product.

Knowing the functions of the parent system, each of its subsystem, and each component can be allocated and analyzed to determine how each subsystem, its components, and interfaces need to be designed. Problems are identified

Understanding and Managing Interfaces

iteratively, and modifications (or even major redesigns) are undertaken. If the responsibilities to design and deliver subsystems or components are assigned to different individuals (teams or suppliers), then they will need to understand how their deliverables are affected by other components or subsystems that interface with their entity.

Constant checks by independent experts, structured walkthroughs, design reviews, or even simulation models (or prototyping) of the interface operations are useful in ensuring that all the interfaces are designed to meet their requirements and allocated functions. Interface control process and documents (described in the section on "Establishment of Interface Control") help monitor and control the complex interface development work.

ESTABLISHMENT OF INTERFACE CONTROL

The basic steps involved in establishment of an interface control process during the product concept definition typically involve (1) assign basic functional areas of responsibility to different functional teams; (2) define design responsibilities of each team; (3) identify and categorize all interfaces; (4) define interfaces to be controlled; (5) establish formal interface control procedures; (6) disseminate scheme, framework, and traceability of requirements and allocated functions to the teams; and (7) monitor the team outputs to ensure that interface control procedures are followed (NASA, 2007).

The interface control process suggested for NASA projects is described in a report by Lalli et al. (1997). The purpose of interface control is to define interface requirements so as to ensure compatibility between interrelated pieces of equipment and to provide an authoritative means of controlling the design of interfaces. Interface design can be controlled by an interface control document (ICD). These documents are useful for the following: (1) control the interface design of the equipment to prevent any changes to characteristics that would affect compatibility with other equipment, (2) define and illustrate physical and functional characteristics of every piece of equipment in sufficient detail to ensure compatibility of the interface, so that this compatibility can be determined from the information in the ICD alone, (3) identify missing interface data and control the submission of these data, (4) communicate coordinated design decisions and design changes to program participants, and (5) identify the source of each interface component.

CONCLUDING REMARKS

A complex product cannot be created without interfaces. Interfaces are very important because they allow connections and transfer of materials, energy, and signals between different entities. Engineers and designers must understand the functions of the entities, characteristics of the interfaces, and trade-offs between various parameters of the interfaces. Interface diagrams and interface matrices are simple but effective techniques in visualizing and understanding the interfaces and issues related to their design and operation.

REFERENCES

Lalli, V. R., Kastner, R. E. and H. N. Hartt. 1997. *Training Manual for Elements of Interface Definition and Control*. NASA Reference Publication no. 1370. Washington, DC: National Aeronautics and Space Administration.

National Aeronautics and Space Administration. 2007. *NASA Systems Engineering Handbook*. Report no. NASA/SP-2007-6105 Rev1. NASA Headquarters, Washington, DC 20546. http://ntrs.nasa.gov/archive/nasa/casi.ntrs.nasa.gov/20080008301_2008008500.pdf (accessed October 15, 2012).

Pimmler, T. U. and S. D. Eppinger. 1994. Integration Analysis of Product Decompositions. In *Proc. ASME 6th Int. Conf. on Design Theory and Methodology*, Minneapolis, MN.

Sacka, M. L. 2008. A Systems Engineering Approach to Improving Vehicle NVH Attribute Management. Master's Thesis for M. S. degree in Engineering Management at the Massachusetts Institute of Technology, Cambridge, MA.

Yassine, A. 2004. An Introduction to Modeling and Analyzing Complex Product Development Processes Using the Design Structure Matrix (DSM) Method. *Quaderni di Management* (Italian Management Review), 9: 71–88.

Zakarian, A. 2008. A New Non-Binary Matrix Clustering Algorithm for Development of System Architectures. *IEEE Transactions on Systems, Man and Cybernetics-Part C*, 38(1): 135–141.

6 Detailed Engineering Design during Product Development

INTRODUCTION

Many steps need to be performed in the detailed engineering design (DED) phase of any complex product development program. Most of these steps can be performed concurrently involving personnel from different disciplines and departments. The sequence of steps and outputs of the steps are generally identified in the systems engineering management plan, product development process, and/or engineering design manuals developed by the company developing the product. The detailed engineering design process generally begins after the concept selection phase. However, some early engineering design work is generally performed during concept development to ensure that the product concept will be technically and economically feasible, and its components could be manufactured and assembled using equipment available within the company or its supplier facilities.

Completion of the DED phase means all components within all systems of the complex product are designed (or developed) which include (a) engineering drawings (usually using a CAD software showing different views and dimensions of a solid model for each component or a system, (b) CAE analyses and evaluations (e.g., structural analysis, thermal analysis, aerodynamic/fluid flow analysis, electrical and electronic circuits analyses), (c) material specifications (i.e., material or composition of metals, plastics, rubbers, etc.) to be used to produce each component including any special treatments such as heat treatment and surface coating, (d) applications of principles related to design for manufacture, design for assembly, and design for sustainability, and (e) evaluations and verification tests (usually engineering tests using computer models and/or physical tests) conducted to ensure that each component meets its stated functional, manufacturing, assembly, safety, and cost requirements (i.e., all attribute requirements cascaded down from the product level). The developed product thus meets the "balanced product design" shown in the systems engineering process (see Chapter 2, Figure 2.2).

ENGINEERING DESIGN

Engineering design requires integration (i.e., simultaneous consideration) of many issues such as (a) meeting many requirements—based on customer needs, governments regulations, and company/industry standards; (b) involving multiple disciplines to ensure consideration of issues and applications of techniques from different disciplines such as electrical, mechanical, materials, software, chemical, safety, manufacturing, assembly, and sustainability; (c) safety analyses and evaluations to reduce or eliminate possible accidents and injuries that can be caused by the product and the possibilities of product liability cases during its use; (d) conducting many design iterations to consider many product attribute requirements and trade-offs between different attributes in various combinations of configurations (called design synthesis in the SE process, see Figure 2.2); and (f) a lot of engineering analyses, design reviews, and testing (subjective and objective) to verify that the components within all lower level systems and the whole product will meet their stated attribute requirements.

Customer needs, attributes, attribute requirements (product level), concept design, system requirements, systems design, subsystem design, requirements and specifications, and design of lower-level systems down to components are all considered during the design activities of team members involved in engineering design. Their output thus is a fully functional product that meets all the requirements.

There is no unique process to accomplish this complex engineering design phase. But the process typically involves the following steps: (Note: The order and number of steps can vary between organizations and products.)

1. Documenting and communicating the specifications of the product to be developed during the early phases (e.g., advanced design, market research and concept design) to the team members, refining the specifications and ensuring that those collectively define the product to be engineered. The product specifications include its characteristics, such as dimensions, functionality, performance, and features, that cover its all attribute requirements.
2. Collecting customer needs data from marketing research and benchmarking data from engineering analysis of competitors' products.
3. Searching for all government requirements (federal, state, and local) applicable to the product during its entire life cycle.
4. Compiling and reviewing of customer feedback and warranty data on existing products available in the market to meet the same customer needs.
5. Conducting functional analysis and requirements analysis for the product.
6. Development and documentation of product attributes and attribute requirements.
7. Crating alternate product concepts designs including CAD drawings and models.
8. Refining product configuration and creating decomposition trees showing systems and lower-level systems down to the component level.
9. Cascading requirements from the product level to all lower-level systems down to components.
10. Identifying all interfaces (between systems and their lower-level systems and components) and developing requirements on all interfaces.
11. Using the systems engineering "V" model to ensure that design of systems, subsystems and components are developed using the top-down approach. Any carryover components to be used should be included in the analyses.
12. Conducting detailed engineering analyses involving material selection, evaluation of functional capabilities using available models and CAE techniques (e.g., applying design for manufacture and assembly techniques and

Detailed Engineering Design during Product Development 123

guidelines for design for sustainability), and building prototype parts and testing to verify that the design of each component meets its stated requirements. CAE analyses can involve evaluation of designs of mechanical structures including CAD modeling and applications of finite element analyses (e.g., stress tests and structural simulations, aerodynamics and thermodynamic analyses, thermo-mechanical analyses, fluid flows [computational fluid dynamics], fatigue and durability analyses, acoustics/sound and noise, vibrations and harshness analyses, electrical and electronic circuits, control systems analyses, dynamic simulations and analysis of moving components), biomechanical and ergonomics analyses, toxic exposures, life cycle costing, and so forth.
13. Conducting Failure Modes and Effects Analyses (FMEA) before the design is released for production. The results of the FMEAs are reviewed with the management and changes are made to the designs to meet the accepted level of risk priority number.
14. Assembling verified components into lower-level systems and lower-level systems into higher-level systems, and tests are performed to ensure that all the assembled subsystems and systems meet their respective requirements. Results of the tests are reviewed with engineering management personnel.
15. Finally, the assembled systems are further assembled to create the whole product.
16. A number of design reviews are conducted during various stages of the above design steps.
17. The whole product is further tested to verify that it meets all stated requirements and validated to ensure that the customers like and are satisfied with the product before it is released for production.

The above steps can be better understood by studying engineering design examples of actual complex products. Thus, the following section provides descriptions of steps applied to six different complex products.

SIX PRODUCT EXAMPLES

The tasks that are performed during the detailed engineering phase depend upon the product, its characteristics and manufacturing, construction and assembly processes. Six different complex products are presented here to illustrate their design-related issues and similarities in engineering design processes.

The following six complex products are included in Table 6.1. The major systems of these products are listed below.

1. <u>Washing Machine</u>: Body system, clothes holding tub and drive system, water and detergent management system, washing cycle control system, controls and display system
2. <u>Refrigerator</u>: Body system, exterior panel and trim system, interior trim and food storage system, refrigeration system, electrical power and control system, water management and ice-making system, and interior lighting system

3. Laptop Computer: Chassis system, microprocessor system, data storage and retrieval system, data input system (keyboard, mouse, and microphone), data output system (display and audio system), wireless communication system, power system, and cooling system
4. Automotive Product: Body system, chassis, steering and suspension system, powertrain system, cooling system, braking system, electrical system, fuel system, exhaust system, climate control system, airflow management system, driver interface system, seating system, lighting system, safety system, and luggage/cargo storage system
5. Wind Turbine: Foundation system, tower structure system, nacelle system, wind turbine (blades and rotor) system, gearbox system, generator system, step-up transformer system, air flow direction and wind velocity sensing system, turbine control system, electrical system, and safety systems including steps, tethering, handholds, and ladders
6. Natural Gas Fired Power Plant: Natural gas storage system, gas and air input and control system, plant housing system, gas turbine system, power generator system, power plant control system, plant lighting system, piping systems, heat exchanger, secondary steam turbine, and generator system, power distribution system, water management system, auxiliary power system, and plant safety/security system.

Table 6.1 also provides some important systems engineering considerations of the products such as variables used to measure their output capacity, special features, product attributes, disciplines involved during design phases, engineering analyses methods used during design, key validation tests to be conducted to determine if the customers will be satisfied with the products, product life cycle, and price.

The customer needs should drive the design of each of the above products. The customer needs can be better understood from the required level of product capacity, special features, and product attributes (see Table 6.1). The overall technical workload during the concept design and detailed engineering phases of these products will depend upon the technologies and complexity involved within the product. The complexity of the product will depend upon the total number of entities (i.e., systems, subsystems, and lower-level systems down to the component level) within the product and the number of interfaces between the entities. In addition, optional features demanded by the customers will include other enhancements added to the basic product design. The capacity (see the second row of Table 6.1) of the product will be estimated from the customer demand. The overall volume of its envelope can be estimated from customer demand (rate of the product output) and the technology used to operate the product.

After the data gathering on customer needs, benchmarking, and documentation of the product specifications, the overall product concept is created. Packaging space needed for creation of the product is usually the first consideration in engineering design where the design engineers along with other team members decide on the shape and size of space required. i.e., the overall envelope of the product. The envelope is created by understanding other existing products available in the market segment by benchmarking. Technological trends and the latest developments are all

TABLE 6.1
Characteristics of Six Complex Products and Disciplines and Analyses Involved in Their Detailed Engineering

Complex Product	Washing Machine	Refrigerator	Laptop Computer	Automotive Product	Wind Turbine	Natural Gas Fired Power Plant
Systems within the product	Body system, clothes holding tub and drive system, water and detergent management system, washing cycle control system, controls and displays system	Body system, exterior panel system, interior trim and food storage system, refrigeration system, electrical power and control system, water management and ice making system, and interior lighting system	Chassis system, microprocessor system, data storage and retrieval system, data input system, data output system, wireless communication system, power system and cooling system	Body system, chassis, steering and suspension system, powertrain system, braking system, electrical system, fuel system, cooling system, exhaust system, climate control system, driver interface system, lighting system, safety system and luggage/cargo storage system	Foundation system, tower structure system, nacelle system, wind turbine (blades and rotor) system, gear box system, generator system, step-up transformer system, air flow direction and wind velocity sensing system, turbine control system, electrical system, and safety system including steps, handholds, and ladders.	Natural gas storage system, gas and air input and control system, gas turbine system, power generator system, power plant control system, plant lighting system, heat exchangers, secondary steam turbine and generator system (for combined cycle plant), power distribution system and plant safety/security system
Capacity	Weight or volume of wash load (3.5 to 5.0 cf)	Volume of refrigerated (6–30 cf) and freezer(1–6 cf) compartments	Processor speed (up to about 3.6 GHz) and data handing capacity, memory capacity (128 GB–1TB SSD, display size (13–16″)	Number of passengers (4–7), volume and weight of load carrying capacity, towing load	1500–3500 kW of electricity produced per hour per turbine	300–1000 MW of electricity produced per hour

(Continued)

TABLE 6.1 (Continued)

Complex Product	Washing Machine	Refrigerator	Laptop Computer	Automotive Product	Wind Turbine	Natural Gas Fired Power Plant
Special Features	Number of settings for fabric types, temperatures, wash cycle time	Adjustable shelves, lighting, auto defrost, garageability	Memory type and storage, processor type and speed, keyboard lighting, touch screen, weight	Interior trim (e.g., leather seats), electronic features, LED lighting, reconfigurable interior and cargo compartments, etc.	Blade defrosters, auto shut-off, elevator in shaft, remote controls, etc.	Emission controls (carbon capture and sequestration), combined cycle, automated control system, etc.
Some Attributes of the Product	Washing Capacity, Packaging Space, Ergonomics, Washing efficiency, features (e.g., settings, displays, top vs. side loading), Noise and Vibrations, Cost/Price	Cooling and Freezer Space, Exterior size, Energy Usage, Features (shelves, adjustability, lighting, Noise and Vibrations, Cost/Price	Weight, Display Size, Overall Size, Processor Capacity, Memory Capacity, Sound Quality and Loudness, Noise, Special Features (e.g., lighted keys, 2 in 1 hinges)	Overall Shape/Body Style, Size, Styling, Performance, Vehicle Dynamics, Aerodynamics, Fuel Economy and Emissions, Costs/Price, Noise, Vibrations and Harshness, Safety, Customer Lifecycle, Product and Process Compatibility	Electricity generation capacity (1.5–3.5 MW), Capacity Factor (28–35%), Exterior and Interior Package, Noise and Vibrations, Safety, Cost/Price	Electricity generation capacity (MW), Capacity Factor (60–87%), Exterior and Interior Package, Noise and Vibrations, Emissions, Safety, Cost/Price
Disciplines Included in Engineering Design Work	Mechanical Engineering, Electrical Engineering, Manufacturing and Assembly, Chemical Engineering, Marketing and Finance	Mechanical Engineering, Electrical Engineering, Manufacturing and Assembly, Chemical Engineering, Marketing and Finance	Mechanical Engineering, Electrical Engineering, Manufacturing and Assembly, Chemical Engineering, Marketing and Finance	Mechanical Engineering, Materials Engineering, Electrical/Electronics Engineering, Manufacturing and Assembly, Chemical Engineering, Safety Engineering, Marketing and Finance	Mechanical Engineering, Civil Engineering, Aerodynamics engineering, Manufacturing and Assembly, Chemical Engineering, Safety Engineering, Marketing and Finance	Mechanical Engineering, Civil Engineering, Electrical Engineering, Manufacturing and Assembly, Chemical Engineering, Safety Engineering, Marketing and Finance

Detailed Engineering Design during Product Development

Specific Examples of Engineering Analyses Performed during Detailed Engineering Design Phase	Packaging and CAD Modeling, Heat and Fluid Flow Analyses, Body Structure and NVH, DFM, DFA, FMEA, LCA, LCCA, Verification and Validation Testing	Packaging and CAD Modeling, Heat and Fluid Flow Analyses, Body Structure and NVH, DFM, DFA, FMEA, LCA, LCCA, Verification and Validation Testing	Electrical and Electronics Architecture, Packaging and CAD Modeling, Heat and Fluid Flow Analyses, Body Structure and NVH, DFM, DFA, FMEA, LCA, LCCA, Verification and Validation Testing	Packaging and CAD Modeling, Heat and Fluid Flow Analyses, Body and Chassis Structure Design and NVH, DFM, DFA, Vehicle Dynamics, FMEA, LCA, LCCA, Verification and Validation Testing, Safety Evaluations	Packaging and CAD Modeling, Heat and Fluid Flow Analyses, Civil Engineering, Structural and NVH, Electrical Systems Architecture and Design, DFM, DFA, FMEA, LCA, LCCA, Verification and Validation Testing	Civil Engineering, Plant Design, Packaging and CAD Modeling, Heat and Fluid Flow Analyses, Chemical Engineering, Body Structure and NVH, DFM, DFA, FMEA, LCA, LCCA, Verification and Validation Testing
Key Validation Tests	Cleanliness of Clothes, Noise level	Temperature Maintenance, Power Consumption	Processing Speed, Uploading/Downloading Speed, Display Readability, Keyboard Comfort	Fun-to-drive, Interior Space, Fuel Consumption, Safety Ratings, Exterior Styling, Ride and Handling	Power Generated, Noise Level, Capacity Factor	Power generated, Capacity Factor, Fuel Consumption
Product Life Cycle (years) (Approximate)	8–15	10–20	3–8	8–14	20–30	25–35
Price ($) (Approximate)	$350–$1200	$500–$3000	$450–$1500	$25,000–$65,000	$1200 (on-shore) – $4300 (off-shore)/kW	$1000–$2500/kW

important considerations to determine improvements in the functionality of the product, its overall configuration, and their effects on other product attributes (e.g., weight, costs, performance).

The engineers make a number of design-related decisions by answering questions such as:

a. How would the systems within the product be configured to fit within the space (envelope)?
b. Can all the systems with their respective subsystems be configured and interfaced within the allocated spaces for each subsystem?
c. Are sufficient clearance spaces provided between systems for the maintenance personnel to service the systems?
d. Would the emissions during processing of materials and use of the product create harmful effects?
e. Would the cost of the product be within the budgeted amount?

As the design of each of the subsystems progresses simultaneously within different teams, the teams meet frequently to ensure that all the subsystems can fit and function within the allocated spaces. Design reviews and verification tests (CAE analyses, simulations, creation of prototype parts, and physical testing of the parts and their assemblies) are conducted to ensure that the components and assembled systems (lower to higher level) meet their respective attribute requirements (see Figure 6.1)

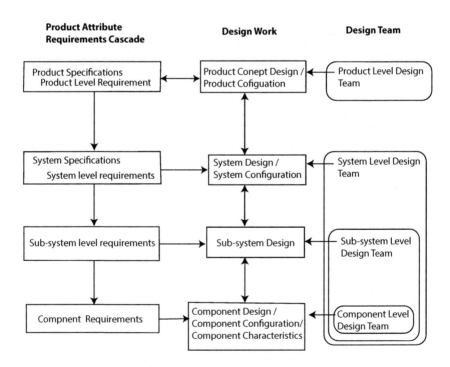

FIGURE 6.1 Product specifications to component design.

Detailed Engineering Design during Product Development

The left side of Figure 6.1 shows that product specifications are developed first to ensure that the product is designed to meet its specifications. Next, the specifications of the systems within the product are developed by cascading the product-level requirements into the system-level requirements. Similarly, the system-level requirements are cascaded down to lower-level systems (subsystems and components). The middle column shows the design work, and the third column shows the teams involved in the design. The teams use CAD to illustrate the shape and size (dimensions) of the components and determine materials for the components. A number of CAE analyses and simulation tests (e.g., stress, strength, and durability evaluations) along with design for manufacturing, design of assembly, and design for sustainability evaluations are conducted to ensure that the components can be assembled into lower-level systems, the lower-level systems can be assembled into the product and the product will function to meet its specifications. (Note that these steps are incorporated in the left side of the systems engineering "V" model.)

ILLUSTRATION OF WIND TURBINE DESIGN USING SYSTEMS ENGINEERING "V" MODEL

The systems engineering "V" model presents all important steps in the life cycle of a product or a system. The model is presented in Chapter 2 and illustrated in Figure 2.4. The model is known as the Systems Engineering "V" Model because the steps are arranged in a "V" shape with succeeding steps shown below or above the preceding steps and staggered in time (see Bhise (2017) and Blanchard and Fabrycky (2011) for more details). The model in Chapter 2 is described for the development of an automotive product. To illustrate the applicability of the model to other products shown in Table 6.1, the following part of this section provides a description in the context of development of a new wind turbine.

The model shows the basic steps of the entire wind turbine program on a horizontal time axis which represents time (t) in months before Job#1. "Job#1" is defined as the event when the first turbine is installed at the work site and begins producing electric power. The wind turbine program begins many months prior to Job#1. The beginning time of the program depends upon the scope and complexity of the program (i.e., the changes in the new turbine as compared to the earlier version) and the state of management's approval to begin the turbine development process.

LEFT SIDE OF THE "V"—DESIGN AND ENGINEERING

In the early stages prior to the official start of the wind turbine program, an advanced design and product planning activity (which usually involves an advanced wind turbine planning department or a special new turbine planning team) determines the wind turbine specifications (i.e., characteristics and its preliminary architecture such as tower type, size, blade material and geometry, type of powertrain, performance characteristics, and so forth. It also provides a list of references and competitors' wind turbines (used for benchmarking) that the new wind turbine may replace or compete with. A small group of engineers and designers (usually about 8 to 12) from the advanced design group are selected and asked to generate a few early wind turbine concepts to understand design and engineering challenges. A business plan including the projected sales volumes, the planned life of the new wind turbine, the wind turbine program timing plan, facilities and tooling plan, manpower plan, and financial plan (including estimates of costs, capital needed, revenue stream, and projected profits) is developed and presented to the senior management along with all other product programs planned by the company (to illustrate how the proposed program will fit within the overall corporate product plan and business strategy).

The wind turbine program typically begins officially after the approval of the business plan by the company management. This program approval event is considered to occur at x-months prior to Job#1 as shown in Figure 2.4. The figure also shows that the advanced design and planning activity begins at (x + y) months prior to Job#1. (Depending upon the scope of the activity, the value of "y" can range from about 3 to 8 months.)

At minus x-months, the chief program manager is selected and each functional group (such as design, structural engineering, aerodynamics engineering, powertrain engineering, electrical engineering, packaging and ergonomics/human factors engineering, safety engineering, manufacturing engineering) within the product development and other related activities (e.g., engineers from the utility companies) are asked to provide personnel to support the wind turbine development work. The personnel are grouped into teams and the teams are organized to design and engineer the wind turbine and its systems and subsystems.

The first major phase after the team formation is to create an overall wind turbine concept (labelled as "Product Concept" in Figure 2.4). During this phase, the designers and the package engineers work with different teams to create the wind turbine concept which involves (a) creating early drawings or CAD models of the proposed wind turbine, (b) creating computer-generated 3-D life-like images and/or videos of the wind, and (c) physical scale model/mock-up of the wind turbine. The images and/or models of the proposed turbine are shown to prospective customers and to the management. Their feedback is used to further refine the product concept.

As the wind turbine concept is being developed, each engineering team decides on how each of the wind turbine systems can be configured to fit within the allocated space (defined by the exterior envelope and interior surfaces of the wind turbine concept—primarily by the nacelle, tower and turbine hub and blades) and how the various systems can be interfaced with other systems to work together to meet all the functional, safety, and other requirements of the product. This step is shown as the "Systems" step in Figure 2.4. Any problems discovered during this phase may require iterating the process back (to the previous phase) to refine or modify the product concept. (This feedback represents the up arrow from the systems to product concept shown in Figure 2.4.)

As the systems are being designed, the next phases involve a more detailed design of the lower-level entities, i.e., design of subsystems of each system and components within each of the subsystems. These subsequent steps, straddled in time to the right, are shown as "Subsystems" and "Components."

The above steps form the left half of the "V" represent the time and activities involved in "Design and Engineering." The up-arrows in the left half of the "V"

Detailed Engineering Design during Product Development 131

indicate the iterative nature of the systems engineering loops covered earlier in Figure 2.2.

RIGHT SIDE OF THE "V"—VERIFICATION, MANUFACTURING, AND ASSEMBLY

The right half of the "V," moving from the bottom to the top, involves manufacturing of components (or lower-level entities) and testing to verify that they meet their functional characteristics and requirements (developed during the left half of the "V"). The components are assembled to form subsystems which are tested to ensure that they meet their functional requirements. Similarly, the subsystems are assembled into systems and tested; and finally, the systems are assembled to create the whole system. At each of the steps, the corresponding assemblies are tested to ensure that they meet the requirements considered during their respective design steps (i.e., the assemblies are verified). These requirements are shown as the horizontal arrows from the left side to the right side of the "V" in Figure 2.4.

The right side of the "V" is thus labelled as "Verification, Manufacturing, and Assembly." It should be noted that down-arrows between various assembly steps in the right half of "V" are not shown in Figure 2.4. The down-arrows would indicate failures in the verification steps. When failures occur, the information is transmitted to the respective design team to incorporate design changes to avoid repetition of such failures.

The engineers and technical experts assigned to various teams in the wind turbine program work throughout all the above steps and continuously evaluate the wind turbine design to verify that the needs of the turbine users are accommodated, and that they will be able to use the wind turbine under all foreseeable usage situations. Early production wind turbines developed just before Job#1 are usually used for additional whole product evaluations for the product validation purposes.

RIGHT SIDE OF THE DIAGRAM—OPERATION AND DISPOSAL

After Job#1, the wind turbines are produced and transported directly to the customer-designated installation site. The model diagram in Figure 2.4 also shows a period called "Operation and Refinement." During this period, the produced wind turbines are purchased, installed, maintained, and operated by the customers and serviced by the manufacturer (or other service providers). The wind turbines may also be refined with some changes (i.e., revised with minor changes during the existing model cycle, or updated as a refreshed new model every few years) during its operational time. When the wind turbine design becomes old and outdated, it is pulled from the market. This marks the end of the production of the wind turbines. At that point, the assembly plant and its equipment are recycled or retooled for the next wind turbine model. As the wind turbines reach the end of their useful life, they are sent to the scrap yards where many of the components may be disassembled. The disassembled components are either recycled for extraction of the materials or sent to the junk yards.

The above systems engineering "V" model thus shows the locations of the "Design and Engineering" and "Verification, Manufacturing, and Assembly" tasks on the time axis along as the left and right sides of the "V," and "Operation and Refinement" and "Retirement and Disposal" tasks on the right side of the "V." The model also shows how the requirements developed during the design process are used for verification of the components, subsystems, systems, and the whole system, as they are assembled. The design process uses the "top-down" approach, i.e., it begins with the development of the product concept on the top left and ends at the top of the "V" on the right side with the assembled and product ready to be installed in a wind farm.

ACTIVITIES IN ENGINEERING DESIGN

Figure 6.2 illustrates the major steps in the detailed engineering design of a component of a product. The top of the diagram in Figure 6.2 shows that the process of designing a product always begins with creating the overall product specifications. The product specifications should be documented and provided to all the team

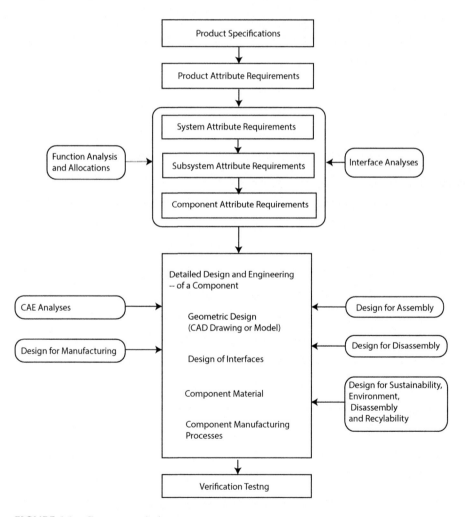

FIGURE 6.2 Component design process.

Detailed Engineering Design during Product Development 133

members to ensure that each member has a clear understanding of details about the product, i.e., what the product should accomplish. For example, if the product is a midsize SUV for the US market the product specifications should provide details such as its overall exterior dimensions, weight, important interior dimensions, powertrain and transmission capabilities, suspension type, and important features. The vehicle attribute managers use the vehicle specifications and develop detailed requirements on all attributes and sub-attributes of the vehicle. The vehicle attribute requirements are cascaded down to vehicle systems (e.g., body system, powertrain system, chassis system, climate control system, and electrical system) and their lower-level systems such as subsystems and sub-subsystems down to the component level. As the vehicle systems are decomposed into lower level systems detailed function analysis is performed to determine how various functions of the vehicle should be allocated (or divided) between vehicle systems. As the configuration of the systems decomposition is being developed, interface analyses (e.g., interface diagrams and interface matrices) are developed to understand interfaces between different levels of systems and components to specify interface requirements.

All the above details about systems decomposition and information generated in requirements analyses are used to conduct detailed design of systems, lower-level systems, and components (see Figure 6.2). Many different design tools such as computer-aided design (CAD), computer-aided engineering (CAE), design of manufacturing (DFM), design for assembly (DFA), design for sustainability (DFS, including design for disassembly and recycling) are applied concurrently by different specialists to ensure that the product will meet its functional requirements and sustainability goals (e.g., reductions in energy usage, emissions, and costs). A number of verification tests are conducted at the component level at progressively higher levels as components are assembled into higher-level systems.

The above-described analyses are conducted in a coordinated and concurrent engineering process that is guided by information obtained from the following activities in the product development process:

1. Product Program Plan (tasks to be performed and timings of different events, i.e., program milestones)
2. Thorough understanding of customer needs: Visiting customers, seeing how they use the product, and interviewing the customers to gather information on what is important to them.
3. Benchmarking—understanding studying existing product designs in terms of how different systems and components in these existing products are designed, e.g., their configuration, materials, unique features, interfaces, manufacturing methods used, costs. Understand their strengths and weaknesses and set goals to come up improved design of the proposed (target) product.
4. Customer feedback and warranty rates on existing and competitive products: Customer likes and dislikes, why, and frequency of warranty problems are understood by team members.
5. Trends in design, technologies, and government regulations are studied.

6. Program management understands and implements important design considerations to guide and control activities related to product configuration, functionality, performance features, so forth, and program timings and budgeting activities
7. Product attribute management is made responsible to ensure that attribute requirements are met during the design activities.
8. Systems engineering management plan (SEMP) is developed early and it is followed to manage design team activities and program management functions.
9. Systems engineers and program management ensures that team members from required disciplines/departments are represented in multi-functional teams, teamwork is managed to follow the SEMP, management of teams and communications between teams (e.g., meetings) is facilitated, manufacturing and assembly personnel are available during the design process, data sources, and suppliers are integrated into the design process.
10. Design team generally create system decomposition trees to understand all subsystems and lower-level systems down to the components (see Figure 2.1 in Chapter 2 for an example of the decomposition tree). QFD charts are created for major systems, subsystems, and components to understand customer needs and cascading the customer needs into engineering specifications. The QFD tool also provides targets for engineering specifications and identified most important functional specifications. Chapter 15 provides more details on the QFD. Interface diagram and interface matrix are also developed to understand interfaces between various systems, subsystems, and components. The requirements for interfaces are developed and trade-offs that need to be considered during designs are identified. Chapter 5 provides more details on the interface analyses.
11. CAD models are developed by package engineers with the help of many specialized engineers. The CAD models and product data are controlled to ensure that the design team members get the latest verified design data. Design changes are communicated to all teams and available within the program management systems.
12. Many design tools such as design for manufacturing (DFM), design for assembly (DFA), design for sustainability (DFS), and failure modes and effects analysis (FMEA) are used to study and improve design details all lower-level and upper-level systems as the design is being refined.
13. Simultaneous (concurrent) design and engineering approach and iterative design (need some iterations but not too many to slow down the progress) steps are incorporated.
14. Attribute managers ensure that team members understand engineering considerations and requirements including test procedures.
15. Design reviews are conducted in a timely manner to meet product functionality and program timings.
16. Evaluations are conducted to ensure that the designed product will meet customer needs by performing required design reviews, and tests (computer and physical).

CONCLUDING REMARKS

The overall phase of detailed engineering of a complex product is very challenging. It involves consideration of all engineering and manufacturing requirements concurrently with the help of inputs from many multidisciplinary teams to ensure that all product requirements, trade-offs between product attributes and risks are studied and the configuration of the whole product along with the design all systems within the product are developed. The complexity of the tasks involved in this phase increases with an increase in the number of systems, lower-level subsystems down to every component in each system within the product and interfaces between the many systems and components. The complex problem is solved by conducting many design iterations with inputs from many engineering analyses, design reviews, and tests to ensure that the design will meet all stated design requirements. It is therefore very important that the personnel working in this phase understand the overall process, their tasks, and available problem-solving techniques by constantly communicating, collaborating, and seeking advice from experts from various engineering specialties including systems engineering which coordinates the activities of different teams through the systems engineering management plan.

REFERENCES

Bhise, V. D. 2017. *Automotive Product Development: A Systems Engineering Implementation.* ISBN: 978-1-4987-0681-0. Boca Raton, FL: CRC Press.

Blanchard, B. S. and W. J. Fabrycky. 2011. *Systems Engineering and Analysis.* Upper Saddle River, NJ: Prentice Hall PTR.

7 Product Evaluation, Verification, and Validation

OBJECTIVES AND INTRODUCTION

The objective of this chapter is to provide the reader with an understanding into what is meant by product evaluation, how evaluations are conducted to verify that the product will meet its stated requirements, and to validate that the right product is developed. The chapter also presents information on types of evaluations and various evaluation issues and covers a number of evaluation methods used in the product verification and validation processes.

WHY EVALUATE, VERIFY, AND VALIDATE?

It is important that we design the right products that will meet the customer needs and the customers will like them. The right products also build customer loyalty and thus the customers will continue to use the products made by the manufacturer and repurchase their next products made by the same manufacturer. The product development team must have a clear focus on meeting the stated customer needs. Otherwise, the designed product may deviate from its goal; for example, the product may become too difficult to use, too heavy, too bulky, consume too much energy, produce much more pollutants as compared to other benchmarked products, too costly, and too boring (not fun-to-use). During the product development process, we must continuously check and ask the question if the product possesses the right characteristics and the product being designed is indeed the right product for the customers.

Testing, Verification, and Validation

Testing is an activity undertaken by using well-established procedures to obtain detailed measurements and data about the performance or characteristics of the product, its systems, subsystems, or components. The collected test data are analyzed to determine if the product, its systems, subsystems, or components meet their stated requirements (i.e., specified during the design process). Thus, a test can be conducted at the product, system, subsystem, or component levels to determine if one or more of the requirements at its corresponding level are met. A test can be performed by using computer models, simulations, prototypes, or physical working samples of the hardware representing the product, its systems, subsystems, or components. Testing methods can be used for verification or validation purposes.

Verification is the process of confirming that the product, its systems, and its components meet their respective requirements. The aim of the verification is to ensure that the tested item (product, its system, subsystem, or component) is built right, that is, it meets its requirements.

Validation, on the contrary, is the process of determining whether the product functions and it possesses the characteristics as expected by its customers when used in its intended environments. The aim of the validation process is to ensure that the right product is designed, and the product can be used and liked by its intended customers.

DISTINCTIONS BETWEEN PRODUCT VERIFICATION AND PRODUCT VALIDATION

From a process perspective, the product verification and product validation processes may be similar in nature, but the objectives are fundamentally different. Verification of a product shows proof of compliance with its requirements—that the product can meet each "shall" statement as proven through performance of a test, analysis, inspection, or demonstration. Validation of a product shows that the product accomplishes the intended purpose in the intended environments—that it meets the expectations of the customer and other stakeholders as shown through performance of a test, analysis, inspection, or demonstration.

Verification testing relates back to the approved requirements set and can be performed at different stages in the product life cycle. The approved specifications, drawings, computer-aided design (CAD) models, physical bucks (or mock-ups), prototypes represent the designs of systems, subsystems, and components and their configuration documentation establish the baseline of that product, which may have to be modified at a later time. Without a verified baseline and appropriate configuration controls, later modifications could be costly or cause major performance problems.

Validation testing is conducted under realistic conditions (or simulated conditions) on the end products for the purpose of determining the effectiveness and suitability of the product for use in the mission operations by typical users. The selection of the verification or validation methods are generally based on the engineering, program management, purchasing, or the certifying agency's judgment as to which is the most effective way to reliably show the product's conformance to requirements or that it will operate as intended.

OVERVIEW ON EVALUATION ISSUES

A product such as an automobile is used by a number of users in a number of different usages. To ensure that the product being designed will meet the needs of its customers, the engineers must conduct evaluations of all product features under all possible usages. A usage can be defined in terms of each task that needs to be performed by the user to meet a certain objective. A task may have many steps or subtasks. For example, the task of getting into a vehicle would involve a user to perform a series of subtasks such as (1) unlocking the door, (2) opening the door, (3) entering the vehicle and sitting in the driver's seat, and (4) closing the door. The evaluations are conducted for a number of purposes such as (1) to determine if the product meets

Product Evaluation, Verification, and Validation

its performance requirements, (2) to determine if the users will be able to use the product and its features, (3) to determine if the product has any unacceptable features that will generate customer complaints after its introduction, (4) to compare the user preferences for a product or its features with other similar or benchmarked products, and (5) to determine if the product will be perceived by the users to be the best in the industry. The purpose of this chapter is to review methods that are useful in the evaluations of products.

The evaluations can be conducted by collecting data in a number of situations. Figure 7.1 presents a simplistic flow diagram illustrating how the customers are involved in the verification and validation processes. Box no. 1 (top left in Figure 7.1) shows that the process begins with understanding the needs of the customers. The list of customer needs is translated into a list of product attributes (Box no. 2). To ensure that the product possesses the attributes, requirements are developed for each product attribute (Box no. 3). The requirements are used during the product development process to create product concepts and product designs; and various systems, subsystems, and components are tested to verify that they meet their respective requirements (Box no. 4). When the entire assembled product is available (usually as an early prototype), it is tested by using a number of customers and experts to validate the product (Box no. 5). After the product is sold to the customers, their feedbacks (through customer clinics, complaints, warranty and repair data, usage experience through dealers, and accident data) are gathered (Box no. 6). The customer feedback data are used in developing (or refining) the list of customer needs (Box no. 1) for incorporating changes in the future models or versions of the product.

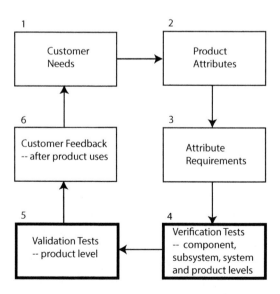

FIGURE 7.1 Verification and validation tests and the customers.

TYPES OF EVALUATIONS

Some examples of types of product evaluations conducted for product verification and validations are given in the following:

1. *Functional evaluations*: These evaluations consist of tests on individual components, subassemblies, and assemblies to ensure that they would perform the required functions or tasks under selected environmental conditions and meet their respective performance specifications.
2. *Durability tests*: The assembled subsystems, systems, and products are tested under most demanding actual situations (e.g., at minimum and maximum operating temperatures, high workloads, high speeds, and maximum electromagnetic fields) over a large number of work cycles (maximum expected work cycles during the life cycle, e.g., 10 years or 150,000 miles for the passenger cars. Note: Many tests may be conducted under accelerated-life-testing conditions).
3. *Assembly evaluations*: Assembly tasks are performed and evaluated under actual work situations or simulated situations (e.g., using 3-D CAD models of the assembly workstations, with the product, tools, and operators/robots) to check that the components can be properly assembled into systems and the systems can be assembled into products with the minimum number of movements and reorientations of the entities. Special attention should be paid on many steps in the tasks such as getting the components to the workstations; transporting or moving the components to the assembly fixtures; orienting and clamping the components in the fixtures; and joining the components using fasteners, welds, and glues, to ensure that sufficient clearance spaces are available for access for easy movements of the components, tools, and arms of a human assembler or a robot. The timings of different events and time required for assembling, inspecting for proper assembly, and so forth, are also checked.
4. *Tests involving human subjects*: Market researchers and human factors engineers conduct tests using representative customers or users. Some products can be operated or used by a single human operator, whereas other complex products such as commercial airplanes and ships require crews where coordination of actions and communications among individuals are important for successful operation of the missions. Some examples of simple test situations involving human subjects are described in the following:
    a. A product (or one or more of its systems, chunks, or features) is shown to a user and the user's responses (e.g., facial expressions and verbal comments) are noted (or recorded). (e.g., this situation occurs when a concept vehicle is displayed in an auto show.)
    b. A product is shown to a user and then responses to questions asked by an interviewer are recorded. (This situation occurs in a market research clinic.)
    c. A customer is asked to use a product and then responses to a number of questions included in a questionnaire or asked by an interviewer are recorded. (This situation can occur in a market research or a human factors test.)

Product Evaluation, Verification, and Validation

    d. A user is asked to use a number of products and the user's performance in completing a set of tasks on each of the products is measured. (This situation can occur in a human factors performance measurement study using a set of products [or alternate designs of the product or its systems] in test situations.)

    e. A user is asked to use a number of products and then asked to rate the products based on a number of criteria (e.g., preference, usability, accommodation, effort, and comfort). (This situation occurs in field evaluations using a number of vehicles—the manufacturer's test vehicle and other competitive vehicles.)

    f. A sample of customers are provided with instrumented products that record the customer behavior and product outputs, for example, video recording of driver behavior and customers' performance as the participants drive where they wish, as they wish, for weeks or months each. This is probably the only valid method to discover what drivers/customers actually do over time in the real world. (This situation occurs in naturalistic driving behavior measurement studies.)

5. *Evaluations after product sales*: Tracking how the product is perceived by its customers based on their product usage experience is a very important approach in validating the product. Manufacturers generally keep track of the customer feedback and repair and warranty data to determine customer satisfaction. The best measure of validation is probably the number of sales in periods after the product introduction. Continued higher sales volumes indicate that customers liked the product and continue to purchase the products. Repurchases of the same product after the original product has been out of service due to reasons such as end of life (or excessive usage or wear), losses in accidents, or simply additional purchases of the same product for an expanded customer base are strong indications of customer acceptance and loyalty.

The aforementioned examples illustrate that an engineer can evaluate a product or its features by using a number of data collection methods and measurements. The evaluations can be conducted by using computer models, simulations, physical models, early prototypes, production prototypes, and the products during and after use. The environment for the evaluations can be simulated in laboratory or field tests can be conducted at selected locations to meet the required environmental conditions (e.g., at different combinations of light, sound, vibrations, static/dynamic, temperature, humidity, and snow/rain levels).

EVALUATION METHODS: AN OVERVIEW

Table 7.1 provides a summary of methods categorized by combinations of types of data collection methods and types of measurements.

    The left-hand column of Table 7.1 shows that the data can be collected by using methods of observation, communication, and experimentation. In the observation method, it is assumed that a product performance can only be observed by a trained observer or an experimenter, or the data can be recorded (e.g., by using a data acquisition system or a video camera) for later observations made by a trained observer. In

TABLE 7.1
Evaluation Methods

Type of Data Collection Method	Type of Measurements	
	Objective Measurements	Subjective Measurements
Observation	Data recorded with instruments or observed by a trained experimenter. Behavior observations of the product during uses (e.g., recordings of product performance measures, customer/user behavior, task durations, errors, difficulties, conflicts, near-accidents).	Observer records and reports are gathered by expert evaluators. Checklists or rating forms completed by observers based on their observations.
Communication	Experimenter reported objective measures (e.g., output levels, response times, speeds, events—displayed by instruments).	Subject reported—detections, identifications of events. Responses in checklists or rating forms. Reporting of problems, difficulties, and errors during operation of equipment.
Experimentation	Measurements with instrumentation: Performance measurements or behavioral measurements.	Data obtained from test subjects: ratings, behavioral measurements (difficulties, errors, etc.)

the communication method, the subject (customer/user) can be asked to report about the problems experienced during the use of the product or asked to provide ratings on his/her impressions about the product while or after performing a given task. In the experimentation method, various test situations are designed by deliberate changes in the combinations of certain independent variables (e.g., parameters of the product, users, or usage situations) and the responses are obtained by using combinations of methods of observation and/or communication. For example, Richards and Bhise (2004) showed how a computer-controlled programmable vehicle buck can be used to create different vehicle packages (configurations) for experimental evaluations and comparisons of different vehicle package parameters (e.g., sitting height; locations of the steering wheel, instrument panel, header, and roof pillars). For more information on many available methods of data collection and their advantages and disadvantages, the reader should refer to Chapanis (1959) and Zikmund and Babin (2009).

The types of measurements can be categorized as objective or subjective as shown in the last two columns of Table 7.1. The objective measures can be defined here as measurements that are not affected by the subject performing the tasks or by the experimenter observing or recording the subject's performance. The objective measures are generally obtained by the use of physical instruments or by unbiased and trained experimenters. The subjective measurements are generally based on the subject's perception and experience during or after performing one or more product usage tasks. The objective measures are generally preferred because they are more precise and unbiased. However, there are many product attributes that cannot be measured without using human subjects as the "measuring instruments." After the users have experienced the product, they are better able to express their perceived impressions about the product and its characteristics by the use of methods of

Product Evaluation, Verification, and Validation

communication. Customer satisfaction (or quality) is probably the most important measure for product validation, which can be assessed better by communicating directly with the customers/users.

The following section provides additional information on the methods of data collection relevant for product evaluations.

METHODS OF DATA COLLECTION AND ANALYSIS

OBSERVATION METHODS

In the observation methods, information is gathered by direct or indirect observations of subjects (experts, customers, and users) and/or the product during the product uses (or operation) to determine if the product is easy or difficult to use or it is operating properly. An observer can directly observe, or a video camera (or other data acquisition instrumentation) can be set up and its recordings can be played back at a later time. The observer needs to be trained to identify and classify different types of predetermined behaviors, events, problems, or errors that a subject commits and/or the product changes (e.g., leaks, noise, excessive heating, emissions/smoke, vibrations, warping, melting, and breaking) during the observation period. The observer can also record durations of different types of events, the number of attempts made to perform an operation, number and sequence of controls used, number of glances made to view certain items and perform a task, and so on.

Some events such as accidents are rare, and they cannot be directly observed due to excessive amount of direct observation time needed until sufficient accident data are collected. However, information about such events can be obtained through reports of near-accidents (i.e., situations where accidents almost occurred but were averted. Note that near-accidents occur at much higher frequencies than accidents) and indirect observations (e.g., through witnesses or from material evidence) gathered after such events (also see Chapter 19). However, the information gathered through indirect observations may not be very reliable due to a number of reasons (e.g., witness may be guessing or even deliberately falsifying, or objects associated with the event of interest may be displaced or removed).

COMMUNICATION METHODS

The communication methods involve asking the user or the customer to provide information about his or her impressions and experiences with the product. The most common technique involves a personal interview where an interviewer asks a series of questions to the user. The questions can be asked before the product usage, during the usage, or after the usage. The user can be asked questions that will require the user to (1) describe the product or the impressions about the product and its attributes, (2) describe the problems experienced while using the product (e.g., difficult to read a label, very hot or cold sitting surfaces, and excessive vibrations or noise), (3) categorize the product using a nominal scale (e.g., acceptable or unacceptable; comfortable or uncomfortable; liked or disliked), (4) rate the product on one or more interval scales describing its characteristics and/or overall impressions (e.g., rating on a 10-point scale for evaluation of comfort during a product usage), or (5) compare the products presented in pairs based on a given attribute.

Some commonly used communication methods in product evaluations involve (1) rating scales: using numeric scales (e.g., a 10-point scale where 10 = excellent and 1 = very poor, scales with adjectives; e.g., acceptance rating and semantic differential scales), (2) paired comparison-based scales [e.g., using Thurstone's Method of Paired Comparisons (Thurstone, 1927; Bhise, 2012) and Analytical Hierarchical Method (Satty, 1980) covered in a section entitled "Paired Comparison-Based Methods" and described in detail in Chapter 3].

EXPERIMENTATION METHODS

The purpose of the experimentation method is to allow the investigator to control the evaluation situation so that causal relationships between response variables (i.e., output or performance of the user or the product) and independent variables that define the product characteristics (e.g., product concept/configuration, location of a control, type of control, size of control, type of display, and operating forces) or operating conditions (e.g., temperatures, sound levels, and light levels) may be evaluated. An experiment includes a series of controlled observations (or measurements of one or more response variables) undertaken in artificial (test) situations with deliberate manipulations of combinations of levels of independent variables to answer one or more hypotheses related to the effect of (or differences due to) the independent variables. Thus, in an experiment, one or more variables (called the independent variables) are manipulated and their effect on another output variable (called the dependent or response variable) is measured, while all other variables that may confound the relationship(s) are eliminated or controlled.

The importance of the experimentation methods is that (1) they help identify the best combination of the independent variables and their levels to be used in designing the product and thus achieve the most desired performance (or effect on the users) and (2) when the competitors' products are included in the experiment along with the manufacturer's product, the superior product can be determined. To assure that this method provides valid information, the researcher designing the experiment needs to ensure that the experimental situation is not missing any critical factor related to the performance of the product or the task being studied. Additional information on the experimental methods can be obtained from Kolarik (1995) or other textbooks on the design of experiments.

OBJECTIVE MEASURES AND DATA ANALYSIS METHODS

The type of objective measurements and the data analysis methods will depend on the type of objective tests (e.g., performance measurement and durability evaluation), existing practices (e.g., test methods and procedures specified in standards), and test/measurement instrumentation available within the manufacturer's facilities and other test laboratories available to the manufacturer. The objective measures can be based on physical measures such as time (taken or elapsed), distance (position or movements in lateral, longitudinal, or vertical directions), velocities, accelerations, and occurrences of predefined outcomes (e.g., product failures) that can be measured to define dependent measures. The dependent measures and their statistics such as means, standard deviations, minimum and/or maximum values, and percentages above and/or below certain preselected levels can be used for statistical analyses based on the experiment design selected for the study.

SUBJECTIVE METHODS AND DATA ANALYSIS

The subjective methods are primarily used by psychologists, market researchers, product evaluators, and ergonomics engineers because in many situations, (1) the subjects are better able to perceive characteristics and issues with the product than the physical measurement instruments, and thus, the subjects can be used as the measurement instruments; (2) suitable objective measurement methods and measures do not exist; and (3) the subjective measures are easier to obtain.

Pew (1993) has pointed out several important points regarding subjective methods. Subjective data must come from the actual user rather than the designer; the user must have an opportunity to experience the conditions to be evaluated before providing opinions; and care must be taken to collect the subjective data independently for each subject; and the final test and evaluation of a system should not be based solely on subjective data.

The two most commonly used subjective measurement methods used during the product development process are (1) rating on a scale and (2) paired comparison-based methods. These two methods are presented in the following.

RATING ON A SCALE

In this method of rating, the subject is first given instructions on the procedure involved in evaluating a given product including explanations on one or more of the product attributes and the rating scales to be used for scaling each attribute. Interval scales are used most commonly. Many different variations in defining the rating scales can be used. The interval scales can differ due to (1) how the end points of the scales are defined, (2) the number of intervals used (Note: odd number of intervals allow use of a midpoint), and (3) how the scale points are specified (e.g., without descriptors vs. with word descriptors and/or numerals).

Figure 7.2 presents eight examples of interval scales. The first four scales (Figure 7.2a through 7.2d) are 5-point numeric scales defined by descriptors (words or adjectives). The last two scales (Figure 7.2g and 7.2h) have 10-point numeric scale and their numeric values range from 1 to 10. Whereas scales in Figure 7.2d through 7.2f have clearly defined midpoints. The three scales also have numbers and adjectives (or descriptors) defining each scale interval. The use of adjectives or descriptors can help the subjects in understanding the levels of the attribute associated with the scale. The use of midpoints (e.g., in scales in Figure 7.2e or f) allows the subject to choose the middle category if the subject is unable to decide if the product attribute in question falls on one or the other side of the scale. The use of a scale such as in Figure 7.2d also allows the subject to first decide if the product was easy or difficult to use and then select the level by using the adjectives "somewhat" or "very." Even the number

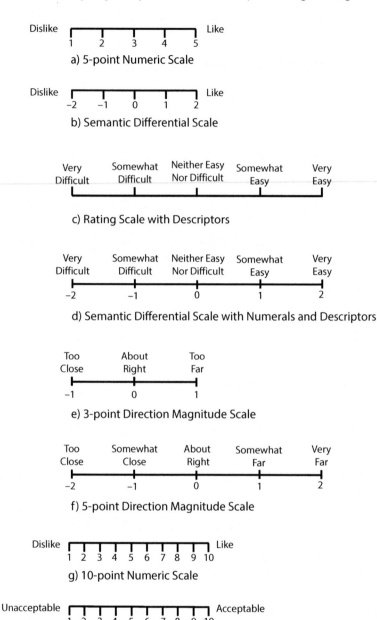

FIGURE 7.2 Examples of 5- and 10-point rating scales. (a) 5-point numeric scale, (b) semantic differential scale, (c) rating scale with descriptors, (d) semantic differential scale with numerals and descriptors, (e) 3-point direction magnitude scale, (f) 5-point direction magnitude scale, (g) 10-point numeric scale, and (h) 10-point acceptance scale.

of intervals can also be used where the subject should be forced to decide between either side of the scale. Thus, the midpoint associated with the inability of the subject to decide between the two sides will be removed. The scales with 5 or less points are easier to use as compared to scales with a larger number of intervals. The direction magnitude scales such as scales in Figure 7.2e and f are particularly useful in evaluating physical dimensions of products. In these scales, the midpoint is defined by the words "about right" and thus a large percentage of responses in this category helps confirm that the evaluated product dimension was designed properly. Whereas a skewed distribution of responses to the left or the right side on the scale will indicate a mismatch in terms of both the direction and magnitude of the problem with the dimension.

Figures 7.3 and 7.4 present 10-point scales with descriptors used to convey the magnitude of the scale values. The scale in Figure 7.4 provides three types of descriptors, namely, (1) acceptability level, (2) customer satisfaction level, and (3) detectability of product problems or defects. In general, a product with ratings below 6 is not introduced in the market as an average customer can detect the problem and hence will be dissatisfied.

Table 7.2 presents a ratings data collection form for a customer evaluation of the exterior and interior characteristics of an automotive product. The form uses two rating scales to evaluate each item. The first scale, a 3-point direction magnitude scale, is used to obtain customer perception of the magnitude of the variable being evaluated. The second, a 10-point acceptance scale, provides customer assessment on the level of acceptance of the rated item.

Scale Value / Rating	1	2	3	4	5	6	7	8	9	10
Descriptor	Dislike Very Much		Dislike Somewhat		Neither Like Nor Dislike		Like Somewhat		Like Very Much	

FIGURE 7.3 A 10-point dislike-like rating scale with descriptors.

Scale Value / Rating	1	2	3	4	5	6	7	8	9	10
Descriptor	Not Acceptable		Poor		Borderline	Acceptable	Fair	Good	Very Good	Excellent
Expected Customer Satisfaction	Very Dissatisfied				Somewhat Dissatisfied		Fairly Well Satisfied	Very Satisfied	Completely Satisfied	
Problem/Product Defect Detectibility	Most Customers				Average Customer		Critical Customers		Trained Observers	Not Perceptible

FIGURE 7.4 A 10-point acceptance scale with descriptors, expected customer satisfaction, and problem detectability.

TABLE 7.2
Illustration of Vehicle Evaluation Using Direction Magnitude and Acceptance Rating Scales

	Item No.	Characteristic to Evaluate	Rating Using Direction Magnitude Scale (Circle One of the Three)			Acceptance Rating (Place an "x" on the 10-Point scale). 1 = Very Unacceptable, 10 = Very Acceptable 1 2 3 4 5 6 7 8 9 10
Exterior characteristics	1	Overall length	Too small	About right	Too long	
	2	Overall width	Too narrow	About right	Too wide	
	3	Overall height	Too low	About right	Too high	
	4	Belt height (lower edge of the side windows)	Too low	About right	Too high	
	5	Windshield rake angle	Too vertical	About right	Too much sloping	
	6	Front overhang	Too small	About right	Too large	
	7	Rear overhang	Too small	About right	Too large	
	8	Ground clearance	Too small	About right	Too large	
	1	Steering wheel longitudinal (fore/aft) location	Too close	About right	Too far	
	2	Steering wheel vertical (up/down) location	Too low	About right	Too high	
	3	Steering wheel diameter	Too small	About right	Too large	
	4	Gas pedal fore/aft location	Too close	About right	Too far	
	5	Gas pedal lateral location	Too much to left	About right	Too much to right	

Interior package characteristics	6	Lateral distance between the gas pedal and the brake pedal	Too small	About right	Too large
	7	Gas pedal to brake pedal lift-off	Too small	About right	Too large
	8	Gearshift lateral location	Too much to left	About right	Too much to right
	9	Gearshift location longitudinal location	Too close	About right	Too far
	10	Height of the top portion of the instrument panel directly in front of the driver	Too low	About right	Too high
	11	Height of the armrest on driver's door	Too low	About right	Too high
	12	Space above the driver's head	Too little	About right	Too generous
	13	Space to the left of the driver's head	Too little	About right	Too generous
	14	Knee space (between instrument panel and right knee with foot on the gas pedal)	Too little	About right	Too generous
	15	Thigh space (between the bottom of the steering wheel and the closest lower surface of the driver's thighs)	Too little	About right	Too generous

The distribution of responses on each direction magnitude scale provides feedback to the designer on how the dimension corresponding to the scale was perceived in terms of its magnitude, and the ratings on the acceptance scale provide the level of acceptability of the dimension. For example, if the ratings on item number 5 in the interior package characteristics part (gas pedal lateral location) in Table 7.2 showed that 80% of the subjects rated the gas pedal location as "too much to the left" on the direction magnitude scale and the average rating on the 10-point acceptance scale was 4.0, the designer can conclude that the gas pedal needs to be moved to the right to improve its acceptability. The author found that such use of dual scales was very helpful in fine-tuning the vehicle dimensions in the early stages of the automotive design process.

ANALYSIS OF 10-POINT RATINGS DATA

Table 7.3 presents a summary of data obtained from 80 subjects who were asked to rate five products (namely, products "R," "W," "T," "Q," and "J") using a 10-point scale. The data for product "R" shows that 5 subjects gave the rating of 10, 6 subjects gave the rating of 9, 12 subjects gave the rating of 8, 17 subjects gave the rating of 7, and so on. Several evaluation measures can be used to analyze the data. The lower part of the table shows that the following evaluation measures were used to compare the products: (1) the number highly preferred, that is, the sum of subjects who rated

TABLE 7.3
Comparison of Five Products Based on 10-Point Ratings Data

	Rating		Number of Rating Responses				
	Product		R	W	T	Q	J
10-point rating scale		10	5	0	16	0	13
		9	6	0	19	0	35
		8	12	5	16	0	16
		7	17	10	12	10	11
		6	11	13	5	7	4
		5	9	18	5	8	1
		4	10	13	2	12	0
		3	6	9	5	22	0
		2	4	5	0	17	0
		1	0	7	0	4	0
	Total		80	80	80	80	80
Evaluation measures	No. highly preferred = no. of ratings ≥ 8		23	5	51	0	64
	No. poorly preferred = no. of ratings ≤ 3		10	21	5	43	0
	No. preferred = no. of ratings ≥ 6		51	28	68	17	79
	No. nonpreferred = no. of ratings ≤ 5		29	52	12	63	1
	Ratio of no. preferred to no. nonpreferred		1.76	0.54	5.67	0.27	79.00
	Log (no. preferred to no. non-preferred)		0.25	−0.27	0.75	−0.57	1.90
	Median value of ratings		6.50	5.00	8.00	3.00	9.00
	Average value of ratings		6.20	4.68	7.76	3.80	8.49
	Standard deviation of ratings		2.14	1.93	1.99	1.77	1.15

Product Evaluation, Verification, and Validation

the product as 8 or higher; (2) number poorly preferred, that is, the sum of subjects who rated the product as 3 or lower; (3) number preferred, that is, the sum of subjects who rated the product as 6 or higher; (4) number nonpreferred, that is, the sum of subjects who rated the product as 5 or lower; (5) ratio of number preferred to number nonpreferred; (6) logarithm of the ratio of number preferred to number nonpreferred; (7) median value of the ratings; (8) average value of the ratings; and (9) standard deviation of the ratings. Several statistical tests can be used to determine statistically significant differences between the products by using any of the evaluation measures (e.g., tests to evaluate differences in means or proportions). The values of the measures presented in the table show that product "J" was most preferred (had highest values of the median, average, and number highly preferred) and product "Q" was least preferred.

The example thus illustrates how the ratings data can be summarized and evaluation measures can be computed to select the preferred product.

PAIRED COMPARISON-BASED METHODS

The method of paired comparison involves evaluating products presented in pairs. In this evaluation method, each subject is essentially asked to compare the two products in each pair using a predefined procedure and is asked to simply identify the better product in the pair on the basis of a given attribute (e.g., comfort or usability). (If the respondent says there is no difference between the two products, the instruction would be to ask the subject to randomly pick one of the products in the pair. The idea is that, if there truly is no difference in that pair among the respondents, the result will average out to 50:50.) The evaluation task of the subject is, thus, easier as compared to the rating on a scale. However, if n products have to be evaluated, then the subject is required to go through each of the $n(n-1)/2$ possible number of pairs and identify the better product in each pair. Thus, if five products need to be evaluated, then the number of possible pairs would be $5(5-1)/2 = 10$. The major advantage of the paired comparison approach is that it makes the subject's evaluation tasks simple and more accurate as the subject has to only compare the two products in each trial and only identify the better product in the pair. The disadvantage of the paired comparison approach is that as the number of products (n) to be evaluated increases, the number of possible paired comparison judgments that each subject needs to make increases rapidly (proportional to the square of n) and the entire evaluation process becomes very time-consuming.

Two commonly used methods based on the paired comparison approach are (1) Thurstone's Method of Paired Comparisons and (2) Analytical Hierarchical Method. Thurstone's Method allows us to develop scale values for each of the n products on a z-scale (z is a normally distributed variable with a mean equal to zero and standard deviation equal to one) of desirability (Thurstone, 1927), whereas the Analytical Hierarchical Method allows us to obtain relative importance weights of each of the n products (Satty, 1980). Both the paired comparison methods are simple and quick to administer and have the potential of providing more reliable evaluation results as compared to other subjective methods where a subject is asked to evaluate (rate) one product at a time using a scale. (It should be noted that rating one product in each trial is more difficult as the subject does not use a reference product to judge the rating value. Furthermore, the rating of a product can be easily influenced by the characteristics of the products rated in preceding trials.) The steps involved in Thurstone's Method are described in Bhise (2012). The Analytical Hierarchical Method is illustrated in Chapter 3.

EVALUATIONS DURING PRODUCT DEVELOPMENT

During the entire product development process, a number of evaluations are conducted to ensure that the product being designed will meet the needs of the customers. The attribute requirements and design issues need to be systematically studied to ensure that all design requirements are considered, and appropriate evaluation methods are used. The results of the evaluations are generally reviewed in the product development process at different milestones with various design and management teams.

Figure 7.5 illustrates the timings of the product verification and validation tests (shown by circles) in the systems engineering "V" model of the product development

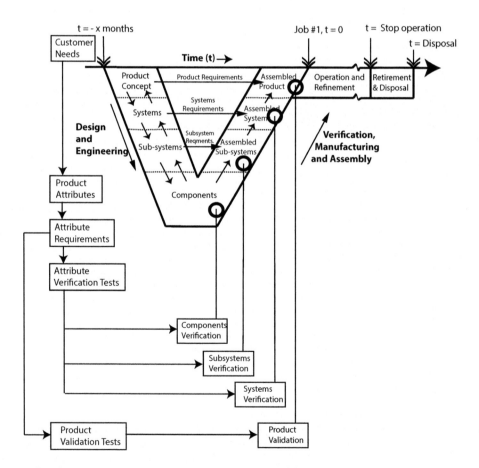

FIGURE 7.5 Verification and validations test during the product development program.

Product Evaluation, Verification, and Validation

process. The figure shows that the customer needs are documented before the program initiation (at $-x$ months). The product attributes are developed from the customer needs. The attribute requirements (developed from the customer needs) are used to create attribute testing and product validation plans. The attribute verification tests are conducted to verify the components, subsystems, and systems. The product validation tests are conducted on fully assembled products available as early prototypes or production units. The verification tests are generally engineering tests using simulations, laboratory, or field tests using test software, test equipment, and test operators. The outputs of the tests (measurements) are compared with the minimum and maximum acceptance criteria values specified in the attribute requirements. The validation tests are usually conducted by using customers, users, and experts to ensure that product meets all the attribute requirements. Appropriate statistical tests are also conducted to assure that the data collected in the tests meet the requirements.

VERIFICATION PLAN AND TESTS

A verification plan is needed to demonstrate that all the product requirements are evaluated, and the targets are met. The product requirements are generally derived from the product specifications and attribute requirements. The requirements can be also selected from available internal (e.g., corporate) manuals of engineering specifications, standards, and brand requirements. The responsibility of documentation of verification plans for various entities is generally assigned to the design release engineers and the attribute engineers.

The planning of the verification testing involves identification of the appropriate methods and test equipment that are needed to verify each requirement. The verification plan must also include the number of product samples that need to be tested to verify each requirement. The verification plan also includes other data such as recommend test dates, specific requirement(s) being verified, the target or acceptance criteria, and the required statistical tests to be performed to support the results of the tests. The engineering managers are usually responsible for approval of the verification test methods and the test plan.

The verification plan also identifies a category for each requirement in terms of how the requirement should be met. The requirements can be categorized as follows:

1. *Must meet*: The verification tests must be performed, and all applicable requirements must be met by the product.
2. *Use carryover*: The verification compliance received in a previous product program (that used the same product design) can be accepted. Additional verification tests may not be needed.
3. *Deviation sign-off*: If the testing results show that the product cannot meet the requirement, a deviation must be written (to allow proceeding without meeting the requirement), approved, and signed-off by the program management.
4. *Does not apply*: The requirement is not relevant to the particular program. Thus, verification tests are not needed.

VALIDATION PLAN AND TESTS

The validation of a complex product also requires a comprehensive validation plan. The validation plan is based on an agreed upon set of tests and evaluations with inputs and product expectations (e.g., quality) of a number of parties (e.g., users, experts, engineers, and market researchers from the company management and purchasing organizations, and government agencies responsible for making and enforcing regulatory requirements on the product).

Table 7.4 presents an evaluation test plan for an automotive product. The test plan is structured to evaluate vehicle attributes by using vehicle (product)-level tests. For validation tests, the product-level tests are conducted using prototype products or early production units. The tests include a combination of evaluation methods involving (1) engineering verification type objective tests that apply to a fully assembled product (i.e., a vehicle) using physical test equipment (e.g., dynamometers); (2) experts (e.g., test drivers, technical experts, and test engineers) who evaluate the product using predefined engineering procedures in an objective manner (i.e., not influenced by individual judgments or biases), and (3) customers who are led to evaluate the product using structured evaluation procedures and provide their subjective impressions (e.g., ratings on a number of product characteristics, their preferences to the products or product features, comments on what they liked or disliked, and reasons to support their ratings or comments).

CAE tests are conducted for verification testing. They are not performed for validation testing as actual products are available for these tests. It should be noted that since the CAE tests do not use actual physical products they do not represent manufacturing variations and assembly issues (e.g., fit, finish, loose parts, and unpredictable vibrations) related to final built entities. Dimensions and coordinates measurements tests are performed on test samples (used for verification and validation testing) to ensure that test samples (i.e., entities at different levels from component to the whole product) meet at the dimensions and tolerances specified in the CAD drawings and models.

The underlying goal in this type of combination of evaluation methods involving physical tests, experts, and customers is to complete the verification and validation within limited time and budget and achieve a high level of objectivity and customer representation. The selection of combinations of objective and subjective evaluations and questions to be included (e.g., who performs the subjective tests: experts vs. customers) are usually determined by the group of decision makers involving engineers, managers, and market researchers who are usually guided by the organizations policies, quality manuals or quality management system, and government regulations. In some cases, the decision makers can be experts, managers, users/operators from the organization that purchases the product, a third party (like a registrar for ISO 9000 certification or independent testing agencies), or the government agencies that enforce regulations on the product (e.g., airline companies that purchase airplanes and the Federal Aviation Administration rules and requirements can help in determining validation procedures for an airplane).

TABLE 7.4
Illustration of a Product-Level Evaluation Plan

Vehicle Attribute	Sub-attribute	Sub-attribute Requirements/Sources	Evaluation Method(s)
Package	Seating Package (Driver & passengers)	Accommodation percentiles and interior dimensions. SAE J 1516, J1517, J4004.	Interior coordinate measurements. Customer Ratings.
	Entry/Exit	Head/torso, knee, thigh, foot space requirements. Distances from Sating Reference Point (SgRP).	Interior coordinate measurements. Customer Ratings.
	Luggage/Cargo Package	Luggage volume requirements. Floor ht to ground.	Interior coordinate measurements. Customer Ratings.
	Fields of View-Visibility	Wiper/defroster zones, mirror fields, pillar obscuration	Interior coordinate measurements. Customer Ratings.
	Powertrain Package	Engine, transmission, and drivetrain envelopes.	Interior coordinate measurements. Eng. Tests.
	Suspensions & Tires Package	Suspension and tire envelopes.	Interior coordinate measurements. Customer Ratings.
	Other Mech Package	Space requirements for fuel tank, electrical, lighting, climate control, etc., systems. FMVSS 108 requirements.	Interior coordinate measurements.
Controls & Displays	Locations-Layout	SAE J1138. Ergonomic requirements.	Interior coordinate measurements.
	Hand & Foot Reach	SAE J287. SAE J1516 and J4004.	Interior coordinate measurements.
	Visibility and Obscurations	FMVSS 111, SAE J1050, J902,903.	Interior coordinate measurements.
	Operability	Ergonomic guidelines, SAE J1139	Ergonomics scorecard.
Safety	Front Impact	FMVSS 204, 208, 212 & 219 requirements.	CAE methods. Sled tests with crash dummies.
	Side Impact	FMVSS 214 requirements.	Engineering tests.
	Rear Impact	FMVSS 301 and 303 requirements.	CAE tests. Eng. Tests.
	Roof Crush	Deformation requirements.	CAE and Eng. Tests
	Sensors, Belts, & Airbags	Anchorage and dummy tests.	CAE and Eng. Tests.
	Other Safety Features	FMVSS 108, SAE lighting stds.	Photometric & Eng. Tests.
Styling/Appearance	Exterior—shape, proportions, etc.	Exterior design guidelines.	Exterior surface measurements. Customer ratings.
	Interior—I/P, Console, trim, etc.	Interior design guidelines	Interior surface measurements. Customer ratings.
	Luggage/Cargo/Storage	Customer requirements.	Customer ratings.
	Underhood Appearance	Design guidelines.	Customer ratings.

(Continued)

TABLE 7.4 (Continued)

Vehicle Attribute	Sub-attribute	Sub-attribute Requirements/Sources	Evaluation Method(s)
	Color/Texture Mastering	Color and texture masters.	Customer ratings.
	Craftsmanship	Craftsmanship guidelines.	Expert and customer ratings. Measurements of mating edges, surfaces and surface finish.
Thermal & Aerodynamics	Aerodynamics	Aero forces, coefficient of drag and noise requirements.	CAE and wind tunnel testing.
	Thermal Management	Temperature guidelines	CAE and heat management tests.
	Water Management	Leak test requirements	Water and air leak tests.
Performance & Drivability	Performance Feel	0-60 mph time. Eng. Requirements	Experts and Customer ratings.
	Fuel Economy	EPA/NHTSA requirements	EPA test procedures
	Long Range Capabilities	Eng. Requirements.	Field tests.
	Drivability	Eng. Requirements.	Field tests.
	Manual Shifting	Eng. Requirements.	Experts and Customer ratings.
Vehicle Dynamics	Ride	Eng. Requirements.	Experts and Customer ratings.
	Steering and Handling	Eng. Requirements.	Experts and Customer ratings.
	Braking	FMVSS 105 requirements.	Field tests.
Noise, Vibrations, & Harshness (NVH)	Road NVH	Eng. Requirements.	Sound measurements. Field tests.
	Powertrain NVH	Eng. Requirements.	Sound measurements. Field tests.
	Wind Noise	Eng. Requirements.	Sound measurements. Field tests.
	Electrical/Mechanical	Eng. Requirements.	Field tests.
	Brake NVH	Eng. Requirements.	Field tests.
	Squeaks & Rattles	Eng. Requirements.	Field tests. Customer ratings.
	Pass by Noise	Eng. Requirements.	Sound measurements. Field tests.
Interior Climate Comfort	Heater Performance	Eng. Requirements.	Field tests. Customer ratings.
	A/C Performance	Eng. Requirements.	Field tests. Customer ratings.
	Water Ingestion	Eng. Requirements.	Field tests.

Category	Subcategory	Source
Security	Vehicle Theft	Eng. Requirements. Expert evaluations.
	Contents/Component Theft	Eng. Requirements. Expert evaluations.
	Personal Security	Eng. Requirements. Expert evaluations.
Emissions	Tailpipe emissions	EPA requirements. Dynamometer and field tests.
	Vapor Emissions	EPA requirements. Dynamometer and field tests
	On-board diagnostics	EPA requirements. Dynamometer and field tests
Weight	Body	Design assumptions. CAE weight predictions & measurements.
	Chassis	Design assumptions. CAE weight predictions & measurements.
	Powertrain	Design assumptions. CAE weight predictions & measurements.
	Climate control	Design assumptions. CAE weight predictions & measurements.
	Electrical	Design assumptions. CAE weight predictions & measurements.
Cost	Cost to the customer	Product planning assumptions. Cost prediction programs.
	Cost to the company	Product planning assumptions. Cost prediction programs.
Customer Life Cycle	Purchase & Service Experience	Historic data and customer feedback.
	Operating Experience	Marketing assumptions. Customer feedback.
	Life Stage Changes	Marketing assumptions. Customer feedback.
	System Upgrading	Marketing assumptions. Customer feedback.
	Disposal/Recyclability	Recycling requirements. Material tracking.
Product/Process Compatibility	Reusability	Reusability requirements. Field data.
	Commonality	Commonality guidelines. Analysis of Component database.
	Carryover	Tooling budget. Analysis of Component database.
	Complexity	Manufacturing budget. Analysis of Component database.
	Tooling/Plant Life Cycle	Manufacturing strategy and budget. Analysis of plant tooling database.

Note: CAE, computer-aided engineering; EPA, U.S. Environmental Protection Agency; NHTSA, National Highway Traffic Safety Administration

CONCLUDING REMARKS

The design teams and their management personnel are responsible for ensuring that their design meets all applicable requirements on the product and the product meets all the customer needs. Thus, evaluations must be conducted to ensure that each entity or the product meets its applicable requirements. The evaluation outcomes should provide support or proof of compliance of the requirements to eliminate future problems. The feedback received from the evaluations is very important to the program management. If certain requirements are not met, then other alternatives (e.g., redesign, retest, or procuring the failed entity from another supplier) need to be immediately considered. Such failures put a lot of pressure on the design teams and the management. If the failed requirement is mandatory by a federal standard or a regulation, then the design must be modified and retested until its compliance to the requirements is verified. A failure to meet a requirement can increase financial risks, reduce customer confidence in the product, and degrade the reputation of the manufacturer. An injury caused by a product defect resulting from a noncompliance of a requirement can also suggest that the manufacturer was negligent and thus support the defendant's claim in a product liability case. On the contrary, meeting all the applicable requirements will confirm that the right product was developed, and it will boost the morale of the design team and the program management.

REFERENCES

Bhise, V. D. 2012. *Ergonomics in the Automotive Design Process*. Boca Raton, FL: The CRC Press.
Chapanis, A. 1959. *Research Techniques in Human Engineering*. Baltimore, MD: The Johns Hopkins Press.
Kolarik, W. J. 1995. *Creating Quality—Concepts, Systems, Strategies, and Tools*. New York: McGraw-Hill.
Pew, R. W. 1993. *Experimental Design Methodology Assessment*. BBN Report no. 7917. Bolt Beranek & Newman, Inc., Cambridge.
Richards, A. and V. Bhise. 2004. Evaluation of the PVM Methodology to Evaluate Vehicle Interior Packages. SAE Paper no. 2004-01-0370. Also published in SAE Report SP-1877, SAE International, Inc., Warrendale, PA.
Satty, T. L. 1980. *The Analytic Hierarchy Process*. New York: McGraw-Hill.
Thurstone, L. L. 1927. The Method of Paired Comparisons for Social Values. *Journal of Abnormal and Social Psychology*, 21: 384–400.
Zikmund, W. G. and B. J. Babin. 2009. *Exploring Market Research*. Ninth Edition. Independence, KY, Cengage Learning.

8 Program Planning and Management

INTRODUCTION

A complex product program requires careful planning and management of its functions. The management needs to ensure that resources (people, capital, and equipment) are provided. And processes and tools are available to multidisciplinary teams to carry out coordinated technical and management functions to develop and produce the products needed by the customers. In addition, collaborative engineering environment, constant reviews of product design and continuously applied management controls, and communication of time schedules, budgets, and costs are provided for successful completion of the product program. A large program typically consists of several projects. The projects are typically managed by project managers, and they report to a program manager who coordinates the progress and accountability of all the projects.

This chapter covers the program and project management functions and tools and the systems engineering management plan (SEMP) required to manage the Systems Engineering (SE) process.

PROGRAM VERSUS PROJECT MANAGEMENT

The life cycle of a complex product can be managed as a program. The program will involve the prime responsibilities of designing the right product, producing it, servicing it during its operating life, and finally closing the production operations and disposing or recycling the products. The entire program is generally divided into several manageable projects such as (1) developing the product, (2) building needed tools and production equipment, (3) building plants and installing equipment to get ready for the production, (4) recruiting and training people to operate the plant, and (5) generating marketing plan and training dealers to sell and service the product. Brown (2008) INCOSE (2012) and NASA (2012) provide additional information on this topic.

A program usually contains many projects. The outputs of projects are used to create the program outcomes. Thus, a program can be either a large project or a group of projects. Each project can have a project manager. The project manager's job is to ensure that his/her project succeeds. The program manager, on the contrary, may not spend much effort on the management of individual projects, but is concerned with the aggregate result or the end state. For example, in an automotive company, a program may include one project to introduce new products to take advantage of rising markets in emerging countries, and another project to protect against the downside of falling markets in developed countries. These projects are opposites with respect to their success conditions, but they fit together in the same program.

Program management thus provides a layer above the management of projects and focuses on selecting the best group of projects, defining them in terms of their objectives, and providing an environment where projects can be run successfully. Program management also emphasizes the coordinating and prioritizing of resources across projects, managing links between the projects, and the overall costs and risks of the program. Program manager should avoid micromanagement of the projects; he should leave the project management to the project managers and concentrate on the success of the overall program.

Program Management Functions

The program management functions typically include the following activities:

1. Projects management: (a) Coordinating projects through a master plan management, (b) status reporting, (c) issues management, and (d) resource management
2. Performance management: (a) Costs measurement, (b) benefits measurement, and (c) analysis of business data
3. Change management: (a) Change facilitation, (b) change communication, and (c) workforce training and transition
4. Knowledge management: (a) Documentation and sharing of lessons learned from past projects and programs, (b) management of standards and best practices, (c) outputs of product and process benchmarking, and (d) customer complaints and feedback data gathering.

Development of Detailed Project Plans

The project development activity requires a lot of inputs. The gathered information is used to develop a project plan. The key project development activities include the following:

1. Collecting inputs from all stakeholders
2. Creating a common understanding of all the projects
3. Preparing documentation of technical plan, management plan, and SEMP (covered in section entitled, "Systems Engineering Management Plan" of this chapter) for each project in the program
4. Supporting the implementation and management of the SE process involving development of requirements, functional analysis and allocation, interface analysis, balanced product design, detailed design, designing and building tools and manufacturing facilities, conduction verification and validation tests, sales, marketing and service, and finally retirement of the product and disposal of facilities.

Project Management

Project management is the discipline of planning, organizing, securing, and managing resources to bring about the successful completion of specific project goals and objectives.

The traditional phased approach involves a sequence of the following five phases to be completed:

1. Project initiation
2. Project planning and design
3. Project execution and construction
4. Project monitoring and controlling systems
5. Project termination

Figure 8.1 presents a flow chart of the aforementioned activities in relation to a project. The project involved a series of tasks to be performed. It is important that the project work must be clearly understood with details about all the tasks to be performed, their sequence, resources (people, equipment, and funds) needed, and time required. Thus, Figure 8.1 shows arrows linking the three primary phases (i.e., phases 2–4) to the tasks to be performed.

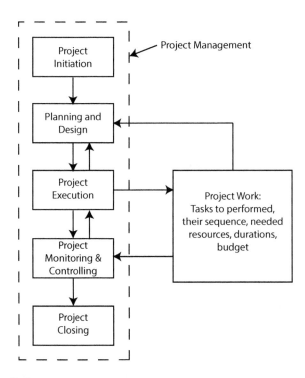

FIGURE 8.1 Project management.

Not all the projects will go through every phase as projects can be terminated before they reach completion. Some projects do not follow a structured planning and/or monitoring stages. Some projects will go through phases 2–4 multiple times. Many industries use variations on these project phases.

STEPS IN PROJECT PLANNING

The basic steps involved in planning a project include the following:

1. Develop a work breakdown structure (WBS) (see the section entitled, "Work Breakdown Structure" in this chapter) of all activities by listing each task in each of the activities. Each task is defined as a group of all steps or actions to be completed to accomplish the task.
2. Identify task inputs, outputs, and deliverables
3. Establish task precedence relationships
4. Determine start and finish time for each task
5. Estimate task duration and resource needs to perform each task. The resource needs include headcount needs by disciplines (e.g., number of designers, number of engineers, and number of test operators), budget to perform the tasks, and special resources (e.g., software applications, training, and product test facilities)
6. Display schedule (e.g., a Gantt chart; see section on "Gantt Chart"). Determine critical path (longest time path of planned activities to the end of the project; see section on "Critical Path Method")
7. Estimate project budget and cash flows (expenses and revenues as functions of time) (see Chapter 9 for more information).

TOOLS USED IN PROJECT PLANNING

GANTT CHART

A Gantt chart is a type of bar chart (with horizontal bar segments on a timescale) that illustrates activities in a project or program schedule. Figure 8.2 illustrates a Gantt chart of a product program. It provides a visual diagram of all activities in the program on a timescale. A Gantt chart illustrates the start and finish dates of all elements or activities in a project or a program. Some Gantt charts also show the dependency (i.e., precedence network) relationships between activities. Gantt charts can be used to show current schedule status using percent-complete by use of shadings or colors of the horizontal bars.

CRITICAL PATH METHOD

The critical path method (CPM) is used for scheduling a set of project activities. The essential technique for using CPM is to construct a model of the project that includes the following:

1. A list of all activities required to complete the project (typically categorized within a WBS)

Program Planning and Management

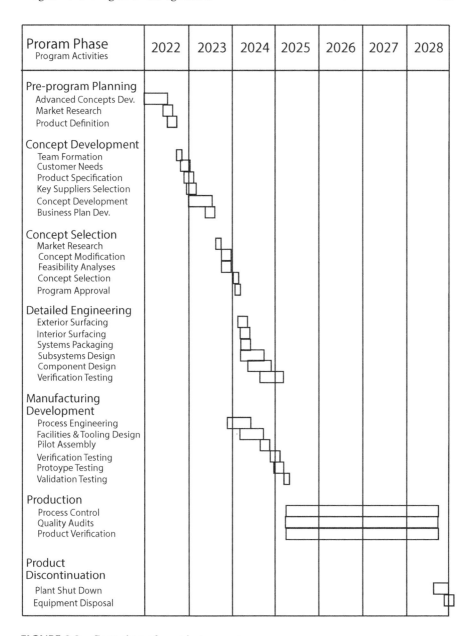

FIGURE 8.2 Gantt chart of a product program.

2. The time (duration) of each activity
3. The dependencies (or sequence of completions) between the activities
4. Project beginning and end dates

Using these values, CPM calculates the longest path of planned activities to the end of the project and the earliest and latest that each activity can start and finish without making the project longer. This process determines which activities are "critical" (i.e., on the longest time path) and which have "total float" (i.e., can be delayed without making the project longer). In project management, a critical path is the sequence of project network activities that add up to the longest overall duration. This determines the shortest time possible to complete the project. Any delay of an activity on the critical path directly impacts the planned project completion date (i.e., there is no float on the critical path). A project can have several, parallel, near-critical paths. An additional parallel path through the network with the total durations shorter than the critical path is called a subcritical or noncritical path.

Program (or Project) Evaluation and Review Technique

The program (or project) evaluation and review technique, commonly abbreviated PERT, is a model for project management designed to analyze and represent the tasks involved in completing a given project. It is commonly used in conjunction with the CPM. PERT is a method to analyze the involved tasks in completing a given project, especially the time needed to complete each task, and identifying the minimum time needed to complete the total project. PERT was developed primarily to simplify the planning and scheduling of large and complex projects. It is able to incorporate uncertainty by making it possible to schedule a project while not knowing precisely the details and durations of all the activities.

The uncertainty in the completion time of each activity is considered by using estimates of optimistic time, most likely time and pessimistic time for each activity. The expected time and variance of time for each activity can be computed as follows:

$$ET_i = \frac{(OT_i + 4MT_i + PS_i)}{6}$$

$$\sigma_i^2 = \left(\frac{(PS_i - OT_i)}{6}\right)^2$$

where ET_i = expected time of the ith activity in the critical path
OT_i = optimistic time estimate to complete ith activity in the critical path
MT_i = most likely time estimate of completing ith activity in the critical path
PS_i = pessimistic time estimate to complete ith activity in the critical path
σ_i^2 = variance of time to complete ith activity in the critical path.

The probability of completion of a project $[P(T \leq k)]$ before a certain date (k), that is, kth day from the project start date, can be estimated by assuming that the total time T of the critical path has a normal distribution with its mean equal to the sum of expected times of all activities (μ_T) and the variance of the total time (σ_T^2) equal to the sum of variances of task completion times of all activities in the critical path as follows:

$$P(T \le k) = \left(\frac{1}{\sigma_T \sqrt{2\pi}}\right) \int_{-\infty}^{k} e^{-Y} dT$$

where

$$Y = \frac{(T - \mu_T)^2}{2\sigma_T^2}$$

$$\mu_T = \sum_i \text{ET}_i$$

$$\sigma_T^2 = \sum_i \sigma_i^2$$

If the project has more than one critical path, then the probabilities of completion of each of the paths before a certain date can be computed. The probability of completion of the project can be computed by multiplying the probabilities of completing all the critical paths before a certain date (if the paths are independent from each other. See the AND gate probability computations as shown in Figure 19.2).

PERT is more of an event-oriented technique rather than start- and completion-oriented and is used more in projects where time, rather than cost, is the major factor. It is applied to very large-scale, one-time, complex, nonroutine infrastructure, and research and development projects.

WORK BREAKDOWN STRUCTURE

A WBS is a tool used in the project management and SE to define and group a project's discrete work elements in a way that helps organize and define the total work scope of the project. A WBS element may be a product, data, a service, or any combination. A WBS also provides the necessary framework for detailed cost estimating and control along with providing guidance for schedule development and control. Additionally, the WBS is a dynamic tool and can be revised and updated as needed by the project manager.

The outputs of the WBS are generally shown in a series of block diagrams using flow charts and tree structures (e.g., with hierarchical levels similar to the decomposition tree shown in Figure 2.1). Each block represents a task and provides many task details and parameters (e.g., time required, dates, costs, and name of the person or department it is assigned to). A WBS typically displays the following: (1) various elements of the project, (2) distribution of work elements of the project in different tasks, (3) distribution of the cost or budget between the elements of the project, and

(4) subdivision of larger work elements into smaller elements. Some versions of the WBS may not consider timings or order of execution of the tasks. (However, many project management software applications used in WBS analysis can create Gantt charts and conduct CPM and PERT analyses.)

PROJECT MANAGEMENT SOFTWARE

Several project management software systems are currently available (e.g., Oracle, Microsoft Project, and Project Standard 2019, developed and sold by the Microsoft Corporation; Microsoft, 2022). The software programs are designed to assist a project manager in developing a plan, assigning resources to tasks, tracking progress, managing the budget, and analyzing workloads. Microsoft Project allows creation of color-enhanced time charts with milestones, tasks, phases, people, and so on, and can share databases with other Microsoft 365 applications (e.g., Microsoft Excel). Many of the software packages allow online sharing of data between project managers and program managers. Thus, all team members have instant access to the project data and many features such as input changes, assign tasks, create personalized dashboards of projects, view calendars, prepare reports, track project issues, and create customized charts and graphs.

OTHER TOOLS

Many other tools are available for specialized analyses such as investment analysis, cost-benefit analyses, expert surveys, simulation models and predictions, risk profile analyses, surcharge calculations, milestone trend analysis, cost trend analysis, target versus actual comparisons of dates, time used, and costs incurred and headcount. These analyses can facilitate communication of project status and improve efficiency and capabilities of project and program managers, especially when these tools are available online and accessible with extensive databases on existing, past, and other similar projects for comparison purposes. The tools also allow managers to create different types of project timing, budget, and progress reports for communications and control of project schedules, cash flows, and problems involving different types of risks.

SYSTEMS ENGINEERING MANAGEMENT PLAN

The SEMP is a higher-level plan (not very detailed) for managing the SE effort to produce a final operational product (or a system) from its initial requirements. Just as a project plan defines how the overall project will be executed, the SEMP defines how the engineering portion of the project will be executed and controlled. The SEMP describes how the efforts of system designers, test engineers, and other engineering and technical disciplines will be integrated, monitored, and controlled during the complete life cycle of the product (DOD, 2012; USDOT, 2007). Figure 8.3 presents a flow chart illustrating the relationship of the SEMP to project work and project management.

Program Planning and Management

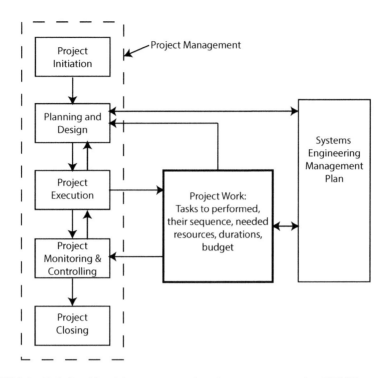

FIGURE 8.3 Relationship of the systems engineering management plan (SEMP) to project work and project management.

For a small project, the SEMP might be included as part of the project plan document, but for any project or program of greater size or complexity, a separate document is recommended. The SEMP provides the communication bridge between the project management team and the technical teams. It also helps coordinate work between and within different technical teams. It establishes the framework to realize the appropriate work (or tasks to be performed) that meet the entry and success criteria of the applicable project phases. The SEMP provides management with necessary information for making systems engineering decisions. It focuses on requirements, design, development (detailed engineering), test, and evaluation; it also addresses traceability of stakeholder requirements and supportability across the project life cycle.

CONTENTS OF SEMP

The purpose of this section is to describe the activities and plans that will act as controls on the project's SE activities. For instance, this section identifies the outputs of each SE activity, such as documentation, meetings, and reviews. This list of required outputs will control the activities of the team and thus will ensure the satisfactory completion of the activities. Some of these plans may be completely defined in the SEMP (in the framework or the complete version). For other plans, the SEMP may only define the requirements for a particular plan. The plan itself is to be prepared as one of the subsequent SE activities, such as may be the case with a verification plan or a deployment plan. Almost any of the plans described in the following discussion may fall into either category. It all depends on the complexity of the particular plan and the amount of up-front SE that can be done at the time the SEMP is prepared.

The first set of required activities relates primarily to the successful management of the project. These activities are likely to have already been included in the project/program plan but may need to be expanded in the SEMP (USDOT, 2007). Generally, they are incorporated into the SEMP, but, on occasion, may be developed as separate documents. The items that can be included in the SEMP are listed as follows. (The items and their descriptions provided in USDOT (2007) were modified to meet the needs of complex product development.)

1. WBS consists of a list of all tasks to be performed on a project, usually broken down to the level of individually budgeted items.
2. *Task inputs*: It is a list of all inputs required for each task in the WBS, such as source requirements documents, interface descriptions, and standards.
3. *Task deliverables*: It is a list of the required deliverables (outputs) of each task in the WBS, including documents, and product configuration including software and hardware.
4. *Task decision gates*: These are a list of critical activities that must be satisfactorily completed before a task is considered completed.
5. *Reviews and meetings*: It is a list of all meetings and reviews of each task in the WBS.
6. *Task resources*: It is identification of resources needed for each task in the WBS, including, for example, personnel, facilities, and support equipment.
7. *Task procurement plan*: It is a list of the procurement activities associated with each task of the WBS, including hardware and software procurement and any contracted or supplier provided services such as SE services or development services.
8. *Critical technical objectives*: It is a summary of the plans for achieving any critical technical objectives that may require special SE activities. It may be that a new software algorithm needs to be developed and its performance verified before it can be used. Or a prototyping effort is needed to develop a user-friendly operator interface. Or a number of real-time operating systems need to be evaluated (verified) before a procurement selection or the level of assembly is made.
9. *Systems engineering schedule*: A schedule of the SE activities that shows the sequencing and duration of these activities. The schedule should show tasks (at least to the level of the WBS), deliverables, important meetings and reviews, and other details needed to control and direct the project. An important management tool is the schedule. It is used to measure the progress of the various teams working on the project and to highlight work areas that need management intervention.
10. *Configuration management plan*: It describes the development team's approach and methods to manage the configuration of the systems within

Program Planning and Management

the products and processes. It will also describe the change control procedures and management of the system's baselines as they evolve.
11. *Data management plan*: It describes how, and which data will be controlled, the methods of documentation, and where the responsibilities for these processes reside.
12. *Verification plan*: It is always required. This plan is written along with the requirements specifications. However, the parts on tests to be conducted can be written earlier.
13. *Verification procedures*: These are developed by the core engineering experts, and they define the step-by-step procedures to conduct verification tests and must be traceable to the verification plan.
14. *Validation plan*: It is required. It assures that the product being designed is the right product and would meet all the customer needs.

The second set of plans is designed to address specific areas of the SE activities (USDOT, 2007). They may be included entirely in the SEMP, or the SEMP may give guidance for their preparation as separate documents. The plans included in the first set listed in the preceding list are generally universally applicable to any project. On the contrary, some of the plans included in this second set are only rarely required. The unique characteristics of a project will dictate their needs. The items that can be included in this second set are described as follows. (The items and their descriptions provided in USDOT (2007) were modified to meet the needs of complex product development.)

1. *Software development plan*: It describes the organization structure, facilities, tools, and processes to be used to produce the project's software. It also describes the plan to produce custom software and procure commercial software products.
2. *Hardware development plan*: It describes the organization structure, facilities, tools, and processes to be used to produce the project's hardware. It describes the plan to produce custom hardware (if any) and to procure commercial hardware products.
3. *Technology plan*: It (if needed) describes the technical and management process to apply new or untried technology. Generally, it addresses performance criteria, assessment of multiple technology solutions, challenges, risks, and fallback options to existing technology.
4. *Interface control plan*: It identifies all important interfaces within and between systems (within the product and external to the product) and identifies the responsibilities of the organizations on both sides of the interfaces.
5. *Technical review plan*: It identifies the purpose, timing, place, presenters and attendees, topics, entrance criteria, and the exit criteria (resolution of all action items) for each technical review to be held for the project/program.
6. *System integration plan*: It defines the sequence of activities that will integrate various product chunks involving components (software and hardware), subsystems, and systems of the product. This plan is especially

important if many subsystems and systems are designed and/or produced by different development teams from different organizations (e.g., suppliers).
7. *Installation plan or deployment plan*: It describes the sequence in which the parts of the product are installed (deployed). This plan is especially important if there are multiple different installations at multiple sites. A critical part of the deployment strategy is to create and maintain a viable operational capability at each site as the deployment progresses.
8. *Operations and maintenance plan*: It defines the actions to be taken to ensure that the product remains operational for its expected lifetime. It defines the maintenance organization and the role of each participant. This plan must cover both hardware and software maintenance.
9. *Training plan*: It describes the training to be provided for both maintenance and operation.
10. *Risk management plan*: It addresses the processes for identifying, assessing, mitigating, and monitoring the risks expected or encountered during a project's life cycle. It identifies the roles and responsibilities of all participating organizations for risk management.
11. *Other plans*: Other plans that might be included are, for example, a safety plan, a security plan, and a resource management plan.

This second list is extensive and by no means exhaustive. These plans should be prepared when they are clearly needed. In general, the need for these plans becomes more important as the number of stakeholders involved in the project increases.

The SEMP must be written in close synchronization with the project plan. Unnecessary duplication between the project plan and the SEMP should be avoided. However, it is often necessary to put further expansion of the SE effort into the SEMP even if they are already described at a higher level in the project plan.

CHECKLIST FOR CRITICAL INFORMATION

The USDOT (2007) guide also provides a checklist to assure that the SEMP includes the following:

1. Technical challenges of the project
2. Description of the processes needed for requirements analysis
3. Description of the design processes and the design analysis steps required for an optimum design
4. Identification and documentation of any necessary supporting technical plans, such as a verification, an integration, and a validation plan
5. Description of stakeholder involvement when it is necessary
6. Identification of all the required technical staff and development teams, and the technical roles to be performed by the system's owner, project staff, stakeholders, and the development teams
7. Description of the interfaces (or interactions) between the various development teams.

Role of Systems Engineers

The role of the systems engineers assigned to the program is essentially to do what is needed to implement the systems engineering process. A carefully developed SEMP will provide a clear roadmap for the systems engineers. They should work closely with all other team members, technical and program planning, to ensure that all basic SE steps are followed (see Figures 2.2. and 2.5). The role of systems engineers is described in more detail in Chapter 2.

Value of Systems Engineering Management Plan

A carefully developed and executed SEMP will enable proper implementation of SE during the program; that is, all the SE steps from obtaining customer needs to the product validation in the product development and subsequent steps during the product operations and disposal stages are completed by the program teams in a timely manner.

The value of the SEMP can be summarized as follows:

1. It will facilitate reducing the risk of schedule and cost overruns and increasing the likelihood that the SE implementation will meet the user's needs.
2. It will engage the right specialists at the right (needed) time (because they will know what needs to be done) and make sure that they perform the right tasks (e.g., analyses or tests), thus resulting in improved stakeholder participation.
3. The product team will be more adaptable, and the developed products and systems will be resilient and meet customer needs.
4. All entities within the product will be verified for functionality and thus the product should have fewer defects.
5. The experience gained and lessons learned during the implementation of the SEMP can be used to create improved SEMP documentation for the next program.

COMPLEXITY IN PROGRAM MANAGEMENT

Programs that require separate program management functions, processes, and people for their management are inherently more complex than simple programs that are generally managed by technical persons responsible for the product development. (See the story on the cyclone grinder development [case study #4] in Chapter 23.) Simple programs do not require additional processes or people to manage the program (the management tasks are generally small and responsibilities in small teams are shared).

Thus, for the management of complex products, the program management should undertake the following:

1. Divide the complex product into a manageable number of "chunks." (Note that a chunk can include one or more systems of the product.)

2. Create an organizational structure with multiple teams (for different systems or chunks of the product) to manage the complex product program.
3. Select team members based on their expertise and capabilities to understand the "big picture," that is, the technical issues related to functioning of the entire product and the interfaces between and within their assigned chunk(s).
4. Train team members to select and apply tools (covered in this book and other tools in specialized disciplines).
5. Require each team to create requirements for assigned chunk based on the customer needs and customer attributes created by the product planning activity.
6. Require each team to provide information on status of its deliverables according to their WBS to the program control team.
7. Require each team to select and apply necessary tools (covered in Chapters 14 through 21) during the product development phases and report results to their parent team during design reviews and program management meetings.

Time Management

To facilitate timely completion of planned activities, the program management should include the following:

1. Gateways/milestones (timely targeted decision points) in the program schedule (see Figures 1.4 and 1.5)
2. Reviews by different specialized areas (by attributes, specialized design and user groups [e.g., technical experts, users, service personnel, and maintenance personnel], peer reviews, subsystems reviews, system reviews, etc.)
3. Definition of work to be completed at each milestone
4. Formal approval plan to proceed to the next phases
5. Plan to handle disapprovals that will involve rework, delays, and workload balancing problems (overtime costs and/or program slippage)
6. Good communication on timing status (ahead of schedule, on schedule vs. behind schedule) and related problems.

Cost Management

The program management should prepare cost and timing charts to communicate and control costs and timings. Various types and formats of charts can be used to control and communicate information on budget levels and comparisons between budgeted costs versus costs incurred and projected costs as functions of time (especially, cost overruns). Figure 8.4 presents an example of a time chart comparing cumulative budgeted cash flow versus actual expenditure.

Challenges in Program Management

Hectic pace and staying on top of all relevant internal and external issues or problems that can affect the project or program pose constant challenges to the program management. Examples of internal factors are failures in meeting verification test

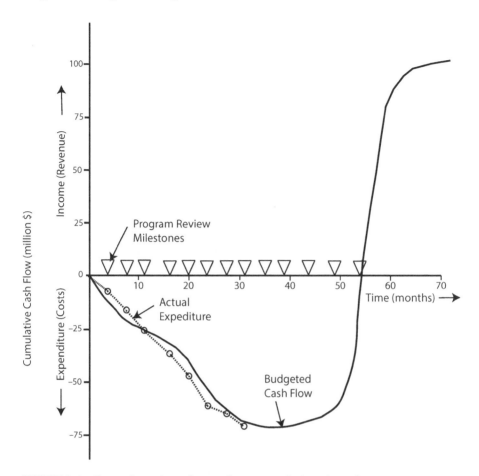

FIGURE 8.4 Comparison of actual expenditure versus budgeted cost flow.

requirements, breakdown of critical test equipment, changes in personnel, bad weather, power outages, and so forth. Examples of external factors are delays caused by supplier problems, changes in the state of economy, changes in the budgets, new technical developments that can change program objectives, political problems in countries affected by the program, and so on. Thus, the program management personnel must be able to handle multiple problems simultaneously, constantly maintain communication with the lower and higher levels of team organization, anticipate problems, and be prepared for various possibilities. Generally, program managers with technical background and familiarity with the technical aspects and issues will be able to handle and foresee possible developing problems quicker than nontechnically oriented program managers.

The component-level problems during the design stage and failures in some verification tests can affect the progress of work on the higher product levels (i.e., risks in not meeting timely deliveries of subsystems, systems and product for verifications and validations, effect on costs due to redesign, rework, retests, etc.). Therefore, it is essential that technical problems and resulting timing and cost problems need to be tracked and communicated immediately through the higher levels so that necessary corrective or precautionary actions can be taken to minimize the program risks. Progress charts on technical issues, timing, and costs need to be kept up-to-date and reviewed through an appropriate reporting structure.

CONCLUDING REMARKS

Project and program management involve a lot of challenges, even in situations where the path to the desired deliverable seems obvious. The difficulties arise due to people, technology, and competition change rapidly. Some examples of sudden changes are (1) the project may be progressing fine until a key team member suddenly resigns, (2) a revolutionary new product is about to be introduced in the market, or (3) a major competitor launches a product almost identical to the manufacturer's new product. The aforementioned situations would force changes in the program. The program complexity will only increase over time as the rates of technological innovations have been increasing rapidly. Furthermore, changes in organizational culture, working environment, economic situation, and scarcity of resources also add substantial challenges in controlling and successful completion of the programs.

REFERENCES

Brown, J. T. 2008. *The Handbook of Program Management: How to Facilitate Project Success with Optimal Program Management*. New York: McGraw-Hill.

Department of Defense (DOD). 2012. *Systems Engineering Plan (SEP) Preparation Guide*. http://www.ndia.org/Divisions/Divisions/SystemsEngineering/Documents/Major%20Current%20Tasks%20and%20Activities/Review%20of%20OSD%20Systems%20Engineering%20Plan/SE_Plan_Prep_Guide_v1.pdf (accessed May 28, 2012).

International Council on Systems Engineering. 2012. *INCOSE Systems Engineering Handbook*. http://www.incose.org (accessed May 1, 2013).

Microsoft Corporation. 2022. *Project 2022*. https://www.microsoft.com/en-us/microsoft-365/project/project-management-software. (accessed: June 13, 2022).

National Aeronautics and Space Administration. 2012. *NASA Systems Engineering Handbook*. http://www.acq.osd.mil/se/docs/NASA-SP-2007-6105-Rev-1-Final-31Dec2007.pdf (accessed May 1, 2013).

U.S. Department of Transportation (USDOT)-Federal Highway Administration. 2007. *Systems Engineering for Intelligent Transportation Systems*. Report no. FHWA-HOP-07-069.

9 Costs and Benefits Considerations and Models

INTRODUCTION

One of the key objectives of all product-producing organizations is to make money unless the organization is a nonprofit or a government entity. The costs of developing, producing, distributing, and maintaining the products are important to all organizations. The goals of the organizations are thus to minimize the total costs during the entire life cycle of the product—from its conception to product disposal. In the early stages, accurate estimates of the costs are required to develop a budget for the product program and to get it approved. The actual expenditure of costs should be continuously compared with the budgeted costs to ensure that the project is meeting its budgetary requirements. Differences between the budgeted costs and actual costs may signal over or under expenditures or errors in estimating the budgeted costs.

The costs are estimated by breaking down a large product program into a series of manageable tasks. Experienced cost estimators, based on the work content in each task and availability of cost information from previously conducted similar tasks and adjustments for the prevailing and future economic and technological conditions, usually develop time and cost estimates to complete the tasks. The costs of all tasks are then added along with some allowances for errors, interest, inflation, and other unknown or unforeseen problems. The project cost estimates are also refined several times in the program as some of the less predictable tasks and unknown issues (e.g., technology development) are resolved or better understood.

The costs are incurred over time. The costs during early product development are nonrecurring; that is, they do not recur or are one-time type of costs associated with product concept development, product design, detailed engineering, testing, building tools, and facilities. Once the production begins, the costs associated with purchasing raw materials, parts purchased from suppliers, plant running costs, direct labor costs, insurance costs, and so forth, are generally proportional to the volume of products manufactured. The cumulative monetary needs decrease as products are sold and the revenues are generated.

The objective of this chapter is to understand different types of costs associated with the various tasks involved during the product life cycle.

TYPES OF COSTS

NONRECURRING AND RECURRING COSTS

The costs are incurred throughout the life cycle of a product program. The total life cycle costs of a product can be divided into (1) nonrecurring costs and (2) recurring costs.

DOI: 10.1201/9781003263357-10

Nonrecurring Costs

These costs represent expenses and investments that are made during the product development, creation of the production systems, and also to retire and dispose the systems after the product is terminated. These costs are incurred before the beginning of production and at the end of production, that is, retirement (disposal) stages in the life cycle of a product. The early costs incurred to reach operational status of the program include product design, development, and refinement costs. The costs include personnel costs (salaries and benefits) of the design team as well as for the development of models, prototypes, market research, verification tests, tools and fixtures design and build, plant and facilities building, equipment/tooling installation, and prove-out. These nonrecurring costs do not vary as a function of the quantity of products produced. Thus, they are also referred to as the "fixed costs."

Recurring Costs

These costs continue to occur and recur throughout the production, sales, and service/maintenance of the products. These costs include personnel costs of the production and distribution (direct and indirect labor), parts and materials purchases, plant and equipment operation and maintenance, utilities, insurance, rents, taxes, licenses, marketing and sales costs, warranty costs, and so forth. The recurring costs vary as a function of the quantity of products produced. Thus, they are also referred to as the "variable costs."

Figure 9.1 shows two charts. The top chart shows various costs (which have negative values, as they represent money spent or lost) as they incur as functions of time during various life cycle stages of a typical product program for a manufactured product. Figure 9.1 also shows revenues. The revenues have positive values as they represent income. They are only generated after the products are sold. (Note: Revenue = Units sold × Unit price.) The lower chart in Figure 9.1 shows the systems engineering "V" model. The timeline of the "V" model is synchronized with the timeline of the upper cost chart.

For cost management purposes, all the costs (negative values) and revenues (positive values) are added, and the cumulative cash flow is frequently reviewed and compared with the budgeted cash flow (i.e., predicted revenue minus budgeted costs). Two cumulative cash flow curves are presented in Figure 9.2. Let us assume that the two cumulative cash flow curves are for two alternative product programs. The alternative 1 incurs much more costs and also extends over a longer duration in the negative cash flow condition than alternative 2. However, the product in the alternative 1 generates revenue at a much higher rate than alternative 2. Understanding of the cumulative cash flow curves (i.e., their levels and timings) is very important before committing to an alternative.

REVENUES BUILDUP OVER TIME AS THE PRODUCT IS SOLD

As the products are sold, the generated revenues (positive values) are tracked and added to the total costs (negative values). The revenues are also affected by a number of factors such as changes in product volumes due to obsolescence of older products and emergence of new trends in design and technologies, introduction and

Costs and Benefits Considerations and Models

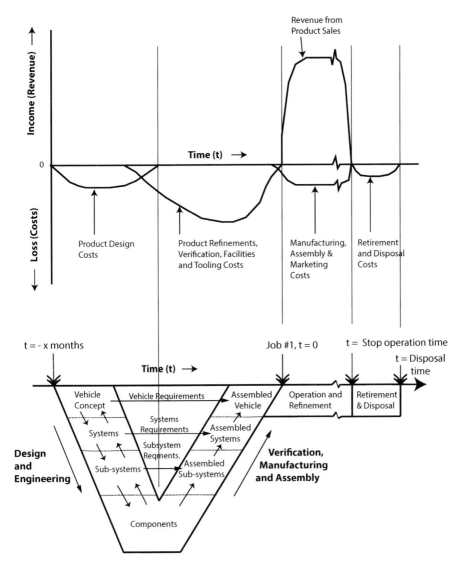

FIGURE 9.1 Costs and revenues in the product life cycle.

availability of new products by the competitors, and changes in economic conditions (e.g., state of economy, interest, inflation, and currency exchange rates).

Make versus Buy Decisions

Most product-producing organizations do not produce all the entities (i.e., systems, subsystems, or components) of the product within their organization. Many of the entities are purchased from other organizations (i.e., suppliers). Typically, standardized components that are common across many similar products are made by

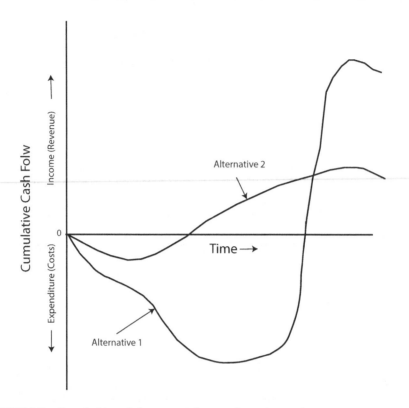

FIGURE 9.2 Cumulative cash flow curves for two alternative product programs.

different organizations. Some examples of standardized components are fasteners (such as nuts, bolts, rivets, clips, and pins), electrical and electronic components (e.g., switches, resistors, transistors, and microprocessors), plumbing supplies (e.g., valves, pipes hoses, and connectors), and so on. Some special components that require unique manufacturing processes and specialized systems, machines, or equipment are also purchased from suppliers with specialized capabilities. For example, major automotive manufacturers typically purchase about 30%–70% of the components (or systems) in the automotive products from their suppliers. The aircraft companies also rely on suppliers to produce most of their components. For example, none of the commercial aircraft manufacturing companies produce jet engines that contribute about 40%–50% of the cost of an airplane. Similarly, specialized systems such as electronic and electrical systems with components such as microprocessors, sensors, actuators, printed circuit boards, and so forth in most complex products are produced by suppliers.

The decision on whether to make or buy an entity depends on many considerations. Some important considerations are given as follows:

1. Availability of in-house manufacturing capability and capacity (e.g., specialized equipment and personnel with unique backgrounds and skills to produce the required product volume)

2. Availability of reliable and low-cost suppliers that can deliver needed volumes of the entities with the required quality requirements
3. Availability of capital required for internal production of the needed entities
4. Need to maintain confidentiality of the competitive information on future product designs or specialized knowledge on some unique processes needed to produce certain entities within the organization to retain competitive advantage.

FIXED VERSUS VARIABLE COSTS

Many organizations organize their total costs into two major categories, namely, fixed costs and variable costs. The fixed costs do not increase or decrease with the output quantity (i.e., production volumes) of products produced. The variable costs are a direct function of the output quantities (i.e., the variable costs increase with an increase in output quantity). The nonrecurring costs are generally treated as the fixed costs and the recurring costs are the variable costs. The cost of any output is the sum of the fixed and the variable costs. The manufacturers should seek to reduce both the fixed and variable costs. However, decreasing the unit cost of an output through increasing product volumes is a much sought-after approach as it spreads the fixed costs over a larger volume. Developing and/or using common components that can be shared across a larger number of products (or models and hence increasing their component volumes) can reduce the total cost of the components substantially. Table 9.1 shows the effect of product volume on three products, namely, A, B, and C. The product cost was computed by using the following simple formula:

$$\text{Product cost} = \left(\frac{\text{Fixed costs}}{\text{Product volume}}\right) + \text{Variable cost per unit}$$

Table 9.1 shows that the unit cost of product A will decrease from $15,000 to $5001.00 as the product volume is increased from 100 units to 1 million units. Similarly, the unit cost of product B can decrease from $105.00 to $5.01 for product volumes of 100 units to 1 million units, respectively. This shows the importance and power of increasing the product volume in reducing the cost of products.

TABLE 9.1
Effect of Product Volume on Product Cost

Product	Fixed Costs	Variable Costs (Unit)	Product Cost ($) Product Volume (Units)					
			10	100	1,000	10,000	100,000	1,000,000
A	$1,000,000.00	$5,000.00	$105,000.00	$15,000.00	$6,000.00	$5,100.00	$5,010.00	$5,001.00
B	$10,000.00	$5.00	$1,005.00	$105.00	$15.00	$6.00	$5.10	$5.01
C	$1,000.00	$2.00	$102.00	$12.00	$3.00	$2.10	$2.01	$2.00

Quality Costs

To ensure that the product being designed will meet the customer needs and satisfy the customers, the organization must perform a number of tasks such as conduct a number of checks, analyses, and evaluations; implement quality control process; honor warranty; and repair or replace failed components. The costs incurred for such tasks can be grouped into the following four categories (Campanella, 1990):

1. *Prevention costs*: These costs are associated with the information gathered and analyses conducted to ensure that the right product is being designed and the product will meet its customer needs. Some examples of the activities involved in this cost category are market research, benchmarking, product performance analyses, design reviews, supplier reviews and ratings, supplier quality planning, training, quality administration, and process validations. Chapters 10 and 17 provide more information on quality issues and techniques.
2. *Appraisal costs*: These costs are related to various appraisals or evaluations conducted to ensure that incoming components and materials and outgoing products will meet quality requirements. Examples of the activities involved in this cost category are purchasing appraisals, maintenance of laboratories with calibrated state-of-the-art testing equipment and trained staff, measurements and tests, inspections, and plant quality audits.
3. *Internal failure costs*: These costs are incurred at the manufacturer's facilities due to product failures during manufacturing, defects observed during testing, troubleshooting and analyzing the failures, rejected and scrapped units (or components), rework, repairs, and so forth.
4. *External failure costs*: These costs are incurred after the product leaves the manufacturer's facilities and is sold to the customers. The costs are due to handling customer complaints, managing returned products, sending replacements, repairing failed products, product recalls, product litigations and liabilities, penalties, lost sales, and so forth.

Manufacturing Costs

The manufacturing costs can be categorized into the following four broad categories:

1. *Costs of parts (components) and subassemblies purchased from the suppliers*: These costs include expenses incurred in purchasing components. Assembled systems and subsystems, standard fasteners from various suppliers.
2. *Costs of parts manufactured internally within the product manufacturer's plants*: These costs are associated with fixed costs for tooling, equipment, and facilities and variable costs associated with purchasing raw materials, expendable tools, processing and operating machines/equipment, inspection, direct labor, coolants, lubricants, utilities, and so on.
3. *Assembly costs*: These include assembly and inspection related to fixed and variable costs of equipment operation (e.g., fixed costs of fixtures and robots

Costs and Benefits Considerations and Models

needed for assembly; variable costs to program and run the assembly robots and/or equipment), direct labor costs, and associated employee benefits.
4. *Overhead costs*: These costs include expenses related to indirect labor (e.g., administrative and plant maintenance personnel and costs of their benefits), employee training, utilities, insurance, property taxes, equipment dismantling, and so on.

It should be noted that all of the preceding four categories have fixed and variable cost components.

Safety Costs

The safety-related costs can be categorized into the following four broad categories (also see Chapter 12):

1. *Accident prevention costs*: These costs represent amounts spent by the organization to avoid or prevent accidents, injuries and adverse health effects from occurring. The accident prevention activities typically include safety analyses (e.g., conducting hazard analyses and failure modes and effects analyses); incorporating engineering changes (e.g., process and equipment improvements to reduce probability of accidents, adding safety devices); conducting safety evaluations/tests and safety reviews; providing safety training to employees; providing/installing and maintaining injury and health protection devices (e.g., hard hats, safety glasses, lockout devices, anti-slip walking surfaces, and providing lifting devices to reduce back injuries). Chapters 12 and 19 provide more information on the accident prevention issues and techniques.
2. *Costs due to accidents*: These costs include losses that an organization incurs due to accidents. The accidents can involve injuries (e.g., medical costs, temporary disability-related costs until an injured person returns to his/her regular job, and costs due to permanent disability), loss of life, damage to facilities and equipment, and/or work stoppages. It should be noted that accident costs are almost always underestimated due to many unreported or unaccounted costs (e.g., loss of production or temporary work slowdowns due to accidents, retraining of replacement workers). In some cases, the incidental costs of accidents have been estimated to be four times as great as the directly accounted costs.
3. *Insurance costs*: This category includes costs to insure (i.e., insurance premiums and workers' compensation costs) against losses due to accidents and injuries, fatalities, and property damage (i.e., repairing or replacing damaged equipment).
4. *Product liability costs*: These are costs incurred in the product liability cases resulting from injuries caused by the product due to defects in the products. These costs include costs to defend cases (e.g., fees charged by lawyers and experts), and compensation or settlement charges paid to the plaintiff, penalties, and fines. (See Chapter 12 for more information on product liability.)

Product Termination Costs

These costs are incurred after the decision is made to terminate the production of the product. These costs include the following:

1. Costs of selling discontinued products at discounted prices or with sales incentives
2. Costs of lost sales of new products due to some customers purchasing the discontinued products at the discounted prices
3. Plant and equipment write-down costs
4. Plant shutdown, equipment removal, and disposal costs
5. Environmental cleanup and site restoration costs
6. Materials recycling costs
7. Continual service, production, and distribution of spare parts for products in service until they are disposed.

Total Life Cycle Costs

These costs include a total of all the aforementioned costs from product conception to end of production and disposal (or recycling) of all products from service and facilities closing. (See Chapter 21 for more details)

EFFECT OF TIME ON COSTS

As the costs are incurred over time, in determining all the aforementioned costs, the effect of time due to factors such as interest rate, inflation rate, and fluctuations in currency exchange rates (if applicable) must be taken into account. Similarly, since the revenues are generated over the selling periods of the products and payments are received over time, the effects of changes in interest rates, inflation, and currency exchange rates should also be considered.

Most complex product programs extend over many years. Therefore, cost computations need to consider the effects of interest and inflation. The computations can be made by using the following variables and the formula.

Let P = value at a time assumed to be the present (called the present value)

i = combined annual interest and inflation rate = $i_r + i_f$
i_r = annual interest rate
i_f = annual inflation rate
n = number of annual interest periods
F = future value after n periods.

With the annual compounding of the combined interest and inflation, the relationship between P and F is as follows (Blanchard and Fabrycky, 2011):

$$F = P(1+i)^n \text{ or } P = F\left\{\frac{1}{(1+i)^n}\right\}$$

Using the preceding formula, the value of $100 today will be $128 in 5 years at 5% combined annual interest and inflation rate. (Note: 128 = 100 (1 + 0.05)^5) This means that $128 spent 5 years from now will be equivalent to $100 today assuming 5% rate of combined interest and inflation.

For a program extending over many periods, the present value of revenues minus the present value of costs can be computed for each period; and the net present values for each period can be summed over the entire duration of the program to obtain the present value of the cumulative cash flow. The present value is generally computed at the beginning of the product program to provide the management the estimate of cash flow over the life of the program.

BENEFITS ESTIMATION

Benefits include income that an organization earns by selling its products and services, investment income, and other incentives it receives through tax breaks, rebates, and special arrangements through state or local governments.

The revenues are the income from selling products and services which are computed by multiplying units sold by the selling price of the unit. The investment income is the interest and dividends received and appreciated value of investments made by the product-producing organization. Incentives typically include reduction in local, state, or federal taxes, special lower interest rates received through government agencies for locating facilities (e.g., a production or assembly plant) of the organization. Other examples of incentives are the value of power buyback provisions paid by the states to power companies using renewable energy, energy-efficient equipment, or reduction of pollutants.

PROJECT FINANCIAL PLAN

AN EXAMPLE: AUTOMOTIVE PRODUCT PROGRAM CASH FLOW

This section presents a simplified cash flow analysis of an automotive product program. The analysis covers a 64-month period—from 40 months before Job #1 to 24 months after Job #1. In the automotive industry, Job #1 represents the time at which first production vehicle rolls out of the assembly plant.

The assumptions used for the costs and revenue computations were as follows:
Program milestones:

1. Program kicks off at –40 months (–40 months represents 40 months before Job #1)
2. Product development team formation begins at –39.5 months
3. Strategic intent confirmation at –34 months
4. Hard-points freeze at –29 months
5. Feasibility sign-off at –27.5 months
6. Program approval at –26 months
7. Surface freeze at –24 months
8. Appearance approval at –19 months

9. Early prototype vehicles available for testing at −14 months
10. Early production prototype vehicles available for testing at −9 months
11. Final prototype vehicles available at −5 months
12. Final sign-off at −2 months
13. Job #1 at 0 month
14. Postproduction at 24 months after Job #1.

The costs estimates used for the preceding illustration are provided in Table 9.1.

Head count cost (i.e., salary plus benefits) = $8000/month
Manufacturing personnel cost = $30/h
Manufacturing plant operation = 2 shifts/day, 25 days/month
Purchased parts and plant overhead costs = $8000/vehicle
Marketing cost = $2000/vehicle
Vehicle sale price = $24,000.

Figure 9.3 presents the cumulative cash flow curve for the vehicle program. The cumulative cash curve was obtained by summing all the costs (negative values) and revenues from product sales (positive values). The spreadsheet used to compute the costs can be downloaded from the website of this book (see: http://www.crcpres) for a file called "Program Cost Flow by Months"). Table 9.2 provides an output of the program. It should be noted that the cost and the revenue values in Figure 9.3 and Table 9.2 were considered without any effects of interest and inflation.

Table 9.2 shows that the point of maximum cumulative expenditure occurred at Job #1 (at month 0). The maximum cumulative expenditure in the program was $1.64 billion at Job #1. The maximum cash flow at 24 months after Job #1 was $6.16 billion.

Another useful Excel spreadsheet for computation of program costs and revenues by quarters is also provided in the website of this book.

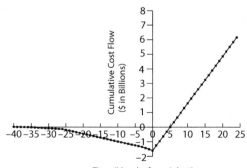

FIGURE 9.3 Cumulative cash flow in an automotive product program.

TABLE 9.2
Costs, Revenue, and Cash Flow in an Automotive Product Program

Months before Job #1	Product Development Headcount	Prod. Dev. Manpower Costs	Services and Supplies Costs	Facilities and Tooling Costs	Product Development Costs Subtotal	Mfg. Headcount	Mfg. Manpower Costs	Number of Vehicles Produced	Parts, Materials & Overhead Costs	Manufacturing Costs Subtotal	Revenue from Vehicle Sales Minus Marketing Costs	Cumulative Cash Flow
−40	50	$400,000	$140,000		$540,000							−$540,000
−39	100	$800,000	$280,000		$1,080,000							−$1,620,000
−38	200	$1,600,000	$560,000		$2,160,000							−$3,780,000
−37	500	$4,000,000	$1,400,000		$5,400,000							−$9,180,000
−36	800	$6,400,000	$2,240,000		$8,640,000							−$17,820,000
−35	1000	$8,000,000	$2,800,000		$10,800,000							−$28,620,000
−34	1000	$8,000,000	$2,800,000		$10,800,000							−$39,420,000
−33	1200	$9,600,000	$3,360,000		$12,960,000							−$52,380,000
−32	1200	$9,600,000	$3,360,000		$12,960,000							−$65,340,000
−31	1200	$9,600,000	$3,360,000		$12,960,000							−$78,300,000
−30	1200	$9,600,000	$3,360,000		$12,960,000							−$91,260,000
−29	1200	$9,600,000	$3,360,000		$12,960,000							−$104,220,000
−28	1200	$9,600,000	$3,360,000		$12,960,000							−$117,180,000
−27	1200	$9,600,000	$3,360,000		$12,960,000							−$130,140,000
−26	1200	$9,600,000	$3,360,000		$12,960,000							−$143,100,000
−25	1200	$9,600,000	$3,360,000	$40,000,000	$52,960,000							−$196,060,000
−24	1200	$9,600,000	$3,360,000	$40,000,000	$52,960,000							−$249,020,000
−23	1200	$9,600,000	$3,360,000	$40,000,000	$52,960,000							−$301,980,000
−22	1200	$9,600,000	$3,360,000	$40,000,000	$52,960,000							−$354,940,000

(Continued)

TABLE 9.2 (Continued)

Months before Job #1	Product Development Headcount	Prod. Dev. Manpower Costs	Services and Supplies Costs	Facilities and Tooling Costs	Product Development Costs Subtotal	Mfg. Headcount	Mfg. Manpower Costs	Number of Vehicles Produced	Parts, Materials & Overhead Costs	Manufacturing Costs Subtotal	Revenue from Vehicle Sales Minus Marketing Costs	Cumulative Cash Flow
-21	1200	$9,600,000	$3,360,000	$40,000,000	$52,960,000							-$407,900,000
-20	1200	$9,600,000	$3,360,000	$40,000,000	$52,960,000							-$460,860,000
-19	1200	$9,600,000	$3,360,000	$40,000,000	$52,960,000							-$513,820,000
-18	1200	$9,600,000	$3,360,000	$40,000,000	$52,960,000							-$566,780,000
-17	1200	$9,600,000	$3,360,000	$40,000,000	$52,960,000							-$619,740,000
-16	1200	$9,600,000	$3,360,000	$40,000,000	$52,960,000							-$672,700,000
-15	1200	$9,600,000	$3,360,000	$40,000,000	$52,960,000							-$725,660,000
-14	1200	$9,600,000	$3,360,000	$40,000,000	$52,960,000							-$778,620,000
-13	1200	$9,600,000	$3,360,000	$40,000,000	$52,960,000							-$831,580,000
-12	1200	$9,600,000	$3,360,000	$40,000,000	$52,960,000							-$884,540,000
-11	1200	$9,600,000	$3,360,000	$40,000,000	$52,960,000							-$937,500,000
-10	1200	$9,600,000	$3,360,000	$40,000,000	$52,960,000							-$990,460,000
-9	1200	$9,600,000	$3,360,000	$40,000,000	$52,960,000							-$1,043,420,000
-8	1200	$9,600,000	$3,360,000	$40,000,000	$52,960,000							-$1,096,380,000
-7	1200	$9,600,000	$3,360,000	$40,000,000	$52,960,000							-$1,149,340,000
-6	1000	$8,000,000	$2,800,000	$40,000,000	$50,800,000							-$1,200,140,000
-5	1000	$8,000,000	$2,800,000	$40,000,000	$50,800,000							-$1,250,940,000
-4	800	$6,400,000	$2,240,000	$40,000,000	$48,640,000							-$1,299,580,000
-3	600	$4,800,000	$1,680,000	$40,000,000	$46,480,000	500	$6,000,000	1000	$8,000,000	$60,480,000		-$1,360,060,000
-2	400	$3,200,000	$1,120,000	$40,000,000	$44,320,000	600	$7,200,000	4000	$32,000,000	$83,520,000		-$1,443,580,000

-1	300	$2,400,000	$840,000	$3,240,000	800	$9,600,000	8000	$64,000,000	$76,840,000		-$1,520,420,000
0	200	$1,600,000	$560,000	$2,160,000	1000	$12,000,000	10000	$80,000,000	$94,160,000		-$1,614,580,000
1					1000	$12,000,000	24000	$192,000,000	$204,000,000	$528,000,000	-$1,290,580,000
2					1000	$12,000,000	24000	$192,000,000	$204,000,000	$528,000,000	-$966,580,000
3					1000	$12,000,000	24000	$192,000,000	$204,000,000	$528,000,000	-$642,580,000
4					1000	$12,000,000	24000	$192,000,000	$204,000,000	$528,000,000	-$318,580,000
5					1000	$12,000,000	24000	$192,000,000	$204,000,000	$528,000,000	$5,420,000
6					1000	$12,000,000	24000	$192,000,000	$204,000,000	$528,000,000	$329,420,000
7					1000	$12,000,000	24000	$192,000,000	$204,000,000	$528,000,000	$653,420,000
8					1000	$12,000,000	24000	$192,000,000	$204,000,000	$528,000,000	$977,420,000
9					1000	$12,000,000	24000	$192,000,000	$204,000,000	$528,000,000	$1,301,420,000
10					1000	$12,000,000	24000	$192,000,000	$204,000,000	$528,000,000	$1,625,420,000
11					1000	$12,000,000	24000	$192,000,000	$204,000,000	$528,000,000	$1,949,420,000
12					1000	$12,000,000	24000	$192,000,000	$204,000,000	$528,000,000	$2,273,420,000
13					1000	$12,000,000	24000	$192,000,000	$204,000,000	$528,000,000	$2,597,420,000
14					1000	$12,000,000	24000	$192,000,000	$204,000,000	$528,000,000	$2,921,420,000
15					1000	$12,000,000	24000	$192,000,000	$204,000,000	$528,000,000	$3,245,420,000
16					1000	$12,000,000	24000	$192,000,000	$204,000,000	$528,000,000	$3,569,420,000
17					1000	$12,000,000	24000	$192,000,000	$204,000,000	$528,000,000	$3,893,420,000
18					1000	$12,000,000	24000	$192,000,000	$204,000,000	$528,000,000	$4,217,420,000
19					1000	$12,000,000	24000	$192,000,000	$204,000,000	$528,000,000	$4,541,420,000
20					1000	$12,000,000	24000	$192,000,000	$204,000,000	$528,000,000	$4,865,420,000
21					1000	$12,000,000	24000	$192,000,000	$204,000,000	$528,000,000	$5,189,420,000
22					1000	$12,000,000	24000	$192,000,000	$204,000,000	$528,000,000	$5,513,420,000
23					1000	$12,000,000	24000	$192,000,000	$204,000,000	$528,000,000	$5,837,420,000
24					1000	$12,000,000	24000	$192,000,000	$204,000,000	$528,000,000	$6,161,420,000

EFFECT OF INTEREST AND/OR INFLATION

The cost curve presented in Figure 9.3 when converted into the present value at the beginning of the program at 10% combined annual interest and inflation rate will be lower. Figure 9.4 shows both curves, first curve (the curve in Figure 9.3) that assumes zero interest and zero inflation and the second curve with 10% combined annual effects of interest and inflation. The second curve was obtained by summing the present value of revenues minus the present value of the costs for each month (period) from the beginning of the program. The website of this book (http://www.crcpress) contains a downloadable Excel spreadsheet illustrating a "V" model of an automotive product development. It shows a Gantt chart with various activities and cost estimates by month (shown in Figures 9.3 and 9.4).

Figure 9.4 thus shows that depending on the magnitude of the interest and inflection rates, their effects can introduce very significant differences in the estimated costs and revenues of the program. The effect will be different at a different base period at which the present value of the program will be computed.

PRODUCT PRICING APPROACHES

TRADITIONAL COSTS-PLUS APPROACH

The traditional approach in determining the product price is to add all the costs per unit (of product) and required profit per unit to come up with the price for the unit. This approach generally does not provide strong incentives to reduce costs as the profits for the manufacturer are assured. The approach also assumes that customers are willing to pay the price (i.e., it is the producer's market—the producer sets the price without regard to the customers). This approach has worked well in the past when the customers had a very limited number of choices in the market for selecting a product.

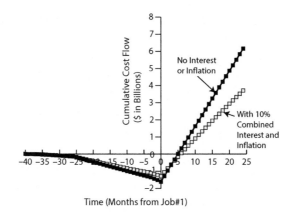

FIGURE 9.4 Illustration of effect of 10% combined interest and inflation on cash flow.

MARKET PRICE-MINUS PROFIT APPROACH

In this approach, the producer determines the lowest price based on the prices of other similar products sold in the markets, then subtracts his dealer margin and expected profit, and the balance is considered as the target cost for producing the product. The target cost is then divided and assigned to each entity in the product. And all internal and external suppliers are asked to meet their respective target costs by improving the product design, manufacturing processes, and operations. For example, in determining the price of a low-cost vehicle for the U.S. market, Hussain and Randive (2010) surveyed the prices of low-cost vehicles sold in the U.S. market. They found that the lowest price of small economy vehicles sold in the U.S. market during 2010–2011 was about $10,000. Thus, they set the target manufacturer's retail price of $8000 (20% below the lowest price sold in the U.S. market). Assuming the dealer margin of 10% ($800) and the manufacturer's profit of $200 (2.78% of factory cost), they set the target cost at $7000 per vehicle and then proceeded to develop target cost for each vehicle system (see Figure 9.5). This assumes that they challenged their suppliers to deliver the systems at the target costs. This approach was also used during the development of the Tata Nano, the lowest-cost vehicle sold for about INR 100,000 (USD 2000) in India (Hussain and Randive, 2011).

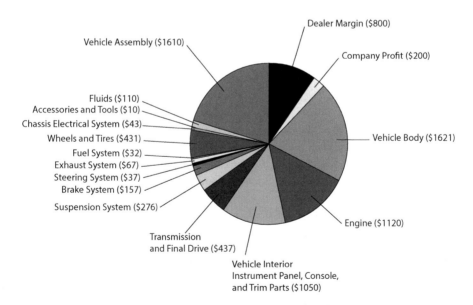

FIGURE 9.5 Low-cost vehicle target cost breakdown. (Data from Hussain, T., S. Randive, *Defining a Low Cost Vehicle for the U.S. Market*, Institute for Advanced Vehicle Systems, College of Engineering and Computer Science, the University of Michigan-Dearborn, Dearborn, MI, 2010.)

SOFTWARE APPLICATIONS

Many different software applications are available to perform product life cycle costing and to create various reports (e.g., by systems, program phases, months; comparisons with budgeted costs). Many of the applications are integrated with other functions such as management information systems, product planning, and supply chain management. The software systems are used for production scheduling, component ordering, inventory control, product control, shop floor management, cost accounting, and so forth. Some examples of such software systems are manufacturing resource planning (MRP) and enterprise resource planning (ERP). The software systems are available from a number of developers (e.g., SAP, Oracle, Microsoft, EPICOR, and Sage).

TRADE-OFFS AND RISKS

Programs and projects involving development of complex products encounter a number of developmental problems and challenges. Many problems involve trade-offs between different attribute requirements and trade-offs between a number of design and manufacturing issues. The costs and timings are directly affected by how the trade-off issues are resolved (see Chapter 3). The design teams deal with these issues constantly during various design stages. Many of these problems are not sufficiently known in the early stages and hence the budgets prepared during the early stages need to be constantly reviewed and some changes in target costs and timings may need to be incorporated in subsequent budgets and milestones in the program.

CONCLUDING REMARKS

Bringing the product to the market at the right time and right price are both very important. Therefore, costs and timings are important parameters used by the program and project managers to evaluate and control their progress. Both these parameters affect the profitability of the organization and its competitive position in the market. Since the initial estimates of these parameters made during the early planning stages are usually not very accurate, they need to be adjusted to account for problems and challenges encountered during the program. Costs and time overruns are universally hated by the management. On the contrary, completions of the program before its planned end date and under its budget are reasons for celebrations of the great accomplishments and deserve special recognition of the program teams.

REFERENCES

Blanchard, B. S. and W. J. Fabrycky. 2011. *Systems Engineering and Analysis*. Fifth Edition. Upper Saddle River, NJ: Prentice Hall.

Campanella, J. 1990. *Principles of Quality Costs*. Second Edition. Milwaukee, WI: ASQC Quality Press.

Hussain, T. and S. Randive. 2010. *Defining a Low Cost Vehicle for the U.S. Market*. Dearborn, MI: Institute for Advanced Vehicle Systems, College of Engineering and Computer Science, the University of Michigan-Dearborn. http://www.engin.umd.umich.edu/IAVS/books/A_Low_Cost_Vehicle_Concept_for_the_U.S._Market.pdf (accessed August 3, 2012).

Part II

Quality, Human Factors, Safety, and Sustainability Approaches

10 Quality Management and Six-Sigma Initiatives

INTRODUCTION

Quality should be built into the product very early in its design stage in a proactive manner than reacting to any defects that occur after the product is launched in the market. This proactive approach is called "Creating Quality" (Kolarik, 1995). Thus, all product design teams should strive to create quality by designing the product that the customer would want. The Systems Engineering (SE) process discussed earlier (see Chapters 1 and 2) clearly implements this approach.

In the Quality Engineering and the Quality Management fields, there are following well-developed approaches to implement quality in products and processes: (1) Total Quality Management (TQM), (2) ISO 9000 Series Standards for organizations, (3) Criteria for Malcolm Baldridge Quality Award (performance criteria to evaluate organizations based on their quality-related accomplishments), (4) Six-Sigma Improvement Methodology (using DMAIC, i.e., *D*efine, *M*easure, *A*nalyze, *I*mprove, and *C*ontrol approach), and (5) Design for Six-Sigma (DFSS) (using IDOV, i.e., *I*dentify, *D*esign, *O*ptimize, and *V*erify approach). The above five approaches are complementary in their implementation in the sense that an organization seeking to improve quality can implement anyone or more of the approaches.

The objectives of this chapter are to provide general background into the above quality approaches and to discuss their importance in creating quality products.

DEFINITION OF QUALITY

Quality is determined by the customers. Therefore, there is no unique definition for quality. The customer needs and customer perceptions of a product can vary widely. And thus, various definitions can be used to define quality. Kolarik (1995) provides an excellent discussion on classical definitions of quality. Some commonly referred definitions of quality and basic considerations in quality are given below.

1. Quality is an inherent feature, a peculiar, identifying, or essential characteristic, a degree of excellence that constitutes the basic nature of a thing or one of its distinguishing attributes (Webster's, 1980).
2. Quality is the totality of features and characteristics of a product or service that bear on its ability to satisfy stated or implied needs now and in the future (ANSI/ASQC, 1987).
3. Quality is customer satisfaction and loyalty. Quality is customer focused (Gryna et al., 2007).
4. Quality is fitness for use, meeting and exceeding customer expectations (Juran and Godfrey, 1999).

5. Quality is conformance to requirements (clearly stated) (Crosby, 1979).
6. Quality should be aimed at the needs of the consumer, present and future (Deming, 1986).
7. Quality $(Q) = P/E$; Where $P = $ Performance of the product, and $E = $ Expectations of the customers about the product.
8. A widely quoted definition of product quality was provided by Garvin (1984, 1988). He suggested that the quality of a manufactured product should be based on the following eight dimensions:
   a. Performance: Product's ability to perform its basic functions
   b. Features: Product features that add to its usefulness, comfort, and convenience
   c. Reliability: Product's ability to perform without failure over time
   d. Conformance: Degree to which the product meets its applicable codes (or regulations/standards) of state or community
   e. Durability: The length of time the product will last before it is retired from service.
   f. Serviceability: Ability to make repairs quickly and at reasonable costs
   g. Aesthetics: The look, feel, and design/styling or appeal of the product
   h. Perceived quality: The impression of quality created by the product in the customer's mind.
9. Other dimensions of quality suggested in the literature are perfection, consistency, eliminating waste, speed of delivery, compliance with policies and procedures, providing a good usable product, "Doing it right the first time," delighting or pleasing the customers, and total customer service and satisfaction.

KEY CONCEPTS IN QUALITY MANAGEMENT

QUALITY GURUS AND THEIR FINDINGS

Many practitioners and researchers in the Quality Management field have provided useful insights into understanding quality-related issues and concepts. Some important observations from the literature are provided below.

1. Quality is a business issue. It is no longer only the quality manager's job. Quality relates to everything in an organization—all individuals and all processes must be involved. (Thus, all individuals and all processes cover "T" for "Total" in TQM.) Quality management demands specialized knowledge, training, and applications of special tools. Implementation of the TQM approach leads to successful organizations. Organization must have a well-developed Quality Management System (QMS) and it must be followed by all in the organization (Kolarik, 1995; Gryna et al., 2007; ISO, 2012).
2. About 85% of poor quality is attributable to the system (management) and only 15% is attributed to the worker (Deming, 1986). Thus, the management must take an active role in creating quality.

3. Understand the purpose of inspection. Inspection is for improvement of processes and reduction of cost (Deming, 1986). (It is not just to scrap nonconforming parts.)
4. Juran's trilogy showed that (1) quality planning, (2) quality control, and (3) quality improvements are three interrelated processes in quality management (Gryna et al., 2007 Juran and Godfrey, 1999). First, the quality actions must be planned through understanding customer needs. Second, the processes that produce products must be controlled to ensure that they deliver what is important to the customers. And third, improvements must be made from the measurements (data) obtained during the quality control processes.
5. The system for causing quality is prevention of defects, not appraisal. The performance standard is zero defects. Quality is free (Crosby, 1979)—when it is implemented early during the product or process development phases.
6. Systems approach must be implemented to improve quality (Feigenbaum, 2004). Quality problems must be approached by studying the whole system; that is, all aspects of the system affecting the customer satisfaction must be studied.
7. Product and process design should be immune to uncontrollable factors; that is, products should be designed by considering "robustness." (Performance of "Robust" products is not affected by factors that we cannot control.) Quality must be designed into the product (Taguchi, 1986; Phadke, 1989).
8. The degree of match between the "true" and "substitute" quality characteristics ultimately determines customer satisfaction (Ishikawa, 1985). A measure (variable) used to evaluate the quality (labeled as "substitute" quality characteristic by Ishikawa) must be selected carefully to ensure that it is truly related to the customer's perception of quality.

PRODUCT QUALITY MEASUREMENTS

Many possible measures (variables) can be used to measure product quality and its attributes. Since the ultimate evaluator of product quality is the customer, most measures of quality are subjective (i.e., they are based on the judgments or ratings made by the customers based on their exposure/experience with the product). Objective measures based on measurements of certain physical characteristics of products (such as dimensions, temperature, airflow, surface finish or roughness, compressibility, and power consumption) can also be used if such measures can be related to the perception of quality or customer satisfaction. (Such measures are used as the "substitute" quality characteristics [Ishikawa, 1985]; see point 8 in the earlier list.)

Some commonly used measures of quality of automotive products are as follows:

1. Percentage of customers satisfied with the vehicle (product)
2. Number of defects observed by the customers per vehicle
3. Number of things gone wrong (TGWs) per 100 vehicles sold
4. Number of things gone right (TGRs) per 100 vehicles sold
5. Number of products returned for repair

6. Ratings on customer satisfaction or how well they liked or disliked the product (e.g., using a 10-point scale, where 10 = liked very much and 1 = disliked very much).

The counts of TGW and TGR can be obtained from customer feedback from mail surveys, dealer received complaints, repair/warranty data, and market research clinics, where owners provided face-to-face feedback on "what they liked or disliked" about the product and "what didn't work well."

Other product quality measures that are used are based on measurements of physical product characteristics, product functions, and product performance under various specified operating conditions. Some examples of physical measures used are (1) values of engineering parameters, for example, use of a CMM (coordinate measurement machines) to measure critical dimensions of parts, (2) performance measures that can be measured by test equipment (e.g., use of engine dynamometers to measure engine output torque vs. speed relationships), and (3) recordings using sensors/meters and/or video equipment (e.g., strength measurements, sound/noise measurements, temperature measurements, fluid flow and velocity measurements, and deceleration levels of human dummies in vehicle crash tests).

Cost-based measures of quality are also used in the industry. They are based on the following four types of quality-related costs: (1) prevention costs (e.g., costs related to design reviews, training, and analyses), (2) appraisal costs (e.g., inspection, testing, and auditing costs), (3) internal failure costs (e.g., scrap, rework, and repair related costs when the product is in the manufacturer's facilities), and (4) external failure costs (e.g., returned products, handling customer complaints, and re-inspection/repair related costs when the product is used by the customers) (see Chapter 8 and Campanella [1990]).

Other performance measures of quality of an organization can be based on (1) ISO 9001 certification (quality can be inferred based on whether the product-producing organization is certified or not), (2) findings during quality audits, and (3) scores on the Baldridge award criteria (NIST, 2013). (Note: ISO 9000 standards and Baldridge award criteria are covered in a later section of this chapter.)

CUSTOMER SATISFACTION AND THE KANO MODEL OF QUALITY

Customer satisfaction with a product is dependent on provision of various features related to different attributes of the product. Figure 10.1 presents three figures to explain the relationship of the product features to customer satisfaction. The abscissa of the figures represents how a desired product attribute is achieved in terms of percent achieved (i.e., from dysfunctional to functional implementation of the product attribute). The ordinate of the figures represents customer satisfaction in terms of the percentage of customers satisfied.

Certain features are "basic." The customer expects these features to be standard (i.e., always expected or available in the product –"it must be designed that way"). These are considered as "unspoken wants." Absence of such basic features will dissatisfy the customers. However, provision of these features will not increase the percentage of satisfied customers beyond some level (e.g., 50%; see Figure 10.1a).

Quality Management and Six-Sigma Initiatives

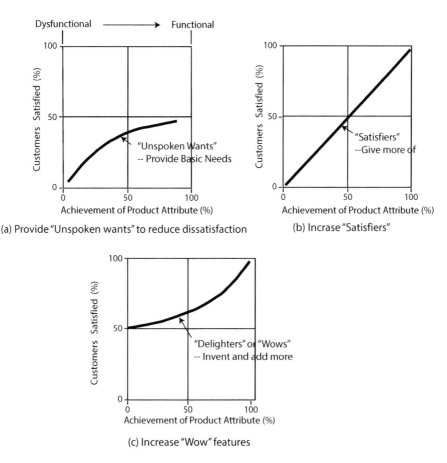

FIGURE 10.1 Three types of features to increase the percentage of satisfied customers. (a) Provide "Unspoken wants" to reduce dissatisfaction. (b) Increase "Satisfiers." (c) Increase "Wow" features.

However, customers like and want more of certain other types of features. Thus, the number of such features should be increased to improve customer satisfaction (i.e., "give me more of such features"). The curve for such features is shown in Figure 10.1b and labeled as the "satisfiers." And the third types of features are the ones that create impressions of "Wow" because the customers have never expected or seen such features in such a product, and they are excited and delighted to have these features. The curve for such features is shown in Figure 10.1c.

These three types of features described above are the three basic components of the Kano Model of Quality. Kano, a Japanese quality researcher, conceptualized that customer satisfaction is affected by three types of product features, namely: (1) the removal of the "dissatisfiers" (i.e., providing the "unspoken wants," otherwise dissatisfaction arises if the product does not provide what the customer expects), (2) increasing "satisfiers" (giving more of these features increases satisfaction with the product), and (3) adding "delighters" (that create "Wows" when present but do not cause dissatisfaction if not incorporated in the product) (Yang and El-Haik, 2003).

A careful study of all the three types of product features (either available in the market currently or new features that could be developed by studying new design and technological trends) and their incorporation in the product are essential to enhance product quality and customer satisfaction. These three types of features should be considered during the early stages of product development and should be included during the development of the product specifications and attribute requirements.

QUALITY INITIATIVES

TOTAL QUALITY MANAGEMENT

Total quality management (TQM) is defined as both a philosophy and a set of guiding principles that represent the foundation of a continuously improving organization. The purpose of the TQM is to provide quality products and/or services to its customers. Quality products, in turn, will increase the organization's productivity and lower its costs (Besterfield et al., 2003).

The guiding principles of the TQM involve the following steps:

1. Continuously improving organization
2. Application of quantitative methods (covered in Chapter 17)
3. Application of human resource and training
4. Improving all processes within an organization
5. Exceeding customer needs now and in the future.

The TQM is also based on the following six basic concepts:

1. Committed and involved management
2. Unwavering focus on customers
3. Effective involvement and utilization of the entire workforce
4. Continuous improvement of all business and production processes
5. Treating suppliers as partners
6. Establishing performance measures for processes.

The letter "T" in the TQM stands for "Total" which means that everyone in the organization, all functions, and all processes must be subject to application of quality principles such as (1) continuously improving all operations performed in the organization, (2) customer focus in all operations performed by all in the organization, and (3) application of quality methods in improving all processes. The second letter "Q" stands for quality that means "customer satisfaction." Thus, everyone in the organization must understand their roles in how to satisfy their customers and strive continuously to improve customer satisfaction. The third letter "M" in the TQM stands for management, which means that quality must be actively managed as a business process by collecting data on the performance of all of their processes, applying quality tools, and monitoring all processes to improve customer satisfaction continuously.

ISO 9000

The ISO 9000 is a series of Quality Standards developed by the International Organization for Standards (ISO) to facilitate international trade by establishing a common set of standards. The standards identify the key elements of a quality system and require that organizations develop a quality system and follow the system to improve the quality of their products and services.

The quality system comprised all the organization's policies, procedures, plans, resources, processes, and delineation of responsibility and authority, all deliberately aimed at achieving product or service quality levels consistent with customer satisfaction and the organization's objectives. When these policies, procedures, plans, and others are taken together, they define how the organization works, and how quality is managed (Goetsch and Stanley, 1998).

Organizations find that they have to achieve the ISO 9000 certification to meet both domestic and international competition. The revised ISO 9000:2000, 9001:2008, and 9004:2009 documents define the eight-quality management principles on which the QMS standards of the series are based (ISO, 2012). These principles can be used by senior management as a framework to guide their organizations toward improved performance.

Brief descriptions of the eight principles as provided by ISO (2012) are presented:

1. *Customer focus*: Organizations depend on their customers and therefore should understand current and future customer needs, should meet customer requirements, and strive to exceed customer expectations.
2. *Leadership*: Leaders establish unity of purpose and direction of the organization. They should create and maintain the internal environment in which people can become fully involved in achieving the organization's objectives.
3. *Involvement of people*: People at all levels are the essence of an organization and their full involvement enables their abilities to be used for the organization's benefit.
4. *Process approach*: A desired result is achieved more efficiently when activities and related resources are managed as a process.
5. *System approach to management*: Identifying, understanding, and managing interrelated processes as a system contributes to the organization's effectiveness and efficiency in achieving its objectives.
6. *Continual improvement*: Continual improvement of the organization's overall performance should be a permanent objective of the organization.
7. *Factual approach to decision making*: Effective decisions are based on the analysis of data and information.
8. *Mutually beneficial supplier relationships*: An organization and its suppliers are interdependent, and a mutually beneficial relationship enhances the ability of both to create value.

MALCOLM BALDRIDGE AWARD CRITERIA

Malcolm Baldrige award is administered by the National Institute of Standards and Technology (NIST) of the U.S. Department of Commerce and given annually by the President of the United States to small and large businesses in manufacturing and service, and to education and healthcare organizations that apply and are judged to be outstanding in the following seven areas: (1) leadership; (2) strategic planning; (3) customer focus; (4) measurements, analysis, and knowledge management; (5) workforce focus; (6) process management; and (7) results (NIST, 2013).

The criteria are used by thousands of organizations of all kinds for self-assessment and training and as a tool to develop performance and business processes. Many organizations have found that using the criteria results in better employee relations, higher productivity, greater customer satisfaction, increased market share, and improved profitability.

The purpose of the award is to help improve quality and productivity of the U.S. businesses and industries by: (1) stimulating American companies to improve quality and productivity, (2) recognizing the achievements of those companies that improve the quality of their goods and services and providing them as examples to others, (3) establishing guidelines and criteria that can be used by business, industrial, governmental, and other organizations in evaluating their own quality improvement efforts, and (4) providing specific guidance for other American organizations that wish to learn how to manage for high quality by making available detailed information on how winning organizations were able to change their cultures and achieve eminence (NIST, 2013).

SIX-SIGMA METHODOLOGIES

Six-Sigma is a systematic methodology for creating sustained improvements to six-sigma levels in important production and other repetitive processes. The "Six-Sigma" term also refers to a philosophy, goal, and/or methodology utilized to drive out waste and improve the quality, cost, and time performance of any business. DFSS is a systematic methodology to design products and processes that will meet customer expectations at a six-sigma level (Creveling et al., 2003; Yang and El-Haik, 2003). The steps and tools used in the six-sigma methodologies are provided in Tables 14.2 and 14.3; and quality tools used in the six-sigma applications are described in detail in Chapter 17.

The six-sigma philosophy is used to reduce the defect rates to very low levels, particularly at or above the six-sigma level. To achieve the six-sigma level, defect rate is equivalent to reducing the variability of each product (or service) parameter (i.e., critical to customer satisfaction) such that its 12 standard deviation ($\pm 6\sigma$) tolerance will be smaller than its customer acceptance region (i.e., the difference between the upper and lower specification levels of the product parameter). This requirement along with a $\pm 1.5\sigma$ range of allowance on the location of the mean value (μ) of the parameter maintains the defect rate below 3.4 defects per million opportunities. This is equivalent to meeting the product parameter acceptance level at 99.999660%. Such a high level of quality (i.e., very low-level defect rates) are particularly important for developing complex products that have many entities (systems, subsystems, and components), many product attributes, many attribute requirements, and many production operations. This issue is further discussed in Chapter 19 and illustrated in Table 19.1 by considering combinations of different levels of reliability and the number of entities or production steps.

DFSS is a systematic methodology to design products and processes that should meet customer expectations at the six-sigma level. The four phases in the DFSS approach can be remembered by the acronym IDOV. The phases involve (1) *I*dentify requirements, (2) *D*esign using inventive approaches, (3) *O*ptimize the design, and (4) *V*erify the design. The approaches and methods used in the four phases of the approach are described in Table 14.3.

Phase I of this approach typically involves identification of requirements by (1) forming a cross-functional design team, (2) determining customer expectations (e.g., by application of QFD; see Chapter 15), and (3) determining product functionality based on customer requirements, technological capabilities, and economic realities.

Phase II of the approach involves creating a design by (1) generating design alternatives, (2) focusing on product and process performance issues and variability factors that affect variables that are critical to quality (CTQ), and (3) evaluating design alternatives.

Phase III involves optimizing the design by seeking to minimize the impact of variations in production (e.g. tolerances of critical product dimensions and tolerance stack-up when parts are assembled) and uses (i.e., when the product is used by a variety of users in different usage situations and environmental conditions) and thus, creating a "robust" design (Phadke, 1989) (see Chapter 17 for Taguchi's concepts).

And phase IV involves verifying the design by particular attention to variables that are CTQ by use of various evaluation methods (see Chapter 7). It should be noted that the above process has many similarities with the systems engineering process described in Chapter 2.

OVERVIEW OF TOOLS USED IN QUALITY MANAGEMENT

The tools used to analyze problems in the quality-related fields can be grouped into the following three categories:

1. Seven traditional or basic quality tools (e.g., cause and effect diagram, check sheet, histogram, scatter diagram, stratification, Pareto chart, and control charts). These traditional tools are used primarily in reactive mode (i.e., after the product or its entities are produced) to improve their quality. They are mostly quantitative in nature; that is, they handle quantitative data. Their format allows visualization of quality issues/problems. They are used primarily on the shop floor. Any of the tools can be used with any other tools.
2. Seven new quality tools (i.e., relations diagram, affinity diagram, tree diagram, matrix diagram, matrix data, arrow chart, and process decision program chart). These tools are mainly used in a proactive manner (to create quality during the early stages of product development). They are used

primarily for planning and by management (for brainstorming). They are qualitative in nature and more effective when used by multifunctional teams. They are also visual, that is, they help in visualizing problems, issues, and relationships. Any of the tools can be used with any other tools.
3. Other product/process development tools include Benchmarking, quality function deployment (QFD), Pugh diagram, failure modes and effects analysis (FMEA), fault tree analysis (FTA), reliability analyses, experiment design, computer-aided design, computer-aided engineering, prototyping, computer-assisted process planning, computer-assisted manufacturing, and so forth.

The above quality tools are described in Chapter 17. Other tools (e.g., QFD, FMEA, FTA, and reliability analysis) are covered in Chapters 15 and 19.

CONCLUDING REMARKS

This chapter provided an overview on the definitions quality, concepts considered in quality, and three major approaches used to improve quality, namely, Total Quality Management, ISO 9000 Certification, and Six-Sigma Methodologies. Product quality is a very important area. Measurement of perception of quality is a difficult area because quality is subjective, and the customers can differ widely on what constitutes quality. However, customer-based quality measures are useful in improving customer satisfaction. Many currently used measures of quality were reviewed. The Kano model of quality helps understand how the product features can be categorized into unspoken wants, satisfiers, and delighters and their relationship to customer satisfaction. Many useful tools used in the quality area are presented in Chapter 17.

REFERENCES

ANSI/ASQC. 1987. *Quality Systems Terminology*. American National Standard A3-1987, Washington, DC: American National Standards Institute.
Besterfield, D. H., Besterfield-Michna, C., Besterfield, G. H. and M. Besterfield-Scare. 2003. *Total Quality Management*. Third Edition. Upper Saddle River, NJ: Prentice Hall. ISBN 0-13-099306-9.
Campanella, J. 1990. *Principles of Quality Costs*. Second Edition. Milwaukee, WI: ASQC Quality Press.
Crosby, P. 1979. *Quality Is Free*. New York: McGraw-Hill.
Creveling, C. M., Slutsky, J. L. and D. Antis, Jr. 2003. *Design for Six Sigma—In Technology and Product Development*. Upper Saddle River, NJ: Prentice Hall PTR.
Deming, W. E. 1986. *Out of Crisis*. Cambridge, MA: MIT Center for Advanced Engineering Studies.
Feigenbaum, A. V. 2004. *Total Quality Control*. Fourth Edition, revised. New York: McGraw-Hill.
Garvin, D. A. 1984. What Does Product Quality Really Mean? *Sloan Management Review*, 26: 25–43.
Garvin, D. A. 1988. *Managing Quality: The Strategic and Competitive Edge*. New York: Free Press.
Goetsch, D. L. and D. Stanley. 1998. *Understanding and Implementing ISO 9000 and ISO Standards*. Upper Saddle River, NJ: Prentice Hall.

Gryna, F. W., Chaua, R. C. H. and J. A. DeFeo. 2007. *Juran's Quality Planning & Analysis for Enterprise Quality*. Fifth Edition. New York: McGraw-Hill.

International Organization for Standardization (ISO). 2012. *ISO 9000 Standards*. Website: http://www.iso.org/iso/iso_catalogue/management_and_leadership_standards/quality_management.htm (accessed February 24, 2012).

Ishikawa, K. 1985. *What Is Total Quality Control?* Upper Saddle River, NJ: Prentice-Hall.

Juran, J. M. and A. B. Godfrey. (Eds). 1999. *Juran's Handbook on Quality*. Fifth Edition. New York: McGraw-Hill.

Kolarik, W. J. 1995. *Creating Quality—Concepts, Systems, Strategies, and Tools*. New York, NY: McGraw-Hill.

National Institute of Standards and Technology (NIST). 2013. *The Malcolm Baldrige National Quality Award Program—Criteria for Performance Excellence*. http://www.nist.gov/bald-rige/publications/business_nonprofit_criteria.cfm (accessed: May 2, 2013).

Phadke, M. S. 1989. *Quality Engineering Using Robust Design*. Upper Saddle River, NJ: P T R Prentice-Hall, Inc.

Taguchi, G. 1986. *Introduction to Quality Engineering: Deigning Quality into Products and Processes*. White Plains, NY: Krus International, UNIPUB (Asian Productivity Organization).

Webster's. 1980. *New Collegiate Dictionary*. Springfield, MA: G. & C. Merriam Company.

Yang, K. and B. El-Haik. 2003. *Design for Six Sigma—A Roadmap for Product Development*. New York, NY: McGraw-Hill.

11 Human Factors Engineering in Product Design

INTRODUCTION

People are involved in working on various product issues throughout the life cycle of a product. And implementation of Systems Engineering is about working with people. Teamwork is used in all phases of complex product development. Thus, it is important that all human tasks must be made easier, simpler, and compatible with human abilities and capabilities. Otherwise, people can make errors in performing the tasks, may not be able to perform the tasks as planned, or may even dislike doing the tasks. Human Factors Engineering discipline involves development and application of knowledge about people (i.e., their characteristics, capabilities, and limitations) in designing the products and systems so that the products and systems are easy, comfortable, convenient, and safe to use.

This chapter covers human factors approach in applying important human characteristics, capabilities, and limitations throughout the life cycle of the product. During the product life cycle, people from all different disciplines interact and work together; and the human factors engineers need to work with the team members from other disciplines to provide the necessary information in the right format about people at the right time. For example, during the early product conceptual design, the human factors engineer must ask questions such as: Who would use the product? Would the people be able to use the product as being designed? What problems would they experience with the product? How should the product be modified? Would people enjoy using the product as designed? Later during the production process, questions will arise related to how the workstations and production equipment should be designed and organized to facilitate production tasks performed by human operators in various processing steps such as manufacturing, assembly, sealing, painting, testing, and inspection. Later as the product is being distributed and sold, human tasks—involved in packing, transportation, unpacking, setting up for sale, and in subsequent operations related to its maintenance—all need to be designed for comfort, convenience, and safe operation.

The product–user interface issues such as how would the users and operators orient, position, and use of the product must consider characteristics such as human anthropometric dimensions, biomechanical characteristics related to body joint angles and postures, human abilities to produce forces, and information processing capabilities. Some human factors design considerations include the following: (1) compatibility with human expectations and capabilities, (2) accommodating a large percentage of the user population, (3) reducing learning time in performing a series of tasks, (4) reducing operator workload, (5) minimizing task completion times and errors while interfacing/using the product, and finally, (6) designing the tasks such that people will enjoy in performing them as compared with complaining about and even avoiding boring, monotonous, or difficult tasks.

This chapter covers the following: (1) definition of human factors engineering, (2) importance of human factors engineering, (3) human limitations and capabilities, (4) human performance measurement, and (5) human factors engineer's role and tasks in the product life cycle.

HUMAN FACTORS ENGINEERING

What Is It?

Human Factors Engineering or Ergonomics is a multidisciplinary science involving fields that have information about people (i.e., psychology, anthropometry, biomechanics, anatomy, physiology, psychophysics, and cognitive psychology). It involves studying human characteristics, capabilities, and limitations, and applying this information to design and evaluate the products, equipment, and systems that people use.

The basic goal of Human Factors Engineering is to design equipment that will achieve the best possible fit between the users and the equipment (product) such that the users' safety (freedom from harm, injury, and loss), comfort, convenience, performance, and efficiency (productivity or increasing the ratio of output to input) are improved. Therefore, the human factors engineer's work involves understanding the needs of the users, measuring their behavior and performance in using the products and analyzing the collected data to determine important variables and their levels that affect the user's performance. The collected data are also used to develop models to predict the behavior and performance of users with different characteristics in completing different tasks.

The field of Human Factors Engineering is also called Human Engineering, Ergonomics, Engineering Psychology, Man–Machine Systems, or Human–Machine Interface Design. After World War II, the field of Human Factors emerged in the United States, mainly among the psychologists, to study the equipment and process design problems primarily from the human information processing viewpoint. The field of ergonomics emerged in European countries around 1949 to improve workplaces and jobs in the industries with an emphasis on biomechanical applications. The word "ergonomics," the science of work laws (or the science of applying natural laws to design work), was coined by joining two Greek words "ergon" (work) and "nomos" (laws) (Jastrzebowski, 1857; Murrell, 1958). Over the past 35 years, the field covers both the physical and information processing aspects and is more commonly known as "Human Factors Engineering" or "Ergonomics" with about equal preference in the use of either name for the field.

Human Factors Engineering Approach

Fitting the Product to the Users: Human Factors Engineering involves "Fitting the product (or equipment) to the people (or users)." This means that the products should be designed such that people (population of users) can fit comfortably (naturally) to the product and they can use the product without any awkward body postures, movements, or errors.

It should be noted that Human Factors Engineering is NOT about fitting the people to the product (i.e., the product should not be designed first and then people are simply asked to somehow adapt or "force-fit" to use them). In some cases, the product is designed such that only people with certain characteristics can fit (accommodate) or use them (which normally involve personnel selection strategy, i.e., placing restrictions on the "type" of people who can use the product).

Designing for the Most: Ergonomics involves "Designing for the Most" (i.e., to ensure that most users within the intended population of the users of the product can fit with the product). It should be noted that if we use other design strategies, like "designing for the average" or "designing for the extreme," only a few individuals within the user population will find the product to be "just right" (or fit very well) for them. Thus, designing for the most will involve making sure that the designer knows what the user population is and knows the distributions of characteristics, capabilities, and limitations of the individuals in that population. For example, in designing a car, the designer makes sure that the seat track should be made long enough and placed from the pedals at distance such that most drivers (from shortest female to tallest male drivers in the population where the vehicle will be marketed) can sit in their preferred driving posture.

Systems Approach: Another important consideration involves "Human as a Systems Component." This means that the designer must treat the human as a component of the system that is being designed. The process for designing a product should thus involve the considerations of the following major components: (1) the user, (2) the product (e.g., a vehicle), and (3) the environment (see Figure 11.1). The characteristics of all the three components in the system must be considered in designing the product. The product design should not only involve designing all the physical components that fit and function well, but also make sure that the user is considered as a human component and the user's characteristics are measured and used in designing the product—to ensure that the product would meet the users' needs related to product use, comfort, convenience, and safety. It should be noted that during the design process of a physical product, the engineer designs each component of the product by paying attention to all its properties (e.g., dimensions, material, hardness, color, surface appearance and touch feel, and how it fits/works with other components). Similarly, when the human is involved as an operator or the user of the product, all relevant human characteristics must be studied and used in designing the product.

Thus, in designing a product, a thorough understanding of the intended user population and the operating environment (e.g., usage conditions such as light levels, temperatures, and vibrations) of the product must be considered. Figure 11.1 shows that when a user uses a product in an environment, the ergonomics engineer must consider the characteristics of all the components in the system and evaluate: (1) how the user will perform various tasks, (2) the user's preferences in using the product, and (3) the pleasing perceptions created by experiencing the product such as the quality, craftsmanship, emotions evoked by the product, and its resulting brand image. Bailey (1996) has proposed another but somewhat similar approach to

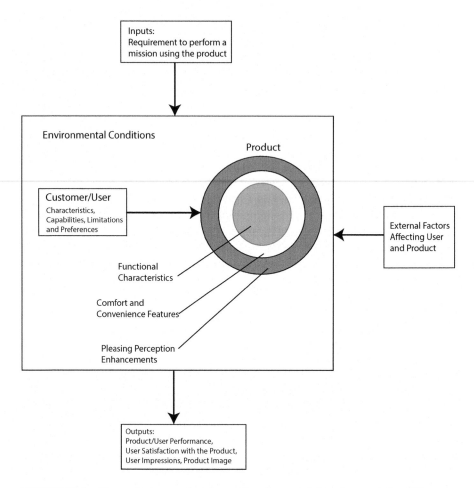

FIGURE 11.1 Human factors engineer's considerations related to characteristics of the user, the product, and the environment, and their relationship to user performance, preference, and perception.

conceptualize ergonomic problems. Bailey's approach involves taking into account the user, the activity (type of operation or usage), and the context (usage situation) in designing a system.

In Figure 11.1, the product is shown here by three concentric circles. The inner circle (or the core) represents the functional aspects of the product. The next ring (surrounding the functional characteristics) represents the comfort and convenience features of the product. And the outer ring represents the amount of pleasing perceptions created by adding craftsmanship and image-enhancing features. The ring model of desirability of a product was suggested by Bhise (2012). In this model, the overall desirability of the product is considered to be proportional to the size of the outer ring. The model suggests that the overall desirability of a product is influenced by the additive effects of three desirable characteristics, namely (1) functionality (size of the inner core), (2) comfort and convenience (thickness of the middle ring), and

Human Factors Engineering in Product Design

(3) quality perceptions (thickness of the out ring). Thus, a designer can decide on how much content to allocate to functions and other product features to increase the overall desirability (i.e., the size of the outer diameter) of the product.

It should be noted that the three components of the ring model are somewhat similar to the three components of the Kano model shown in Figure 10.1. The functional characteristics are like the unspoken wants (basic features that make the product functional) in the Kano model. The comfort and convenience features are like the satisfiers in the Kano model. And the pleasing perceptions can create impressions of excellence in the attention to details (e.g., selection of good-looking materials with pleasing touch feel, excellent fit and finish, better lighting, smoothness in movement of controls, crispness in switch activation feel, and so forth—which contribute to the feeling of craftsmanship), and thus, they contribute to the creation of the "Wow" features in the Kano model.

Figure 11.1 also shows inputs, outputs, and external factors affecting the product and its users (e.g., operator workload due to basic product usage tasks and other additional tasks, climate, lighting, and noise levels in the working environment). The inputs dictate the requirements of the tasks, that is, what the user needs to do with the product. The outputs can be measured to determine how well the user completes the task (i.e., the user performance and also how the product performs or affects the user's performance) and the user's impressions about the product (e.g., Was the experience in using the product "fun to use" or a "pain to use"?).

It is important to realize that the human factors engineers working with other specialty disciplines during the product development are in a unique situation to help in improving the sizes of both the outer rings of the desirability model. The two outer rings are also largely responsible for increasing the product appeal and brand differentiation.

HUMAN FACTORS RESEARCH STUDIES

Most research studies available in the Human Factors field can be categorized into the following three types or their combinations:

1. *Descriptive research*: This type of research generally provides data describing human characteristics of different populations (e.g., means, standard deviations, and distributions of characteristics such as anthropometric dimensions, strength measurements, detection thresholds for visual or auditory signals, information processing rates, decision, and response times).
2. *Experimental research*: This type of research generally involves experiments conducted to determine the effects of different combinations of independent variables on certain response variables under carefully manipulated and controlled experimental situations (e.g., determining the effects of different vehicle parameters on comfort/ease/difficulty ratings or performance measures during an entry into a vehicle, seating, operating driver interface equipment, and loading/unloading luggage in the trunk or cargo areas).

3. *Evaluative research*: This type of research generally involves comparisons of user performance (and/or preference) in using different products (e.g., determining which of the four proposed cell phone designs would be the most convenient to use and preferred by their users).

HUMAN FACTORS ENGINEER'S RESPONSIBILITIES IN DESIGNING COMPLEX PRODUCTS

The inclusion of the human factors or ergonomics engineers as a part of the product development team is now an accepted practice in most complex product-producing industries (e.g., computer, automotive, aviation, maritime, and nuclear). The ergonomics engineers work from the earliest stages of new product concepts creation to the periods when the customers use the product, dispose it, and are ready to purchase their next product.

The ergonomics engineer's major tasks during the product development phases are as follows:

1. Provide the product design teams with needed ergonomics design guidelines, requirements, information and data from past studies, results of analyses, scorecards, and recommendations for product decisions at the right time (called the gateways or milestones) in front of the right level of decision makers (involving design teams, program managers, chief engineers, senior management, and others).
2. Apply available methods, models, and procedures to address issues raised during the product development process (e.g., conduct field of view analyses to determine visible and obscured areas within the user's visual field).
3. Conduct quick-react studies (i.e., experiments or surveys) to answer questions raised during the product development process.
4. Evaluate product/program assumptions, concepts, sketches, drawings, CAD models, physical models/mock-ups/bucks, mechanical prototypes, prototype products, and production units made by the manufacturer and its competitors.
5. Participate in the design and data collection phases of customer clinics and market research clinics involving the product concept and other existing leading products as comparators (or controls).
6. Obtain, review, and act on the customer feedback data from complaints, warranty, customer satisfaction surveys, market research data (e.g., J. D. Powers survey data [J. D. Power and Associates, 2013]), inspection surveys with owners, magazines related to consumer, and other specialized products (e.g., cars, trucks, appliances, manufacturing equipment).
7. Provide ergonomics consultations to members of the product development teams.
8. Long term: Conduct research, translate research results into design guidelines, and develop design tools for designing future products.

IMPORTANCE OF HUMAN FACTORS ENGINEERING

CHARACTERISTICS OF ERGONOMICALLY DESIGNED PRODUCTS

In general, an ergonomically well-designed product (that meets all applicable ergonomics requirements) will be liked by its users and their responses can be described by realizing the following:

1. An ergonomically designed product should "fit" people well (like a well-fitting suit). (Note: One would use a well-fitting suit much more often than a poor-fitting one.) Thus, when it is time to replace an old product, a customer will most likely purchase a newer version of the same product that fits him/her well. This suggests that ergonomically designed products, more likely, will be repurchased.
2. An ergonomically designed product can be used with minimal mental and/or physical work. Thus, as the product usage increases, the customers will realize the ease, comfort/convenience features, and the absence of problems while using the product.
3. An ergonomically designed product is "easy-to-learn." (Note: Owner's manuals of easy-to-learn products are seldom used. Easy-to-learn products work in an "expected" manner.)
4. A product with usability problems (i.e., the absence of ergonomics) can be quickly noticed—usually after use. Thus, ergonomic characteristics of many products are not generally noticed in the showrooms where the customer does not have an opportunity to use them.
5. Ergonomically designed products are generally more efficient (productive) and safer (less injurious).

WHY APPLY HUMAN FACTORS ENGINEERING?

Complex products involve people during their design and uses. Thus, application of human factors knowledge is essential during product development. Further, because of the fierce completion in the market and tighter program schedules and budgets, the products need to be designed "right the first time" without costly redesigns. Therefore, the human factors engineers should be included early in the product development process to achieve the following:

1. Create functionally superior products, processes, or systems.
2. Reduce costly and time-consuming redesigns (with early incorporation of human factors inputs in the design process, superior products, or systems can be developed without additional design iterations).
3. Achieve outstanding designs early. (There are thousands of ways to design a product, but only a few designs are truly outstanding. You want to find those "outstanding" designs quickly).

Human Factors Engineering Is Not Commonsense

1. Commonsense ideas/solutions are often wrong. For example, a designer wanted to create an instrument panel illuminated with "deep red" lighting for a new hot sports car. The human factors engineer reminded him that about 8% of the males have a color deficiency in perceiving "the color red." The designer said, "but the air force used red-colored instrument panels in airplanes." The human factors engineer reminded the designer again saying "color-deficient persons cannot get a pilot's license, but a car is a consumer product and you don't want to annoy these color-deficient males in using your vehicle. If you want red then we should add some yellow in it and make it orangish-red so that the red-color deficient people can still read the instruments" (Bhise, 2012).
2. Knowledge-based decisions are superior as they minimize usability problems. (The ergonomist brings his specialized knowledge and data about the users at an early point in the design process.)

A BRIEF OVERVIEW OF HUMAN CHARACTERISTICS AND CAPABILITIES

The human characteristics and capabilities used in product (equipment) design can be classified as follows.

Physical Capabilities

These capabilities can be measured by using physical instruments (e.g., anthropometers, measuring tapes, rulers, calipers, goniometers, optical scanners, weighing scales, strength/force measuring gauges, and force platforms).

Anthropometric characteristics (which involve measurements of human body dimensions): The measurements made when a human subject is stationary (not moving) are called "static" dimensions, which are generally taken when a subject is standing erect or sitting in an anthropometric measurement chair (with the torso and lower legs vertical, and upper legs horizontal). Some static anthropometric dimensions are as follows: standing height, standing shoulder width, standing eye height, standing elbow height, sitting height, sitting eye height, sitting shoulder width, sitting elbow-to-elbow width, sitting hip width, sitting knee height, buttock-to-knee distance (sitting), elbow-to-fingertip distance, foot length, foot breadth, hand length, hand breadth, hand depth, hand length, breadth and depth of fingers measured at different joints, body weight, and so on. The dimensions are measured under standardized standing or sitting postures so that data for different populations can be compared. Other measurements of human body (and body segments) such as surface areas, volumes, centers of gravity, and weights of different body segments are also considered as a part of anthropometry (science of human body dimensions).

The human body dimensions measured when a subject is in a work posture (e.g., sitting in a car seat and with arm extended to reach a push button or grasp a control located at a given lateral and vertical distance with respect to the sitting position (measured from a seating reference point), operating equipment or performing some work in a workplace) are called "functional" anthropometric dimensions. The functional anthropometric dimensions cannot be accurately predicted from the static anthropometric dimensions because body tissues deflect differently in different postures and the movements around the body joints are complex (i.e., they cannot be predicted well by using simple approximations of pin joints). Further, the postures that people assume in a given work situation can vary between individuals and within an individual (i.e., when the same individual is measured at a different time). The variations also depend on other factors such as preferences, state of health, age, gender, clothing, available space, and clearances. In many equipment design-related issues, it is important to measure the relevant anthropometric dimensions and posture angles at predetermined instances with respect to certain reference points in the workplaces under actual dynamic working conditions (e.g., by video recording or by using a motion analysis system to digitize trajectories of different points on the body parts). The data from distributions of such measurements can be used to determine how most people (e.g., 95% of the population) are accommodated (i.e., fitted within a workspace).

Biomechanical characteristics (ability to produce forces/strength and make body movements): The human strength measurements are made by using different force measurement devices (e.g., hand dynamometer for grip strength measurements, force platforms for forces exerted while pushing or pulling) as functions of time and body posture angles (measured by video recording or use of goniometers). The maximum strength capabilities under various hand, foot, and body movements under different directions (e.g., pushing or pulling in different directions such as up, down, left or right, or at different joint angles) and for lifting and carrying loads over different vertical and horizontal distances (with respect to certain human body points, e.g., ankle point) for males and females have been measured by different researchers. The data are available in different human factors textbooks and handbooks (Kroemer et al., 1994; Konz and Johnson, 2004; Salvendy, 2012).

INFORMATION PROCESSING CAPABILITIES

These are mental (cognitive) capabilities involving the acquisition of information through various sensors (in the human eyes, ears, joints, vestibular tissues, etc.), transmitting this sensed information to the brain, recalling information stored in the memory, processing the information to make decisions (e.g., detecting, recognizing, comparing, selecting, and estimating), and making responses (e.g., motor action—generating a body movement, activating a control, or making a verbal response).

Detection (or recognition) times for different signals, detection rates (percent signals detected), reaction times (or response time), information processing speeds (measured in bits per second), memory capacities (number of items stored in the memory), time to complete tasks, and error rates (errors per 1000 opportunities) in completing tasks are examples of commonly used measures to evaluate human information processing capabilities. Various human information processing models have been developed to understand how humans process information and make decisions under different situations (Wickens et al., 1998). These models and data available from the literature are generally used by the human factors engineers to evaluate human information processing issues related to product design. Laboratory, simulator, and/or field research studies using prototype products or early working products are also commonly conducted to evaluate the information processing performance of selected subjects under different product usage situations.

OTHER FACTORS AFFECTING HUMAN CAPABILITIES

In general, most human abilities degrade as people get older. The degradation in most human abilities is about 5%–10% per decade after about 25 years of age. With practice, humans can perform complex tasks with very little or no conscious effort. However, humans are not very consistent and precise in performing tasks like machines. Thus, human performance in performing most tasks varies considerably. The variability in the same subject performing the same task is called the "within subject variability," whereas the "between subject variability" is based on differences in the performance of different subjects performing the same task. There are many other factors such as fatigue, monotony (or boredom), attention, and motivation that can affect human capabilities and hence their performance.

PERCENTILE VALUES

When designing a product, the human factors engineers must make sure that most users in the population can perform the tasks associated with the product. The criteria for human capabilities or characteristics are generally expressed in terms of percentile values. Thus, a 95th percentile capability of a given characteristic means that 95% of people in the population can meet or have the capability specified in the characteristic. For example, 95th percentile reach distance in a specified work situation means at least 95% of people in that population will be capable of reaching a specified location in a given work configuration.

HUMAN ERRORS

Human operators will make errors in operating equipment due to a number of reasons. For example, Fitts and Jones (1947a, 1947b) measured errors that pilots made during operating aircraft controls and reading aircraft displays. We are interested here in understanding how a product can be designed to minimize the possibilities of user errors. Most human errors occur due to information processing failures. For example, Treat (1980) studied over 13,500 police-reported traffic accidents and found that in over 70% of the accidents, human errors were identified as the definite causes of accidents. The definite causes were defined as the accidents in which the in-depth accident investigating multidisciplinary team members were over 95% confident that the accidents were caused by human factors (e.g., recognition errors, decision errors, and performance errors).

Human Factors Engineering in Product Design

DEFINITION OF AN ERROR

An error can be defined as an act or action that is not a part of normal pattern of behavior (or a series of actions) required to perform a task. The following three definitions of an error illustrate that an error involves an abnormal act or behavior.

1. An act involving an unintentional deviation from the truth or accuracy (Webster's Dictionary, 1980)
2. An out-of-tolerance action
3. An inconsistency with normal programmed behavioral pattern that differs from prescribed procedures.

In general, when a human operator's capabilities are overloaded (or exceeded) or when the operator is inattentive, an error is possible. The likelihood of human error will increase when human factors, principles, or guidelines are violated, that is, when the equipment is not designed with proper consideration of operator characteristics, capabilities, and limitations. Humans will always make some errors because they are not like machines. The basic error rate data available in the literature (Gertman and Blackman, 2001) show that even for the best "human factored" equipment, the error rates of the order of 1 error in 10,000 operations to 1 error in 1,000 operations are quite common.

TYPES OF HUMAN ERRORS

Human errors can be classified as follows:

1. *Detection error*—failure to detect a signal or a target (e.g., a driver fails to see a pedestrian). The detection error can occur due to a number of reasons such as: (1) if the signal was weak in relation to its background (on noise), (2) the signal occurrence violated the user's expectancy (spatial or temporal), or (3) the user was not attentive (alert) or distracted.
2. *Discrimination error*—failure to discriminate an item from other items that differ in some characteristics. Some examples of the discrimination error are as follows: (1) a driver failed to discriminate a red stop lamp signal from a red tail lamp signal, (2) a driver failed to discriminate a red traffic signal from other lamps in his visual field, or (3) a driver failed to recognize and discriminate a push button from other push buttons in an instrument panel.
3. *Interpretation error*—failure to recognize or interpret a situation, a signal, a hazard, or a scale (e.g., a tachometer reading was interpreted as the speed reading by a driver).
4. *Omission error*—failure (or forgetting) to perform a required action (e.g., forgetting to look in the side view mirror before changing lane).
5. *Commission error*—performing a function that should not have been performed. A commission error can involve:

a. *Extraneous act error*—introducing a step or a task that should not have been performed (e.g., a driver changed radio station while turning on the windshield defrost function).
b. *Sequential error*—performing a step or a task out of a required sequence (e.g., a driver changed radio station band after selecting a radio preset station).
c. *Time error*—failing to perform within an allotted time or excessive time spent (e.g., a driver required a long glance or more than a single glance while using a control or a display).
6. *Substitution error*—using or substituting another item (e.g., control or display) instead of the desired one (e.g., a driver pressed the accelerator pedal instead of the brake pedal).
7. *Reversal error*—reversing the direction of activation or interpreting a displayed signal in the opposite direction (e.g., increased temperature instead of decreasing).
8. *Inadequate response error*—insufficient control action or movement or insufficient force applied during a control activation (e.g., insufficient brake pedal movement during a stopping maneuver), or error in response due to an error in judgment or estimation of signal magnitude, distance, speed, and location).
9. *Legibility error*—error related to not being able to read a display (due to factors such as small font size, low contrast of the letter or numeral against its background, insufficient light, excessive glare, or parallax)
10. *Recovered error*—an error has occurred but the operator could correct the error after some elapsed time. (Note: Many errors made by the human operators are recovered [i.e., corrected] and thus undesired events [e.g., accidents] are avoided.)
11. *Unrecovered error*—an error that the operator fails to correct (or will not or cannot correct).

HUMAN INTERFACE

During the design of any product involving a human user, the human factors engineer needs to understand the following four types of human interfaces:

1. *Accommodation of the human body* (i.e., positioning and supporting the human body in the workspace [e.g., a seat installed in a cockpit or a workstation, armrests, footrests, and headrest] and making sure that adequate clearances are provided for body motions with consideration of extra space for clothes, shoes, hats, special insulating suits, and masks)
2. *Protecting human body from the environment* (i.e., protection from heat, rain, falling/flying objects, toxic substances [e.g., helmet, protective suit, gloves, or enclosed workplace such as a vehicle body], energy absorbing padding, or insulating materials)
3. *Providing inputs to the equipment* (i.e., through controls—selection, location, and design of controls such as hand/finger-operated controls [push

buttons, keyboards, touch pads], foot controls [pedals, floor switches], voice controls, proximity switches, and gesture sensing controls)
4. *Receiving inputs from the equipment* (i.e., through displays—selection, location, and design of visual displays [e.g., analog and digital gauges, or screens], sound/auditory displays [e.g., bells, whistles, beeps, and voice commands], tactile displays [e.g., feel of shapes, surfaces, textures, parting lines, and vibratory signals], and olfactory displays [sensory perception of smells—pleasant or unpleasant]).

All the above four types of interfaces need to be designed with considerations of characteristics, capabilities, and limitations of the intended population of users. The following examples will illustrate this point.

1. Human anthropometric dimensions are needed to design dimensions of workspaces, heights of work surfaces, reach distances to various work locations in the 3-dimensions, grasp areas of handles, armrests, footrests, step heights, and step widths.
2. Human biomechanical characteristics are needed to design seats, exertion of forces to operate controls, lifting and carrying items, designing walking, sitting, and resting surfaces.
3. Human information processing capabilities need to be considered to design controls, to displays, to simplify decision-making in operating equipment. Humans are also very good at sensing changes in situations and are very adaptable to changing situations.

However, limitations on human capabilities and how the capabilities are affected by other factors such as gender, age, workload, psychological and physiological stress, and fatigue are also very important in designing products to ensure that a large percentage of users (i.e., "most" users) can be accommodated under all operating environments and the users will enjoy using the equipment.

Many textbooks, handbooks, and research papers provide valuable information on human characteristics from descriptive, experimental, and evaluative research studies (Bhise, 2012; Bridger, 2008; Pheasant and Haslegrave, 2006; Kroemer et al., 1994; Wickens et al., 1998; SAE, 2010; Salvendy, 2012; Van Cott and Kinkade, 1972).

USER PERFORMANCE MEASUREMENTS

User's performance while using a product can be measured by using a number of measures such as time taken to complete a task, the number of errors made during the performance of the task, subjective ratings on ease or difficulty during the use, increase in the user's workload, and increase in the user's heart rate. The problem of "what to measure" depends on the researcher's understanding of the user's tasks, the purpose of the research, and resources available. Chapter 7 on Product Evaluations also includes information on methods for performance evaluations.

Types and Categories of User Performance Measures

User performance measures can be categorized as follows:

1. *Behavioral measures*: These measures are derived from measurements of user behavior and/or choices made by the user while performing a task (e.g., "what did he/she do" by recording his/her eye movements, body movements, sequences in performing different movements and tasks, decisions made, and time taken).
2. *Physical measures*: These measures are obtained from the measurements made by physical instruments of parameters such as time, distance, speed, acceleration, and forces exerted.
3. *Subjective measures*: These measures are based on the judgments of the user (or of an observing experimenter), for example, ratings, preferences, judgments, thresholds of perception, and detection of signals.
4. *Physiological measures*: These measures are based on the recordings of the physiological state of the user obtained from measurements of changes in the user's heart rate, sweat rate, oxygen intake, galvanic skin resistance, electrical activities in different skeletal muscles (electromyogram [EMG]) or the heart (electrocardiogram [ECG or EKG]).
5. *Accident-based safety performance measures*: These measures are computed from searching through various accident databases. Some examples of measures are number of accidents of a given type, accident rates (e.g., number of accidents per hours worked, number of accidents per travel miles), and accident severity (e.g., amount of property damage [dollar loss], injury level, workdays lost due to an injury) (see Chapters 12 and 19 for more information).
6. *Equivalency-based measures*: These measures are based on comparisons of performance or user judgments under two different conditions to determine if they are equal or different (e.g., sound quality of radio A equal to the sound quality of radio B, worse, or better).
7. *Monitory measures*: These measures are based on the dollar value of costs or benefits such as usage costs, energy consumption, or savings per use.

Characteristics of Effective Performance Measures

For any measure to be acceptable and useful, it should meet certain key characteristics. The following characteristics were based on the characteristics of effective safety performance measures presented by Tarrants (1980).

1. *Administrative feasibility*: The measuring system or measuring instrumentation used to obtain the value of the performance measure must be practical, that is, one must be able to construct it and use it quickly and easily without excessive costs. Thus, a product evaluator should be able to use the measurement system to obtain the necessary measurements. And the

product development team should be able to set targets by using the measure and determine if the product or its attribute goals have been reached.
2. *Interval scale*: The measurement system should be able to provide values of the measure by using at least an interval scale. The interval scale should be graduated with equal and linear units, that is, the difference between any two successive point values should be the same throughout the scale.

   It should be noted that there are four types (or orders) of measurement scales: nominal, ordinal, interval, and ratio scales—with ascending order of power to perform mathematical operations and applications of statistical inference techniques.
   a. The nominal scale (the most primitive) is used for categorizing or naming (e.g., football jersey numbers, car model numbers) of items. One can only analyze the data obtained by using the nominal scale by counting frequencies or percentages of values in each category.
   b. The ordinal scale is used to order or rank items. But the distance measured on a scale between ranked items may not be equal. Thus, in ranked items, one can only conclude that an item with a higher rank is better than another item with a lower rank (e.g., the Mohs hardness scale in geology).
   c. The interval scale has equal-length intervals but the zero point on the interval scale is arbitrary (e.g., like the Fahrenheit or Celsius temperature scales). On an interval scale, the difference between any two items measured on a scale can be determined by difference between their two respective scale values.
   d. The ratio scale is the most informative (or quantitative) as the ratios of quantities defined by the scale values can be constructed. For example, a 10-lb weight is two times heavier than a 5-lb weight. The ratio scale also contains an absolute zero (i.e., the point of "no amount").

   Thus, we should make sure that the measure that we select should use the highest possible order of scale—with the interval scale as the minimum order of acceptable scale.
3. *Quantifiable*: A quantitative measure will allow comparison between any two values in terms of at least a difference on an interval scale. The quantitative measure should permit application of more sensitive statistical inference techniques. (Note: A nonquantitative, i.e., a qualitative, measure limits statistical inference [due to use of data on nominal and/or ordinal scales] and opens the way for individual interpretation. For example, if the result of a speedometer comparison study states that "the analog speedometers are better than digital speedometers," the reader does not have sufficient information on the magnitude of improvement gained by the use of an analog speedometer over the digital speedometer.)
4. *Sensitivity*: The measurement technique should be sensitive enough to detect changes in a characteristic of the product or the user's performance to serve as a criterion for evaluation. (Note: A tiny diamond cannot be measured on a cattle scale.)

5. *Reliability*: The measurement technique should be capable of providing the same results for successive applications in the same situations.
6. *Stability*: If a process does not change, the performance level obtained from the measure at any other time should remain unchanged.
7. *Validity*: The measure should produce information that is representative of what is to be inferred in the real world. This is particularly important because in product evaluation many different types of measures can be used. For example, the measures can vary from indirect, intermediate, and surrogate measures to direct measures related to the final outcome (e.g., in an automotive braking performance test, the temperature of the brake pads can be considered as an intermediate or surrogate measure as compared to the stopping distance as a direct measure of the braking performance).
8. *Error-free results*: An ideal measuring instrument should yield results that are free from errors. However, in general, any measurement will have some constant and random errors. The product evaluator needs to understand the sources of such errors and minimize the errors. The errors can also be statistically isolated and estimated by their sources (or effects of the sources such as measuring instruments, test operator, test method, and test environment).

HUMAN FACTORS METHODS: AN OVERVIEW

The methods used in human factors engineering analyses and studies can be categorized into the following four broad categories.

1. *Human factors checklists and scorecards*: The human factors experts can conduct evaluations by completing checklists and/or scorecards. The checklists can involve an expert's judgment on how the product meets human factors guidelines, design considerations, and requirements. The scorecards are generally used to summarize the findings of human factors evaluations.
2. *Task analysis*: It analyzes a given task by breaking it down into subtasks and determining operational problems (e.g., errors) by evaluating the effects of demands placed by each subtask on the human operator and capabilities needed to perform each of the subtasks.
3. *Human performance models*: Available human factors models can be applied to predict performance of different users using a product in a given situation.
4. *Laboratory and field tests*: Data can be obtained from measurements of the following: (a) observations of a subject performing a task using the product; (b) applying methods of communication, for example, by asking the subject to provide ratings on selected product features and/or asking the subject to describe problems and difficulties in using the product; (c) measuring

the subject's performance by using physical instruments; (d) measuring the physiological state of the subject while completing different tasks; and (e) measuring the user's workload in performing one or more tasks. The evaluations can be conducted by using task simulators, early prototypes, or production versions of the product.

The above methods are described in greater details in Chapter 18.

CONSIDERATIONS IN THE APPLICATIONS OF HUMAN FACTORS GUIDELINES

Human performance in a task can be affected by a number of factors related to the product, the user, and the usage situations. Thus, in designing a product, each item such as a control, a display, a seat, or a handle that can affect the user's performance must be carefully evaluated to ensure that all applicable and relevant human factors principles are met. Thus, during the design stages of a complex product, the human factors engineer's work can be very challenging. The challenges are due to simultaneous considerations of many design guidelines, many requirements in different standards, and many trade-offs between different product attributes and issues.

Many human factors design guides and standards generally provide design guidelines in addition to design or performance requirements. The design guidelines are statements that provide information on how to design. The guidelines are generally derived based on research studies, past experiences (e.g., lessons learned from past product development programs), and customer feedback. The guidelines thus help the designer in incorporating the best practices and avoiding certain problems. The guidelines generally should be treated differently than the requirements (see Chapter 4 for more information on different types of requirements) because any failure in meeting a requirement may disqualify the product from getting a certification of compliance to the standard. On the other hand, a guideline could be violated if a better design approach or method is available.

Table 11.1 illustrates human factors considerations, human factors advice (guidelines or principles) and comments, and references that are generally analyzed during different product design steps.

CONCLUDING REMARKS

Human Factors Engineering is an important discipline and it offers knowledge about people to design products so that people will find them easy and enjoyable to learn and use. Easy-to-use products can also reduce human errors and thus will be safe (i.e., free from accidents) during their uses. Products that are difficult to use are generally not liked by their users, and therefore, such products will most likely not be repurchased.

TABLE 11.1
Illustrations of Human Factors Considerations in Product Design Applications

Design Step	Human Factors Considerations	Statements of Human Factors Advice/Guidelines/Principles	Comments and References
1 Determine user population	Geographic location of product usages, male-to-female ratio of users, age of users, market segment characteristics (education, income level of users)	Select population that will be representative of the users of the product. Understand and measure user characteristics relevant to the product usage situations. Determine user needs.	Conduct literature survey. Measure anthropometric characteristics of subjects invited to participant in market research clinic (Pheasant and Haslegrave, 2006).
2 Benchmark similar products	Determine product usages. Evaluate products and develop human factors score cards.	Refine list of product usage situations and customer needs. Understand ergonomic strengths and weaknesses of the competitive products. Discuss examples of ergonomically good and poor features in the competitive products with the design team.	Benchmarking helps the team members in understanding human factors issues.
3 Determine occupant postures in the product, clearances, and space required to operate/use the product.	Percentage of users that can be accommodated (or "fitted"). Hand, foot, and body movements during usages. Eye locations of users from displays and product openings or apertures for field of view evaluation.	Let a large man (95th percentile) fit within the user space (e.g., cockpit) and a small woman (5th percentile) reach the controls. Study visibility of short and tall users. Use seat design principles.	Early product concepts must be evaluated for space and visibility considerations (Bhise, 2012).
4 Determine the range of environmental conditions during product usage	Effect of ambient conditions (e.g., light levels, noise levels, temperatures, air flows, social interactions with other users, etc.) on user performance and preferences.	Evaluate human performance issues of product usages under combinations of extreme environmental conditions.	Some examples: legibility of displays under sun glare and night usages. Ability to hear auditory signals or voice messages under noisy conditions. Ability to process information under extended periods of thermal stress.

5	Entry and egress in the product workspace	Ease in entry into the product (e.g., a vehicle) and egress from the product space must be considered. Steps, handholds, seat height, door opening spaces, lighting and visibility of interfering components.	Evaluate needs of extreme users (e.g., tall, short, small, obese). Study foot, hand, and body movements. The user must be at least in three-point contacts (e.g., two hand and one foot contact or one hand and two foot contacts) with the product during the entire maneuver.	Accidental hitting of body parts and possibility of slips and falls during entry/exit must be evaluated. Use full-size mock up for evaluations. Use task analysis (Bhise, 2012).
6	Product operation	Visibility of required field of views, visibility of displays and controls, legibility of displays, interpretability of displayed information, reach distances to controls, expectancy of control locations, direction of motion stereotypes, feedback from control activations, tactile feel of controls, etc.	Locate controls and displays in expected locations. All major displays and controls required during product operation must be visible and located close (less than 35°) to normal line of sight. Locate all major controls within maximum and minimum comfortable reach distances. Locate controls and displays by considering importance, frequency of usage, function grouping, sequence of eye and hand motions, associations of controls to displays.	Apply control and display design guidelines. Evaluate user workload and simplify tasks (redesign user interface) to avoid chances of errors (Bhise, 2012).
7	Loading/unloading and/or carrying/lifting equipment, cargo, etc.	Determine weights, sizes, heights, distances, types of handles, frequencies of lifting/carrying tasks.	Apply lifting and carrying guidelines. Some examples: Evaluate force, frequency of force applications, and joint deviations for possible cumulative trauma injuries. Design so that weakest (5th percentile female) can perform these tasks. Hold load close to body. Provide hand support during lifting.	Konz and Johnson (2004) and Chaffin et al. (1999).

REFERENCES

Bailey, R. W. 1996. *Human Performance Engineering*. Upper Saddle River, NJ: Prentice Hall PTR.
Bhise, V. D. 2012. *Ergonomics in the Automotive Design Process*. Boca Raton, FL: CRC Press.
Bridger, R. S. 2008. *Introduction to Ergonomics*. Third Edition. Boca Raton, FL: CRC Press.
Chaffin, D. B., Andersson, G. B. J. and B. J. Martin. 1999. *Occupational Biomechanics*. New York: John Wiley & Sons, Inc.
Fitts, P. M. and R. E. Jones. 1947a. Analysis of Factors Contributing to 460 "Pilot Errors" Experiences in Operating Aircraft Controls. Memorandum Report TSEAA-694-12. Aero Medical Laboratory, Air Materiel Command, Wright-Patterson Air Force Base, Dayton, Ohio. In *Selected Papers on Human Factors in the Design and Use of Control Systems* (Ed.) Wallace Sanaiko, H. (pp. 332–358). New York: Dover Publications, Inc.
Fitts, P. M. and R. E. Jones. 1947b. Psychological Aspects of Instrument Display. I: Analysis of Factors Contributing to 270 "Pilot Errors" Experiences in Reading and Interpreting Aircraft Instruments. Memorandum Report TSEAA-694-12A. Aero Medical Laboratory, Air Materiel Command, Wright-Patterson Air Force Base, Dayton, Ohio. In *Selected Papers on Human Factors in the Design and Use of Control Systems* (Ed.) Wallace Sanaiko, H. (pp. 359–396) New York: Dover Publications, Inc.
Gertman, D. L. and H. S. Blackman. 2001. *Human Reliability and Safety Analysis Data Handbook*. New York, NY: John Wiley.
J. D. Power and Associates. 2013. Customer Surveys on Initial Quality, In-Service and Product Appeal. https://www.jdpower.com/autos/car-ratings/ (accessed May 2, 2013).
Jastrzebowski, B. W. 1857. *An Outline of Ergonomics, or the Science of Work Based upon the Truth Drawn from the Science of Nature*. Przyrodo i Premyst, 29–32, Proznan. Warszawa: Central Institute for Labour Protection. (Reprinted as Commemorative edition published on the occasion of the *XIVth Triennial Congress of the International Ergonomic Association and the 44th Annual Meeting of the Human Factors and Ergonomics Society* [San Diego, CA], 2000.)
Konz, S. and S. Johnson. 2004. *Work Design—Occupational Ergonomics*. Sixth Edition. Scottsdale, AZ: Holcomb Hathaway, Publishers, Inc. IBSN# 1-890871-48-6.
Kroemer, K. H. E., Kroemer, H. B. and K. E. Kroemer-Elbert. 1994. *Ergonomics—How to Design for Ease and Efficiency*. Englewood Cliffs, NJ: Prentice Hall.
Murrell, K. F. H. 1958. The Term "Ergonomics". *American Psychologist*, 13(10): 602.
Pheasant, S. and C. M. Haslegrave. 2006. *BODYSPACE: Anthropometry, Ergonomics and the Design of Work*. Third Edition. London: CRC Press, Taylor and Francis Group.
Salvendy, G. (Ed.). 2012. *Handbook of Human Factors and Ergonomics*. Hoboken, NJ: John Wiley & Sons, Inc.
Society of Automotive Engineers, Inc. 2010. *SAE Handbook*. Warrendale, PA: SAE.
Tarrants, W. E. 1980. *The Measurement of Safety Performance*. New York: Garland STPM Press.
Treat, J. R. 1980. *A Study of Precrash Factors Involved in Traffic Accidents*. Highway Safety Research Institute Report. University of Michigan, Ann Arbor, MI.
Van Cott, H. P. and R. G. Kinkade (Eds.). 1972. *Human Engineering Guide to Equipment Design*. Sponsored by the Joint Army-Navy-Air Force Steering Committee, Washington, DC: McGraw-Hill, Inc./U. S. Government Printing Press.
Webster's. 1980. *New Collegiate Dictionary*. Springfield, MA: G. & C. Merriam Company.
Wickens, C., Gordon, S. E. and Y. Liu. 1998. *An Introduction to Human Factors Engineering*. Upper Saddle River, NJ: Pearson Prentice Hall (Addison Wesley Longman, Inc.).

12 Safety Engineering in Product Design

INTRODUCTION

Products should be safe to use. This means that every product should be designed such that it does not cause injuries and generate any adverse health effects (e.g., exposure to harmful conditions or toxic substances) in people during the product life cycle. Accidents can cause injuries, loss of lives, loss of work, delays in work, property damage, and also can incur costs. Designers should ensure that a safe product is created in the early conceptual and detailed design stages so that failures of the product during its uses are minimized. A safe product should have characteristics and features that prevent accidents from occurring and prevent injuries in cases if accidents occur. Just like the proactive approach to quality, safety should be "built in" the product in its early design stages, and safety evaluations (or monitoring) must be conducted as the product advances through its life stages.

Customers are very conscious of safety and they demand that products should be safe to use, and the users and others should be free from any harm that can result from the uses and even some foreseeable misuses of the product. Thus, safety is an important product attribute and it should be considered at the earliest time in the conceptualization of any new product. Product liability is also another important consideration during the product design. Safety requirements should be developed early to ensure that safe products can be developed and potential for the product liability related to costly litigations are minimized. Special surveys of new safety technologies, benchmarking of safety features in other similar competitive products, safety reviews, and safety analyses should be conducted during the early design stages.

The objective of this chapter is to provide basic background into Safety Engineering, related issues, considerations, methodologies, and safety-related costs to ensure that safe products are developed. The topics covered in this chapter include the following: (1) definition accident, (2) accident causation theories, (3) safety performance measurement, (4) product liability principles, and (5) approaches and methodologies used in solving safety-related problems.

BACKGROUND: SAFETY ENGINEERING

DEFINITION OF SAFETY ENGINEERING

Safety Engineering is a specialized engineering field that deals with application of multidisciplinary concepts and techniques to design and evaluate products, systems, and processes with the primary objective of improving safety and providing healthful working environments.

SAFETY PROBLEMS

People build work systems (where different types of work gets done—from producing products to providing services) involving workstations that include equipment (products), people (workers, operators, or users), and environment. Some workplaces may have unsafe working conditions and some people may commit unsafe acts. The products could also be used by people in unsafe conditions and/or while committing unsafe acts. The unsafe acts (committed by people) and unsafe conditions (also caused by unsafe acts committed by people) can cause accidents; and the accidents can cause injuries and losses (incur costs). Accidents can also lead to work stoppages and inefficiencies in work processes. Many research studies involving analyses of the accident data and accident causation factors have shown that the majority of accidents are preventable (Heinrich et al., 1980). Product manufacturers incur safety costs due to (1) litigations and liabilities resulting from the accidents caused by their products, (2) extra efforts undertaken to create safer products through accident prevention actions (e.g., incorporation of safety features in the products and conducting special design reviews and tests during product development to ensure that the products are safe), and (3) occurrences of accidents during the life cycle of the products. The safety-related costs of a product manufacturer can amount to about 5% to 15% of the revenues generated by the products (e.g., add all the costs in conducting literature surveys, benchmarking, safety analyses, safety reviews, purchasing, testing, installing safety devices, and defending litigations).

The need for safety engineering generally becomes obvious when any of the following problems occur: (1) increase in the number of accidents and injuries, (2) increase in the costs or losses resulting from the accidents, (3) employee turnover, (4) employee complaints, and/or (5) increased product litigations.

Figure 12.1 illustrates the safety engineering approach in dealing with existing or new products. The safety problem is generally noticed with the mounting accident statistics obtained from the accident databases, changes in the safety regulations, increases in customer complaints, and/or product liability litigations. The data from these sources are reviewed by the product designers and engineers to determine how various safety problems can occur. Many accident causation theories are considered, and accident situations are analyzed to determine the causes of the accidents. Several hazard analysis techniques (see Chapter 19) are also applied to predict safety-critical situations before any accidents occur. The analyses uncover many potential causes for future accidents that could be eliminated by undertaking one or more accident preventive countermeasures. Cost-benefit analyses are usually performed to determine the countermeasures that can provide higher ratios of the benefit (i.e., reduction in costs due to preventable accidents) to the cost of implementation of the countermeasures. The selected countermeasures are implemented, and the safety performance is monitored (e.g., by maintaining control charts of the accident data) to determine the effectiveness of the countermeasures. If the countermeasures are found to be ineffective, then further changes in safety prevention strategies are considered by iterating the whole process as shown by the feedback loop in Figure 12.1.

Safety Engineering in Product Design

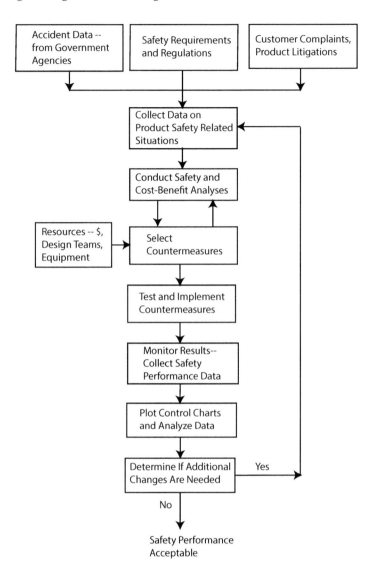

FIGURE 12.1 Safety engineering approach.

IMPORTANCE AND NEED OF SAFETY ENGINEERING

Safety engineering is an important field as it seeks to: (1) design safer products and (2) build and operate safer systems. Safer products and systems are generally efficient to use as they tend to reduce costs and time losses due to accidents. Thus, safety engineering should be considered as one of the important disciplines during the entire product life cycle, and it is especially useful during the early phases of product and system development.

Goetsch (2007) has pointed out that need for safety engineering becomes more acute when any one or more of the above problems are observed in an organization: (1) rapidly increasing safety-related costs, (2) need to meet new safety regulations, (3) increasing litigations, (4) growing interest in ethics and corporate responsibility, (5) increased pressure from labor organizations and employees, (6) realization that safer products and systems are more productive, (7) realizing that safety and quality are closely related, (8) greater awareness and professionalization of health and safety, and (9) new hazards due to faster pace of technological changes.

3Es OF SAFETY ENGINEERING AND COUNTERMEASURES

In developing safe products, engineers should always think about the "3Es" of safety engineering, which include the following:

1. First "E" for *E*ngineering: Products should be designed and engineered with safety in mind.
2. Second "E" for *E*ducation: Safety education will help the designers and users to understand the importance of safety.
3. Third "E" for *E*nforcement: Safety requirements and safety practices must be enforced (through approaches such as training, incentives, inspections, audits, regulations, fines or penalties, and product recalls) to ensure that people act responsibly.

Thus, commonly considered safety countermeasures include the following:

1. Engineering solutions (e.g., incorporation of fail-safe designs, lock-outs, and alterations to products and processes to minimize accidents and injuries)
2. Administrative solutions (e.g., screening employees, limiting exposures to unsafe/toxic environments, rotating people, establishing policies, regulations/laws, practices/procedures, enforcement, training, and awareness of hazardous situations)
3. Personal protection (e.g., isolation of people from hazards, providing hard hats, safety goggles, masks, and seat belts)

METHODS USED IN SAFETY ENGINEERING

The techniques of safety engineering should be applied by engineers in designing complex products and systems to minimize the probability of safety-critical failures. The "Systems Safety Engineering" function helps to identify "safety hazards" in emerging designs and can assist with the techniques to "mitigate" the effects of potentially hazardous conditions that cannot be designed out of systems.

The methods used by safety engineers are listed as follows:

1. Critical incident technique
2. Behavioral sampling
3. Checklists to identify hazards

Safety Engineering in Product Design

4. Hazard analysis (or methods safety analysis)
5. Fish diagram (cause and effect diagram)
6. Failure modes and effects analysis (FMEA) and failure modes, effects, and criticality analysis (FMECA)
7. Logical analysis and fault tree analysis (FTA)
8. Reliability analysis
9. Risk analysis
10. Cost-benefit analysis
11. Accident data analysis (e.g., data gathering and statistical analysis of accident frequency, rates, severity, and accident costs)
12. Accident investigation
13. Accident reconstruction and accident simulation
14. Control charts (of different types for monitoring occurrences of hazards, unsafe acts, unsafe conditions, and accidents)
15. Experimental studies (e.g., to determine the effects of countermeasures on near-accident and accident rates by comparing "before" the countermeasures to "after" the countermeasures data)

The first 10 of the methods listed above can be conducted without waiting for accidents to occur. The accident data-based methods (#11 and above) can be applied after the accidents have occurred. The control charts and experimental studies can be based on measurements of non-accidents (e.g., unsafe acts, unsafe conditions, hazards, or near-accidents) or accidents. The above methods are described in Chapters 14 and 19.

HISTORIC BACKGROUND

Some historic points in the evolution of safety engineering are briefly described below.

1. *Circa 1800*: Before the industrial revolution in Britain and the United States (circa 1800), claims of injury or damages were predicated on the laws of contracts/privities and trespass (Hammer, 1980). (The rule of privity can be described as follows: A seller is liable for injury by his product only to the party with whom he has contracted to supply the product.)
2. *1830–1910*: Industrial Revolution (after circa 1830), Factory Inspections (1835–1875) (e.g., guarding of dangerous machines), 1880 Boiler Standards and 1910 Workmen's Compensation Acts played major roles in advancing safety in the industry. (The workmen's compensation can be described as: "No fault" liability which limits damages an employer has to pay for an injury at work and limits the right of employees to sue their employers.)
3. *1925–1960*: Safety Research was primarily conducted by the insurance companies (responsible for settling injury- and accident-related claims covered by the insurance policies).
4. *1963–1970*: A number of government safety laws were enacted by Congress (e.g., the Consumer Products Safety Act and the Motor Vehicle Safety

Act). Many federal safety agencies (e.g., the Consumer Products Safety Commission and the National Highway Traffic Safety Administration) were created to develop and enforce safety regulations.
5. *1963–1973*: A number of product safety and liability cases and decisions (e.g., strict liability, negligence and implied warranty). (Note: Product Liability concepts are presented in a later section entitled "Product Safety and Liability" of this chapter.)
6. *1973–present*: Strong awareness of safety issues due to presence of federal and state safety regulations, enforcement activities, and well-maintained accident databases by government agencies and private organizations. Thus, safer products are now expected by all.

DEFINITION OF AN ACCIDENT

A number of safety researchers have provided definitions to describe an accident, that is, when a situation would be called an accident. The definitions also help in understanding the concepts of an accident and issues related to accident prevention. A few commonly referred accident definitions are provided below.

1. An accident is a set of complex events involving sequence, human actions/behavior (unsafe acts), unsafe conditions, and some degrees of the following characteristics (Petersen and Goodale, 1980):
   a. *Degree of unexpectedness*—the less an event could have been anticipated, the more it is likely to be called an accident.
   b. *Degree of avoidability*—the less the event could have been avoided, the more likely it is to be called an accident.
   c. *Degree of intention*—the less the event resulted from a deliberate action or lack of an action, the more likely it is to be called an accident.
   d. *Degree of warning*—the less warning, the more likely it is to be called an accident.
   e. *Duration of occurrence*—the more quickly it happens, the more likely it is to be called an accident.
   f. *Degree of negligence*—the more reckless or carelessness involved, the less likely it is to be called an accident.
   g. *Degree of misjudgment*—the more mistakes in judgment involved, the less likely it is to be called an accident.
2. An accident is any unplanned and uncontrolled event caused by human, situational, or environmental factors, or any combination of these factors that interrupts the work process, which may or may not result in injury, illness, death, property damage, or other undesired events, but which has a potential to do so (Colling, 1990).
3. An accident is an unplanned and uncontrolled event in which the action or reaction of an object, substance, person, or radiation results in personal injury or probability thereof (Heinrich et al., 1980).
4. Accident is an unplanned, not necessarily injurious or damaging event, which interrupts the completion of an activity, is invariably preceded by an

unsafe act and or an unsafe condition or some combination of unsafe acts and/or unsafe conditions (Tarrants, 1980).

ACCIDENT CAUSATION THEORIES

Many researchers have proposed theories to explain how accidents occur. It is important to understand the theories so that countermeasures can be generated to reduce the occurrences of the accidents. These theories also help in undertaking accident-preventing actions during the development of safe products and processes. This section provides brief descriptions of several accident causation theories that are useful in designing safe products. More detailed descriptions of the theories are provided by Petersen and Goodale (1980).

1. *Act of God* (Demons or other supernatural forces): This theory is recognized in the legal literature and by the insurance industry that some accidents can only be explained as an "act of God." The theory assumes that such acts are bad happenings outside an individual's control (or reasons such as the victims were presumably marked for punishment because of some unknown quality, the devil did his handiwork, etc.).
2. *Accidents are "rare and random" events*: This theory recognizes that accidents are very low probability events, and they can happen and do happen to anyone. Here, an accident is considered as a "lottery"—an event whose outcome seems to be determined by chance. In early 1900s, the accidents to a large extent were considered uncontrollable and unpredictable. Minimum thought was given to design of environments that could reduce probability of an accident or harm. It also suggests that repeated violations of common-sense safe practices eventually and invariably will lead to an accident.
3. *Accident-prone theory*: This theory assumes that accidents are caused by some invariant human characteristics identified as "accident proneness." Accident proneness may be defined as the continuing or consistent tendency of a person to have accidents as a result of his or her stable characteristics (or response tendencies). Such accident-prone people are also called accident "repeaters," that is, they are involved in repeated number of accidents (or are "over-involved" in accident occurrences as compared to normal individuals).

    (Note: The accident proneness theory can be tested by analyzing accident data. Compare an "observed" distribution of the number of individuals involved in x number of accidents in a population with a "fitted" distribution of the expected number of individuals involved in x number of accidents assuming Poisson distribution ($P[x]$, where x is the number of accidents occurring to an individual and $x = 0, 1, 2, 3, \ldots$). Statistically significant difference between the "observed" and "fitted" (expected) distributions suggests the presence of accident proneness. This suggests that individuals with certain accident-prone characteristics are "over-involved" in accidents; that is, accident-prone individuals have larger values of x than others with smaller values of x.)

4. *Chain of multiple events*: This model assumes the existence of many factors influencing accidents rather than any key cause. The probability of an accident (P) that will occur in a given unit of activity (A) is assumed to be a function of a whole set of factors and conditions. If these factors are designated as $x_1, x_2, x_3, x_4, x_5, \ldots, x_k$, the probability of an accident in the activity would be a function of the factors (i.e., $P_A = f[x_1, x_2, x_3, x_4, x_5, \ldots, x_k]$).
5. *Energy exchange model*: Most accidents are caused by unplanned or unwanted release of excessive amounts of energy (e.g., mechanical, electrical, chemical, thermal, and ionizing radiation) or hazardous materials (e.g., carbon monoxide, carbon dioxide, hydrogen sulfide, methane, and water). (However, with a few exceptions, these accidents due to energy releases can also be explained by the unsafe acts theory where an unsafe act may trigger the release of large amounts of energy or a hazardous material, which in turn causes the accident.)
6. *Epidemiological model*: Epidemiology is the study of causal relationship between environmental factors and disease. The epidemiological theory suggests that the models used for studying diseases can also be used to study causal relationships between environmental factors and accidents. The model assumes that the key components are (a) predisposition characteristics (i.e., a susceptible host, a disease-producing agent or a virus, and a hazardous environment) and (b) situational characteristics (e.g., not wearing sufficiently warm [low insulation] clothing in a cold environment, risk-taking, and an untrained host). The predisposition characteristics are assumed to create a disease-producing condition or an accident-like condition (i.e., unexpected, unavoidable, or unintentional), which in combination with the situational characteristics results in an accident that causes injuries or damage (see Figure 12.2).
7. *Domino theory*: An accident is caused by a sequence of events. Each event can be assumed to be represented by a domino.
    The domino sequence is as follows (see Figure 12.3):
    a. Faults of persons are created by the environment or acquired by inheritance.
    b. Unsafe acts and conditions are caused by faults of persons.
    c. Accidents are caused by unsafe acts of persons and/or exposure to unsafe conditions.
    d. Injury results from an accident.
    The Domino theory thus suggests that the above four-events sequence involved in the causation of an accident can be broken by removing any one

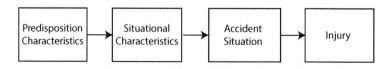

FIGURE 12.2 Sequence of events in the epidemiological model of accident causation.

Safety Engineering in Product Design

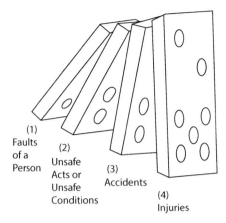

FIGURE 12.3 Sequence of dominos in the domino theory of accident causation.

of the events (like removal of a Domino in the chain) to prevent an accident from occurring.

8. *Unsafe acts theory*: Accidents occur primarily due to unsafe acts of people. The unsafe acts occur due to the following reasons: (1) misunderstanding of instructions, (2) lack of knowledge or training, (3) recklessness or violent temper, and/or (4) actions that exceed human capabilities and limitations (e.g., speed of response needed was beyond operator's capability to react).

9. *Human factors or human error models*: There are several human factors models that can be related to accident causation. The models postulate that human failures occur due to reasons such as (1) task demand exceeds operator capabilities, (2) operator experiences information overload, (3) operator's attention is diverted/distracted, (4) operator is not consistent in his response (i.e., variability in human operator's output is too high), (5) operator is under stress, and (6) operator fails to get the right information at the right time [or fails to process the information] needed to make the decision and thus does not make the right response needed to avert the accident. Most human failures can also be explained as a result of one or more human errors. Thus, the human factors models can also be considered to originate from an occurrence of a human error.

Many of the human failures are due to information processing errors where the human operator fails to make the right decision at the right point in time, and thus, the accidents are caused by human errors. Some examples of human errors are as follows: (1) interpretational errors (i.e., errors in interpreting situations or signal interpretation; for example, the operator misunderstood the meaning of the "red" flashing light on his instrument panel), (2) substitution errors (i.e., substituted a different action instead of the intended action; for example, the driver pressed the gas pedal instead of the brake pedal—substituted a wrong control), (3) reversal errors (i.e., operator responded with an action in the opposite direction instead of the intended direction; for example, turned a control clockwise instead of counterclockwise), (4) legibility errors (e.g., operator could not read a display due to small font size, poorly lit display at night, or sunlight reflection glare from the display lens), (5) forgetting errors or omission errors (i.e., forgot to perform an action; for example, pilot took-off without reading the fuel gage), (6) commission errors (i.e., performed a task or step when not required or in a different sequence), (7) other errors (i.e., control operational errors due to violation of one or more human factors principles in the design of the control; for example, a control is not located at an expected location, the direction of control motion violates its direction-of-motion stereotype, the control not located with other controls of the same functional group, the control was not located close to its associated display, etc.)

SAFETY PERFORMANCE MEASURES

Two types of measures (i.e., variables) are used to measure safety performance. They are (1) accident-based measures (i.e., based on accident data) and (2) non-accident measures (i.e., based on measurements of data from events other than accidents. For example, behaviors exhibited during the use of a product, unsafe acts committed, or errors made in operating equipment). The accident-based measures are more believable (or "hard") as compared with non-accident measures that are regarded as "soft" measures of safety. The advantages of the non-accident measures are that they can be obtained without waiting for accidents to occur and the non-accident events occur at much higher rates (more frequent) than accidents that in general occur very rarely. On the other hand, the accident-based measures are "hard" or ultimate measures of safety and have higher face validity.

Why Measure Safety Performance?

Measurements are essential to determine the level of a problem or effects of changes in the design of a product or a system. The measurement of safety performance allows us to assess the following types of problems:

1. State of safety, that is, accurately determining the level of safety in an operation or effectiveness of a product and/or process in achieving safety objectives
2. Assist in business planning and safety improvement activities
3. Allow us to evaluate, compare, or calibrate accident prevention initiatives
4. Provide feedback on past safety actions
5. Predict future safety costs.

Currently Used Accident Measures

Some examples of currently used accident-based safety performance measures are:

1. Number of accidents
2. Number of injury accidents
3. Number of persons injured in accidents

Safety Engineering in Product Design

4. Number of disabling injury accidents
5. Number of fatal accidents
6. Number of fatalities in accidents
7. Incident rates (e.g., lost time [disabling] injury frequency rate, number of accidents per 200,000 work hours)
8. Lost workdays (number of workdays lost due to accidents)
9. Accident costs.

The above measures are computed over a pre-defined exposure (e.g., time duration or number of product usage cycles). The accidents can be also categorized by using a number of classification criteria, such as accident type, accident severity, accident location, characteristics of a person involved in the accident, type of equipment, and type of operation or environment involved in the accident.

ACCIDENT-BASED INCIDENT RATES

Incident rates are popularly used to measure safety performance in work-related industries (Tarrants, 1980; Goetsch, 2007; NSC, 2012). The commonly used incident rates based on incidents of injuries, illnesses, or fatalities are defined as follows:

$$IR = \frac{(N \times 200,000)}{T}$$

where IR = Total injury and illness incident rate

N = Number of injuries, illnesses, and fatalities resulting from the accidents
T = Total hours worked by all employees during the period in question
200,000 hours = 100 employees working 40 hours/week times 50 weeks in 1 year.

The incident rates based on the 200,000 hours of exposure are as follows:

1. Injury rate
2. Illness rate
3. Fatality rate
4. Lost workday cases rate (accident cases where at least one workday was lost due to an accident)
5. Number of lost workdays rate
6. Specific hazard rate
7. Lost workday injury rate

The problems and issues with the use of the incident rates are as follows:

1. Since they are based on accidents, they are postmortem. Thus, one has to wait for accidents to occur before the value of the measure can be computed. Thus, long periods of time need to elapse before reliable estimates of the safety measure can be obtained.

2. Since accidents are rare events, they are unreliable for small work organizations (i.e., they do not accumulate a large number of work hours).
3. They are not very useful to predict the effectiveness of safety countermeasures in shorter time periods.

Advantages and Disadvantages of Current Accident-Based Measures

Some advantages of the currently used accident-based measures are as follows:

1. Quick acceptance (i.e., they are an "accepted standard") as compared with non-accident data, which may be regarded as questionable by many decision-makers
2. Motivate management (i.e., they get management's attention and motivate them to take prompt actions)
3. Long history of use
4. Used by government agencies (e.g., U.S. Occupational Health and Safety Administration) and industry associations
5. Easy to calculate
6. Indicate trends in performance
7. Good for self-comparison.

Some disadvantages of the currently used accident-based measures are as follows:

1. They can only be computed after the accidents have occurred (i.e., they are reactive or postmortem).
2. The numbers can be easily manipulated as many unreported accidents (intentionally or unintentionally) can cause underestimation of the safety problem.
3. They may be biased (due to management attitude to restricted work, doctor influence on reporting, worker attitude to light duties, compensation system, motivation to achieve safety awards, and competitions between organizations based on safety performance).
4. The measured number of accidents is typically low making it difficult to establish trends (accidents, in general, are rare events).
5. The accidents differ in severity (i.e., the severity of injuries, amount of property damage, and losses are different in different accidents). Thus, comparisons based only on the number of accidents can be misleading.
6. Some managers or safety specialists may regard an accident as a "once off or a freak" event. (Thus, may disregard the accident data.)

Non-accident Measures

Unlike the accident-based measures, the non-accident measures are not standardized by government agencies or industries. However, methods have been used (e.g., Behavior Sampling and Critical Incident Technique, see Chapter 19) to measure unsafe acts, unsafe conditions, and errors. The frequency and occurrence rates of such incidences have been used to evaluate safety performance (Tarrants, 1980).

Safety Engineering in Product Design

Safety costs have been routinely tracked by many organizations. They include (1) costs due to accidents and (2) accident prevention costs. The costs due to accidents are generally underestimated due to unreported or unaccounted accidents. Incidental costs of accidents have been estimated to be four times as great as the actual costs. The accident prevention costs include costs of safety analyses, engineering changes, evaluations/tests, reviews, training, protection devices, and so forth. (Note: Safety costs are summarized in a later section of this chapter.)

SAFETY ANALYSIS METHODOLOGIES

Two Possibilities: Accident versus Hazard

The distinction between accident and hazard can be understood by considering the following two considerations:

1. *Accident*: Accident is an event in which damage to property or injury to personnel has occurred (or accidents are occurring, i.e., accident cases and data are accumulating).
2. *Hazard*: Hazard is a real or potential condition that could cause damage to property or injury to personnel but has not occurred so.

The following two possibilities need to be considered prior to deciding on the type of safety analysis and methods to use (Hammer 1980, 1989; Colling, 1990).

1. *An accident has occurred or accidents are occurring*: This possibility leads to (a) accident investigation and (b) accident analysis. An accident investigation usually precedes an accident analysis.
2. *Not waiting for accidents to occur*: This possibility leads to conducting hazard identification and hazard analysis. The hazard analysis involves the following: (a) hazard identification, (b) determining whether controls are in place to prevent occurrences of hazards, (c) formulate countermeasures, and (d) select the best countermeasures to implement to avoid future accidents.

Accident Analysis Methods

Accident analysis can be considered to include the following methods: (1) Accident investigation, (2) accident analysis, and (3) accident data analysis. These three methods are not distinctively different and there is considerable overlap between their contents. The applications of the methods can also vary depending on the accident researcher involved in performing the analyses. These three methods are described below.

1. *Accident investigation*: The accident investigation involves reading the accident report, visiting the accident site, talking to the witnesses, gathering all facts about a particular accident such as who was involved, how the accident occurred, what the injuries and losses were, and so on. A detailed accident investigation typically involves: (a) reading individual accident

reports, (b) sending independent accident investigators (or a team of multidisciplinary experts—in case of detailed investigation) to verify the details of the accident, (c) reconstruct the accident (i.e., describe how the chain of events led to the accident), and (d) preparing a detailed report on each accident case.

2. *Accident analysis*: An accident analysis usually involves more than one accident of a given type (e.g., accidents involving a particular product model under a certain type of situation, while performing a certain task or a maneuver). The accident analysis involves: (a) collecting and analyzing accident data, (b) determining the causes and circumstances of the accidents, (c) investigating possible chains of events that led to the accidents, and (d) creating a model of the accident situation to reconstruct and illustrate details about the behavior of various elements or events that led to the accidents.

3. *Accident data analysis*: The accident data analysis usually involves the following: (a) securing access to one or more accident databases, (b) understanding variables and categories used in creating the database, (c) evaluating completeness of the data (i.e., understanding missing data or uncategorized variables that are generally categorized as "other" or "not available"), (d) creating tabular summaries of accident data based on relevant variables of interest, and (e) conducting statistical tests to determine if any of the differences due to variables and their categories are statistically significant.

The methods used for accident analysis are described in Chapter 19.

HAZARD ANALYSIS METHODS

The methods used for hazard analysis also have some overlap in content and differences in formats and details depending on the individuals or organizations conducting the analyses. The methods used for hazard analysis are listed as follows:

1. General hazard analysis
2. Detailed hazard analysis
3. Methods safety analysis (like operations analysis)
4. Job safety analysis/job hazard analysis (to uncover hazards in a job)
5. Failure modes and effects analysis (FMEA)
6. Failure modes and effects and criticality analysis (FMECA)
7. Fault tree analysis (FTA)
8. Error analysis
9. Human reliability analysis.

The methods used for hazard analysis are described in Chapter 19. More information on the above-listed hazard analysis methods can be obtained from Brown (1976), Hammer (1980), Colling (1990), Roland and Moriarty (1990), and Goetsch (2007).

PRODUCT SAFETY AND LIABILITY

Product manufacturers are sued in large numbers by users, misusers, and even abusers of their products. Injuries resulting from the use (or often misuse) are the basis for an increasing number of product liability lawsuits (Hammer, 1980; Goetsch, 2007). The suits cost the industries millions of dollars each year. Therefore, the objectives of this section are to provide the reader with background into basic concepts and issues related to the product liability, to understand the role and importance of safety analyses in product and systems design, and to help a product engineer communicate with a product liability lawyer.

The best way a manufacturer can prevent or defend such claims is by manufacturing a reasonably safe and reliable product (or system or process), and where necessary by providing instructions for its proper use. The key to achieve a reasonably safe and reliable product and to reduce the product liability exposure is to "build in" product safety during the early design stages of the product.

Terms and Principles Used in Product Litigations

The following terms used in product litigations will help the reader understand the issues (Hammer, 1980; Goetsch, 2007).

1. *Liability*: It can be defined as an obligation to rectify or to recompense for any injury or damage for which the liable person has been held responsible or for failure of a person to meet a warranty. Here, the product user is the loser (or injured) and is assumed to be demanding compensation for the injury, losses, and/or sufferings caused by the product.
2. *Plaintiff*: A person (or a party) who starts a legal case against another to obtain a remedy for an injury caused by the product.
3. *Defendant*: A manufacturer (or the seller) who is faced with proving that the product is safe.
4. *Three major legal principles*: The following three principles are generally considered in establishing liability: (a) negligence, (b) strict liability, and (c) breach of warranty (Hammer, 1980).
   a. The negligence principle involves failure of the defendant to exercise a reasonable amount of care in the design and manufacture of his product (or to carry out a legal duty) so that injury or property damage does not occur to a user or other person. Thus, here the focus is on the conduct of defendant (manufacturer), i.e., his duty and/or care. The plaintiff must prove that the defendant's conduct involved: (i) an unreasonably great risk of causing damage, (ii) defendant failed to exercise ordinary or reasonable care, and/or (iii) not using available knowledge (e.g., new developments, design methods, safety practices, and safety devices) that would have decreased the level of risk.
   b. The principle of strict liability is based on the concept that a manufacturer of a product is liable for injury due to a defect, without the necessity for a plaintiff to show negligence or fault. Here, the plaintiff must

only prove that the product was defective, unreasonably dangerous, and the proximate cause of the harm. Thus, in the strict liability, the focus is on the quality of product—rather than the fault of the manufacturer (i.e., regardless of whether or not the manufacturer acted reasonably). The manufacturer is said to be "strictly liable" because his liability does not depend on his own conduct or care. Therefore, defense is particularly difficult and frustrating to the manufacturer. This is the basic cause for what is called the product liability crisis.
  c. The breach of warranty can involve the following two principles: (i) implied warranty and (ii) expressed warranty and misrepresentation.

The principle of implied warranty involves an implication by a manufacturer or dealer that a product is suitable for a specific purpose or use, is in good condition, or is safe by placing it on sale. The implied warranty of safety is the principle that any product by being placed on sale is implied to be safe. The implied warranty of merchantability implies that the product sold is in as good condition as other products of its type. The implied warranty of fitness implies that the product is suitable for the purpose for which it is sold.

The principle of expressed warranty involves a statement by a manufacturer or dealer, either in writing or orally, that his product will perform in a specific way, is suitable for a specific purpose, or contains specific safeguards.

Product Defects

The first step in product liability (in cases where no express warranty or misrepresentation is involved) is to prove that the product was defective. (Note: It is not sufficient to establish that the product was dangerous (e.g., even a knife can be dangerous). Thus, the product must have a design defect (i.e., a defect in its basic design) or a manufacturing defect (e.g., a flaw in the manufacturing process).

Some examples of design defects are as follows: (1) a concealed danger created by the design, for example, a sharp edge after a collapse in an accident, (2) needed safety devices not included, and (3) involved materials of inadequate strength or failed to comply with accepted standards.

Some examples of manufacturing defects are as follows: (1) poor quality material used for structural components, (2) failed to meet required material hardness (e.g., failure in the heat treating process), (3) a sharp edge or flash left on a grasp handle (e.g., operator forgot to grind the sharp edge, which can lead to an injury), and (4) improper assembly (e.g., misaligned parts, loose parts, missing parts, wrong electrical connections [transposed wires]).

Warnings

The courts have recognized the concept of the product manufacturer's duty to warn the users about potential safety-related considerations and problems. The general position of the courts is to always use warnings. The cost of supplying warnings is low as generally only the printing costs are involved (e.g., warning labels, warning

Safety Engineering in Product Design

messages included in screens or user's manuals). In many cases, the manufacturers have been found to be liable for failure to warn, even when the (missing) warning would have been of dubious value. From the Human Factors perspective, too many warnings are ineffective as people will generally disregard frequent occurrences of warnings. But courts view the provision of warnings as desirable and thus it may have some effect on reducing the manufacturer's negligence (due to failure to warn). Thus, the presence or absence of warnings is generally an issue as compared with their effectiveness.

Engineering and management, thus, can be vulnerable in the following areas: (1) product design; (2) product manufacturing and materials selection; (3) packaging, installation, and application/use of the product (i.e., operation); and (4) warning labels. Failure to comply with the regulations (i.e., applicable standards) can mean that the manufacturer is negligent. Compliance to government standards (which are generally minimum standards) may provide some protection against negligence-related cases but it offers no protection in strict liability cases.

SAFETY COSTS

Safety costs are all costs associated in incorporating safety in the product design and ensuring that the products operate safely during their operational life. These costs include the following:

1. Costs incurred in creating safe products:
   a. Costs associated in gathering data on safety regulations, past accidents, and litigations-related data
   b. Costs associated in developing safety requirements, cascading requirements to various systems, subsystems, and components of the product (Note: Safety is a product attribute.)
   c. Costs to design and implement safety features in the product (e.g., costs associated in conducting safety analyses, product safety design reviews, meetings with experts, management, government agency experts, and lawyers)
   d. Costs associated in following special safety precautions (e.g., checks, inspections) during manufacturing and assembling
   e. Verification and validation costs (e.g., safety testing costs)
2. Costs incurred during product uses/operations
   a. Costs associated in gathering data on safety-related incidences (e.g., meeting with customers, users, dealers, repair shop personnel, government agency personnel, and lawyers; investigating product failures and accidents)
   b. Conducting safety analyses and tests
   c. Providing technical and legal support on product litigations, recalls, repairs, fines, customer relations campaigns, and so on.
   d. Costs associated with fixing the defects (i.e., product recalls, repairs, retests)

3. Safety costs related to product discontinuation and disposal
   a. Disposal or recycling of retired products
   b. Disposal of plant equipment and hazardous/toxic substances (e.g., toxicity tests after disposal).

SECURITY CONSIDERATIONS IN PRODUCT DESIGN

Increased awareness of security issues with computers, databases, and terrorist attacks (e.g., cyberattacks) has raised many issues with the security of large complex products (e.g., commercial aircraft, cruise ships, and power plants).

Some important security issues involve: (1) lockout of products from unauthorized users, (2) shielding products from viruses and security threats, (3) resilience from threats (how to make the product performance insensitive to threats), (4) system shutdown or operating under high threat levels, and (5) the abilities of the product to function during and after threat incidences.

Product designers, especially with embedded computerized control systems, need to include the security issues along with the safety issues discussed in this chapter.

CONCLUDING REMARKS

Product safety is an important discipline, and the product designers must make sure that their products are safe. Safety requirements must be incorporated very early in the product design and product attribute engineering process. Safety reviews must be conducted during both the design and manufacturing phases to ensure that the products do not have any design and manufacturing defects that will increase risks to the users as well as the manufacturer. It should be realized that during the product liability cases, the manufacturer is considered to be an expert and very knowledgeable about the product safety requirements, available safety devices, safety technologies, and safety-related design and manufacturing considerations. Furthermore, the manufacturer is also expected to provide warnings to the users about any potential hazard associated with the uses of the product. The design engineers should maintain proper records on safety analyses and safety-related decisions based on potential benefits of the product to the customers versus costs incurred to make the product reasonably safe for defending their decisions if challenged during any future product reviews and liability cases.

REFERENCES

Brown, D. B. 1976. *Systems Analysis and Design for Safety-Safety Systems Engineering.* Englewood Cliffs, NJ: Prentice-Hall, Inc.

Colling, D. A. 1990. *Industrial Safety Management and Technology.* Englewood Cliffs, NJ: Prentice Hall.

Goetsch, D. L. 2007. *Occupational Safety and Health for Technologist, Engineers, and Managers.* Sixth Edition. ISBN: 0132397609. Englewood Cliffs, NJ: Prentice Hall.

Hammer, W. 1980. *Product Safety Management and Engineering.* Englewood Cliffs, NJ: Prentice-Hall, Inc.

Hammer, W. 1989. *Occupational Safety Management and Engineering*. Fourth Edition. Englewood Cliffs, NJ: Prentice Hall.

Heinrich, H. W., Petersen, D. and N. Roos. 1980. *Industrial Accident Prevention*. Fifth Edition. New York: McGraw-Hill, Inc.

National Safety Council (NSC). 2012. *Injury Facts*. 2012 Edition. Spring Lake Drive, Itasca, IL: National Safety Council 1121. Website: http://shop.nsc.org/Reference-Injury-Facts-2012-Book-P124.aspx (accessed October 25, 2012).

Petersen, D. and J. Goodale. 1980. *Readings in Industrial Accident Prevention*. New York: McGraw-Hill, Inc.

Roland, H. E. and B. Moriarty. 1990. *System Safety Engineering and Management*. Second Edition. New York: John Wiley & Sons, Inc.

Tarrants, W. E. 1980. *The Measurement of Safety Performance*. New York: Garland Publishing, Inc.

13 Design for Sustainability

INTRODUCTION

The concept of sustainability as applied to products and systems has expanded now to how the products or systems affect energy usage, environment, resource use efficiency, and social responsibility. Thus, the product or a system designer has to select or create a design that is socially responsible in terms of least impact on the environment (i.e., polluting potential), energy consumption, and economic effects by considering factors such as safety, health, economic benefits (e.g., job creation), and reduced use of resources (e.g., reduced water consumption). The manufacturers of products are finding that they are increasingly held responsible for the products that they create because of the effects of the products on the society. Thus, this chapter is about how sustainability can be assessed and evaluations can be used to judge how the product (or system) is developed.

ISO 14001 also recognizes the need for environmental management and provides guidelines for development of an effective Environment Management System (EMS) which can give companies a significant competitive advantage (ISO, 2021). ISO 14001 certification will ensure that the company's business is focused on its environmental impact, supported by effective management processes.

WHAT IS SUSTAINABILITY?

Design for Sustainability (DFS): It is a strategy that focuses on incorporating systematic changes in design thinking to foster sustainable patterns in production and consumption. This can be achieved by focusing on enhancing the functionality, innovation, life span, and resource efficiency of products and services by considering environmental, social, and economic aspects (Niinimäki and Hassi, 2011).

Design for sustainability can be also considered as an approach that puts the well-being of people and the sustainability of the environment first. To date, sustainable design has generally been a reaction to global environmental crises and focuses on resource efficiency. Whilst efficiency is important, growth in goods and services consistently outpaces efficiency gains.

Definition of Sustainable Design: Sustainable design means product design is intended to minimize negative environmental impacts, to promote the health and comfort of users, and to take measures to reduce consumption of nonrenewable resources, minimize waste, and to create healthy, productive environments.

Sustainable design principles include the ability to (a) optimize site potential (where a new system will be installed), (b) minimize non-renewable energy consumption, (c) use environmentally preferable products, (d) protect and conserve water, (e) enhance indoor environmental quality, and (f) optimize operational and maintenance practices.

The growing popularity of sustainability and its implications for the practice of engineering, particularly for the built environment involves two particular methodologies (a) life cycle assessment and (b) leadership in energy and environmental design. Design for Sustainability, thus, is the response to this challenge. It includes but goes beyond what Design for the Environment (DFE) or eco-design provides, by integrating social, economic, environmental, and institutional aspects and by offering opportunities to get involved, express one's own (or organization's) identity beyond consuming standardized mass products.

What Is a Sustainable Product?

Considering the broader definition of sustainability that includes many above aspects, a sustainable product or a system should have the following characteristics:

1. Uses less energy—It is energy efficient and reduces energy consumption and waste.
2. Eco-friendly (Eco-products)—It minimizes negative impacts on the environment (or designed for environment). It is green, natural, non-toxic, compostable, recyclable, less polluting, and less waste producing, and produces healthy living and operating conditions.
3. Lean—It uses fewer resources (money, materials, and energy) and recycles to reuse materials.
4. Using energy from renewable sources—It avoids production of pollutants (e.g., CO_2, methane, and nitrous oxides) generated by using coal-fired electric power plants.
5. Socially responsible—It cares about society and its well-being.
6. Safe, comfortable, and productive—It does not cause injuries, discomfort, and waste of materials or energy.
7. Long lasting with increased life span—It promotes durability.
8. Designed for disassembly—Products and systems can be disassembled easily to remove recyclable materials.

Sustainable products are those products that provide environmental, social, and economic benefits while protecting public health and environment over their whole life cycle, from the extraction of raw materials (to produce the products) until their final disposal.

Life Cycle Consideration

All products have an effect on the environment and society, which may occur at any or all stages of their life cycle—from raw material acquisition, manufacture, distribution, use, and retirement/disposal (see Chapter 21). Design for the life cycle is a systematic way of considering the entire life or life cycle of a product from the beginning of its concept development to its disposal through recycling or scrappage (i.e., from birth-to-death, or cradle-to-grave considerations). The goal is to reduce environmental effects (reduce toxic substances), avoid materials that cannot be reused or

Design for Sustainability

recycled economically, reduce use of resources (e.g., materials, energy, and costs) in producing products, and produce products that are safe (i.e., non-injury causing), comfortable (easy to use), durable (long-lasting—reducing unnecessary production, consumption, and disposal problems (e.g., use bio-degradable materials)).

TOOLS/METHODS USED FOR SUSTAINABILITY ANALYSES

1. Life cycle analysis (LCA) (Evaluating costs and benefits over the entire life cycle of the product [or the system] being analyzed)
2. Eco-Design (eliminating or reducing pollution toxic substances, using more bio-degradable materials), and neutrality toward environment or design for environment (DFE). Eco-logical design can reverse adverse effects on health, safety, comfort, and productivity due to pollution, waste, and other ills.
3. Design for durability
4. Design for manufacturing and assembly (DFM and DFA techniques for evaluating and reducing costs of manufacturing and assembly. See Chapter 16)
5. Design for disassembly (reducing the cost of separating components of different materials for recycling/reuse)
6. Design for waste reduction and reusability
7. Design for recyclability
8. Socially responsible design (i.e., products and systems have socially acceptable characteristics).

In implementing the above analysis techniques in creating sustainable products and systems, the companies become proactive as compared to reactive and in doing so introduce innovation in designing their products and systems (Martin, 2016). Value is created across stakeholders through sustainability, including product marketing via product differentiation and brand enhancement, research and development with innovation and design optimization, and finance and operations with identified areas for cost reduction. For example, by collecting data for a life cycle assessment (LCA) study, a company learned that it was using both hot water and chlorine in a sanitizing step. Only one method was necessary to safely sanitize, and the LCA identified that costly redundancy. In addition, value can be created by identifying and acting on potential areas for risk and regulation.

DESIGN FOR ENVIRONMENT

Design for the Environment (DFE) is a design approach to reduce the overall human health and environmental impact of a product, process, or service, where impacts are considered across its entire life cycle. Different software tools have been developed to assist designers in finding opportunities to improve products or processes/services.

<u>EPA's Design for Environment Program</u>: EPA's DFE program encourages the industries to move to safer alternatives, complements regulatory action by showing

that safer and higher functioning alternatives are available, or points out the limitations to chemical substitution for a particular use (EPA, 2021).

ISO 14001 Standard for Managing Environmental Programs: ISO 14001 Environmental Management Standard (EMS) establishes a framework that businesses can follow to develop an effective environmental management system (ISO, 2021; EPA, 2021). ISO 14001 covers environmental management system requirements such as documentation, training, auditing, defining environmental aspects and their impact, performance evaluation, life cycle assessments, leadership, and continuous improvement. This standard intends to protect the earth's environment while permitting the expansion of international trade and commerce.

ISO 14001 international standard is intended to help companies to reduce their carbon footprint and lower costs. It allows the companies to set up a fully integrated and effective EMS that will streamline operations, allocate time and resources more effectively, and increase overall efficiency.

The ISO 14001:2015 document provides the following:

1. It specifies the requirements for an environmental management system that an organization can use to enhance its environmental performance. ISO 14001:2015 is intended for use by an organization seeking to manage its environmental responsibilities in a systematic manner that contributes to the environmental pillar of sustainability.
2. It helps an organization achieve the intended outcomes of its environmental management system, which provide value for the environment, the organization itself, and interested parties. Consistent with the organization's environmental policy, the intended outcomes of an environmental management system include (a) enhancement of environmental performance, (b) fulfillment of compliance obligations, and (c) achievement of environmental objectives.
3. It is applicable to any organization, regardless of size, type, and nature, and applies to the environmental aspects of its activities, products, and services that the organization determines it can either control or influence considering a life cycle perspective.
4. It does not state specific environmental performance criteria.
5. It can be used in whole or in part to systematically improve environmental management. Claims of conformity to ISO 14001:2015, however, are not acceptable unless all its requirements are incorporated into an organization's environmental management system and fulfilled without exclusion.

For example, Dell (2018) follows ISO 14001 for managing environmental programs. Dell (a computer company) considers environmental opportunities and challenges at every stage of the product life cycle—from design and development, manufacturing and operations, to product use and recovery. Product design efforts are guided by corporate environmental policy and governance set to continuously improve their performance. Thus, their products are designed to include environmentally responsible

Design for Sustainability

materials, using efficient designs that require fewer materials and maximize reusability and recyclability throughout the product lifecycle.

Design for Environment also included the following objectives:

1. Design to minimize material usage (e.g., reducing packaging material, avoiding painting by using colored plastic panels)
2. Design for disassembly (e.g., using mechanical fasteners instead of glue)
3. Design for recycling and use of recycled materials (e.g., using cardboard made up of paper product waste)
4. Design for remanufacturing (e.g., refilling used printer cartridges)
5. Design to minimize hazardous materials (e.g., using lead-free solder)
6. Design for energy efficiency (e.g., using automobile equipped with hybrid vs. internal combustion powertrain)
7. Design to meet regulations and standards (e.g., meeting government automotive emission and fuel economy requirements).

DESIGN FOR DISASSEMBLY

DFD involves designing a product to be disassembled for easier maintenance, repair, recovery, remanufacture, and reuse of components or materials. DFD is a part of Design for the Environment (DFE) and sustainable product design. DFD is becoming increasingly recognized as an effective tool by designers, manufacturers, and legislative boards alike.

Implementation of DFD is not only to meet legal requirements set up by WEEE (Waste Electrical and Electronic Equipment Directive) and RoHS (Restriction of Hazard Substance) (European Commission, 2022) but for the following benefits such as:

a. Reducing waste in the manufacturing and recovery processes
b. Reduce production costs and allow for greater technical efficiency
c. Provide greater flexibility during product development, shorter development time scales, and reduced development costs
d. Allows the product and its components to be better suited for re-use or recycling when it has reached its end of life.

Goal of DFD

The goal of DFD is to close the production loop by conceiving, developing, and building a product with a long-term view of how its components can be refurbished and reused, or disposed of safely at the end of the product's life. Three important factors to be considered during the early design process include:

a. Selection and use of materials
b. Geometric design of components and product architecture
c. Selection and use of fasteners.

DFD AND DFA GUIDELINES

DFD through component design and product architecture shares many of the principles used in DFA (Design for Assembly, see Chapter 16). Dowie-Bhamra (1996) provided a number of guidelines for DFD. The relevant guidelines and principles from Dowie-Bhamra (1996) and Fauldi (2015) are summarized below.

The guidelines for DFD are as follows:

1. Minimize the number of components used in an assembly, either by integrating parts or through system re-design. Reducing the number of components in an assembly will in general reduce total disassembly time.
2. Minimize the number of material types used in an assembly
3. Separate working components into modular sub-assemblies
4. Construct sub-assemblies in planes which do not affect the function of the components
5. Avoid using laminates which require separation prior to re-use
6. Avoid painting parts as only a small percentage of paint can contaminate and prevent an entire batch of plastic from being recycled
7. Choose recycling-compatible materials
8. Avoid using materials which require separating before recycling
9. Use as few components and component types as possible (without compromising the structural integrity or function of the product)
10. Integrate components (which relate to the same function) where possible
11. Standardize the use of fasteners—use commonly available parts and maintain consistency within the design
12. Make components easily separable
13. Apply non-contaminating markings (e.g., through etching or molding) to materials for ease of sorting
14. Maintain good access (clearances for hand-fingers and tools, and visibility) to components and fasteners. Consider making the plane of access to components the same for all components
15. Do not paint plastic parts or other coatings which may contaminate other plastics when recycled
16. Consider the use of ADSM technology (Active Disassembly using Smart Materials) that dissolves at higher temperatures. (See later section for non-temperature-critical products).

The above DFD assembly guidelines can have different effects while considering assembly tasks. Therefore, the DFD guidelines that have positive, negative, and little effect during assembly considerations were sorted in three separate groups. The sorted lists are provided below.

DFD Guidelines Having Positive Effects on Assembly

1. Reduce the number of components
2. Reduce the number of separate fasteners
3. Provide open access and visibility for separation points

Design for Sustainability

4. Avoid orientation changes during disassembly
5. Avoid non-rigid (flexible) parts
6. Ensure that disassembly can be done with common tools and equipment
7. Design for ease of handling and cleaning of all components.

DFD Guidelines Having Negative Effects on Assembly

1. Design two-way snap fits or break-points on snap fits
2. Use joining elements that are detachable or easy to destroy
3. Design for ease of separation of components
4. Use water-soluble adhesives.

DFD Guidelines Having Relatively Little Effect on Assembly

1. Design products for reuse
2. Eliminate the need to separate parts
3. Reduce the number of different materials.
4. Enable simultaneous separation and disassembly
5. Place components in logical groups according to recycling and disassembly sequences
6. Identify separation points and materials
7. Facilitate the sorting of noncompatible materials
8. Use molded-in material identification in multiple locations
9. Provide a technique to safely dispose of hazardous waste
10. Select an efficient disassembly sequence.

USE OF FASTENERS

Since fasteners play an integral part in the joining of components and subassemblies, designers should consider the following:

a. Minimize the number of fasteners used within an assembly
b. Minimize the types of fastener used within an assembly
c. Standardize the fasteners used
d. Do not compromise the structural qualities of the assembly by using too few or inadequate fasteners
e. Use snap-fits where possible to eliminate the need for a fastener
f. Consider work-hardening, fracture, fatigue failure and general wear when designing snap-fits
g. Consider the use of destructive fasteners or those incorporating ADSM technologies (Active Disassembly using Smart Materials; see later section of this chapter).

RECYCLING AND MATERIAL RECOVERY

Methods for recycling and recovery of materials are also important design considerations. Here, the basic rule considered is that the material removal rate (MRR) has to be economical to disassemble a material and recycle it by manual or mechanical disassembly. MRR is the amount of material (grams) that can be removed per minute for recycling to be cost-neutral for manual disassembly.

SELECTION AND USE OF MATERIALS

Metals are generally easier to recycle, but the following guidelines should be considered:

a. Non-plated metals are more recyclable than plated ones.
b. Low alloy metals are more recyclable than high alloy ones.
c. Most cast irons are easily recycled.
d. Aluminum alloys, steel, and magnesium alloys are readily separated and recycled from automotive shredder output.
e. Contamination of iron or steel with copper, tin, zinc, lead, or aluminum reduces recyclability.
f. Contamination of aluminum with iron, steel, chromium, zinc, lead, copper, or magnesium reduces recyclability.
g. Contamination of zinc with iron, steel, lead, tin, or cadmium reduces recyclability.

Thus, the following material selection rules are considered during the product design phase:

a. The selection of materials should in no way compromise the structural requirements of the design.
b. Regulated/restricted materials often have legislation stating that they must be recycled, it is more economical to avoid these materials if possible.
c. Materials should be marked according to standards (e.g., ISO 1043) for identification purposes.
d. If more than one material is used within an assembly, they should be made from a similar material or at least be easily separable, so that they can be recycled individually.
e. The use of materials with different properties (e.g., magnetic, or specific gravity) can be beneficial during the separation/sorting process.
f. Use recyclable materials based on technical and economical characteristics.
g. Use recycled materials, where possible to increase the recycled content.
h. Standardize material types for the cooperation and its suppliers.
i. Reduce the number of material types.
j. Use compatible materials (if different materials are needed, select one material that is most preferred).
k. Eliminate incompatible laminated/non-separable materials.

PRODUCT DESIGN GUIDELINES FOR RECYCLING

General guidelines that should be considered during product design stages are as follows:

Design for Sustainability

  a. At the early phases of detailed engineering of a new product program, the design engineers, product engineers, purchasing and supply personnel, and the component supplier should discuss recycling issues associated with the product concept and determine the "best fit" materials and processes for specific applications.
  b. Suppliers should be encouraged to demonstrate recyclability and to take materials back for recycling at the end of the product's useful life to be recycled in other applications.
  c. The use of materials which have been recycled is desirable where it is economically viable.

DESIGN FOR ACTIVE DISASSEMBLY

Active Disassembly (AD) involves the disassembly of components using an all-encompassing stimulus, rather than a fastener-specific tool or machine. It uses smart materials which undergo self-disassembly when exposed to specific temperatures. Shape Memory Polymers (SMPs) and Shape Memory Alloys (SMAs) form the majority of the smart materials used.

AD fasteners change their form to a pre-set shape when exposed to a specific trigger temperature, which can range from approximately 65 degrees Celsius to 120 degrees Celsius, depending on the material. They change from threaded fasteners to bolts without threads. Thus, they fall out of the hole without using tools.

CONCLUDING REMARKS

Applying DFS involves special attention by the design team during detailed engineering of systems and components. It helps in reducing life cycle costs, pollution, and improves safety, comfort, and well-being of users. And it is a socially responsible goal for the organizations to achieve. It also provides organizations with a competitive advantage and leadership in achieving these higher-level goals. Product should be also designed for disassembly, recyclability, and reuse to reduce waste and costs. A number of design guidelines were provided in this chapter to achieve these goals.

REFERENCES

Dell, Inc. 2018. Design for Environment – White Paper. Website: https://i.dell.com/sites/content/corporate/corp-comm/en/Documents/design-for-environment.pdf (Accessed: May 5, 2022).

Dowie-Bhamra, T. 1996. Design for Disassembly. Co-design: The Interdisciplinary. Journal of Design and Contextual Studies, No. 5-6. (Also in Website: https://www.engen.org.au/index_htm_files/DFD-guidelines.pdf; Accessed: February 20, 2022).

EPA. 2021. Frequent Questions About Environmental Management Systems. (Website: https://www.epa.gov/ems/frequent-questions-about-environmental-managment-systems; Accessed: December 24, 2021).

European Commission. 2022. Restriction of Hazardous Substances in Electrical and Electronic Equipment (RoHS). Website: https://ec.europa.eu/environment/topics/waste-and-recycling/rohs-directive_en (Accessed: May 5, 2022).

Fauldi, Jeremy. 2015. How to Design for Disassembly and Recycling (Video). Autodesk Sustainability Workshop. Website: https://www.youtube.com/watch?v=vcFRvuOnWQ8 (Accessed: February 20, 2022).

ISO. 2021. ISO 14001:2015 Environmental management systems — Requirements with guidance for use. (Website: https://www.iso.org/standard/60857.html; Accessed: December 24, 2021).

Martin, Shelly. 2016. Design for Sustainability. February 15, 2016. (Website: https://www.textileworld.com/textile-world/features/2016/02/design-for-sustainability/; Accessed: December 24, 2021).

Niinimäki, Kirsi and Lotta Hassi. 2011. Emerging design strategies in sustainable production and consumption of textiles and clothing. *Journal of Cleaner Production* 19(16): 1876–1883, May 2011.

Part III

Tools Used in Product Development, Quality, Human Factors, and Safety Engineering

14 Methods and Toolbox

INTRODUCTION

The objectives of this chapter are to provide an overview of the methods used in the product life cycle phases and to provide information on problem-solving approaches used for product evaluations. The methods (also called techniques or tools) provide the capabilities to organize, visualize, and analyze the collected data and to understand relationships between different variables such as the product characteristics, product attributes, performance measures, customer ratings, and customer satisfaction measures. The relationships shown by the methods and supported by statistical inference methods provide important information to make decisions regarding product configurations, function allocations, trade-offs between different variables related to product attributes, material selection, technology development, process selection, costs, and so forth.

OVERVIEW OF METHODS

Implementation of Systems Engineering and Project Management in the product programs involves use of many methods. In general, the methods facilitate performing the following tasks: (1) organize or sort the collected data, (2) display the data, (3) analyze the data, and (4) facilitate applications of statistical data analysis techniques. The statistical data analysis techniques provide support in evaluating problems such as determining (1) underlying statistical distributions of variables in the collected data, (2) relationships between independent variables and response variables, (3) cause and effect relationships, (4) groups (or clusters) of variables that have common issues, and (5) combinations and levels of variables that have the largest or least amount of effect on certain dependent variables.

Most of these methods can be used to solve many problems that are encountered during the entire life cycle of the products. Some methods are especially useful in the early phases of the product development process such as product conceptualization and detailed engineering. Other specialized engineering analysis tools such as structural analysis, aerodynamic analysis, thermal analysis, electrical analysis, and control systems analysis are especially useful to solve some specialized design issues. Whereas other tools in production planning, control, and product distribution are useful in later stages of the product program.

Table 14.1 presents a summary of various methods used in the product programs. The methods thus form a "toolbox" for the systems engineers and related fields covered in the preceding chapters. The major fields or areas (where the methods originate and are applied) are presented in the first column of the table. The second column presents the names of the methods (or tools) used in these fields. The third and the succeeding columns present phases in the product life cycle—from pre-concept

DOI: 10.1201/9781003263357-17

planning to product retirement/disposal. The "x" marks in the cells of the table indicate application possibilities of the methods. The methods shown in this table are described in Chapters 15 through 21. The specialized engineering analysis methods such as those used in mechanical engineering, production engineering, electrical engineering, chemical engineering, and materials engineering are outside the scope of this book and hence not covered in this book. Here, the emphasis is placed primarily on the tools used to perform data-driven decision-making tasks involving product planning, product quality, human factors, safety engineering, systems engineering, and program management.

CLASSIFICATION OF METHODS

In addition to the classification of the methods by fields as shown in Table 14.1, the methods can be also classified by how the data are gathered (i.e., their data gathering methods) and data presentation formats. The data gathering methods can be classified as follows: (1) observation methods, (2) communication methods, and (3) experimentation methods. The data presentation methods are as follows: (1) charting methods and (2) plotting methods. (Note that it is important to realize here that: (1) data are needed for use in the application of all the methods and (2) the data are processed, organized, and results obtained from applications of all the methods need to be displayed and presented to facilitate decision-making. See also Chapter 7, which presents methods of evaluations [that generate data] and Table 7.1 that organizes the methods of data collection and types of measurements.) The following part of this section provides description of the methods of data gathering and data presentation methods.

OBSERVATION METHODS

In the observation methods, information is gathered by direct or indirect observations of the product usage situations to determine how the product performs and how the product users react to the product. One or more observers can directly observe, or video cameras can be set up (or data acquisition systems can be installed in a product being evaluated) and their recordings can be played back at a later time for observation and analysis. The observers need to be trained to identify and classify different types of predetermined behaviors of the product and its users, events, problems, or errors that the users commit during the observation period. The observers can also record durations of different types of events, number of attempts made to perform an operation, number and sequence of controls used, number of glances made, and so forth. Some events such as accidents are rare, and they cannot be measured through direct observations due to the excessive amount of direct observation time needed until sufficient numbers of accidents occur and the accident data are collected. However, information about such events can be obtained "indirectly," for example, through reports of near accidents (i.e., situations where accidents almost occurred but were averted) and observations gathered after such events through witnesses or from material evidence (e.g., spillage of materials, damaged materials, skid marks). Therefore, the information gathered through indirect observations may not be very reliable due to a number of reasons (e.g., a witness may be guessing or even

Methods and Toolbox

TABLE 14.1
Methods Used in Various Fields to Solve Problems in Different Phases of the Product Life Cycle

Major Field/Area	Method (or Tool)	Pre-concept Planning	Product Concept Design	Detailed Design	Facilities and Tooling Design	Product Testing	Product Verification	Production Planning	Production	Product Distribution	Product Usage/Service	Product Retirement/Disposal
Product Development	QFD	x	x	x	x			x				
	Benchmarking	x	x	x	x	x		x	x		x	x
Tools	Pugh diagram	x	x	x	x			x	x			
	CAD/Solid Modeling Software		x	x	x							
	CAE analyses (e.g., finite element analysis, computational fluid mechanics, thermal analysis)		x	x	x	x						
	Parametric analysis		x	x	x							
	Rapid prototyping (hardware and software)		x	x	x			x				
	Design for manufacture/assembly/Disassembly/service		x	x	x	x		x	x	x	x	x
	Life cycle analysis and life cycle cost analysis	x	x	x								
	FMEA		x	x					x	x	x	x
	Model-based Systems Engineering SysML/SML	x	x	x	x							
New Quality Tools	Relations diagram (flow/interface diagram)	x	x	x	x				x	x			
	Affinity diagram	x	x	x	x				x	x			
	Systematic diagram (tree diagram)	x	x	x	x				x	x			
	Matrix diagram (interface matrix)	x	x	x	x				x				
	Matrix data analysis	x	x	x	x				x				
	Process decision program chart, process chart	x	x	x	x				x	x	x	x	x
	Arrow diagram	x	x	x	x				x	x	x	x	x
Traditional Quality Tools	Pareto chart	x		x					x	x	x	x	
	Cause and effect diagram (Fish diagram)	x	x	x	x	x			x	x	x	x	
	Check sheets				x		x		x	x	x	x	
	Histogram					x				x	x	x	
	Scatter diagram		x			x	x		x	x	x	x	
	Stratification		x		x	x	x		x	x	x	x	

(Continued)

TABLE 14.1 (Continued)

Major Field/Area	Method (or Tool)	Pre-concept Planning	Product Concept Design	Detailed Design	Facilities and Tooling Design	Product Testing	Product Verification	Production Planning	Production	Product Distribution	Product Usage/Service	Product Retirement/Disposal
	Control charts								x	x	x	
	Experiment design		x	x		x	x		x	x	x	
Human Factors Engineering	Task analysis		x	x				x	x	x	x	
	Anthropometric and Biomechanical Human Models			x	x			x	x	x		
Tools	Human Performance Measurements and Models		x	x		x	x	x	x	x	x	
	Prototyping and simulators		x	x					x	x		
Safety Engineering	Hazard analysis	x	x	x	x			x	x	x	x	x
	FMEA		x	x	x			x	x	x	x	x
Tools	Fault tree analysis		x	x	x			x	x	x	x	
	Accident analysis	x		x					x	x	x	
Project/Program	Timing charts with milestones	x	x	x	x	x	x	x	x	x	x	x
	Cost tables	x	x	x	x	x	x	x	x	x	x	x
Planning and Management	Progress and problem charts	x	x	x	x	x	x	x	x	x	x	x
	Cost-Benefit Analysis	x	x		x							
Production	CAPP			x	x			x	x	x	x	
Planning and	MRP				x			x	x	x	x	
Management	ERP				x			x	x	x	x	

Methods and Toolbox

deliberately falsifying, or objects associated with the event of interest may have been displaced or removed).

COMMUNICATION METHODS

The communication methods involve asking the users or the customers to provide information about their impressions and experiences with the product or a process. The most common technique involves a personal interview where an interviewer asks each user a series of questions. The questions can be asked before, during, or after the product usage. The user can be asked questions that will require the user to (1) describe the product or the impressions about the product and its attributes (e.g., usability), (2) describe the problems experienced while using the product (e.g., could not locate or view a critical item), (3) categorize the product or its performance using a nominal scale (e.g., acceptable or unacceptable, comfortable or uncomfortable), (4) rate the product on one or more scales describing its characteristics and/or overall impressions (e.g., workload ratings), or (5) compare the product with other competitors' products presented in pairs based on a given attribute (e.g., ease of use, comfort). Interviews can also be conducted with a group of individuals, such as in a focus group, that includes about 8–12 individuals with a similar background and are led by an interviewer/moderator to brainstorm through a series of questions, and the participants are asked to provide opinions or suggest issues related to one or more products.

Some commonly used tools in communication methods in product evaluations include the following: (1) rating scales: using numeric scales, scales with adjectives (e.g., acceptance ratings and semantic differential scales) and (2) paired comparison-based scales (e.g., using Thurstone's method of paired comparisons and analytical hierarchical method). These tools are described in Chapter 7.

In addition, many tools used in fields such as Industrial Engineering, Quality Engineering and Design for Six Sigma, and Safety Engineering can be used. Some examples of such tools are process charts, task analysis, arrow diagrams, interface diagrams, matrix diagrams, quality function deployment (QFD), Pugh analysis, failure modes and effects analysis (FMEA), and fault tree analysis. The above-mentioned tools rely heavily on the information obtained through the methods of observation and communication involving the users/customers and members of the multifunctional design teams. Additional information on many of these tools is presented in Chapters 15 through 21 and also in other books such as those by Kolarik (1995), Besterfield et al. (2003), Creveling et al. (2003), and Yang and El-Haik (2003).

EXPERIMENTATION METHODS

The purpose of experimental research is to allow the investigator to control a research situation (e.g., selecting a product design, performing a task or a test condition) so that causal relationships between the response variables and independent variables may be evaluated. An experiment includes a series of controlled observations (or measurements of response variables) undertaken in artificial (test) situations with deliberate manipulations of combinations of independent variables in order to answer one or more hypotheses related to the effect of (or differences due to) the independent variables. Thus, in an experiment, one or more variables (called independent variables) are manipulated, and their effect on another variable (called dependent or response variable) is measured, while all other variables that may confound the relationships are eliminated or controlled.

The importance of the experimental methods is as follows: (1) they help identify the best combination of independent variables and their levels to be used in designing the product and thus provide the most desired effect on the users, and (2) when the competitors' products are included in the experiment along with the manufacturer's product, the superior product can be determined. To assure that this method provides valid information, the researcher designing the experiment needs to ensure that the experimental situation is not missing any critical factor related to performance of the product or the task being studied. Additional information on the experimental methods can be obtained from Chapter 17, Kolarik (1995) or other textbooks on Design of Experiments.

Experiments can be also conducted by exercising computer models using various combinations of input variables (or configurations). The computer modeling methods can be classified as follows: (1) mathematical models, (2) simulation models, (3) visualization or animation models, and (4) prototyping using a combination of hardware and software.

DATA PRESENTATION METHODS

The data presentation methods allow us to organize and visualize important aspects (e.g., trends, groups, relationships) within the data. These methods can be classified as follows:

1. Charting methods (e.g., flow diagram, process charts, fish diagram, interface diagram, time charts, tree diagrams [event trees, fault trees, decision trees])
2. Plotting methods (e.g., histograms, scatter diagrams, pie charts, Pareto charts, polar plots, 3D charts, control charts)
3. Tabular formatted methods (e.g., spreadsheet, matrix diagrams, Pugh diagram, design structure matrix, FMEA).

METHODS IN PRODUCT DEVELOPMENT, QUALITY, HUMAN FACTORS, SAFETY, AND PROGRAM MANAGEMENT

The important methods used during product development (see the top part of Table 14.1) covered in Chapter 15 are as follows: (1) Benchmarking and breakthrough, (2) Pugh diagram, (3) QFD, (4) FMEA, (5) program status chart, and (6) business plan. Design for X methods, where X stands for sustainability, environment, manufacture, and assembly, are covered in Chapters 13 and 16. Life cycle analysis (LCA) and life cycle cost analysis (LCCA) methods are covered in Chapter 21.

The methods used in the Quality Engineering area are categorized as New Quality Tools and Traditional Quality Tools. The tools are listed in the second and third sections of Table 14.1. Chapter 17 presents descriptions and examples of these tools.

The methods used in the Human Factors Engineering area listed in Table 14.1 are presented in Chapter 18. Similarly, methods used in the Safety Engineering area listed in Table 14.1 are presented in Chapter 19.

Other production planning and enterprise management tools such as computer-assisted design (CAD), computer-assisted process planning (CAPP), materials requirements planning (MRP), and enterprise resource planning (ERP) are outside the scope of this book and therefore not covered in this book. They are typically covered in computer-assisted manufacturing books such as the one by Groover (2008).

It should be noted that methods used in Systems Engineering implementation and management are covered in Section I of this book.

INTEGRATION OF TOOLS IN APPLICATIONS

Most of the tools covered in this chapter and described in Chapters 13 through 21 can be used with any other tools. The selection of tools and their sequence of applications largely depends on the past experiences and backgrounds of the individuals involved in the product programs. However, there is a common underlying sequence of steps in problem solving (see Chapter 3, Table 3.1), which typically includes (1) defining the problem, (2) collecting data to understand issues and variables affecting the problem, (3) developing alternate solutions, (4) evaluating the solutions (by use of a model or an experiment), and (5) applying the selected solution.

Tables 14.2 and 14.3 present various methods that can be used in each of the phases of the Design, Measure, Analyze, Improve, and Control (DMAIC) and

TABLE 14.2
DMAIC Process Used Solving Six-Sigma Improvement Projects

Phases of the DMAIC Process	Tools, Methods and Models
Phase I: Define the Problem Describe the problem in operational terms; Precisely specify the process or problem to be evaluated; Identify customers and critical to quality; Boundaries, Timeframe, etc.	Process Flow Diagram, Process Charts, Relations Diagram, Affinity Diagram, Cause-Effect Diagram, Pareto Chart, FMEA
Phase II: Measure Allow quantification of inputs and outputs; Measure internal processes that affect critical-to-customer; Measure process capability	Process observation; Operator Interviews and Feedback; Statistical Process Control Tools: Histograms, Gauge Studies, Control Charts
Phase III: Analyze Identify key variables that cause errors, defects, or variations; Experiment; Plot data; Determine relationships	P-charts, Scatter Diagrams, Box plots, Correlations, C-E diagrams, Experiment Design, Regression Analysis
Phase IV: Improve Generate ideas for improvement; Determine combinations of factor levels that produce significant improvements; Determine optimum product parameters/process settings	Design of Experiments; Taguchi Methods, Control charts, Verification Tests
Phase IV: Control Establish methods/procedures to control and maintain optimum settings of product/processes	Define and install Quality System—procedures, responsibilities, training; Change control system; Control plans; Audits

TABLE 14.3
IDOV Process Used in Design for Six Sigma (DFSS)

Phases of the IDOV Process	Tools, Methods, and Models
Phase I: Identify Requirements Form cross-functional team. Determine customer expectations. Determine product functionality based on customer requirements, technological capabilities, and economic constraints	Customer Surveys Benchmarking QFD Kano Model Ring Model Risk Analysis
Phase II: Characterize the Design (Invent, Innovate)	CAD/CAE Pugh Matrix Interface Diagram/Matrix FMEA Design for X
Phase III: Optimize the Design	Simulation Tools Robustness Design of Experiements Reliability Analysis Tolerancing
Phase IV: Verify the Design Pilot Tests Refine and Verify Continuously Ensure Production System Capability (to meet desired sigma level)	Reliability Testing Error Proofing Measurement System Evaluation, Process Capability Assessments Process Control Training

Identify, Design, Optimize, and Verify (IDOV) processes of the Six-sigma improvement projects and the Design for Six-Sigma projects, respectively. The methods presented in the second columns of these tables are described in Chapter 17. A trained Black belt (a professional certified to apply various quality engineering tools) usually selects a set of tools that can best help in accomplishing tasks to be performed to solve the problem in different phases of the quality-related projects.

The typical sequence of tools used in the DMAIC process includes cause-and-effect diagram, Pareto chart, and experiment design. Whereas the IDOV process uses benchmarking, QFD, FMEA, Pugh diagram, experiment design, and requirements compliance testing (verification). Some examples of the applications of the tools and the problem-solving processes are provided in Chapters 22–24.

CONCLUDING REMARKS

Use of specialized and proven methods is very important in understanding and solving many problems encountered in the product programs. The methods help in identifying and estimating the strengths of effects and relationships between different variables, and thus, provide information for decision-making. Succeeding chapters in Section III of the book provide descriptions of the methods/tools along with a few examples illustrating their applications. Section IV provides several case studies where many of the tools provided important information for making key decisions.

REFERENCES

Besterfield, D. H., Besterfield-Michna, C., Besterfield, G. H. and M. Besterfield-Scare. 2003. *Total Quality Management*. Third Edition. ISBN 0-13-099306-9. Upper Saddle River, NJ: Prentice Hall.

Creveling, C. M., Slutsky, J. L. and D. Antis, Jr. 2003. *Design for Six Sigma—In Technology and Product Development*. Upper Saddle River, NJ: Prentice Hall PTR.

Groover, M. P. 2008. *Automation, Production Systems, and Computer-Integrated Manufacturing*. Third Edition. Upper Saddle River, NJ: Pearson-Prentice Hall.

Kolarik, W. J. 1995. *Creating Quality—Concepts, Systems, Strategies, and Tools*. New York: McGraw-Hill, Inc.

Yang, K. and B. El-Haik. 2003. *Design for Six Sigma*. ISBN 0071412085. New York: McGraw-Hill.

15 Product Development Tools

INTRODUCTION

This chapter describes applications of the important tools used during the early stages of product development. The tools covered include (1) benchmarking and breakthrough, (2) Pugh diagram, (3) quality function deployment (QFD), (4) failure modes and effects analysis (FMEA), and (5) other product development tools such as business plan, program status charts, computer-aided design (CAD), and model-based systems engineering (MBSE).

Benchmarking is used to compare currently available competitors' products with manufacturer's current product concepts to understand the gaps between the product concepts and the benchmarked products. The knowledge gained thus can be used to incorporate some of the best ideas learned from the competitors. Breakthrough methodology allows us to gain large improvements in product designs by thinking beyond the present product design and production capabilities. The Pugh diagram is used to select a product concept by using a number of product attributes to compare various alternate product concepts with a reference called the "datum." QFD is used to translate customer needs into engineering functional specifications and to determine the engineering product specifications that are critical to customer satisfaction. The FMEA method is used to improve the reliability and safety of the product by tabulating its failure modes and their effects on other entities within the product by prioritizing the uncovered product problems (failure modes) by applying a risk priority number (RPN) assessment method. Business plan, program status charts, and CAD are used in the communication of product and program information within the design, engineering, and program management activities. Model-based systems engineering is a software-based approach to implement systems engineering to increase speed and accuracy in product (system) development.

BENCHMARKING AND BREAKTHROUGH

Benchmarking and breakthrough methods are generally used during the very early stages of a product development program. From the information gathered during the benchmarking exercises, the product designers can realize the gaps between the characteristics and capabilities of the products of their leading competitors and their new product concepts. The breakthrough approach forces the design teams to look beyond the existing products and technologies and think about developing a totally new product or new features and achieve a major improvement step over the existing product designs.

BENCHMARKING

Benchmarking is a process of measuring products, services, or practices against the toughest competitors, or the companies recognized as industry leaders. Thus, it is a search for the industry's best products or practices that can lead to superior performance.

A multifunctional team within the manufacturer's product development staff is usually selected to perform product benchmarking. The benchmarking exercise typically starts with identifying the toughest competitors (e.g., very successful and recognized brands as the industry leaders) and their products (models) that serve similar customer needs to the manufacturer's proposed product. The selected competitor products are compared against the target product. The target product is the product considered by the manufacturer to be its future product (or an existing model of the future product). The team gathers all important competitive products and information of the products and compares the competitors' products to their target product through a set of evaluations (e.g., measurements of product characteristics, tearing down [i.e., disassembling] the products into their lower-level entities for close observations, evaluations by experts on issues such as performance, capabilities, and unique features, materials, and manufacturing processes used by the competitors, and estimates of costs to produce the benchmarked products). The information gathered from the comparative evaluations is usually very detailed. However, the depth of evaluations included in benchmarking can vary between problem applications and companies. For example, the benchmarking of an automotive disk brake may involve comparisons based on part dimensions, weights, materials used, surface characteristics, strength characteristics, heat dissipation characteristics, processes needed for its production, estimated production costs, features that would be "liked very much" by the customers, features that can be "hated" by the customers, features that create a "wow" reaction among the team members and potential customers, special performance tests such as part temperatures during severe braking torque applications, brake "squealing" sound, and so on. In addition, digital pictures and videos can be taken to help visualize the differences in benchmarked products.

The gathered information is generally summarized in a tabular format with product characteristics listed as rows and different benchmarked products represented in columns. Table 15.1 presents an example of a table created by Hale and Pelowski (2006) to compare the dimensions (or parameters) of 10 different motorcycles. The dimensions are described in the left-hand column of the table and the values of dimensions of the 10 motorcycles are provided in the subsequent columns. Such data are useful in establishing ranges of values of different parameters and in setting target values of each of the parameters of the product being designed.

Figures 15.1 and 15.2 present photographs collected during benchmarking exercises to show the differences in various headlamps and hand controls used in different brands and types of motorcycles (Hale and Pelowski, 2006). For example, the photographs in Figure 15.1 provide quick insights into the types of available configurations of motorcycle headlamps with characteristics such as single round headlamps, dual round headlamps, multi-compartment-styled headlamps, and lamp mountings under the handlebar versus mountings inside the shroud holding the

TABLE 15.1
Benchmarking of Motorcycle Dimensions*

Bike Brand	Honda	Suzuki	Yamaha	Harley Davidson	Yamaha	Ducati	Harley-Davidson	Harley-Davidson	BMW	Ducati
Bike Model	CBR 600 RR	GSX-R750	Vstar 1100	Electraglide	FJR 1300	ST4S	1200C	Fatboy	R1200RT	Monster S2R
Type	Sport	Sport	Cruiser	Touring	Touring	Touring	Cruiser	Cruiser	Touring	Sport
Origin	AS	AS	AS	NA	AS	EU	NA	NA	EU	EU
Footrest ht from ground	0.0	0.0	0.0	0.0	0.0	0.0	0.0	0.0	0.0	0.0
Footrest from the front of bike	125.4	129.9	97.9	94.0	132.4	134.3	91.3	94.0	123.7	130.0
Handgrip ht from ground	0.0	0.0	0.0	0.0	0.0	0.0	0.0	0.0	0.0	0.0
Handgrip from the front of bike	58.9	60.3	102.1	95.9	80.6	68.4	96.5	95.9	84.5	65.7
Seat ht from the ground	0.0	0.0	0.0	0.0	0.0	0.0	0.0	0.0	0.0	0.0
Seat from the front of bike	0.0	0.0	0.0	0.0	0.0	0.0	0.0	0.0	0.0	0.0
Seat length	0.0	0.0	0.0	0.0	0.0	0.0	0.0	0.0	0.0	0.0
Seat width	35.6	35.1	46.7	41.9	31.4	31.6	31.8	37.1	34.9	26.0
Seat adjust	0.0	0.0	0.0	0.0	0.0	0.0	0.0	0.0	0.0	0.0
Footrest length	7.6	10.2	12.7	12.1	8.9	10.6	10.2	10.8	9.5	7.6
Footrest width	2.1	2.9	29.8	30.5	3.2	3.2	3.8	29.5	4.4	2.5
Footrest stance	52.7	47.6	66.0	60.3	54.0	49.8	67.9	62.5	52.7	44.5
Rear brake ht from ground	33.2	34.2	42.3	44.6	26.7	32.5	38.7	39.7	30.9	30.1
Rear brake	113.5	117.2	90.3	79.1	117.5	120.7	80.2	81.0	110.7	117.8
From front of bike Rear brake	4.8	5.1	8.3	11.7	5.7	4.1	5.9	12.2	4.4	3.8
Length Rear brake	2.1	1.7	5.4	6.7	3.2	1.9	3.8	7.0	1.7	2.1
Width Rear brake	0.0	3.5	8.3	9.7	3.8	1.4	3.3	8.1	4.4	1.9
Offset from the footrest Rear brake	N	Y	Y	N	N	Y	N	N	Y	Y
Adjust Shifter ht from	32.7	33.1	41.7	41.8	25.9	30.1	38.9	38.2	32.6	30.1
The ground Shifter ht from	113.0	117.8	85.7	80.0	118.6	120.8	79.9	79.4	111.0	117.3
The front of bike Shifter length	3.8	4.1	6.4	5.1	3.8	4.4	5.7	5.1	4.1	3.8
Shifter width	1.9	1.7	2.5	2.7	2.1	2.4	2.5	2.5	1.9	2.1
Shifter offset from the footrest	1.0	4.1	5.1	3.2	1.9	4.4	3.2	4.1	5.1	3.8

(Continued)

TABLE 15.1 (Continued)

Bike Brand	Honda	Suzuki	Yamaha	Harley Davidson	Yamaha	Ducati	Harley-Davidson	Harley-Davidson	BMW	Ducati
Bike Model	CBR 600 RR	GSX-R750	Vstar 1100	Electraglide	FJR 1300	ST4S	1200C	Fatboy	R1200RT	Monster S2R
Type	Sport	Sport	Cruiser	Touring	Touring	Touring	Cruiser	Cruiser	Touring	Sport
Origin	AS	AS	AS	NA	AS	EU	NA	NA	EU	EU
Shifter adjust	Y	Y	Y	N	N	Y	Y	Y	N	Y
Grip stance	54.6	55.2	75.2	72.0	59.0	61.0	62.2	75.5	61.8	63.8
Grip diameter	3.2	3.1	3.5	3.6	3.3	3.1	3.3	3.6	3.6	3.1
Grip length	11.4	11.4	14.1	11.4	12.7	11.7	11.4	11.4	11.4	12.1
Grip adjust	N	N	N	N	N	N	N	N	N	Y
Clutch lever from grip center	9.8	9.5	9.2	9.8	9.8	9.4	8.9	9.5	8.6	9.5
Clutch adjust	N	N	N	N	Y	N	N	N	Y	Y
Brake lever from grip center	8.4	8.4	8.9	8.9	9.0	9.5	8.9	8.9	8.6	9.5
Brake adjust	Y	Y	N	N	Y	N	N	N	Y	Y
Inside grip to lo hi beam	2.5	2.5	3.4	5.0	2.0	2.0	5.0	5.0	3.0	2.3
Lo hi width	1.8	1.8	2.0	1.4	1.5	1.6	1.4	1.4	1.8	1.5
Lo hi ht	2.0	2.5	1.8	2.5	2.2	2.4	2.7	2.7	2.3	2.0
Inside grip to turn left	2.3	2.5	2.9	3.5	2.5	2.5	4.0	4.0	3.2	2.6
Turn width left	0.9	1.2	1.7	3.0	1.1	0.8	2.9	2.9	3.5	0.8
Turn ht left	1.3	1.2	1.1	1.9	1.0	1.0	2.0	2.0	1.5	1.0
Inside grip to horn	2.3	3.5	4.0	2.5	2.5	1.1	2.5	2.5	4.5	2.6
Horn width	1.8	2.5	2.3	1.4	2.2	2.2	1.4	1.4	3.0	2.2
Horn ht	1.4	1.2	1.2	2.5	1.1	1.1	2.6	2.6	1.7	1.2
Inside grip to on off	3.6	4.0	3.3	2.5	4.5	5.4	3.0	3.0	3.0	5.0
On off width	1.8	1.8	2.0	1.4	1.3	1.3	1.4	1.4	0.5	1.5
On off ht	2.3	2.5	1.8	2.5	2.1	2.4	2.4	2.4	1.2	2.0

Product Development Tools

Attribute										
Inside grip to starter	2.9	3.5	4.0	4.0	5.0	4.8	5.0	5.0	3.0	5.0
Starter width	2.5	1.7	2.0	1.4	2.0	1.0	1.4	1.4	1.5	0.9
Starter ht	1.2	1.1	1.0	2.5	0.9	1.0	2.5	2.5	1.0	0.9
Inside grip to turn right	—	—	—	4.0	—	—	4.0	4.0	3.0	—
Turn width right	—	—	—	3.0	—	—	2.9	2.9	3.5	—
Turn ht right	—	—	—	2.0	—	—	2.0	2.0	1.5	—
Inside grip to turn deac	—	—	—	—	—	—	—	—	4.0	—
Turn deac width	—	—	—	—	—	—	—	—	3.0	—
Turn deac height	—	—	—	—	—	—	—	—	1.7	—
Speedo dia	4.0	3.0	11.9	8.0	8.3	8.0	8.0	10.3	9.0	7.7
Speedo font	1.5	—	0.6	0.5	0.5	0.2	0.4	0.5	0.6	0.5
Speedo ht from ground	91.5	93.6	92.2	118.1	104.6	98.9	104.2	97.7	109.0	101.3
Speedo from front of bike	37.7	38.8	100.4	73.0	46.3	42.2	94.5	97.3	50.8	52.2
Speedo offset from the center of bike	4.0	5.0	0.0	6.0	0.0	6.4	0.0	0.0	8.0	4.0
Tach dia	7.0	7.6	—	8.0	7.5	8.0	—	—	9.0	7.7
Tach font ht	0.4	0.4	—	0.5	0.8	0.4	—	—	0.6	0.7
Fuel dia	2.0	—	—	1.5	4.0	5.1	—	3.1	1.0	—
Fuel font ht	—	—	—	0.5	1.0	2.2	—	—	—	—
Coolant dia	—	1.0	—	1.5	2.5	2.7	—	—	1.0	—
Coolant font ht	—	—	—	0.5	0.5	1.1	—	—	0.6	—
Warning font icon ht	0.5	0.6	—	1.0	0.8	1.0	—	0.4	—	0.7
Mirror ht from the ground	99.7	109.0	115.3	121.2	113.5	105.9	119.2	112.6	103.9	117.8
Mirror from front of bike	39.5	44.9	99.5	91.4	43.8	47.5	93.8	92.4	57.5	62.1
Mirror length	14.0	13.5	15.2	13.0	14.0	15.2	13.3	13.0	11.4	12.1
Mirror ht	8.4	8.4	7.6	8.3	8.6	8.3	8.3	8.3	13.3	8.3
Mirror center to center	66.0	66.0	89.9	76.7	78.7	75.2	67.9	87.6	72.4	76.2
Windshield ht from the ground	110.2	114.5	—	152.4	132.1	118.5	—	—	142.8	113.6
Windshield from the front of bike	48.7	52.5	—	85.7	61.0	50.5	—	—	61.9	53.2
Windshield ht	19.1	22.2	—	26.0	27.9	22.9	—	—	47.0	14.0
Windshield width	28.3	31.4	—	53.3	49.8	30.2	—	—	50.8	24.9

(Continued)

TABLE 15.1 (Continued)

Bike Brand	Honda	Suzuki	Yamaha	Harley Davidson	Yamaha	Ducati	Harley-Davidson	Harley-Davidson	BMW	Ducati
Bike Model	CBR 600 RR	GSX-R750	Vstar 1100	Electraglide	FJR 1300	ST4S	1200C	Fatboy	R1200RT	Monster S2R
Type	Sport	Sport	Cruiser	Touring	Touring	Touring	Cruiser	Cruiser	Touring	Sport
Origin	AS	AS	AS	NA	AS	EU	NA	NA	EU	EU
Headlamp ht from the ground	76.2	76.1	89.4	93.8	83.1	83.4	93.1	90.4	89.0	85.8
Headlamp from the front of bike	22.4	18.1	49.4	39.1	24.8	21.9	53.5	47.3	18.1	36.2
Headlamp length	15.9	6.4	18.1	16.5	15.2	26.0	13.7	17.1	10.2	18.4
Headlamp ht	7.6	6.4	18.1	16.5	11.4	11.4	13.7	17.1	9.8	18.4
Tail lamp ht from the ground	100.3	98.0	52.4	57.5	83.8	83.8	53.2	56.5	87.0	86.8
Tail lamp width	16.8	16.5	10.2	10.8	18.4	14.6	10.8	11.4	13.2	14.0
Tail lamp ht	4.1	7.6	14.3	7.6	8.9	7.0	8.3	8.3	7.0	4.4
Front turn lamp ht from the ground	64.3	69.5	76.0	76.2	164.9	85.6	104.4	99.2	100.3	86.9
Front turn lamp from the front of bike	30.0	28.6	50.3	45.7	44.5	36.4	80.2	76.2	44.5	43.2
Front turn lamp length	7.9	7.6	8.3	8.3	5.7	5.1	5.7	6.0	10.2	7.0
Front turn lamp ht	6.0	4.4	8.3	8.3	7.6	8.6	5.7	6.0	12.7	4.1
Front turn lamp stance	61.0	55.2	50.0	41.9	48.9	44.1	41.4	64.1	74.9	43.2
Rear turn lamp ht from the ground	83.6	85.8	51.3	48.4	80.1	76.5	50.6	65.3	87.2	81.8
Rear turn lamp from the front of bike	192.7	189.4	238.9	229.9	207.3	208.1	202.1	222.4	212.4	192.1
Rear turn lamp length	7.9	7.6	8.3	8.3	5.4	6.0	5.7	6.0	14.0	7.0
Rear turn lamp ht	6.0	4.4	8.3	8.3	6.4	4.4	5.7	6.0	5.1	4.1
Rear turn lamp stance	29.8	38.1	43.8	43.2	28.9	48.9	41.4	36.8	31.1	43.2
Wheelbase	139.9	139.9	164.5	162.0	150.5	144.0	158.8	163.5	147.8	144.8
Stand	side	side	side	side	center	side	side	side	both	side
Seat to handgrip x	92.6	97.8	66.4	62.2	77.5	83.5	65.2	63.5	70.2	81.8
Seat to handgrip z	−0.4	0.2	33.7	33.9	25.3	13.3	42.3	31.6	29.6	13.0

Product Development Tools

Seat to footrest x	26.0	28.3	70.5	64.1	25.7	17.6	70.5	65.4	31.0	17.5
Seat to footrest z	-46.4	-46.2	-48.0	-44.4	-46.7	-44.6	-35.5	-42.4	-46.1	-46.7
Footrest to shifter x	12.4	12.1	12.2	14.0	13.8	13.5	11.4	14.6	12.7	12.7
Footrest to shifter z	-6.3	-7.6	19.5	14.0	-4.6	-6.2	8.8	11.9	-2.9	-6.7
Footrest to brake x	11.9	12.7	7.6	14.9	14.9	13.7	11.1	13.0	13.0	12.2
Footrest to brake z	-5.8	-6.5	20.1	16.9	-3.8	-3.8	8.6	13.4	-4.7	-6.7
Eye pt x	93.9	93.0	154.0	134.0	123.0	104.0	139.8	140.6	129.0	106.0
Eye pt z	149.5	147.0	143.5	146.5	147.5	143.0	147.5	144.0	149.5	150.2
Shoulder pt x	112.5	111.0	168.0	145.0	135.0	120.0	153.7	155.0	143.0	124.5
Shoulder pt z	133.5	131.2	124.8	125.0	127.0	128.0	127.0	124.5	131.5	132.0
H pt x	136.0	136.0	166.0	147.5	143.0	140.0	156.7	154.8	149.0	140.0
H pt z	95.0	92.0	82.5	85.5	84.0	87.0	81.5	82.0	89.5	90.0
Torso angle	59.0	58.5	88.2	86.0	80.0	64.5	84.0	86.2	80.0	66.9
Torso upper leg angle	82.5	86.0	94.1	97.5	99.0	86.0	96.0	98.2	96.5	85.3
Upper leg Lower leg angle	73.0	79.0	112.3	119.5	83.0	76.5	115.3	123.0	85.7	77.0

Note: All dimensions are in cms. The following abbreviations are used in the descriptions of the motorcycle dimensions: x = longitudinal fore/aft direction, y = lateral left/right direction, z = vertical direction, H = H-point (Hip point) used to locate rider seating position, ht = height, lo = low beam, hi = high beam, Lo hi = low/high beam switch, Speedo = speedometer, Tach = tachometer, beam = headlamp beam switch, turn = turn signal switch, deac = turn signal deactivation switch, and dia = diameter.

Source: Hale, A. and D. Pelowski, *Ergonomic Study of Asian, European, and North American Motorcycles: Commonality and Differences between Cruiser, Sport, and Touring Categories*, University of Michigan-Dearborn, Dearborn, MI, 2006.

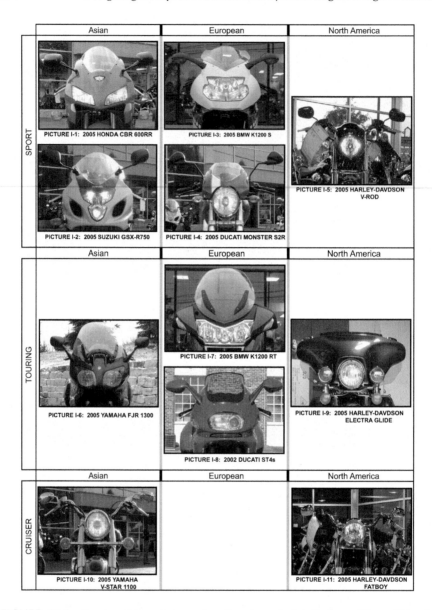

FIGURE 15.1 Benchmarking photographs of motorcycle headlamps. (From Hale, A. and D. Pelowski, *Ergonomic Study of Asian, European, and North American Motorcycles: Commonality and Differences between Cruiser, Sport, and Touring Categories*, University of Michigan-Dearborn, Dearborn, MI, 2006.)

windshield. The photographs in Figure 15.2 show different shapes and styles in designing hand grips and clutch squeeze handlebars with chrome plating as well as dark high-strength plastic materials. The benchmarking information was used by Hale et al. (2007) to determine commonality and differences between cruiser, sport,

Product Development Tools

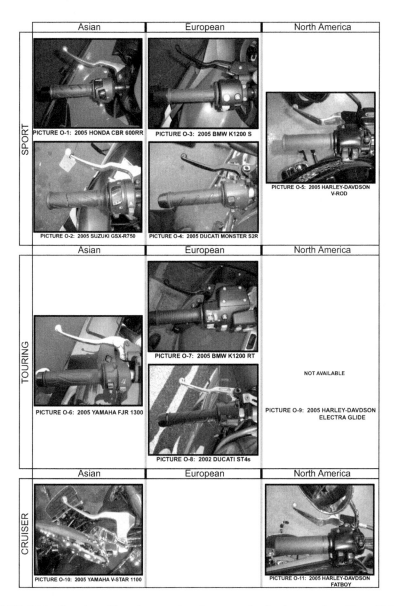

FIGURE 15.2 Benchmarking photographs of motorcycle clutch controls. (From Hale, A. and D. Pelowski, *Ergonomic Study of Asian, European, and North American Motorcycles: Commonality and Differences between Cruiser, Sport, and Touring Categories*, University of Michigan-Dearborn, Dearborn, MI, 2006.)

and touring categories of motorcycles produced by Asian, European, and North American motorcycle manufacturers. Thus, benchmarking is a powerful tool in giving opportunities to the design team to learn from the currently available products and gain new ideas or insights for improving their product.

The data from the benchmarking studies can also be entered into computers as a relational database and later sorted for use by different functional areas as subsets. The information is used to determine the gaps between the characteristics of the competitors' products and those of the manufacturer's target product. An action plan can be developed to close the gaps by determining how the target product can be improved by implementing many of the good ideas used in the competitors' designs and avoiding the problems uncovered in the poor designs. Thus, benchmarking can reveal one or more best competitors' products that can be used as reference products during subsequent product development phases. In addition, some innovative changes can be made to further improve the best design found during the benchmarking exercises.

Peters and Waterman (1982) called the benchmarking exercise "creative swiping from your best competitor." Many companies have benchmarking laboratories where many products produced by different competitors are collected and the products (or their systems, subsystems, or components) are displayed along with the gathered information. Such laboratories are excellent learning tools for designers, engineers, and product planners. The greatest advantage of benchmarking is that it allows team members to understand the competitors and learn from their products and processes needed to create the products within a very short period.

Benchmarking, thus, can help in reducing the gaps between the manufacturer's target product and its best competitors. However, merely designing as well as the best competitors is not sufficient because these best competitors will also be continuously improving their future products. Thus, the manufacturer's target product should have capabilities that extend well beyond the best benchmarked products. Simply selecting the best design based on the best set of characteristics among the benchmarked products may not produce an overall best product because trade-offs exist between different product characteristics related to issues such as costs, performance, customer preferences, manufacturing methods, and so forth.

Benchmarking is an important technique for quality improvements, and it is recognized in the Baldrige Quality Award criteria and ISO 9000 requirements (NIST, 2012; ISO, 2012). A quality producer must continuously compare its product with its best competitors and use the information to improve its processes and products continuously. Interestingly, the process of benchmarking is not new—it is like keeping up with the "Joneses" (friends, neighbors, or colleagues). We find out what others have done (e.g., observe, or ask others, or conduct literature surveys), compare our situation with the findings reported by others, and then decide on the next course of action.

Another example of a benchmarking study and product evaluations is provided in Chapter 23. Case study 2 in Chapter 23 shows how an automotive supplier benchmarked many automotive steering wheels and used the information to design a study to evaluate different steering wheel designs to improve the touch–feel characteristics of a steering wheel for a future automotive product.

BREAKTHROUGH

The breakthrough approach involves throwing away all the existing product designs (and processes) and brainstorming to develop a totally new design to obtain huge potential gains in terms of performance improvements, costs, and added customer satisfaction. Breakthrough designs typically require radically new thought dimensions and lead to the adoption of new technologies. Thus, implementation of a breakthrough design creates new problems in systems integration and management. Some examples of breakthrough product designs are as follows: (1) the old cathode ray tube computer screens and televisions replaced by liquid crystal plasma technology displays, (2) tungsten light sources replaced by compact fluorescent lamps and light-emitting-diode lamps, and (3) fuel systems of gasoline engines with carburetors replaced by direct fuel injection along with turbo-boost technologies.

DIFFERENCES BETWEEN BENCHMARKING AND BREAKTHROUGH

The differences between benchmarking and breakthrough are as follows:

1. Benchmarking is a quick/short-term process to seek ideas for product (or process) improvements, whereas the breakthrough process takes a longer time for its implementation.
2. Benchmarking generally has a narrower focus (over a smaller set of changes) than breakthrough, which involves an expanded focus (or a complete redesign).
3. Benchmarking produces smaller improvements compared to breakthroughs, which involve dramatically improved performance and radically new dimensions that usually require new technologies.
4. The benchmarking exercise should be conducted before brainstorming to generate the breakthrough designs.

Creative breakthrough thinking, the theory of inventive problem solving (Altshuller, 1997), the IDEALS concept (Nadler, 1967; Nadler and Hibino, 1998), and reengineering processes (Champy, 1995) have suggested several approaches and principles in developing improved products and processes. Such approaches can be very useful in developing breakthrough product concepts.

PUGH DIAGRAM

Pugh diagram is a tabular formatted tool consisting of a matrix of product attributes (or characteristics) and alternate product concepts along with a benchmark (reference) product called the datum. The diagram helps to undertake a structured concept selection process and is generally created by a multidisciplinary team to converge on a superior product concept. The process involves creation of the matrix by inputs from all the team members. The rows of the matrix consist of product attributes based on customer needs, and the columns represent different alternate product concepts.

The evaluations of each product concept on each attribute are made with respect to a datum (a reference product, usually an existing product of the manufacturer). The process uses classification metrics of "same as the datum" (S), "better than the datum" (+), and "worse than the datum" (−). The scores for each product concept are obtained by simply adding the number of "+" and "−" signs in each column. The product concept with the highest net score ("sum of pluses" minus "sum of minuses") is considered to be the preferred product concept. Several iterations are used to improve product superiority by combining the best features of highly ranked concepts until a superior concept emerges and becomes the new benchmark.

AN EXAMPLE OF PUGH DIAGRAM APPLICATION

An automotive power train engineer wanted to determine if the performance of a transient turbo-charged gasoline engine can be improved over the gasoline turbo direct injection (GTDI) methodology by using the following three concepts: (1) concept 1—an electric turbo boost (e-Turbo), (2) concept 2—a hybrid turbo using an electric motor assisting in parallel with the turbo operated by the exhaust gases, or (3) concept 3—use of an electrical compressor only (Black, 2011). The engineer created a Pugh diagram to compare the aforementioned three technologies with the GDTI as the datum. Table 15.2 presents the Pugh diagram. The product attributes used to compare the aforementioned three technologies are presented in the second column from the left of the table. The four right-hand columns represent product concepts 1, 2, and 3 and the datum as the last column.

All three product concepts improve the "performance" attribute (attribute 4) compared to the datum by eliminating turbo lag (a transient condition during quick accelerations). This is shown by the + signs in all three product concepts (columns) of the row corresponding to attribute 4. However, they introduce additional negatives into the system due to additional cost (attribute 2), weight (attribute 10), noise (attribute 11), electrical and electronics (attribute 9), and life cycle durability (attribute 1) challenges. The bottom row of "net score" (sum of pluses minus sum of minuses) shows that none of the three product concepts were better than the datum, since the net scores of all are negative. The concept 2 (hybrid turbo) is the least negative based on

TABLE 15.2
Pugh Diagram for Product Concept Selection

Attribute No.	Customer-based Product Attribute	Product Concept #1	Product Concept #2	Product Concept #3	Datum
1	Life-Cycle Durability	−	−	−	
2	Cost	−	−	−	
3	Package and Ergonomics	-	+	+	
4	Performance	+	+	+	
5	Fuel Economy	S	+	S	
6	Safety/Security	−	-	-	
7	Vehicle Dynamics	+	+	+	
8	Emissions	+	+	+	
9	Electrical and Electronics	−	−	−	
10	Weight	−	−	−	
11	Noise, Vibrations, and Harshness	−	−	−	
	Sum of Pluses	3	5	4	
	Sum of Minuses	6	6	6	
	Net Score	−3	−1	−2	

the net score. Life cycle durability, cost, safety/security, electrical and electronics, weight, noise, vibrations, and harshness (NVH) are all additional problem issues with concept 2 compared to the datum (traditional turbo [GTDI]). In the aforementioned analysis, all the product attributes were considered to have equal weight, that is, the number of + and − signs were simply added to obtain the net score.

Table 3.7 in Chapter 3 showed a method of using different weights for the attributes. This method can also be used in the Pugh selection process. Table 15.3 presents an additional analysis on the aforementioned problem using importance weighting for each of the product attributes. The importance of each product attribute was rated by using a 10-point scale, where 10 = most important and 1 = least important (see the column with the heading "Importance Rating" in Table 15.3). The importance scores were converted to "Importance Weight" (by dividing the importance rating of each attribute by the sum of all importance ratings). The importance weights are shown in the column to the right side of the importance rating column. Each product concept was evaluated with respect to the datum (the current product GTDI system) for each attribute by using another 10-point scale ranging from −5 to +5. Here, the +5 score indicates that a given product concept is very much better than the datum and the −5 score indicates that the product concept is very much inferior to the datum. The sum of the weighted scores of each product concept was obtained by summing the multiplied values of importance weight and product rating over the entire set of product attributes. The weighted sums of the three product concepts are −0.46, −0.33, and −0.97 (see the last line of Table 15.3). Concept 2 had the largest value of the weight sum (−0.33). Thus, product concept 2 (hybrid turbo) emerged as the winner among the three concepts. However, it is still worse than the datum (because the weighted sum value is negative). If fuel economy becomes more important in the future, then this concept has the potential to be implemented.

TABLE 15.3
Pugh Analysis with Ratings

Customer-Based Product Attribute	Importance Rating	Importance Weight	Preference Ratings Using −5 to +5 Scale Compared to the Datum			
			Product Concept 1	Product Concept 2	Product Concept 3	Datum
Customer life cycle durability	5	0.06	−3	−3	−3	—
Cost	10	0.13	−3	−5	−5	—
Package and ergonomics	3	0.04	−3	3	3	—
Performance	10	0.13	5	5	5	—
Fuel economy	10	0.13	0	5	0	—
Safety/security	10	0.13	−3	−3	−3	—
Vehicle dynamics	8	0.10	5	5	5	—
Emissions	3	0.04	5	5	5	—
Electrical and electronics	5	0.06	−3	−5	−5	—
Weight	5	0.06	−3	−5	−5	—
NVH	9	0.12	−3	−5	−5	—
Sum or weighted sum	78	1.00	−0.46	−0.33	−0.97	—

The benefit of the hybrid turbo is that it enables completely independent intakes and exhausts and many modes of operation including additional fuel savings. This also helps eliminate some of the air intake system routing and packaging issues. The penalties in the trade-offs are due to (1) higher electrical load to drive the electric motor–driven compressor, (2) poor reliability and durability of the electrically driven compressor, (3) added complexity resulting from extra parts, and (4) additional costs of extra hardware.

QUALITY FUNCTION DEPLOYMENT

Quality function deployment (QFD) is a technique used to understand customer needs (voice of the customer) and to transform the customer needs into engineering characteristics of products or processes in terms of functional or design requirements. It relates "what" (what are the customer needs?) to "how" (how should an engineer design to meet the customer needs?), "how much" (the magnitude of design variables, i.e., their target values), and also provides competitive benchmarking information, all in one diagram. QFD was originally developed by Dr. Yoji Akao of Japan in 1966 (Akao, 1991). QFD is applied in a wide variety of applications and is considered a key tool in Design for Six-Sigma projects (Yang and El-Haik, 2003). It is also known as the "house of quality" because the correlations matrix drawn on top of the QFD matrix diagram resembles the roof of a house (Kolarik, 1995; Besterfield et al., 2003).

Figure 15.3 shows the basic structure (or regions) of the QFD. The contents of each region of the QFD are described as follows:

Customer needs (what): the needs of the customers as expressed by what the customer wants in the product are listed sequentially in the rows of this region. Each customer need should be described by using the customer's words (as the customer would describe it in his or her own words, e.g., give me a product that lasts for a long time, a product that looks good, a product that works well, an energy-efficient product, a product that is fun to use, etc.). The needs can also be listed in a hierarchical order such as primary needs, secondary needs (within each primary need), and tertiary needs (within each secondary need). The ith customer need is defined as C_i. (The mathematical definitions of all the QFD variables are described in the later pages of this section.)

Importance ratings: this column provides importance ratings for each of the customer needs. The importance ratings (or weights) can be defined by using several different rating techniques. However, a 10-point rating scale is commonly used, where 10 = extremely important and 1 = not at all important. The importance rating for ith customer need is defined as W_i.

Functional specifications (how): the functional specifications are created by the engineers involved in product development to define how the product "should function" or "should be designed to meet its customer requirements." The functional specifications describe how the engineers will address the customer needs. Thus, these specifications should be described

Product Development Tools

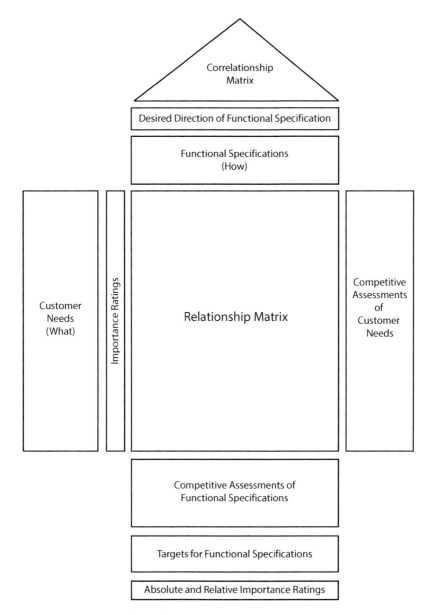

FIGURE 15.3 Structure of a quality function deployment diagram showing its regions and contents.

using technical terms and variables that are used and selected by the engineers such as materials, dimensions, strengths, manufacturing processes, test requirements, and so forth. Here, the engineer should list each functional (or engineering) specification in a separate column. The jth functional specification is defined as F_j.

The functional specifications should describe engineering issues, methods, or variables that need to be considered during the product development. Some examples of variables for functional specifications are the type of construction (e.g., welded vs. assembled using fasteners), material to be considered to make the entity (e.g., steel, high-strength steel, aluminum, and carbon fiber), type of production process to be used (e.g., extrusion vs. cast), locations (e.g., locations expected by the customers for operation and service), physical space (e.g., size and volume) needed for the specified entity, product characteristics (e.g., maximum achievable acceleration), capacity or capabilities of the entity, and durability (e.g., works without a failure for 100,000 cycles under specified conditions).

Relationship matrix: the relationship matrix is formed by the customer needs as its rows and the functional specifications as its columns. Each cell of the matrix represents the strength of the relationship between the customer need and the functional specification defining the cell. The weights of 9, 3, and 1 are commonly used to define strong, medium, and low relationship, respectively. The following coded symbols are used to show the strengths of relationships: two concentric circles (for 9 = strong), one open circle (for 3 = medium), and a triangle (for 1 = low). A cell is left blank when no relationship exists between the customer need and functional specification defining the cell. The relationship matrix is defined by its cells as R_{ij}s.

Desired direction of functional specification: this row (placed above the functional specifications row) shows an up arrow, a down arrow, or a 0 to indicate the desired direction of the value of the engineering specification. The up arrow indicates that a higher value is desirable. The down arrow indicates that a lower value is desirable. And 0 indicates that the functional specification is not dependent on either increase or decrease in its value. Thus, a quick visual scan of this row gives graphic information on whether the desired values of functional specifications need to be larger, smaller, or not dependent on their values.

Correlation matrix (roof): the correlation (or interrelationship) matrix is formed by relationships between combinations of any two engineering specifications defined by the cells of the matrix. The direction and strength of the relationship is indicated in the cell by a positive or a negative number (defined as I_{jk}) in the cell. Coded symbols are also used to indicate the direction and strength of the relationships. Only half of the correlation matrix above the diagonal is the roof of the QFD chart (see the roof in Figure 15.4).

Absolute importance ratings of functional specifications: the absolute importance rating of each functional specification (defined as A_j) is computed by summing weighted relationships between customer needs and the functional specification. The weighting of relationships is based on the importance rating (W_i) of each customer need and the strength of the relationships (R_{ij}s). The expression for the computation of values of A_js is given in the next part of this section.

Relative importance ratings of functional specifications: the relative importance rating (defined as V_j) of each functional specification is expressed as

Product Development Tools

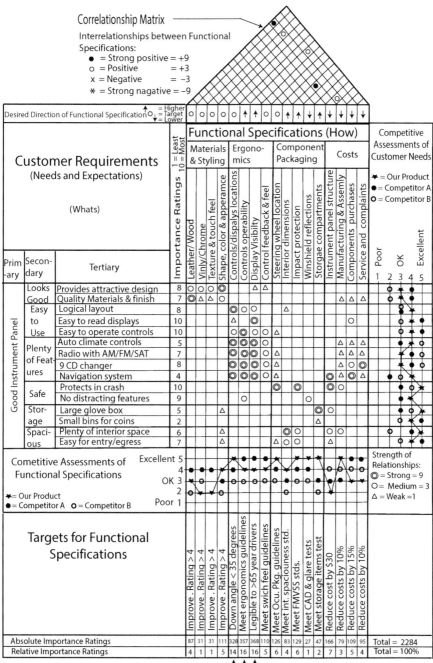

FIGURE 15.4 A quality function deployment chart for an automotive instrument panel.

the percentage of the ratio of its contribution A_j to the sum of all A_js. The expression for computation of values of V_js is given in the next part of this section.

Competitive assessment of customer needs: each product used in benchmarking (along with the manufacturer's target product) is rated to determine how well each customer need (C_i) is satisfied by the product. The rating is usually given by the customers (or by the product development team members after they are very familiar with the customer needs and the products) by using a 5-point scale, where 1 = poor (the product poorly meets the customer need) and 5 = excellent (the product meets the customer need at a high excellence level). The ratings of each product are plotted on the right side of the relationship matrix.

Competitive assessment of functional specifications: each product used in benchmarking (along with the manufacturer's target product) is rated to determine how well each functional specification is satisfied by the product. The rating is usually provided by technical experts or the product development team members (after they are very familiar with each functional specification and the products) by using a 5-point scale, where 1 = poor (the product poorly meets the functional specification) and 5 = excellent (the product meets the functional specification at a high excellence level). The ratings of each product are plotted just below the relationship matrix.

Targets for functional specification: the target value for each functional specification (column of QFD) determined by the team (based on the data shown in the entire QFD chart) is provided below the competitive assessment of the functional specification. The targets are determined after extensive discussions within the team on how the specified high-level product requirements should be achieved (e.g., best in class, among the leaders, slightly above average, average, or below average among all the current products in its class). The target value can be also provided by indicating the rating level (used in the competitive assessments of the functional specification) that should be achieved. The target value should be precisely specified using a quantitative value (e.g., rating value greater than or equal to 4 on a 5-point scale, reduce costs by 10%, achieve an operating temperature of 100 deg, F, must have breaking strength over 300 lbs.). Use of words such as achieving proper, adequate, maximum, minimum, improve, or reduce should be avoided in specifying the target value as any of these words do not provide precise information and their values cannot be interpreted at the same level by any two engineers.

The following mathematical definitions will clarify the aforementioned variables and their relationships.

Let us assume the following:

C_i = ith customer need, where $i = 1, 2, \ldots, m$
F_j = jth functional specification, where $j = 1, 2, \ldots, n$

W_i = importance rating of ith customer need, where $i = 1, 2, \ldots, m$ (The value of W can range from 1 to 10, where 10 = extremely important and 1 = not at all important.)

R_{ij} = relationship between ith customer need and jth functional specification (The values assigned to R_{ij} will be 9, 3, or 1 to define strong, medium, and low relationships, respectively. The ijth cell in the relationship matrix will be left unfilled [i.e., blank] if there is no relationship between C_i and F_j.)

I_{jk} = Relationship between jth functional specification and kth functional specification where j and k are equal to $1, 2, \ldots, n$, and $k \neq j$. (Thus, it is the interrelationship between two functional specifications shown in the roof of the QFD diagram.)

The values of I_{jk} can be as follows:

+9 = Strong positive relationship between jth functional specification and kth functional specification
+3 = Positive relationship between jth functional specification and kth functional specification
0 = No relationship between jth functional specification and kth functional specification
−3 = Negative relationship between jth functional specification and kth functional specification
−9 = Strong negative relationship between jth functional specification and kth functional specification
A_j = Absolute importance rating of jth functional specification

$$= \sum_i \left[W_i \times R_{ij} \right]$$

V_j = Relative importance rating of jth functional specification (percentage)

$$= \frac{100 \times A_j}{\left[\sum_j A_j \right]}$$

AN EXAMPLE OF THE QUALITY FUNCTION DEPLOYMENT CHART

An instrument panel had to be designed for a new mid-sized four-door sedan. The manufacturer formed a team of interior designers, package engineers, body engineers, electrical engineers, ergonomics engineers, market researchers, and engineers from the suppliers of the instrument panel, instrument cluster, radio, climate controls, and navigation system. The team interviewed over 75 customers (i.e., the owners of their current vehicle model and two of their best competitors) and asked them about their needs and expectations in the instrument panel of their future vehicle. The customers first started telling the team that they wanted a good instrument panel.

The team members kept on asking several probing questions such as the following: What do you mean by a good instrument panel? What would you like in the instrument panel? The customers responded saying that a good instrument panel means (1) it should look good, (2) it should be easy to use, (3) it should have plenty of features, (4) it should be safe, (5) it should have enough storage capacity, and (6) finally it should make the vehicle interior look spacious (it must not be crammed with little clearances, and its outer surfaces should not be too close to their knees). The aforementioned needs were considered as the secondary needs. The team then probed more into each of the secondary needs and created lists of tertiary needs. The primary, secondary, and tertiary needs were listed on the left side as "whats" (i.e., what the customers wanted) in the QFD shown in Figure 15.4.

The team members also asked customers to rate each of the tertiary needs on a 10-point importance scale, where 1 = not at all important and 10 = extremely important. The importance ratings are provided to the right side of the customer requirements column in Figure 15.4.

The team brainstormed and created a list of "hows," that is, how would they design the instrument panel? How should it function? What would be their technical descriptors or functional specifications of the instrument panel? The functional specifications developed by the team were listed as column headings in the QFD chart. The columns are placed immediately to the right of the importance ratings.

Next, the team discussed every combination of customer needs and functional specifications and provided the strengths of their relationships using the following scale: (1) strong relation (weight = 9), (2) medium relation (weight = 3), and (3) low relation (weight = 1). The symbols corresponding to the weights were placed in the cells of the relationship matrix (Figure 15.4). Similarly, the relationships between every pair of functional specifications were discussed by the team and symbols corresponding to very positive to very negative relationships were placed in the interrelationship (or correlation) matrix shown on the top of the QFD chart as its roof.

Based on the information gathered during the customer interviews, the team members rated the instrument panels in each of the three vehicles (their current product and two competitors called competitor A and competitor B) and plotted the ratings on each of the tertiary customer needs and functional specifications. The plot of the competitive assessment of customer needs is provided on the right side of the relationship matrix. The plot of competitive assessment of functional specifications is provided below the relationship matrix (i.e., the matrix of customer needs and functional specifications) (Figure 15.4).

The team developed targets (i.e., what target values or rating levels, guidelines, and/or requirements to use for designing their future instrument panel) for each of the functional specifications by discussing how their product compared with the two competitors and their marketing goals. The targets were shown in a section below the competitive assessments of the functional specifications in the QFD. Some targets are specified to meet the Federal Motor Vehicle Safety Standards (FMVSS) (e.g., impact protection targets in Figure 15.4).

Finally, the absolute and relative importance ratings of each of the functional specifications were computed and entered in the last two rows of the QFD chart. The three functional specifications that received the highest importance ratings were (1) display visibility (i.e., displays visible to the driver without any obscurations and legible labels and graphics), (2) control operability (i.e., how the controls are configured, and their operating motions), and (3) locations of the controls and displays (i.e., placement of the controls within the driver's reach distances and displays located high in the instrument panel). These three functional specifications must get very high priority in designing the instrument panel.

CASCADING QUALITY FUNCTION DEPLOYMENTS

The QFD technique can be cascaded in multiple steps to link the customer needs to the production specifications of the components (i.e., from its product-level customer needs to its component-level production specifications). The cascading is shown in Figure 15.5. The figure shows a series of five QFD charts linked such that outputs of a preceding chart become inputs to the next or succeeding QFD chart. Here, the first QFD chart on the left translates the customer needs (labeled as A) into the functional specifications of the product (labeled as B). The second QFD chart takes the functional specifications as inputs (described in rows and labeled as B) and translates them into systems specifications (shown in the columns and labeled as C). (The systems are defined here as those that form the complex product.) The third QFD translates the system specifications (labeled as C) into specifications of their components (labeled as D). The fourth QFD translates a component's specifications (labeled as D) into its manufacturing process specifications, that is, how the component should be produced using manufacturing processes and machines (labeled as E). And the last (fifth) QFD translates the manufacturing process specifications (labeled as E) into component production specifications, that is, characteristics of the component after it is produced (labeled as F). Thus, the component production specifications can be traced back to the original customer needs. Such a series of QFD cascades ensures that components when produced will indeed function to meet the customer needs of the product. The number of cascading steps required is determined by the engineers to make sure that all important levels of entities and processes are considered.

ADVANTAGES AND DISADVANTAGES OF QUALITY FUNCTION DEPLOYMENT

Developing a single QFD chart can be time-consuming as it takes many hours of teamwork involving meetings, discussions, customer visits, benchmarking of competitive products, development of targets, and so on. The advantage is that it exposes the entire product design team to all aspects of the decisions to be made during the product development process. The process of developing a QFD thus educates the team, documents the collected information, and prioritizes the information needed during its development. Thus, when the team begins developing the product, the subsequent decisions generally take much less time as the team members have already discussed all the issues and they are very aware of most of the interfacing and trade-off considerations. The product developed using the QFD, therefore, would have a better chance of designing the right product and satisfying its customers.

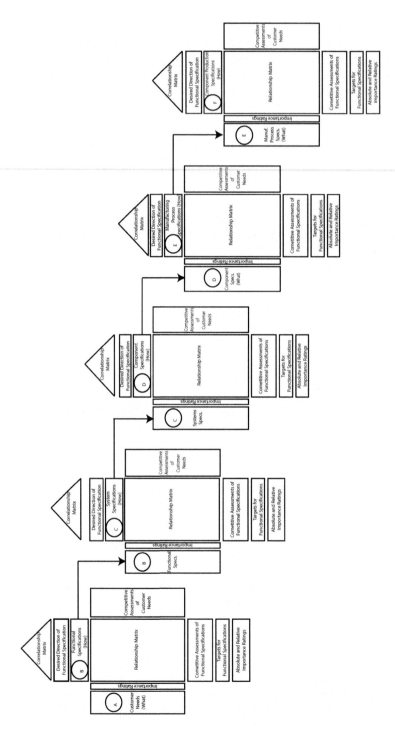

FIGURE 15.5 Using a series of quality function deployments to cascade customer requirements to component manufacturing process specifications.

FAILURE MODES AND EFFECTS ANALYSIS

The FMEA method was initially developed in the 1960s as a systems safety analysis tool. It was used in the early days in defense and aerospace systems design to ensure that the product (e.g., an aircraft, spaceship, or missile) was designed to minimize probabilities of all major failures by brainstorming and evaluating all possible failures that could occur and to act on the resulting prioritized list of corrective actions. For over the past 25 years, the method has been routinely used by product design and process design engineers to reduce the risk of failures in the designs of products and processes used in production, operation, and maintenance of various systems used in many industries (e.g., automotive, aviation, utilities, and construction). The FMEA conducted by product design engineers is typically referred to as DFMEA (the first letter "D" stands for design), and the FMEA conducted by process designers is referred to as PFMEA (the first letter "P" stands for process).

FMEA is a proactive and qualitative tool used by quality, safety, and product/process engineers to improve reliability (i.e., eliminate failures and thus improve quality and customer satisfaction) (Kolarik, 1995; Besterfield et al., 2003). Development of an FMEA involves the following basic tasks: (1) identify possible failure modes and failure mechanisms, (2) determine the effects or consequences that the failures may have on the product and/or process performance, (3) determine methods of detecting the identified failure modes, (4) determine possible means for prevention of the failure modes, and (5) develop an action plan to reduce the risks due to the identified failures. FMEA is very effective when performed early in product or process development and conducted by experienced multifunctional team members as a team exercise.

The method involves creating a table with each row representing a possible failure mode of the given product (or process) and providing information about the failure mode in the following 12 columns of the FMEA table:

1. Description of a system, subsystem, or component.
2. Description of a potential failure mode of the system, subsystem, or component.
3. Description of potential effects of the failure on the product/system and its subsystems, components, or other systems.
4. Potential causes of the failure.
5. Severity rating of the effect due to the failure.
6. Occurrence rating of the failure.
7. Detection rating of the failure or its causes.
8. Risk priority number (RPN): it is the multiplication of the three ratings in points 5, 6, and 7 above.
9. Recommended actions to eliminate the failures with higher RPNs.
10. Responsibility of the persons or activities assigned to undertake the recommended actions and target completion date.
11. Description of the actions taken.
12. Resulting ratings (severity, occurrence, and detection) and RPNs (after the action is taken) of the identified failures in point 2 above.

Examples of rating scales used for severity, occurrence, and detection are presented in Tables 15.4 through 15.6, respectively. The definitions of the scales generally vary between different organizations depending on the type of industry, product or process, nature of the failures, associated risks to humans, and costs due to the failures.

AN EXAMPLE OF A FAILURE MODES AND EFFECTS ANALYSIS

An automatic transmission in an automobile will not operate properly if the transmission fluid leaks out. An engineer designing a transmission fluid hose conducted

TABLE 15.4
An Example of a Rating Scale for Severity

Rating	Effect	Criteria: Severity of Effect
10	Hazardous—without warning	Very high severity rating when potential failure mode affects safe product operation and/or involves noncompliance with government regulations without warning.
9	Hazardous—with warning	Very high severity rating when potential failure mode affects safe product operation and/or involves noncompliance with government regulations with warning.
8	Very high	Product inoperable with loss of primary function.
7	High	Product operable with reduced level of performance. Customer dissatisfied.
6	Moderate	Product operable but usage with reduced level of comfort or convenience. Customer experiences discomfort.
5	Low	Product operable but usage without comfort or convenience. Customer experiences discomfort.
4	Very low	Minor product defect (e.g., noise, vibrations, poor surface finish) only noticed by most customers.
3	Minor	Minor product defect only noticed by average customer.
2	Very minor	Minor product defect only noticed by discriminating customer.
1	None	No effect.

TABLE 15.5
An Example of a Rating Scale for Occurrence

Rating	Probability of Failure	Possible Failure Rates
10	Very high: Failure is almost inevitable	≥1 in 2
9		1 in 3
8	High: Repeated failures	1 in 8
7		1 in 20
6	Moderate: Occasional failures	1 in 80
5		1 in 800
4	Low: Relatively low failures	1 in 2,000
3		1 in 15,000
2	Remote: Failure is unlikely	1 in 150,000
1		≤1 in 1,500,000

Product Development Tools

TABLE 15.6
An Example of a Rating Scale for Detection

Rating	Detection	Criteria: Likelihood of Detection by Design Control
10	Absolutely uncertain	Design control will not and/or cannot detect a potential cause or mechanism for the failure mode, or there is no design control.
9	Very remote	Very remote chance that the design control will detect a potential cause/mechanism for the failure mode.
8	Remote	Remote chance that the design control will detect a potential cause/mechanism for the failure mode.
7	Very low	Very low chance that the design control will detect a potential cause/mechanism for the failure mode.
6	Low	Low chance that the design control will detect a potential cause/mechanism for the failure mode.
5	Moderate	Moderate chance that the design control will detect a potential cause/mechanism for the failure mode.
4	Moderately high	Moderately high chance that the design control will detect a potential cause/mechanism for the failure mode.
3	High	High chance that the design control will detect a potential cause/mechanism for the failure mode.
2	Very high	Very high chance that the design control will detect a potential cause/mechanism for the failure mode.
1	Almost certain	Design control will almost certainly detect a potential cause/mechanism for the failure mode.

an FMEA to evaluate possible failures caused by the hose. This FMEA is presented in Table 15.7. The hose involved in this example consists of a nylon tube with connectors inserted into each end. Ferrules are crimped onto each end to help hold the connectors on. A conduit made of plastic covers the hose to protect it from heat and moving parts near the engine. The hose carries the transmission fluid from the reservoir to the clutch actuation system. The transmission fluid in the hose is pressurized during the operation to about 6.5 b (94.3 psi). The hose must also be able to withstand temperatures of 60°C (140°F). During the development phase, the design itself is proved in a series of tests referred to as the Design Verification Plan and Report (DVP&R). The DVP&R happens once for each design, so it does not take into account all the sources of variability that the product and materials are exposed to during the life of the part. There were a few items with RPN above 50 in the FMEA. The actions taken by the engineer reduced the RPN for two of the three failure modes (see Table 15.7).

FAILURE MODES AND EFFECTS AND CRITICALITY ANALYSIS

The failure modes and effects and criticality analysis (FMECA) are very similar in format and content to the FMEA described earlier. It contains an additional column of criticality. The criticality column provides a rating showing the level of criticality of the failure (in each row) in accomplishing the major goal (or mission) of the product. The technique is also called failure modes and criticality analysis (Hammer, 1980).

TABLE 15.7
An Example of FMEA

Part Number: 1008-91

Original Date: 2/15/2012

Description: Powertrain Fluid Tube Assembly
Model Year: 2012

Date Revised: 6/20/2012
Prepared by: T. Jack

No.	Item/Function	Potential Failure Mode	Potential Effects of Failure	Severity Rating (S)	Potential Causes of Failure	Occurrence Failure Rating (O)	Current Design Controls	Detection Rating (D)	RPN	Recommended Action	Responsibility	Action	Action Results S	O	D	RPN
1	Hose—allows transmission fluid to travel and exert pressure to point of use	Leak in the hose	A slow leak would not be noticed immediately. The driver would notice gear shifting delays over time and finally, the vehicle will stop shifting gears	8	Hole in the hose. Defective hose material	3	100% leak test with air	1	24	No action required						
2		Hose burst at low pressure	A quick and fast leak with immediate loss of shifting function	8	Defective hose material	3	Material burst test	7	168	Investigate if variations in the burst test results	T. Jack 07/20/12	Reduced hose extrusion temperature for consistency	8	1	7	56
3	Connectors—attach a hose to pressurized oil reservoir	Leaks around connector	A slow leak would not be noticed immediately. The driver would notice gear shifting delays over time and finally the vehicle will stop shifting gears	8	Loose connector or omitted washer or uneven connector mating surface	3	100% leak test with air	1	24							

#	Item/Function	Failure Mode	Failure Effect	S	Cause	O	Current Controls	D	RPN	Recommended Action	Responsibility	Action Taken	S	O	D	RPN
4	Connector disconnected from hose or reservoir	A quick and fast leak with immediate loss of shifting function	8	Connector not fully seated in the assembly	2	Pull out testing at final inspection. 100% visual checks during service	4	64	Investigate if a sensor could detect if the connector was fully seated at the hose end	T. Jack 07/20/12	Sensor not possible currently	8	2	4	64
5	Conduit—protects hose during operation from moving parts and stones sprayed by tires	Insufficient conduit wear strength	Possible damage to the tube in service	6	Conduit material defective	1	Design verified during sign-off	7	42	No action required						
6	Ferrule—holds the connector on to the hose	Connector disconnected from hose or reservoir	A quick and fast leak with immediate loss of shifting function	8	Ferrule or hose not properly positioned in the crimping die or press failure	2	Pull out testing at final inspection. 100% visual checks during service	4	64	Investigate if parts presence sensor can be located on the die	T. Jack 07/20/12	Sensor installed on the die and press stops if the sensor is not activated	8	1	2	16
7	Barcode label—lot and part traceability	Missing label	Loss of traceability and identification	4	Label fell off, never affixed, not affixed	4	Visual inspection of label	4	32	No action required						
8	Packing—protect the part while shipping properly	Damaged part	Returned by the customer	4	Insufficient packing strength	1	Packaging standard used for all parts shipping	4	16	No action required						

Criticality can be rated by using different scales for different products. The criticality ratings typically cover a range from low criticality involving the stoppage of equipment (requiring minor maintenance) to high criticality involving failures resulting in the potential loss of life.

OTHER PRODUCT DEVELOPMENT TOOLS

During the product development process, tools from many areas such as systems engineering, specialty engineering areas, and program and project management are used to manage both technical and business activities of the program. The systems engineering and program management tools are covered in Section I of this book. Sections II and III of the book cover approaches and tools used in product development, quality, safety, and human factors engineering areas. Among other remaining tools, the following part of this section covers the business plan, program status chart, design standards, CAD tools, and other evaluation tools. These tools are covered in this chapter because they provide important information in management and technical decision-making activities during the product development process.

BUSINESS PLAN

Business plan is generally a proposal for creating or developing a new product. It is usually prepared internally within a company to obtain concurrence from the top management to approve the product program. The business plan is thus a document prepared to describe the details of a proposed product, product program timing plan, and corporate resource needs to develop the product. It is typically prepared jointly by the company's product planning, engineering, marketing, and finance activities.

The business plan should include the following:

1. Description of the proposed product:
    a. Product configuration (e.g., for an automotive product, the body style of the vehicle such as a sedan, a coupe, a crossover, a sports utility vehicle, a pickup, or a multi-passenger vehicle and its variations)
    b. Size class
    c. Markets where the product will be sold and used (countries)
    d. Market segment: luxury, entry luxury, or economy
    e. Manufacturer's suggested retail price and price range with different optional equipment
    f. Production capacity and estimated sales volumes over the product life cycle
    g. Makes, models, and prices of leading competitors in the proposed market segment
2. Attribute rankings (i.e., how the product would be positioned in its market segment such as best in class, above the class average, and average in the class by considering each product attribute)
3. Pugh diagram showing product attributes and changes in the proposed concept with respect to the datum (selected reference product) and competitors

4. Dimensions and options:
   a. Overall exterior dimensions (e.g., product package envelope with length, width, height, wheelbase, and cargo/storage volume)
   b. Interior dimensions (e.g., people package with legroom, headroom, shoulder room, number of seating locations, and so forth)
   c. Major changes in the product's systems as compared to the previous (outgoing) model, for example, drive options (front-wheel drive, rear-wheel drive, or all-wheel drive), power trains (types, sizes, and capacities of engines/motors and transmissions), and descriptions of unique features (e.g., type of suspensions) and optional equipment
5. A one-paragraph description of the proposed product with several adjectives to describe its image, stance, and styling characteristics (e.g., futuristic, traditional, retro, fast, dynamic, aerodynamic, tough, or chunky—like a Tonka truck)
6. Program schedule:
   a. Program kick-off date, timings of major milestones
   b. Job #1 date (i.e., date when the first production unit comes out of the manufacturing plant) and model year
7. Projected sales volumes:
   a. Quarterly or yearly sales estimates of each model in each market segment
8. Financial analysis:
   a. Curves of estimated cumulative costs and revenues during product life cycle for different scenarios (best case, average, and worse case)
   b. Anticipated quarterly funding needed during product development and during revenue build-up
   c. Anticipated date of break-even point
9. Product life cycle:
   a. Estimated life span
   b. Possible product refreshments, future models, and variations
   c. Recycling of the plant and products
10. Proposed plant location and plant investments
11. Description of risks involved in undertaking the program such as technical risks, time/schedule risks and costs related risks (i.e., running over the budget)
12. Justification (reasons) for approval of the proposed plan versus other alternatives.

PROGRAM STATUS CHART

Product planners and program managers keep track of progress on product development programs by using a number of different techniques such as program timing charts and Gantt charts, which are presented in Chapter 8, and cash flows, which are covered in Chapter 9. One chart that is very popular is the program status chart, which is typically used to track the status of problems encountered during a program (or a project).

The status charts are also called the "Red–Yellow–Green" charts as they indicate problem status by the use of colors as follows: (1) Red indicates that the problem is not yet solved and it is a "job stopper" (i.e., it will stop the progress on the entire program until it is solved), (2) yellow indicates that the problem can introduce significant delays in the program unless it is solved quickly, and (3) green indicates that the problem is no longer a timing threat to the program.

Table 15.8 presents an illustration of a program status chart. The status column in the chart uses letters R, Y, and G to indicate red, yellow, and green colors when colored charts can be made on a normal "black-ink-only" printer. Such charts are typically used in senior management–level meetings to draw attention and get fast resolution on the job-stopper problems.

STANDARDS

Product design standards serve as very useful tools in saving time required to make design decisions. Properly developed design standards that incorporate rationale and assumptions used in their development can provide basic knowledge on whether a standard can be applied during the design process. When the standard meets the needs of the customers for the product, the use of the design requirements and design procedures provided in the standards can reduce the time required to get the necessary information and decide on how to design.

Some standards may only specify how the product should perform and thus provide design flexibility (i.e., the designer can design by using any appropriate solution as long as it meets the required performance). Such performance standards can promote innovative product designs as they will not be restricting compliance to any given design specifications. The advantages and disadvantages of design versus performance standards along with other types of standards and issues related to the standards are covered in Chapter 4.

MODEL-BASED SYSTEMS ENGINEERING

Model-based systems engineering (MBSE) is now rapidly gaining interest among the SE and product design community. It is a systems engineering methodology that focuses on creating and using predeveloped models as the primary means of information exchange between engineers, rather than on document-based information exchange. It helps to support system requirements, design, analysis, verification, and validation activities beginning in the conceptual design phase and continuing throughout development and later life cycle phases. The MBSE methodology is used in many industries such as aerospace, defense, rail, and automotive.

MBSE is the formalized application of modeling to support system or product development activities involving system requirements, design, analysis, verification, and validation beginning in the conceptual design phase and continuing throughout the development and later life cycle phases of the system (or product). It is, thus, an implementation of the systems engineering process that focuses on the use of a software application (i.e., a model) as the primary means of information exchange, rather than on document-based information exchange. It helps in developing and managing

TABLE 15.8
Program Status Chart

Program: XM25　　　　Program Manager: RJW　　　　Date: 08/30/11

Problem No.	Problem Description	Status[a]	Target Date	Expected Completion Date	System/Subsystem/Component	Product Attribute	Responsibility: Organization and Manager
1	Gloss level on the top resurface of the instrument panel should be reduced to meet the veiling glare standard	Y	9/15/2011	10/30/2011	Instrument Panel	Safety	Body and Trim; JBM
2	Squeaking noise during braking	Y	11/10/2011	12/30/2011	Braking system—brake pad material	Safety	Chassis-Brakes; WLV
3	Premature wear in front bearings of turbo boosters	R	11/15/2011	Unresolved	Power train—turbo boosters	Performance	Powertrain; EER
4	Noticeable jerk during transmission shifting	Y	11/25/2011	12/30/2011	Power train—X5 transmission	Performance	Transmission; JJT
5	Wiper flutter at speeds over 70 mph	G	7/15/2011	8/26/2011	Body—visibility	Safety	Body Electrical; RGK

[a] R, red: critical issue (job stopper)—must be resolved before next program review meeting; Y, yellow: important customer need—must be resolved by next milestone; G, green: issue resolved—no action needed.

requirements and interfaces between systems with complex functionality and capabilities. Model-based approach as compared to the traditional document-based approach is considered superior because of the ability of the software to obtain inputs from many team members and to maintain a detailed database on relationships and traceability created by the software applications. The model-based approach is also found to reduce the time taken to develop product/system specifications and it is also more accurate and error-free.

SysML is a general-purpose modeling language for systems engineering applications (SysML.org, 2021. It is used in developing models used for applications of MBSE. It supports the specification, analysis, design, verification, and validation of a broad range of systems and systems-of-systems. The MBSE is intended to produce an integrated system model (using SysML) which reflects multiple views of the whole product (or system) to flow (or cascade) requirements to lower levels or systems and to specify the interaction and interconnection of its components, and their functions. As the requirements are specified, projects are created and handed off to downstream engineering, which includes interdisciplinary subsystem teams containing team members who specialize in software, electronics, mechanical, and other design disciplines/departments.

The outputs of the SE projects include, but not limited to, the following (Douglass, 2021):

a. Requirements specification
b. Analysis of requirements, whether this is done with use cases or user stories
c. System architectural specifications
d. System interface specifications
e. Trace relations between the elements of the different work products
f. Safety, reliability, and security (and resulting requirements) analyses
g. Architectural design trade studies

The MBSE approach implemented in IBM's Rhapsody has been claimed to be very useful in meeting these objectives (IBM, 2022a and 2022b). IBM Rhapsody can be easily accessed through a web browser so any stakeholder can view requirements and make changes. All team members can rely upon a single point of truth. All changes are automatically tracked to ensure each change can be traced to its originator. When a change is made to a model, it is reflected in the code, as well as the reverse. This traceability is necessary for compliance. The process is completely automated to ensure speed and accuracy. Diagrams in an IBM Rhapsody model are interrelated, so changes to an element in one diagram are automatically propagated throughout the model, enhancing data consistency across systems. Simulation provides early validation. IBM Rhapsody tests models to expose problems early in the process. This helps protect against costly systems failures following release. When a defect is discovered, the software can investigate, test, and help make the needed correction. Models can be reused within the same initiative or applied to other initiatives, speeding up the process and reducing costs.

Douglass (2021) also illustrates how the MBSE software with its packages can help perform the following (a) Actor Package: This package holds the actors—elements that represent objects that interact with the system, (b) Design Synthesis Package: This holds all design-related information for the systems model, (c) Architectural Analysis Package: This holds architectural analyses, such as trade studies, usually in one nested package per analysis, (d) Architectural Design Package: This holds the architectural design, the system and subsystem blocks, and their relations, (e) Functional Analysis Package: This holds the use case analyses, one (nested) package per use case analyzed, (f) Interfaces Package: This holds the logical system and subsystem interfaces as well as the logical data schema for data and flows passed via those interfaces, (g) Requirements Analysis Package: This holds all the requirements information, (h) Requirements Package: This holds the requirements, (i) Use Case Diagrams Package: This holds the system use cases and use case diagrams. In addition, the software allows creation of plans for risk management, verification (to ensure that the product meets its requirements at different levels) and validation, and allows conducting simulations.

Advantages of MBSE are as follows:

a. Encourage collaboration
b. Maintain regulatory compliance
c. Mitigate ambiguous specifications
d. Reduce costs
e. Improve response times
f. Accelerate time to market
g. Remove data inconsistencies and rework.

MBSE is considered a highly efficient engineering practice that uses very descriptive, precise system model artifacts to improve processes. It helps solve the collaboration, cost, and calendar problems that team members must face and overcome each day. The model-based process and its results are much more reliable than traditional approaches that MBSE solutions are recommended by the International Council of Systems Engineering (INCOSE) and the US Department of Defense.

COMPUTER-AIDED DESIGN TOOLS

A number of CAD tools are used to create three-dimensional solid models using software such as AutoCAD, CATIA, Pro/Engineer, SolidWorks, Rhino, and so forth. These tools not only perform the traditional engineering drawing and drafting work but also allow the visualization of the product model from different eye points to evaluate issues such as exterior and interior appearance (e.g., shapes, continuity/discontinuity between adjacent surfaces, tangents, and reflections), spaces (clearances) and postures of human occupants/operators (with digital human models) in products (e.g., cars, airplanes, and boats) or workplaces, the feeling of interior spaciousness and storage spaces, layouts of hardware placements (mechanical packaging), comparisons of alternate designs (by superimposition or side-by-side viewing of different product concepts and competitive products), assembly analyses to evaluate assembly feasibility (e.g., by detecting interferences between parts being assembled), alternate assembly methods and fit (e.g., gaps) between parts, and so on. The newer CAD models can also simulate movements of parts within the product and movements of the product in its work environment to visualize how the product will look and fit within other existing systems.

CAD models are also very useful for communications between different design studios, product engineering offices, and supplier facilities. The CAD files for the products can also be used as inputs to various other sophisticated computer-aided engineering analyses to evaluate structural/mechanical (e.g., strength, dynamic forces, deflections, and vibrations in simulated operating environments), aerodynamic (using computer-aided fluid dynamics), and thermal (temperature, heat build-up, and heat transfer) aspects of the products. The CAD files can also be used to facilitate manufacturing. For example, CAD files serve as inputs to computer-aided process planning as well as in creating machining programs for the computer numerical control machines.

CAD has become an especially important technology within the scope of computer-aided technologies, with benefits such as lower product development costs and a greatly shortened design cycle. CAD enables designers to create layouts and develop their work on a display screen, print it out, and save it for future editing, thus saving time in creating variations in designs and their drawings.

PROTOTYPING AND SIMULATION

Virtual and physical prototyped parts can be created for visual evaluations and physical mock-ups for use in design reviews. Many computer simulation systems are also available for human factors testing of user interfaces. Three-dimensional parametric solid modeling requires the design engineer to input values of key parameters, what can be referred to as the "design intent." The objects and features created can be shown to the customers for their feedback and be adjusted by creating many design iterations until an acceptable design is achieved. Further, any future modifications can be easily made by inputting parameter changes in a computer-controlled prototype. Many automotive manufacturers use computer-controlled adjustable vehicle models (or programmable vehicle bucks) during early concept phases to compare and evaluate several automotive designs by quick changes in many key vehicle parameters (Prefix Corporation, 2012).

A number of specialized computer software systems are increasingly used to simulate product testing and evaluation. CAD is also used to create accurate photo simulations that are often required in the preparation of environmental impact reports, in which CADs of intended buildings, vehicles, and other products are superimposed into photographs or videos of existing environments to represent what such locales will be like when the proposed facilities are allowed to be built. Potential blockage of view, corridors, and shadow studies are also frequently made through the use of CAD.

PHYSICAL MOCK-UPS

Physical mock-ups of product concepts of products such as cars, trucks, airplanes, and boats are useful during design reviews to get a better feel of the size, space, and configuration of the product in its early phases. The mock-ups can also be shown to

Product Development Tools 301

potential customers and users for their feedback during informal quick evaluations as well as structured market research clinics.

TECHNOLOGY ASSESSMENT TOOLS

Using new technologies to improve product designs has been a continuous process to gain improvements in performance, efficiencies, safety, and costs. However, most new technologies cannot be immediately applied. It takes many years or even decades to solve problems in bringing some new technology applications to a state of readiness and implementation. The technical experts in various specialized areas generally follow advances in new technologies. The progress in the most promising technologies is closely followed, and research departments are asked to perform evaluations and undertake development projects to improve the technologies so that they can be quickly implemented in future products.

Many methods to assess technologies have been developed. Forgie and Evans (2011) have provided an excellent review of available techniques for technology assessments.

CONCLUDING REMARKS

The product development process involves the integration of many ideas, product features, and technologies. It is important to use tools that can help in searching, developing, and evaluating ideas that can be implemented to develop balanced product concepts. Systems engineering along with other specialized engineering and management disciplines allows simultaneous consideration of many inputs from multidisciplinary teams. Developing a right product at the right time is important. The tools presented in this chapter and their applications would thus aid in selecting the right product concept and refining the concept during the early stages of the product development.

An example of how a series of QFD charts can be used to cascade customer needs down to specifications for the component manufacturing process.

REFERENCES

Akao, Y. 1991. *Development History of Quality Function Deployment. The Customer Driven Approach to Quality Planning and Deployment.* ISBN 92-833-1121-3. Minato, Tokyo, Japan: Asian Productivity Organization.

Altshuller, G. S. 1997. *40 Principles: TRIZ Keys to Technical Innovation.* Translated by Shulyak, L. and S. Rodman. Worcester, MA: Technical Innovation Center.

Besterfield, D. H., Besterfield-Michna, C., Besterfield, G. H. and M. Besterfield-Scare. 2003. *Total Quality Management.* Third Edition. ISBN 0-13-099306-9. Upper Saddle River, NJ: Prentice Hall.

Black, S. 2011. *e-Turbo Concepts to Eliminate Turbo Lag and Enable Engine Downsizing. Term Project submitted for the Automotive Systems Engineering Class (AE 500) in Winter Term 2011.* Dearborn, MI: University of Michigan-Dearborn.

Champy, J. 1995. *Reengineering Management.* Business Books. New York: HarperCollins Publishers.

Douglass, Bruce Powel. 2021. *Agile Model-Based Systems Engineering Cookbook: Improve System Development by Applying Proven Recipes for Effective Agile Systems Engineering*. Birmingham, UK: Packt Publishing Ltd. Kindle Edition.

Forgie, C. C. and G. W. Evans. 2011. Assessing Technology Maturity as an Indicator of Systems Development Risk. In *Systems Engineering Tools and Methods* (Eds.) Kamrani, A. K. and M. Azimi (pp. 111–134). Boca Raton, FL: CRC Press.

Hale, A. and D. Pelowski. 2006. *Ergonomic Study of Asian, European, and North American Motorcycles—Commonality and Differences between Cruiser, Sport, and Touring Categories*. Advisor: Prof. V. D. Bhise. Capstone Project Report, Automotive Systems Engineering Program. Dearborn, MI: College of Engineering and Computer Science, University of Michigan-Dearborn.

Hale, A., Pelowski, D. and V. D. Bhise. 2007. Commonality and Differences between Cruiser, Sport, and Touring Motorcycles: An Ergonomics Study. SAE Paper no. 2007-01-0438. Presented at the SAE 2007 World Congress, Detroit, MI.

Hammer, W. 1980. *Product Safety Management and Engineering*. Englewood Cliffs, NJ: Prentice Hall.

IBM. 2022a. IBM Engineering Systems Design. (Website: https://www.ibm.com/internet-of-things/learn/mbse-smart-paper/index.html?chapter-01; Accessed: January 4, 2022)

IBM. 2022b. IBM Engineering Systems Design Rhapsody. (Website: https://www.ibm.com/products/systems-design-rhapsody?utm_content=SRCWW&p1=Search&p4=43700050290364589&p5=e&gclid=EAIaIQobChMIh4OMn9yY9QIV_HRvBB0YUwH1EAAYASAAEgKijvD_BwE&gclsrc=aw.ds; Accessed: January 4, 2022).

International Organization for Standardization (ISO). 2012. *ISO 9000 Standards*. Website: http://www.iso.org/iso/iso_catalogue/management_and_leadership_standards/quality_management.htm (accessed February 24, 2012).

Kolarik, W. J. 1995. *Creating Quality—Concepts, Systems, Strategies, and Tools*. New York: McGraw-Hill.

Nadler, G. 1967. *Work Systems Design: The IDEALS Concept*. Homewood, IL: Irwin.

Nadler, J. and S. Hibino. 1998. *Breakthrough Thinking: The Seven Principles of Creative Problem Solving*. Roseville, CA: Prima Publishing and Communications.

NIST, 2012. *Baldridge Quality Award Criteria*, National Institute of Standards and Technology (NIST). Website: http://www.nist.gov/baldrige/publications/business_nonprofit_criteria.cfm (accessed: May 2, 2013).

Peters, T. and R. H. Waterman, Jr. 1982. *In Search of Excellence*. First Harper Business Essentials. New York: HarperCollins Publishers.

Prefix Corporation. 2012. Programmable Vehicle Model (PV). Website: http://www.prefix.com/PVM/ (accessed June 20, 2012).

SysML.org. 2021. What is the relation between SysML and UML? Website: https://sysml.org/sysml-faq/what-is-relation-between-sysml-and-uml.html (Accessed: October 5, 2021)

Yang, K. and B. El-Haik. 2003. *Design for Six Sigma*. ISBN 0071412085. New York: McGraw-Hill.

16 Design for Manufacturing and Assembly

INTRODUCTION

During the detailed design phase, engineers design the components to ensure that the components function as specified in their cascaded attribute requirements. The components are also evaluated to ensure that they can be manufactured and assembled easily and economically. Thus, during the detailed design and engineering phase of any component, it is very important to ensure that (a) the component will meet its functional characteristics, (b) the component can be manufactured easily using the best and most economical method, and (c) the component can be easily assembled into its system (i.e., the system to which it belongs to meet the function of the system) with least amount of assembly cost and time. The components should be also designed to meet sustainability goals for their life cycle (see Chapter 13). Thus, this chapter provides manufacturing and assembly considerations and methods that can be used during the early design phase of the components.

DESIGN, FUNCTIONING, MANUFACTURING, AND ASSEMBLY

Design of any component involves studying requirements on the system within which the component belongs. The requirements on every component must be cascaded down from the system within which it belongs because the component exists to provide one or more of the functions of the system (Note: These are systems engineering considerations.).

The functioning of a component (i.e., how a component serves its function within a system in which it belongs) involves determining its configuration (e.g., shape, dimensions, features), materials, and special treatments such as heat treating and surface coating to improve its durability. The function of a component will also depend upon how functions of its parent system are allocated to different components within the system and how the components are interfaced with each other.

Manufacturing here refers to production of individual part (or component) from raw materials. Manufacturing involves use of operations such as stamping, sawing, forging, casting, and machining during its production.

Assembly refers to the addition and/or joining of parts to form sub-assemblies or assemblies. Assembly typically involves human assemblers, special purpose automated machines, and/or robots that perform operations such as moving, positioning, inserting, and aligning of components, and fastening them to a base component or an assembly by use of methods such as welding, riveting, nailing, or fastening screws/ nuts/bolts, or applying adhesives.

Design for Manufacturing (DFM) and Design for Assembly (DFA) are primary tools for improving costs and plant operations. DFM and DFA are considered as the integration of product design and process planning into one common activity because product design must include assumptions about functions of each component, its materials, and manufacturing and assembly processes used in its production. The goal is to design a product that can be easily and economically manufactured. The importance of designing for manufacturing is underlined by the fact that about 70% of manufacturing costs of a product are determined by its design decisions (cost of materials, processing, and assembly), with production decisions responsible for about only 20% (such as process planning or machine tool selection) (Boothroyd et al., 2011).

The component design involves understanding all the requirements related to functions, fits and interfaces of the system in which it belongs. Thus, its shape, size, and dimensions, material, strength, and other functional characteristics (e.g., speeds, forces, operating temperatures, other substances (e.g., corrosives) that will be in the operating environment, and so forth) are determined by conducting many different analyses such as (a) customer needs analyses, (b) mechanical analyses involving stress, strength, deflection, and vibrations, (c) thermal and airflow analyses, (d) electrical analyses involving resistance, current flow, voltages, and capacitances, (e) chemical analyses involving corrosion resistance and toxicity, and (f) durability in withstanding required number of duty cycles under stated loads and environmental conditions. In addition, the manufacturing and assembly requirements of the components must be considered concurrently. Otherwise, the component may not meet the cost, time, and other manufacturing and assembly requirements which mostly are dependent on variables related to manufacturing processes, manufacturing equipment, workstation layouts, plant layout, assembly line configuration, and so forth.

By applying a DFA tool, the communication between manufacturing, assembly, and design engineering is improved, and ideas and reasons considered during the design process become well-documented for future reference.

PRINCIPLES OF DFMA

DFMA should be applied early in product development along with considerations of its production process, tooling, and the assembly process design. Doing so will make manufacture less time-consuming, which will reduce overall costs and increase the ease of manufacturing. The exact process of DFMA will depend on what product (i.e., its functions) is being designed and how it is produced.

General principles of DFMA include (a) selection of materials, (b) designing components for efficient assembly, (c) standardization of components, (d) reducing the number of components, and (e) minimizing the amount of manufacturing and assembly operations. Other basics of effective DFMA include product and process design simplicity by reducing complexities due to considerations of details such as the number of design features such as slots, steps or groves to handle, orient, and feed components, number of steps and workstations in manufacturing and assembling components, number of tools, fixtures, machines/robots, and grippers, and setup time required for tools and machines.

MATERIALS, MANUFACTURING AND ASSEMBLY CONSIDERATIONS: AN EXAMPLE OF IC ENGINE PISTON

Designing a piston in an internal combustion (IC) engine requires consideration of its attributes to ensure its functionality and selection of the right combination of materials, manufacturing, and assembly processes.

The attributes of a piston are:

a. Weight (if too heavy, the engine loses power)
b. Withstands combustion temperatures
c. Withstands high crankshaft speeds (RPMs) and accelerations/decelerations
d. Shape (e.g., long skirt increases friction)
e. Scratch resistant surfaces
f. Smoother outside surfaces
g. Precise dimensions (tolerances with respect to the cylinder wall, piston pins, and piston rings)
h. Low friction coefficient with the cylinder wall
i. Low expansion coefficient (allows tighter piston-to-cylinder-wall clearances)
j. Costs.

Manufacturing processes in producing the piston are assumed to be (a) casting, diecasting, or forging, and (b) machining (for precise surface dimensions). Forged pistons have more complex manufacturing processes and are more expensive to make. Instead of a simple mold for casting, the forging process needs a large press, which rams the aluminum into a complex mold under high pressure. Machining forged pistons is also time-consuming and expensive. The forged piston advantage is greater strength, harder surfaces, more predictable expansion properties, and virtually no porosity. Another advantage to forged pistons is the ability to make them lighter and with a smaller skirt. It can be designed this way because the forged pistons are stronger. They can be machined to remove more material out of them without suffering structural losses. Forged pistons have a distinctive look, with an extra-hard surface and reducing machining marks. The forged pistons are thus used when high rpm and high temperatures are expected especially in supercharged and turbocharged IC engines.

Further, the assembly of the piston into an engine involves (a) attaching a connecting rod to the piston using a piston pin, (b) attaching piston rings to the piston, (c) inserting the piston into an engine cylinder, and (d) attaching the connecting rod to the crankshaft. It should be noted that for a proper functioning of the engine tolerances between the parts (i.e., piston and piston rings, piston cylinder, piston pin, and connecting rod) must be achieved to required levels.

Thus, the functioning of the piston, its manufacturing, and its assembly all need to be considered during its design process to ensure that it will meet the requirements of the engine.

MANUFACTURING AND ASSEMBLY CONSIDERATIONS

The following considerations must be carefully studied during the selection of manufacturing processes:

1. Most components are not produced by a single process, but they require a sequence of different processes to achieve all the required attributes of the final part. Selection of the manufacturing process for any component will depend upon many factors, such as function of the component, operating environment and usage cycles, and materials used to produce the product. Understanding of manufacturing processes, materials, benchmarking of other existing products performing similar functions and produced by leading manufacturers and advances in new technologies are important before a manufacturing process is selected.
2. Shape and other attributes of the component (e.g., strength, weight, conductivity, ability to withstand temperatures, optical properties, texture, color, and so forth), the capabilities of manufacturing processes to produce products with consistent output quality and properties of the materials
3. Number of components, use of standardized components, required quantities of components, and rates of production
4. Reducing the overall manufacturing costs.

The following considerations must be carefully studied during the selection of assembly processes:

1. The process of assembly can be divided into two separate types of elements a) handling (acquiring, orienting, and moving the components), and b) insertion and fastening (mating a part to another component or group of components).
2. Time required for assembling each component which in-turn depends upon:
    a. Feeding/presenting method (component orientation, direction, and speed)
    b. Handling/carrying/getting into position and inserting without damage, collisions, fumbling, and jamming
    c. Method of joining/fastening components (e.g., welding, riveting, screwing, plastic deformation, and adhesive bonding)
    d. The total number of components in the assembly: Reducing part count (originally driven by local economic analysis, now driven by part cost itself) by considering:
        i. Function of the component: Serves as a carrier or can be combined with other parts.
        ii. Movements of adjacent parts after assembly
        iii. Materials of adjacent parts
        iv. Disassembly needs (repair, inspection, upgrade)
        v. Type and number of fasteners.

Design for Manufacturing and Assembly

MANUFACTURING AND ASSMBLY COSTS

The manufacturing and assembly costs include:

a. Cost of raw materials required for producing the component
b. Cost of the manufacturing process (i.e., capital and operating costs of processing equipment at required production rates in processes such as casting, forging, stamping, machining, and painting)
c. Assembly cost to install the component (includes costs of bringing the component, orienting, and positioning the component on a base component or an assembly of components, and fastening the component (e.g., riveting, screwing, plastic deformation of material, welding, adhesive bonding)
d. Direct labor costs
e. Indirect labor costs (e.g., maintenance)
f. Costs of parts purchased from suppliers
g. Overhead costs (e.g., insurance, supervisory/management personnel).

The cost of assembly is determined by the number of components in the assembly, and the ease with which the components can be handled, inserted, and fastened. Thus, the main goal in assembly cost reduction should be to reduce the number of components in an assembly. While reducing the number of components additional analyses should be conducted to ensure that the complexity and costs of the remaining components do not increase over the original design.

The desirable features for ease of assembly are:

a. Component is easy to align
b. Component is easy to insert and grip
c. Enough clearance exists between assembled components
d. Assembly requires a few fasteners
e. Limited need for fixtures
f. Components arrive at the assembly station in the correct orientation.

The assembly method will depend upon (a) the characteristics (or features) of the component, (b) the base component (or assembly) on which the component is to be assembled, and (c) desired production rate.

ASSEMBLY ENGINEER'S RECOMMENDATIONS TO COMPONENT DESIGNERS FOR ASSEMBLY COST REDUCTIONS

Based on the above considerations, the recommendations that the assembly engineers can make to the component designers can be described as follows:

1. <u>Include a multidisciplinary design team</u>. About 70% of manufacturing costs of a product are determined by the decision of the design engineering team. Therefore, a multidisciplinary team approach involving design, engineering, manufacturing, assembly, marketing, and finance personnel is

very important to ensure that requirements and considerations of all product attributes and systems in the complex product are considered during design of all components in the product.

2. <u>Reduce number of components</u>. If components can be combined, the number of steps in installation and the installation time can be reduced. There are drawbacks to consider. This can affect serviceability and increase to the cost of remaining components. It can also make assembling the product more complicated if not designed well. Evaluating alternative designs and assembly methods by using the method of Boothroyd et al. (2011) or similar methods is important before selecting a component design.

3. <u>Reduce the number of fasteners</u>. Fasteners take a lot of time to install. They add installation and maintenance costs. In the case of nuts and bolts, they improve serviceability (e.g., rivets or adhesive bonds are more difficult to disassemble and damage components during disassembly), so any change in the number of fasteners should consider serviceability.

4. <u>Change the type of fastener</u>. If components replacement or serviceability is not an issue, then nuts and bolts could be replaced with rivets. If the components are welded, then replacing the welds with fasteners or using a chemical adhesive may improve the assembly time under manual assembly. If the assembly was done in a robot cell, welding the component might be faster than all of the above depending on the workstation layout. Boothroyd tables should be used for time/cost estimation when considering a design change (e.g., type of installation process and whether manual, automated, or robotic must be taken into consideration).

5. <u>Make components easy to orient</u>. Symmetry helps with feeding and handling the component since it can be installed in more than one way (by orienting and aligning). Symmetric component will reduce assembly time as compared to an asymmetric component. End-to-end and rotational symmetry about the axis of insertion improve the ease at which a component can be inserted into the assembly. If symmetrical design is not possible, choose (or incorporate) pronounced asymmetrical features to help in quickly identifying the asymmetry and thus simplify the component orientation and alignment.

6. <u>Avoid components that are very small or large</u>. Components that are very small or large are difficult to grasp, feed, and handle, and they take a longer time to assemble. Operators have a difficult time picking up just one very small part, but mechanical feeders will have less difficulty. Both operators and feeders will have difficulty handling very large parts. Avoid very small and slippery component geometries and surfaces (hard to handle).

7. <u>Avoid flexible components</u> that are hard to feed, grasp, and insert.

8. <u>Reduce components that can tangle</u>. In the case of components that tangle very easily (e.g., wiring harnesses, springs or snap rings), it would be a good idea to apply cable and zip ties in as many places as possible so that wires do not catch during assembly and increase assembly time. This will also ensure that operators do not damage components during assembly.

Design for Manufacturing and Assembly

9. Provide chamfers for insertion holes to make things easy to align and insert. Any components that need to be inserted into a hole should also have relief cut into the ends to make those parts easier to align.
10. Provide air relief in blind holes. Blind holes with tight fits should have air relief the passage to reduce insertion efforts because the air pressure in the stuck hole will increase the insertion force.
11. Provide very obvious aligning aids (features) for aligning and inserting asymmetric parts. Otherwise, assembly time and cost will increase. If it is installed incorrectly, the assembly may require reworking or repairing jammed parts.
12. For components that could nest and jam up, provide bosses or webbing on the component to prevent nesting with other components in bins. For components with thin lips that can stack in feeders, make those lips thicker so that they cannot overlap. This will improve handling and feeding times. Add small design features that help components not-jamming-up during transport and feeding.
13. Keep the insertion forces as low as possible. Both robots and operators will need more time if component insertion is difficult. An additional step (motion and force provided by another source) may be required if the component is beyond the robot or the human operator's ability to insert.
14. The component should be designed so that it does not need to be held down during assembly. This will remove component holding difficulty (added step to hold the component) after inserting and the holding time-related costs for manual and robotic assembly.
15. Provide self-locating features. Thus, components do not have to be held in place during assembly.
16. Add guide pins and special bosses to components when not secured immediately so that they self-align.
17. Provide aligning features to help robots easily align components. This will also help operators in manual assembly tasks involving aligning of components. Notches, slots, holes, bosses, etc. will all help, especially, if the operator does not have good visibility of moving components during assembly.
18. Design the product with the least number of reorientation motions. If the product can be assembled from bottom to top without any component or product orientation motions, that is preferred, as those orientation motions can be time-consuming. Otherwise, an additional fixture or a feature in the fixture may have to be designed to account for the reorientation motions.
19. Standardize simple components like fasteners. Multiple sizes of standardized fasteners will require different feeders and screw-tightening heads or tools.
20. Design components so they can be located first before they are released. This will ensure that the component is not misplaced or mislocated.
21. Reduce the amount of screws where possible and replace them with simpler fasteners such as rivets or snap fits.
22. Provide sufficient clearance space for insertion tools or fingers to reduce insertion time.

23. Allow insertion of components from above (easy to feed by gravity and reduce component holding).
24. Make component design obvious with matching features on the fixtures for easier component orientation for assembly.
25. Simplify component design, so the least number of fixtures are needed. The reduced number of features required to orient, align, and position a component will reduce handling and positioning time.

METHODS TO ESTIMATE ASSEMBLY TIME

Several different methods are used to estimate the time to complete assembly tasks. All the methods are based on pre-determined time estimates obtained from actual time measurements under different combinations of variables affecting assembly time. The methods typically provide time estimates for different work elements in tabular formats with various variables that affect the time estimates. Total assembly time is calculated by adding estimated times of all work motions obtained from the tables. The magnitudes of the estimated times give the analyst feedback and ideas on modifications to the components and assembly methods to reduce assembly time.

METHODS-TIME MEASUREMENT

Methods-Time Measurement (MTM) was developed by Maynard, Stegemerten, and Schwab (1948) in 1946 from motion pictures of manufacturing operations in Westinghouse. Three basic methods, MTM-1, MTM-2, and MTM-3 are available. The MTM-1 is the most detailed, and MTM-2 and MTM-3 are simplified, i.e., use less number of variables to provide time estimates more quickly with a lesser number of steps and hence less precise than MTM-1 estimates (Maynard et al., 1948; Konz and Johnson, 2008). MTM-1 is still used in many companies to estimate times for manufacturing and assembly operations involving manual tasks such as hand and body motions.

The motions are broken down into 10 categories: Reach, Move, Turn, Apply Pressure, Grasp, Position, Release, Disengage, Body Motions (e.g., leg-foot, burn body, bend, sit, kneel on knee), and Eye Motion. Tables are provided for each category of motion with variables that effect the category of motion in columns and rows of the tables. The cell in the table provides time estimates for the motion in time measurement units (TMUs), which is 0.00001 hour (or 1/100,000 of an hour or 0.036 s). The times are for an experienced operator working at a normal pace (100%) and no allowances are included in the times.

The Reach motion is usually with an empty hand or finger while Move is generally movement with an object (weight) in hand. The Reach and Move times are affected by movement distance, the accuracy of movement, and whether the hand was in motion before or after the movement. The Move time is also affected by the weight of the object moved and the acceleration and deceleration involved in the movement. Turn is the movement that rotates the hand, wrist, and forearm about the long axis of the forearm. The amount of time depends upon the number of degrees turned as well as the weight of the object or resistance against the turn made. Apply Pressure is the application of force without resultant movement. Grasp is the motion used to gain control of an object. It is almost always followed by a Reach. Grasp is divided into five types of grasps: (a) pic-up grasp (easily grasped objects, very small or object lying close against a flat surface) and interference grasps of nearly cylindrical objects of different diameters (less than 0.25 inch to greater than 0.5 inch), (b) regrasp (change grasp without relinquishing control), (c) transfer grasp (control transferred from one hand to other), (d) select grasp (objects jumbled with other objects so that search and select occur), and (e) contact, sliding, or hook grasp. Position is a collection of minor hand-finger movements for aligning, orienting, and engaging one object with another object. The Position time depends upon the class of fit, symmetry, and ease of handling. MTM-1 includes 12 tables to obtain time estimates of all motions. Learning the use of MTM-1 requires about a week of classroom lectures and shop training by observing and recording data on motions performed by actual operators performing different operators over several weeks.

BOOTHROYD ET AL. ASSEMBLY EVALUATION METHODS

Boothroyd et al. (2011) have provided methods to evaluate assembly (i.e., to predict either time or costs to assemble) of components using manual (with human operator), equipment involving automatic machines, and robots. The methods involve use of tables developed from actual assembly tasks conducted under different combinations of component characteristics, component handling situations, assembly equipment, and their setups.

BOOTHROYD ET AL. MANUAL ASSEMBLY EVALUATION METHOD

The method is based on a classification system for the assembly process. The classification system is a systematic arrangement of component features that affect the acquisition, movement, orientation, insertion, fastening of the part plus some other operations that are not associated with a specific component such as turning assembly over.

For manual assembly, two classification tables are provided: one for component handling and the second for insertion and fastening.

The handling classification system consists of two-digit codes from 00 to 99 for a total of about 100 classifications. The first digit identifies the row and the second identifies the column in the tables. Each code has a handling time in seconds (range from 1.13 to 9.0 s). Factors that affect the handling time include:

a. Amount of hand turning/manipulation in degrees of the component required in orienting
b. Easy to grasp components or components with handling difficulties
c. Thickness and size of components (for grasping difficulties)
d. Components needing tweezers for grasping and manipulation
e. Components requiring two hands for manipulation

f. Component weight less than 10 lbs or more than 10 lbs for two-hand operation or requiring mechanical assistance.

The manual insertion and fastening timetable contains 63 two-digit code numbers with insertion times that range from 1.5 s to 12.0 s. These times are for small parts and must be scaled up for larger parts.

Factors that affect manual insertion and fastening time include:

a. Whether the component is secured immediately or after other operation
b. Accessibility of the insertion region
c. Ability to use insertion region (hand/finger clearance and visual obstructions)
d. Ease of aligning and positioning of the component
e. Whether a tool is needed
f. Whether the component stays put after being placed or whether the assembler must hold it until other components or fasteners are installed.
g. Simplicity of insertion operation
h. Fastening method (no screwing, plastic deformation, riveting and screwing).

To estimate assembly time using the Boothroyd et al. manual assembly method, a worksheet is created with each component listed in a separate row of the table. The table includes information such as the number of components to be assembled, manual handling code, handling time, manual insertion code, insertion time, total operation time, minimum number of parts, and a brief description of the motion required for assembling the component (see Table 16.2).

OTHER BOOTHROYD ET AL. ASSEMBLY EVALUATION METHODS

Boothroyd et al. (2011) provide assembly evaluation methods for other assembly situations such as use of automated assembly machines, robots, and specialized assemblies such as wiring harnesses and printed circuit boards.

AN EXAMPLE OF APPLICATIONS OF MTM-1 VS. BOOTHROYD'S MANUAL ASSEMBLY TIME ESTIMATING METHODS

A gas valve diaphragm assembly consisting of eight components shown in Figure 16.1 was assembled manually by using a workstation shown in Figure 16.2. The workstation consisted of five bins arranged in a semi-circular configuration so that hand motions in picking up the five types of components (namely valve plates [diaphragms], screws, plate holders, washers, and nuts can be reduced in moving to the assembly work fixture in front of the operator.

The work fixture was designed to first hold the two screws with their heads facing the bottom side of the fixture. The fixture had recessed holes to hold the screw heads and slotted screwdriver pins were mounted inside each hole to hold the screws from rotating. Next the plate holder with positioned over the two screws with its two holes passing through the screws and its two ears facing down. The gas valve plate was

Design for Manufacturing and Assembly 313

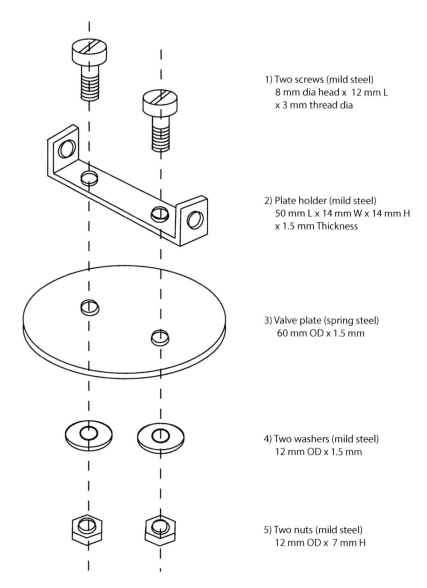

1) Two screws (mild steel)
   8 mm dia head x 12 mm L
   x 3 mm thread dia

2) Plate holder (mild steel)
   50 mm L x 14 mm W x 14 mm H
   x 1.5 mm Thickness

3) Valve plate (spring steel)
   60 mm OD x 1.5 mm

4) Two washers (mild steel)
   12 mm OD x 1.5 mm

5) Two nuts (mild steel)
   12 mm OD x 7 mm H

FIGURE 16.1 Gas valve diaphragm assembly. (Dimensions are in mm).

then positioned over the hole holes and inserted over the two screws. The two washers were placed over the two screws simultaneously using both hands of the operator. And in the last operation, the two nuts were placed and tightened over the two screws simultaneously using both hands of the operator. The entire assembly was lifted by the operator's right hand and placed in the output bin.

The MTM-1 table showing left- and right-hand motions required for the assembly and the MTM-1 codes and time estimates are shown in Table 16.1

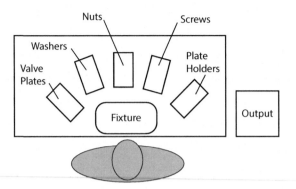

FIGURE 16.2 Layout of the assembly workstation.

It would take 17.3 sec (479.7 TMU) to assemble with both hands.

Estimates of times used to assemble the gas valve diaphragm using the Boothroyd et al. (2011) method for manual assembly are shown in Table 16.2.

Comparing the total assembly times of 17.3 secs obtained by applying the MTM-1 with 57.25 secs obtained by applying the Boothroyd et al. method, shows that MTM-1 provides substantially lower time estimates. The differences in these total assembly times are due to many of the variables related to component characteristics that affect component handing and insertion that are taken into account by the Boothroyd et al. method (which the MTM-1 does not consider).

There are many differences and similarities between the two methods. The following section illustrates some similarities and dissimilarities between the two methods.

SIMILARITIES BETWEEN MTM-1 AND BOOTHROYD ET AL. MANUAL ASSEMBLY METHODS

1. Both methods are systematic approaches to evaluate assembly operations and help estimate and reduce overall assembly time.
2. Both methods help to calculate approximate assembly times before a component is launched—for planning and budgetary reasons.
3. Both methods start out with the current design of components and allow the team to visualize potential issues (by comparing times for other possible assembly motions) throughout the assembly operation.
4. Difficult assembly steps can be evaluated according to the fit and required movements, allowing the team to clearly separate them from more simple assembly operations.
5. Standard times are provided for the most common manual assembly operations.
6. Both methods can only be applied to smaller components, large assemblies need to be considered differently.

Design for Manufacturing and Assembly

TABLE 16.1
MTM-1 Table Showing Codes and Time Estimates for the Assembly Motions

Left-Hand Motion Description	No.	Left-Hand Code	Time (TMU)	Right-Hand Code	Right-Hand No.	Right-Hand Motion Description
Reach to a screw	1	R10C	12.9	R10C	1	Reach to a screw
Grasp a screw	1	G1C2	17.4	G1C2	1	Grasp a screw
Move screw to the fixture	1	M16Cf	37.4	M16Cf	1	Move screw to the fixture
Release screw	1	RL2	2.0	RL2	1	Release screw
			17.0	R16C	1	Reach to plate holder
Reach to a valve plate	1	R16C	17.0			
			8.7	G1C2	1	Grasp plate holder
Grasp a valve plate	1	G1B	3.5			
Regrasp valve plate	1	G2	5.6			
			5.6	G2	1	Regrasp plate holder
			18.7	M16Cf	1	Move the plate holder over the two screws
			25.3	P2SSD	1	Position the plate holder over the two screws
			2.0	RL2	1	Release plate holder
Move the valve plate over the two screws	1	M16Cf	18.7			
Position the plate over the two screws	1	P2SSD	25.3			
Release plate	1	RL2	2.0			
Reach to washer	1	R16C	17.0	R16C	1	Reach to washer
Grasp a washer	1	G4B	9.1	G4B	1	Grasp a washer
Move the washer over a screw	1	M16f	37.4	M16cf	1	Move the washer over a screw
Release the washer	1	RL1	2.0	RL1	1	Release the washer
Reach to nut	1	R16C	17.0	R16C	1	Reach to nut
Grasp a nut	1	G4B	9.1	G4B	1	Grasp a nut
Move the nut over a screw	1	M16cf	37.4	M16cf	1	Move the nut over a screw
Position nut over the screw	1	P2SSD	50.4	P2SSD	1	Position nut over the screw
Turn nut 1/2 inch along the circumference	6	M1/2Bf	12.0	M1/2Bf	6	Turn nut 1/2 inch along the circumference
Release nut	6	RL1	12.0	RL1	6	Release nut
Regrasp nut	6	G1A	12.0	G1A	6	Regrasp nut
			6.7	R3C	1	Reach to assembly
			2.0	G1A	1	Grasp assembly
			18.7	M14Cf	1	Move assembly
			2.0	RL1	1	Release assembly
Reach to the home position	1		15.8	R16B	1	Reach to the home position
		Total TMUs ---->	479.7			
		Total sec ----->	17.2692			

TABLE 16.2
Table Showing Boothroyd et al. Codes and Time Estimates for the Assembly Motions

Component	No. of Components	Manual Handling Code	Handling Time per Component (sec)	Manual Insertion Code	Insertion Time per Component (sec)	Time (sec)	Description of Task
Valve Plate	1	03	1.69	02	2.5	4.19	Place over the plate holder and inserted over the two screws
Plate Holder	1	11	1.8	02	2.5	4.3	Place over two screws
Screw	2	42	4.35	41	7.5	23.7	Held in fixture with head at the bottom
Washer	2	40	3.6	02	2.5	12.2	Insert washer over screw
Nut	2	01	1.43	92	5	12.86	Place nut over screw and tighten
					Total -->	57.25	

7. Both methods help to visualize the importance of certain design features to the product design and engineering team and seek the opinion of the manufacturing engineering team.
8. Both methods provide time estimates. The component features and issues considered will not always be identical, especially for complex assemblies.

DISSIMILARITIES BETWEEN MTM-1 AND BOOTHROYD ET AL. METHODS

1. MTM considers the actual assembly station layouts and reach distances per part, while the Boothroyd et al. method does not.
2. The Boothroyd et al. method is focused on the handling and insertion motions, while MTM considers hand/finger and body movements.
3. Boothroyd methods focuss on the actual part geometry (thickness, symmetry, weight) when calculating approximate assembly times, challenging the team to improve these factors before the final design of the part is launched. MTM considers the weight of components and overall size for grasping movements.
4. The Boothroyd et al. method considers the accessibility for operator hands, fingers, and tools. MTM does not consider these.
5. The Boothroyd et al. method considers the visibility of components for the operator while MTM considers eye movements that need to be included by the analyst.
6. The Boothroyd et al. method provides alternative fastening solutions, such as plastic bending, riveting, screwing, etc. for fastening whereas MTM requires modeling of the hand-finger movements involved in each operation.

Design for Manufacturing and Assembly 317

7. MTM allows for simultaneous operations between left and right hands and evaluates the feasibility of combining these movements to save time whereas the Boothroyd et al. method does not consider simultaneous or sequential operations.
8. Boothroyd et al. consider special situations such as use of tweezers, optical magnification, and holding down parts during assembly. MTM does not consider such situations (but they can be modeled approximately by using other motion elements).

BOOTHROYD ET AL. METHODS FOR ESTIMATING ASSEMBLY TIMES FOR HIGH-SPEED AUTOMATIC ASSEMBLY AND ROBOTIC ASSEMBLY

HIGH-SPEED AUTOMATIC ASSEMBLY

High-speed automatic assembly consists of part feeders, conveyors, and assembly machines. The standard machine bases used for assembly machines include: (a) Rotary indexing machines, (b) In-line indexing machines, and (c) Continuous motion machines (In-line free transfer or asynchronous machines).

The Boothroyd et al. (2011) method for evaluating high-speed automatic assembly involves many variables that are considered through the use of a series of tables to account for various component features affecting feeding and insertion of components. The steps involved in the procedure are as follows.

1. Obtain a five-digit feed code from tables
   a. First digit based on rotational or non-rotational components
   b. Second digit based on part orientation
   c. Third digit based on the feature (e.g., slots, steps, chamfers, flats, etc.) controlling orientation
   d. Fourth digit based on additional feeding difficulty (e.g., overlapping parts, delicate, and flexible)
   e. Fifth digit based on additional feeding difficulty (e.g., tangling, light/not light, sticky parts).
2. Obtain orienting efficiency E and relative feeder cost (C_r) from the table
3. Obtain addition to C_r due to feeding difficulty
4. Obtain relative workhead cost (W_c) due to direction of holding, securing, aligning, and resistance
5. Compute total insertion cost ($C_t = C_f + C_i$), where C_f = feeding cost and C_i = insertion costs. The assembly method is evaluated based on total insertion cost (C_t) and not on assembly time.

Table 16.3 provides an example of outputs for an assembly of 10 components in an automotive starter motor. (All costs are in cents.)

TABLE 16.3
An Illustration of Output Table for Evaluation of High-Speed Automatic Assembly System for an Automotive Starter Motor

Name	Repeat Count	Feed Code	Orientation Efficiency (€)	Relative Feeder Cost (Cr)	Overall Length	Max Feed Rate (Fm)	Required Feed Stations	Feeding Cost (Cf)	Insertion Code	Relative Workhead Cost (Wc)	Required Insertion Stations	Insertion Cost (Ci)	Total Cost (Ct)	Minimum No. of Parts
Armature	1	22020	0.75	2	300	3.75	1	0.12	02	1.5	1	0.18	0.3	1
Field frame with brush	1	25720	0.45	3	150	4.5	1	0.18	02	1.5	1	0.18	0.36	1
Brush plate	1	63020	0.4	2	10	60	1	0.12	02	1.5	1	0.18	0.3	1
End shield	1	01322	0.25	4	80	4.69	1	0.24	02	1.5	1	0.18	0.42	1
End shield bushing	1	00022	0.7	4	20	52.5	1	0.24	01	1.5	1	0.18	0.42	0
Shim	2	00040	0.7	3	20	52.5	1	0.18	00	1	1	0.12	0.3	1
C-washer	1	63046	0.4	5	20	30	1	0.3	41	2.1	1	0.252	0.552	1
Seal	1	00022	0.7	4	25	42	1	0.24	0	1	1	0.12	0.36	0
End shield cap	1	11000	0.3	1	27	16.67	1	0.06	02	1.5	1	0.18	0.24	1
End shield cap screws	2	21000	0.9	1	8	168.75	1	0.06	38	0.8	1	0.096	0.156	0
												Total --->	3.408	

Design for Manufacturing and Assembly

TABLE 16.4
An Illustration of Output Table for Evaluation of Robotic Assembly

	Components	Codes	AR	AG	TP	TG	Total Cost ($) [60,000*AR] +[5000*AG]	Net Total Cost per workstation ($)	Total Time for Operation (sec) [3*TP] +[2*TG]	Net Total Time, sec	Net Total Workstation Time, sec
	Differential Case/Carrier	02	1.0	1.0	1.05		65000		3.15	3.15	
Workstation #01	Side gear	00	1.5	0	1.0		90000	90000	3*2	6	15.75
	Thrust washer (Side Gear)	11	1.0	0	1.1		60000		3.3*2	6.6	
	Spider gear							Cost of labor			
Workstation #02	Thrust washer (Spider Gear)	Manual Assembly Station									12.456
	Cross pin shaft										
Workstation #03	Cross pin bolt	36	1.0	0.5	1.25	TG is	62500	62500	3.75	3.75	3.75
	Carrier bearing	00	1.0	0	1.0	notrequired	60000		3.15*2	6.3	
Workstation #04	Race	30	1.0	0	1.0	as assembly	60000	60000	3*2	6	18.9
	Shim	30	1.0	0	1.0	involves	60000		3.3*2	6.6	
Workstation #05	Ring Gear	00	1.0	0	1.0	Multistation.	60000	62500	3	3	10.5
	Bolt	36	1.0	0.5	1.25		62500		3.75*2	7.5	
Workstation #06	Bearing cap	00	1.0	0	1.0		60000	62500	3	3	6.75
	Bearing bolt	36	1.0	0.5	1.25		62500		3.75	3.75	
	Gasket	00	1.0	0	1.0		60000		3	3	
Workstation #07	Cover	30	1.0	0	1.0		60000	62500	3	3	12.75
	Filler Plug	36	1.0	0.5	1.25		62500		3	3	
	Cover Bolt	34	1.0	0.5	1.0		62500		3.75	3.75	
Workstation #08	Axle shaft	43	1.5	0	1.2		92500	92500	3.6	3.6	3.6
						Total -->	1040000	492500		72	84.456

ROBOTIC ASSEMBLY

Boothroyd et al. (2011) provide a method to estimate costs and times for robotic assembly for: (1) Single-station one-arm robot, (2) Single-station two-arm system, and (3) Multi-station system. Tables are provided to obtain values of the following four variables that are used to obtain the evaluation measures (cost of robots and assembly time): (a) AR = relative robot cost, (b) AG = relative addition gripper or tool cost, (c) TP = relative affective basic operation time, and (d) TG = relative time penalty for gripper or tool change. The above four are multipliers to basic costs and time ($60,000 for robot, $15,000 for gripper; 3 sec time for robot to perform a basic operation and 2 sec for time penalty for gripper of tool change).

The row and columns of the tables are classified with the following digit code:

a. First Digit: The appropriate row is selected based on the direction of insertion
b. Second Digit: The appropriate column is selected based on the needs of a special gripper, clamping temporarily after insertion, and part tendency to align itself during insertion
c. When the row and columns are selected for an operation, the numbers in the box (cell) of the table provide values of AR, AG, TP, and TG for computing robot costs, gripper tool cost, total time for operation, and gripper/tool time penalty.

Table 16.4 provides an example of outputs for an eight-workstation assembly for assembling 19 components (see Case Study 5 in Chapter 24).

CONCLUDING REMARKS

It is important to ensure that at the detailed engineering phase all components of a designed product can be easily and economically manufactured and assembled. Design for Manufacturing and Design for Assembly are a different set of methods that are used to predict time and costs for manufacturing and assembly processes. These methods should be applied iteratively to ensure that product design can meet all the functional, sustainability, manufacturing, and assembly requirements of all components, systems, and the product. During application of the design for manufacturing the materials and manufacturing processes are selected. And during the design for assembly application, the assembly methods and processes are selected to ensure that the production costs and processing times are economical and would meet the goals of the product program.

REFERENCES

Boothroyd, G., Dewhurst, P., and W. A. Knight, *Product Design for Manufacturing and Assembly*. 3rd Edition, Boca Raton, FL: CRC Taylor &Francis Group, 2011.

Konz, S. and S. Johnson, *Work Design: Occupational Ergonomics*, 7th Edition, Scottsdale, Arizona: Holcomb Hathaway, Publishers Inc., 2008.

Maynard, H., Stegemerten, G., and J. Schwab, *Methods Time Measurement*. New York: McGraw-Hill, 1948.

17 Traditional and New Quality Tools

INTRODUCTION

This chapter presents seven traditional quality tools and seven new quality tools used in the Total Quality Management (TQM) field. The seven traditional tools are as follows: (1) Pareto chart, (2) cause-and-effect (C-E) diagram, (3) check sheet, (4) histogram, (5) scatter diagram, (6) stratification, and (7) control chart. The seven new tools are as follows: (1) relations diagram, (2) affinity diagram, (3) systematic diagram, (4) matrix diagram, (5) matrix data analysis, (6) process decision program chart (PDPC), and (7) arrow diagram.

This chapter also includes some basic information on experiment design and its usefulness in selecting the right combinations of levels of product parameters (or independent variables) based on statistical analysis techniques and provides some discussions on Taguchi concepts of quadratic costs, robustness, and Taguchi methods for experimentation.

TRADITIONAL QUALITY TOOLS

The traditional quality tools are quantitative in nature and any of the tools can be used with any other tools. They are called traditional because they have been popularly used on the shop floor by plant personnel to summarize and visualize data obtained from production samples (Besterfield et al., 2003; Kolarik, 1995). They are easy to use and help visualize certain characteristics of the data such as distribution of the data and relationships between variables in the data. The following part of this section presents the purpose, description, and examples of each of the seven traditional quality tools.

Pareto Chart

Purpose
The purpose of a Pareto chart is to provide visual guidance in selecting problems that occur with high frequencies so that large improvements in quality can be obtained by eliminating only a few problems.

Description
A Pareto chart is a series of vertical bars whose heights reflect the frequency of problems (or categories of defects usually related to quality issues). Each bar represents a given category of a problem found in the database. (Note: This bar chart is not a histogram. A histogram presents the frequency of items [height of bars] falling in equal intervals of a continuous variable.) The problems (bars) are arranged in a descending

order of frequency (height of the bar) from the left to the right. This means that the categories represented by the tall bars on the left occur with higher relative frequency than those on the right. This bar chart is used to separate the "vital few" from the "trivial many" categories of problems or issues.

This chart is based on the Pareto principle, which states that a large percentage of the problems come from a relatively low percentage of categories. This principle is also commonly referred to in the industry as the 80/20 rule, where 20% of something is typically responsible for 80% of the results. Pareto charts are extremely useful because they can be used to identify those problems that have the greatest cumulative effect on product quality. A curve of the cumulative percentage of problems is also superimposed on the top of the bars. Thus, the Pareto chart helps in screening out many of the less frequently occurring problems and allows the analyst to focus attention on a few more frequently occurring problems.

The Pareto chart is named after Vilfredo Pareto, a nineteenth-century economist who postulated that a large share of wealth is owned by a small percentage of the population. Pareto noticed that 80% of Italy's land was owned by 20% of the population. This basic principle translates well into quality problems. This tool helps implement the "continuous improvement" principle by helping to select the most frequently occurring problems in each improvement cycle.

Example: Pareto Chart of Customer Complaints

An example of the Pareto chart is presented in Figure 17.1. The figure presents customer complaint data gathered by an automotive power window supplier. The data were categorized by types of complaints received on the power windows. Out of the total 383 complaints, 197 (or 51%) complaints included grinding noise during window operation, 95 (or 25%) complaints stated that the window regulator did not work properly, 82 (or 21%) complaints stated that the glass did not properly open and close while driving, 6 (or 2%) complaints stated that the window regulator did not work at

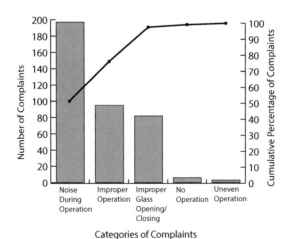

FIGURE 17.1 Pareto chart of window regulator customer complaints.

all after sitting in the parking lot for a few days, and 3 (or 1%) complaints involved uneven (or intermittent) operation of the window regulator system.

The Pareto chart of the data provided in Figure 17.1 shows two vertical axes. The vertical axis on the left provides the scale for the number of complaints presented in the vertical bars. The vertical axis on the right presents the scale for the cumulative percentage of complaints presented by the curve (i.e., the dark lines joining points shown with small squares). The cumulative curve shows that 76% of the complaints came from the first two categories, namely, noise during operation and improper operation.

CAUSE-AND-EFFECT DIAGRAM

Purpose

The purpose of the C-E diagram is to display the causes of a given problem. The causes are grouped by major areas and arranged according to the levels of hierarchy such as primary, secondary, and tertiary causes.

Description

The C-E diagram is also called a Fish diagram (also known as a fishbone diagram, herringbone diagram, Fishikawa, or Ishikawa diagram) (Ishikawa, 1985). The diagram looks like a fish, as shown in Figure 17.2. The C-E diagram is created to illustrate the causation on a single event. The event is written in the head of the fish (as shown in the oval on the right side of the diagrams in Figures 17.2 and 17.3). The

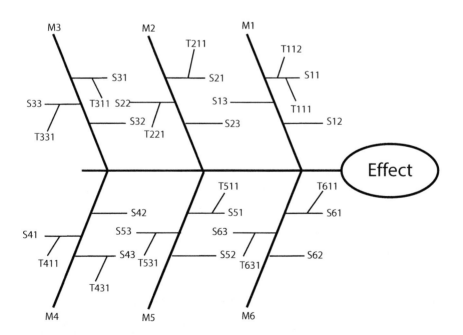

FIGURE 17.2 C-E diagram.

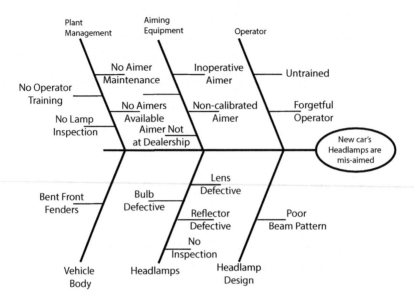

FIGURE 17.3 Example of a C-E diagram.

causes are organized by major categories shown on the major bones (M1 to M6) connected to the backbone in Figure 17.2. The secondary and tertiary causes are shown as S11 to S62 and T111 to T631, respectively.

The major causes are usually grouped, and each group is assigned to a major bone. The major sources of cause categories typically include: (1) People: Any person involved with the process; (2) Methods (or Process): How the process is performed and the specific requirements for doing it, such as policies, procedures, rules, regulations, and laws; (3) Machines: Any equipment, computers, tools, and others required to accomplish the job; (4) Materials: Raw materials, parts, fasteners, lubricants, fluids, and others used to produce the final product; (5) Measurements: Data generated from the process that are used to evaluate its quality; and (6) Environment: The conditions, such as location, time, temperature, and culture in which the process operates, and so forth.

Example: C-E Diagram for Misaimed Headlamps

Some owners of new cars complained that the headlamps of their new cars were not properly aimed (or were misaimed). In order to understand the causes of such complaints, a team of engineers from the car manufacturer's vehicle lighting group, the headlamp assembly operators at the car assembly plant, and a few mechanics at different dealerships met and discussed possible causes for the misaimed headlamps. They categorized various causes of headlamp misaim into the following six major groups: (1) operator (who is responsible for headlamp aiming), (2) aiming equipment, (3) plant management, (4) vehicle body, (5) headlamps, and (6) headlamp design. The six groups are represented by six major bones in the fish diagram in Figure 17.3. The subcauses under each of the groups are shown as secondary bones

Traditional and New Quality Tools

of each major bone. For example, the operator-related problems are as follows: (1) the operator was untrained with the headlamp aiming procedure or (2) the operator forgot to aim the headlamp. The above two causes are shown as secondary bones (or links) attached to the primary bone representing the operator-related causes. Similarly, the headlamp-related major bone (shown in the lower side of the diagram) is shown with its secondary bones as causes related to (1) defective headlamp lens, (2) defective (e.g., warped) reflector, (3) defective bulb (e.g., with crooked filament), and (4) non-inspected headlamp unit.

The fish diagram thus provides a composite view containing all causes related to the event described in the fish head and the causes are categorized by the major bones. The next step in the problem solving generally involves the team to systematically reviewing all causes in the fish diagram and selecting the causes that they think would occur with high frequencies and developing an action plan to make changes to eliminate the causes.

Cause-and-Effect Process Diagram

The C-E process diagram shown in Figure 17.4 is a slightly different format of the fish diagram to describe the causes involved in a process. The steps involved in the process are listed in boxes placed on the backbone of the fish diagram. In Figure 17.4, the steps in the process are as follows: (1) define the project, (2) select a project manager, and (3) select team members. The problem to be investigated (shown in the fish

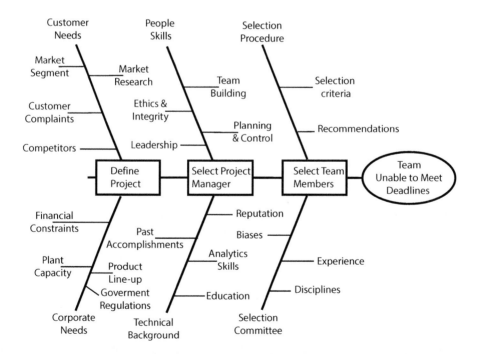

FIGURE 17.4 Example of a C-E process diagram.

head) was "Team Unable to Meet Deadlines." The major causes in each step are categorized and represented as major bones. The minor causes under each major bone are shown as labels attached to the secondary bones. The labels are provided with only one or two words, and they indicate a problem area in a given step represented by each major bone.

For example, the first process step involved "define project" (see the left side of Figure 17.4). The major problems related to defining the project were from "customer needs" and "corporate needs." The problems with the customer needs were due to problems related to (1) understanding the "market segment," (2) conducting the "market research," (3) receiving and understanding the "customer complaints," and (4) understanding the "competitors" and their products. Similarly, each major bone presents problems related to its area in each process step.

The C-E process diagram thus provides all causes associated with each process step and organizes them in categories assigned to each major bone. Thus, collectively the entire C-E process diagram presents all causes for the head event by each process step and in each major area represented by each major bone. The C-E process chart shown in Figure 17.4 has only three steps. A process with many steps can thus have a very long fish diagram. (The author has created some fish diagrams that were over 40 ft. in length [investigated causes of a problem in a manufacturing process with many steps]. One C-E process diagram was created on a long paper roll and the paper was taped around all walls of a large conference room!)

The advantage of the C-E process diagram is that it displays all causes associated with each step in the process. As additional causes are uncovered, they can be simply added at locations corresponding to the step in the process. The diagram thus provides a composite map of possible causes that can be further discussed by a problem-solving team to determine process changes to eliminate many of the causes.

CHECK SHEET

Purpose
The purpose of a check sheet is to maintain running counts of problems (or defects or issues) found during inspections of outputs of a process.

Description
A check sheet generally consists of a piece of paper with a list of problems (or issues) and spaces next to each problem to place check marks as the problems are found during inspections of parts (or products). A thoughtfully constructed check sheet should provide an accurate and a quick understanding of a number of characteristics of the inspected problems. The reviewer of the check sheet should get the information not only on the number of problems of each type, but also additional information on the relative frequencies of the problems and their locations in the inspected units. Thus, the visual layout and appearance of a check sheet is important as it can provide quick visualization and analysis of the data in the check sheet. Check sheets can be created by using tabular and/or pictorial (graphical) formats. Creativity plays a major role in the design of a check sheet.

Example: Checklist for Door Trim Defects

An engineer wanted inspectors to visually check interior door trim panels of cars before they are sent to the assembly line for installation of hardware such as window switches, mirror switches, latches, grab handles, speakers, map pockets, and window glass moving mechanism. The inspectors were asked to record the locations of the defects (e.g., scratches, cuts, wrinkles, and misaligned parts or parting lines) observed in different areas of the trim panel. The areas are shown in the sketch of the driver's door shown in the top portion of Figure 17.5.

Depending on the location of a defect, the inspector placed a check mark ("x") in the row corresponding to the location in the table placed in the lower part of Figure 17.5. The number of check marks provided a graphic view of the relative frequencies of the defects in different areas, and they quickly conveyed the finding that areas S, U, and Q had the highest concentrations of defects.

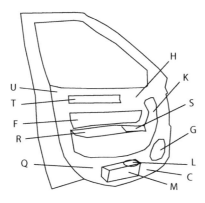

Location of Door Trim Area	Check Marks	No. of Defects
U : Upper Substrate	xxxxxxxxxxx	11
T : Upper Decorative Trim	xx	2
H : Inside Handle Trim Bezel	xxxxx	5
F : Bolster	xxxx	4
R : Armrest		0
S : Window Switch Plate	xxxxxxxxxxxxxxxxxxxxx	21
G : Speaker Grill	xxxxx	5
Q : Lower Substrate	xxxxxxxxxx	10
M : Map Pocket	xxxx	4
K : Grab Handle	xxxxx	5
L : Cup Holder	xx	2
C : Lower Insert	xxxxx	5

FIGURE 17.5 Example of a tabular check sheet with a picture.

Example: Check Sheet for Defects in Painted Car Body

The check sheet for recording defects in the painted car body simply consisted of a picture of the car (see Figure 17.6). The inspectors were asked to place a checkmark in the area where each defect (e.g., uneven paint, dripped paint, or unpainted area) was observed. Visual observation of the frequency of check marks and concentration in different areas of the check sheet provided feedback to the paint department that the rocker panel, doors, cowl, and rear pillar areas received uneven paint. The paint department used the completed check sheets to reprogram the path and speed of the painting robots to avoid excessive paint spray in the identified areas and to apply additional paint in the missed areas.

HISTOGRAM

Purpose

The purpose of a histogram is to display the distribution of frequency of measured values of a continuous variable (e.g., a characteristic of a product such as the diameter of a shaft, surface finish, hardness of the material, or an activation force of a latch—usually obtained from physical measurements of product samples).

Description

A histogram is a graphical representation, showing a visual impression of the distribution of a measured variable. To construct a histogram, a sample of products from a process is obtained and a continuous variable representing a product characteristic is measured from every item in the sample. The measured values are sorted into equal-length intervals on a linear scale representing the measured variable. The frequency values falling in each interval are plotted as vertical bars. Thus, a histogram consists of adjacent vertical bars (or rectangles), erected over discrete intervals (bins), with an area equal to the frequency of the observations (measurements) in the interval.

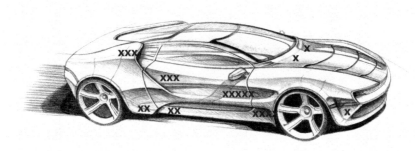

FIGURE 17.6 Example of a pictorial check sheet used in inspection of paint job on a car body.

Note: An inspector places an "x" mark in the region where uneven paint was observed.

Traditional and New Quality Tools

The height of a rectangle is also equal to the frequency density of the interval, that is, the frequency divided by the width of the interval. The total area of the histogram is equal to the number of measurements. A histogram may also be normalized displaying relative frequencies.

A histogram is a bar chart—where each bar represents the frequency of occurrences within the interval represented by the bar. Not all bar charts are histograms. (Note: Bar charts of categorical data are not histograms.) Sample size or observations >30 and the number of intervals between 5 and 15 are generally needed for estimating the shape of the distribution.

Example: Histogram of Resistance of an Electrical Component

Figure 17.7 presents a histogram of resistance (measured in ohms) of an electrical component. Eighty samples of the component were obtained from its production process. The resistance of each sample was measured. The resistance values varied from 61 to 125 ohms. Using the interval of 5 ohms, the number of sample components (having their resistance values) falling in each 5-ohm interval were counted. The frequency count was plotted against the interval value (midpoint of the interval) (see Figure 17.7).

SCATTER DIAGRAM

Purpose

The purpose of a scatter diagram is to show the relationship between two continuous variables (X and Y) by plotting the locations of the data points in the two-dimensional (X–Y) space.

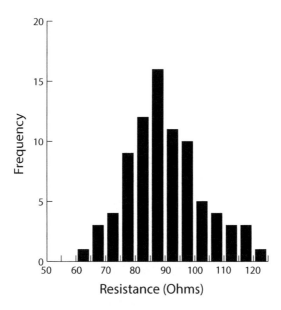

FIGURE 17.7 Example of a histogram showing resistance values of 80 samples.

Description

A scatterplot is a two-dimensional plot of many data points. Each data point is defined by values of two characteristics, both defined and measured as continuous variables. The axes of the scatterplot are orthogonal, and each axis represents a continuous variable (e.g., a product characteristic such as weight, length, temperature, or resistance).

Example: Scatterplot of Sitting Height versus Standing Height of 30 Human Operators

Figure 17.8 presents a scatterplot of two anthropometric dimensions of 30 human operators in an assembly shop. The sitting height and standing height of each operator were measured by use of an anthropometric dimension measuring laser scanner in the standardized sitting and standing postures. Defining the sitting height as the x-axis and the standing height as the y-axis, the locations of 30 data points corresponding to the 30 operators were plotted. The plot is called a "scatter diagram" and is shown in Figure 17.8. The scatter diagram shows the relationship between the two variables. The locations of data points in Figure 17.8 show that the two variables are correlated (i.e., sitting height increases with an increase in standing height, or vice versa).

STRATIFICATION

Purpose

The purpose of stratification is to separate (or categorize) data points based on a certain characteristic that defines its category (class, strata, or belongingness). A stratification process helps in determining differences between data points (e.g., their locations in a plot) due to the effects of certain variables that define their strata.

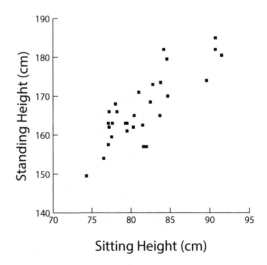

FIGURE 17.8 Scatter diagram showing relationship between standing height and sitting height of 30 human operators in an assembly shop.

Description

The stratification is usually used with other plotting tools such as scatter diagrams, histograms, or bar charts. The data separated (identified by codes) by categories help in visualizing and understanding differences due to the categories. Stratification can be done by data sources (or by variables) such as machines, operators/users, suppliers, locations, time of the day, work shifts, days in the week, and seasons.

Example: Stratification of Anthropometric Data by Gender

Figure 17.9 presents the same data points used in Figure 17.8. However, the data points are coded (stratified) to indicate the gender of the operator. The data points of male operators are shown as black-filled squares and the data points of female operators are shown as unfilled (or open) squares. The data points of males are located higher than that of the females. This indicates that males had larger standing heights than females for the same sitting height.

CONTROL CHARTS

Purpose

The purpose of a control chart is to distinguish between variation in a process resulting from common causes (natural variations) and variation resulting from some special causes. It presents a graphic display of process stability or instability over time. Once the stability of the process is achieved (i.e., when the process is found to be within control limits), the process parameters can be estimated and used to determine

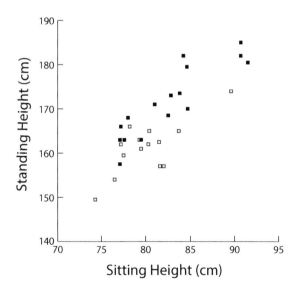

FIGURE 17.9 Scatter diagram with data stratified by gender.

Note: Data points of males and females are in "black filled" and "unfilled" squares, respectively.

the capability of the process (by measures such as C_p, C_{pk}, percent defectives or average number of defects per inspection units; see Appendix 8 for definitions of process capability indices C_p and C_{pk}). If the process capability is not acceptable, then the process should be improved, and new control charts should be created (based on new data after the process changes) for a continued reassessment of the process.

Description

A control chart is a visual plot of the process outputs over time along with a centerline and upper and lower control limits. Control charts can be used to describe process outputs by measuring continuous variables (e.g., a product characteristic such as dimension, weight, temperature, capacitance, or resistance) or by using discrete variables such as the number of defectives (nonconforming parts) or the number of defects (abnormalities such as cracks, scratches, wrinkles, and warps). Control charts of process output data measured using continuous variables are defined as the *Variables Control Charts*. Examples of commonly used variables charts are as follows: mean and range charts (X-bar and R), mean and standard deviation charts (X-bar and S), or X and moving range charts (X and R_M). Control charts based on data using discrete variables are used to create *Attributes Control Charts* (e.g., P chart, NP chart, U chart, and C chart).

To develop a control chart, samples of process outputs (e.g., entities such as products, systems, subsystems, or components) are selected in subgroups. A subgroup can have one or more samples. Many subgroups are obtained at preselected time intervals and their samples are inspected. For variables charts, a subgroup containing one sample can be used to create X and moving range chart. Subgroups of sample size (n) between 2 and 5 are generally used for X-bar and R charts (Note: X-bar denotes arithmetic mean and R denotes the range of measurements from a subgroup); and subgroups with samples of six or more are used to create X-bar and S charts (Note: S denotes the estimate of the standard deviation of measurements in a subgroup). The number of samples (n) in a subgroup for P and NP charts should be large (at least equal to the reciprocal of P. Note: P and NP denote the proportion of defectives and the number of defectives, respectively.). And for U and C charts, the sample size in each subgroup should be large enough to obtain at least a few defects in each subgroup. (Note: U denotes the average number of defects in an inspection sample and C denotes the number of defects in a sample.) The formulas used to compute centerlines and control limits for different control charts are provided in Appendix 8.

When one of the more points in a control chart falls outside the band between lower and upper control limits, the process is considered to be out of control and the point(s) falling outside the control limits are considered to be outliers due to some special or assignable cause that affected the process. Thus, when a process is found to be out of control, the engineer responsible for the process should investigate the reasons for the out-of-control points and fix the process such that the special or assignable causes do not occur. When such out-of-control points are removed and the control limits are recomputed, the remaining points, most likely, will fall within the new control limits. If the process is still found to be out of control, again the new special causes for the out-of-control points need to be investigated and the process

Traditional and New Quality Tools

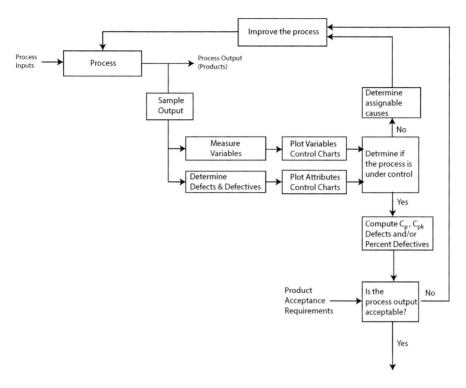

FIGURE 17.10 Flow diagram of process improvement using control charts.

should be fixed again. The iterative process should be repeated until all points fall within the newly recomputed control limits. Thus, when all points fall within the control limits, the process is considered to be stable (i.e., running within its natural variability).

Figure 17.10 presents a flow diagram showing how a process can be improved by sampling process outputs, measuring and/or inspecting the samples, plotting control charts, determining if assignable (or special) causes are present and/or the process output is acceptable.

SOME EXAMPLES OF CONTROL CHARTS

Variables Control Charts

A vehicle lighting engineer wanted to evaluate the process capability of a stop lamp production line. The stop lamps must meet the luminous intensity requirements specified in the Federal Motor Vehicle Safety Standard 108. The standard requires that each manufactured stop lamp must not produce luminous intensity greater than 300 candelas and no less than 80 candelas measured along the lamp axis. The engineer measured the luminous intensity of six randomly drawn stop lamps from each day's production for over 25 consecutive workdays. The data are provided in Table 17.1.

TABLE 17.1
Stop Lamp Intensity Measurements with Subgroup Size of 6 and 25 Subgroups

Day (Subgroup Number)	Stop Lamp Intensity Data (Measured in candelas directed along lamp axis)					
	Sample 1	Sample 2	Sample 3	Sample 4	Sample 5	Sample 6
1	182.75	148.34	181.45	105.33	70.05	156.05
2	153.34	91.93	223.79	138.57	202.46	81.09
3	278.86	320.45	160.90	225.35	214.24	323.67
4	172.20	90.67	184.50	130.90	137.41	180.44
5	188.52	107.98	84.50	108.47	196.33	201.32
6	121.79	155.48	192.32	180.57	119.70	177.41
7	46.36	57.30	101.26	117.99	124.19	65.00
8	237.36	176.90	85.19	179.94	265.46	171.10
9	222.18	140.25	185.93	160.52	152.19	197.98
10	78.68	62.00	58.56	113.76	144.39	69.30
11	176.98	181.28	204.76	76.62	249.64	223.98
12	146.80	162.00	118.40	255.73	162.13	180.07
13	154.08	170.39	160.47	129.98	211.09	174.24
14	234.70	231.18	187.33	158.76	190.50	148.32
15	124.59	211.91	249.66	89.31	138.69	204.59
16	135.36	218.62	186.03	190.14	192.03	160.94
17	132.03	91.67	187.57	243.81	191.90	136.23
18	222.86	97.89	116.96	98.56	123.19	166.34
19	147.68	178.60	232.77	158.24	137.58	127.35
20	289.56	305.60	141.56	56.56	178.36	80.07
21	103.27	180.11	184.72	131.65	80.76	214.18
22	253.02	206.33	185.06	174.44	154.82	207.31
23	82.24	135.19	190.27	189.11	160.28	112.00
24	174.33	151.77	102.39	188.92	257.47	146.31
25	153.70	115.85	154.62	139.47	149.70	169.77

The above data were used to solve the following questions:

1. Determine the 3-sigma X-bar and R-chart parameters, the center line and the control limits, and plot the charts.
2. Do the above charts show that the process is under control and is running naturally? What would you do to improve the process?
3. Compute C_p and C_{pk} (process capability indexes) of the process before and after your improvements and comment on the capability of the process before and after the improvements.

The X-bar and R-charts with centerlines and control limits are shown in Figure 17.11. The formulas used for computations of the centerline and 3-sigma control limits are presented in Appendix 8. The charts display that the process is out of statistical control and not running within natural variations. Points for subgroups 3, 7, and 10 in

Traditional and New Quality Tools

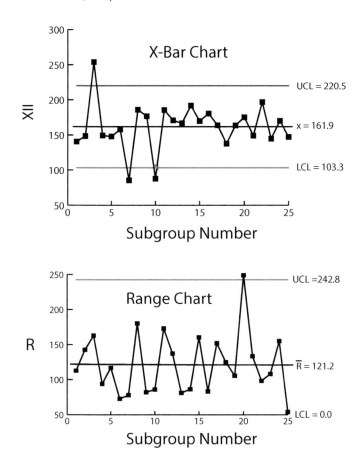

FIGURE 17.11 X-bar and range charts for stop lamp intensity data.

the X-bar chart and point 20 in the R-chart are outside their respective control limits. Assuming that assignable causes exist, and they can be eliminated by the production personnel, the four outlier points were removed and the two control charts were redrawn. The redrawn charts showed that the process was under control. Table 17.2 presents various parameters of the process before (original data) and after the above outlier points were removed.

The values obtained for C_p before and after the improvements (removal of outlier points) were 0.84 and 0.87, respectively, showing slight improvements in the process. The values of C_{pk} also improved from 0.71 to 0.75. The fact that both the process capability parameters are less than 1 showed that both before and after the improvements the process was both potentially and actually incapable. Thus, to obtain additional increases in process capabilities, more dramatic changes in both products (e.g., introduction of filament-free lighting sources such as light-emitting diode) and production processes (e.g., low variability reflector production) would be needed.

TABLE 17.2
Summary of Results from before and after Removal of Points Outside the Control Limits of X-Bar and R Charts

Parameter	Original Data	After Removal of Points Outside the Control Limits
X-bar	161.9	161.1
UCL of X-bar	220.5	220.5
LCL of X-bar	103.3	107.6
R-bar	121.2	116.8
UCL of Range	242.8	234.1
LCL of Range	0.0	0.0
σ-cap (Estimation of sigma)	48.42	45.80
C_p = Potential capability index	0.84	0.87
C_{pk} = Realized capability index	0.71	0.75
Defectives per million opportunities	19,826	13,049

Attributes Control Charts

A manufacturer of die-cast toy cars inspected samples of car bodies from 20 consecutive shifts of production from a die-casting machine. The number of car bodies (sample size) inspected in each shift varied from shift to shift. Each incidence of the following three types of defects was noted: voids (unfilled areas), ejection marks (scratches), and flash (small amount of thin material present on the car bodies at the parting lines between the dies).

Using the data provided in Table 17.3, the answers to the following questions are provided below.

a. Is the die-casting process stable by considering the proportion of defective car bodies?
b. Assuming that 24 car bodies are treated as an inspection unit, construct a U chart and determine whether the average number of defects (considering all defects combined) per inspection unit is under control.

Figure 17.12 presents a P chart for the proportion of defective car bodies. The chart was created by using unequal sample sizes provided in the second column of Table 17.3. The upper control limit of p is not constant as the sample size is different for each subgroup. The die-casting process based on the P chart is not stable because the value of p for the 10th subgroup is outside its upper control limit.

Figure 17.13 presents a U chart based on total number of defects (sum of defects involving voids, marks, and flashes) in each subgroup. The U chart shows that the u-values (average number of total defects per inspection unit of 24 die-cast car bodies) of all the subgroups were within their respective upper and lower control limits. Thus, the die-casting process based on the total number of defects is stable.

TABLE 17.3
Data on Number of Defects and Number of Defective Die-Cast Car Bodies

Shift Number	Sample Size	Number of Defects			Number of Defective Car Bodies
		Voids	Marks	Flashes	
1	150	5	4	2	6
2	132	7	3	3	6
3	100	2	0	1	1
4	105	4	4	8	4
5	60	0	0	3	3
6	126	2	3	2	5
7	132	9	5	2	6
8	162	5	4	9	7
9	156	5	4	3	9
10	190	9	3	5	17
11	210	12	6	6	8
12	167	7	4	3	1
13	134	7	5	4	6
14	125	1	2	0	3
15	110	0	0	5	3
16	134	1	1	3	7
17	167	8	3	7	6
18	190	4	3	4	3
19	123	0	0	2	5
20	154	4	3	6	5

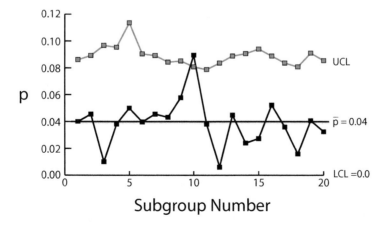

FIGURE 17.12 P chart for the proportion of defective car bodies.

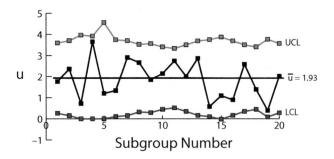

FIGURE 17.13 U chart for the total number of defects per inspection unit of 24 car bodies.

NEW QUALITY TOOLS

These quality tools are mainly used in a proactive manner (to create quality). Thus, they are used primarily for planning (e.g., for brainstorming) and by the management (Kolarik, 1995). The tools are qualitative in nature and more effective when used through active participation of all members of the multifunctional product teams. The format of the tools is visual in nature, that is, they help visualize problems, issues, relationships, and so on. Any of the tools can be used with any other tools. The tools are as follows: (1) relations diagram, (2) affinity diagram, (3) systematic diagram, (4) matrix diagram, (5) matrix data analysis, (6) PDPC, and (7) arrow diagram.

The following part of this section presents purpose, description, and examples of each of the seven new quality tools.

RELATIONS DIAGRAM

Purpose

The purpose of a relationship diagram is to show the causal relationship between different events (or issues) to understand causation of a given event and to determine its root causes. This technique is useful during brainstorming, problem solving, and determining sequences of events leading to a particular problem.

Description

The relationship diagram is a flow diagram with arrows showing the relationship between different events or issues. It is typically used during a brainstorming session held among individuals attempting to define a problem and to understand its causation. As a brainstorming session progresses, a number of issues that could lead to the problem are raised by different individuals. The issues are recorded on an easel or on 5″ × 3″ cards. The cards are pasted (or the issues are rewritten) on a large sheet of paper and the individuals are asked to determine if any relationships exist between any issue to other issues. If a relationship is identified, then an arrow is drawn from an originating issue (which is the cause) to other resulting issues (which is the effect) to indicate the existence of causal relationship. The process is repeated over all issues. The description of any of the issues and their relationships can be modified (combined, removed, or added) as more information is obtained during the discussions. The process is iterated until all individuals feel that all relevant issues

Traditional and New Quality Tools

and relationships between them are documented. The root causes of the problem are generally the earliest occurring causes in the chain of events.

Example: Understanding Causation of Headlamp Misaim

A team of engineers and vehicle assembly plant personnel were gathered to brainstorm and develop a list of issues that can cause shipment of a new vehicle to the end-customer with misaimed headlamps. The 17 issues identified in the brainstorming session are presented in Figure 17.14.

The team discussed the 17 issues, determined relationships between the issues, and created a relationship diagram. The relationship diagram is shown in Figure 17.15. The arrow diagram thus shows the flow of a series of events (causes) leading to the final outcome (customer returned a new car to the dealer to fix misaimed headlamps). The

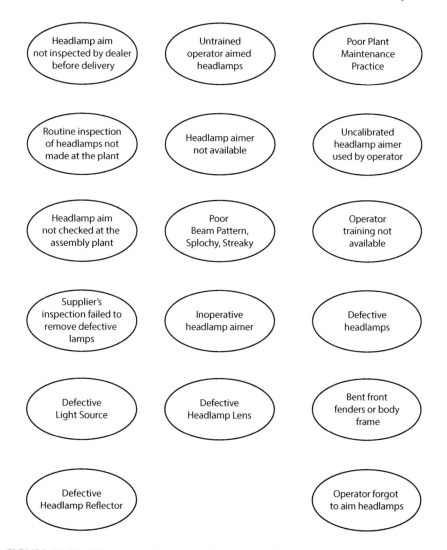

FIGURE 17.14 Illustration of problems discovered during group discussions.

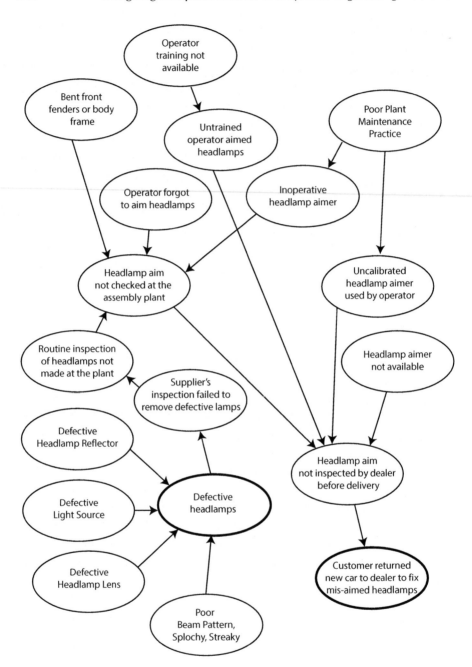

FIGURE 17.15 Example of a relations diagram.

diagram suggests that key issues leading to the misaimed headlamps are as follows: (1) headlamp aim not checked at the plant, (2) untrained operator aimed headlamps, (3) uncalibrated headlamp aimer used for aiming, (4) headlamp aimer not available, and (5) headlamp aim not inspected by the dealer before vehicle delivery to the customer.

Affinity Diagram

Purpose

The purpose of an affinity diagram is to group a list of issues by determining some common characteristics underlying the issues in each group. The common characteristic is considered to create some affinity (or binding) between the issues within each group. The affinity diagram thus reduces a long list of issues into a manageable number of groups of issues.

Description

It is assumed that a list containing issues related to a given problem is available. A deck of blank 3" × 5" cards are obtained, and each issue is written on a separate card. The cards are read to a team of decision makers who are familiar with the problem and asked to sort the cards into groups based on some common underlying considerations.

Example: Grouping Causes of Headlamp Misaim

The issues related to headlamp misaim problem presented in Figure 17.14 were given to a team of vehicle-lighting engineers to create an affinity diagram. Figure 17.16 contains an affinity diagram created by the team. The affinity diagram contains four groups under the following headings: (1) headlamp supplier issues, (2) assembly plant management issues, (3) headlamp aiming workstation issues, and (4) car dealership issues. The issues in each group can be further investigated by people familiar or responsible for work in each group to recommend solutions to the problem.

The affinity diagram is a useful technique for organizing many issues or causes of a problem into a few main groups. A decision maker can use this diagram to eliminate causes one by one or to determine one or more root causes.

Systematic Diagram

Purpose

The purpose of a systematic diagram is to show possible alternatives available to solve a problem by constructing a tree diagram by hierarchical decomposition of the original problem objective into a series of lower-level objectives followed by questions, and ways and means to solve the problem. The possible solutions are categorized into (1) practical (i.e., the solution is practical and thus, the problem can be solved using the ways and means suggested in the branch), (2) uncertain (i.e., sufficient information is not available to determine if the solution suggested in the branch can be implemented), and (3) impractical (i.e., the solution suggested in the branch is not practical and thus, it should be rejected).

Description

The systematic diagram consists of a tree structure presented in a horizontal orientation. The primary objective called the first-level objective is described in a box on the left hand side of the diagram, and the successive branches of the tree describing second-level objectives, third-level objectives, and so forth are developed toward the right side. After the objectives are described sufficiently in their lower levels,

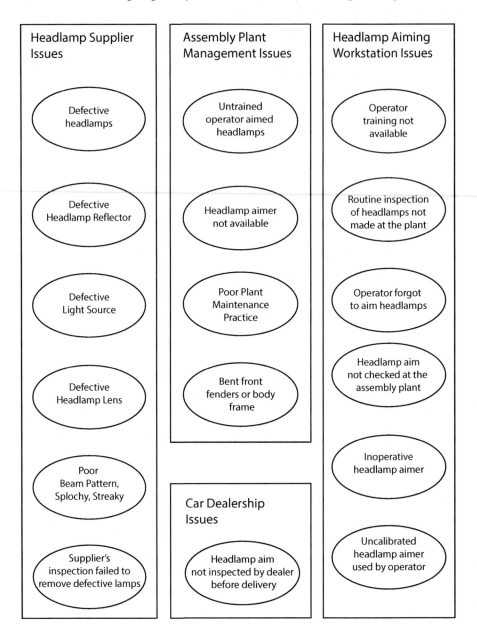

FIGURE 17.16 Affinity diagram showing clusters of causes.

the ways and means to accomplish each objective are described and developed in branches toward the right hand side. When the branches are sufficiently developed to describe possible solutions, the solutions are categorized as follows: (1) practical, (2) uncertain, and (3) impractical. The practical solutions can be recommended for immediate implementation, whereas the uncertain solutions may be considered after more information is available.

Traditional and New Quality Tools

The advantages of the systematic diagram are as follows: (1) it helps systematic development of possible solutions to meet all lower to higher level objectives, (2) when the diagram is developed by brainstorming with the help of team members from all disciplines, the solutions can be applied immediately without additional meetings to obtain the concurrence of all affected parties, and (3) more effective solutions will be realized because all possible major areas will be explored during the implementation of this tool.

Example: Alternatives to Reduce Product Development Time

Figure 17.17 presents a systematic diagram for a product program that required the team to consider reducing the program development time by 50%. Two second-level objectives were considered (1) improving the efficiency of the product development by means of implementing new computer-integrated technologies and (2) expanding current product development capability by either hiring more designers and engineers in the company's U.S. operations or by farming out the development work to outside design and engineering firms.

The final means considered to solve the problems and decisions on each of the alternatives were as follows (see last four columns of Figure 17.17):

1. Implement CATIA software to speed up and communicate product design within all engineering activities (categorized as practical [open circle symbol])
2. Develop a new computer-assisted design system that will be custom designed for the unique needs of the company (categorized as impractical [X symbol] because of the substantial time and cost involved in the development of the software and purchase and implementation of new hardware)
3. Purchasing a new computer-assisted engineering modeling software already available in the market (categorized as uncertain [triangle symbol] due to further testing of the software needed to ensure that it will perform all the necessary analyses efficiently)
4. Training current employees to adopt the new technologies (categorized as uncertain because the existing technical manpower was not capable of learning advanced software and modeling methods)
5. Hiring more designers and engineers within their Detroit office (categorized as uncertain because of substantial costs and increase in the budget)
6. Increasing manpower in the Los Angeles office (categorized as impractical because of long distance and limitations of existing facilities)
7. Contracting work to General Design Company (categorized as impractical because of unsatisfactory work performed by that company in their previous project)
8. Contracting work to Fast Design Work (categorized as practical because the organization was highly regarded in terms of their technical capabilities and past experience)
9. Developing a new engineering center in Europe (categorized as impractical because of high labor costs)

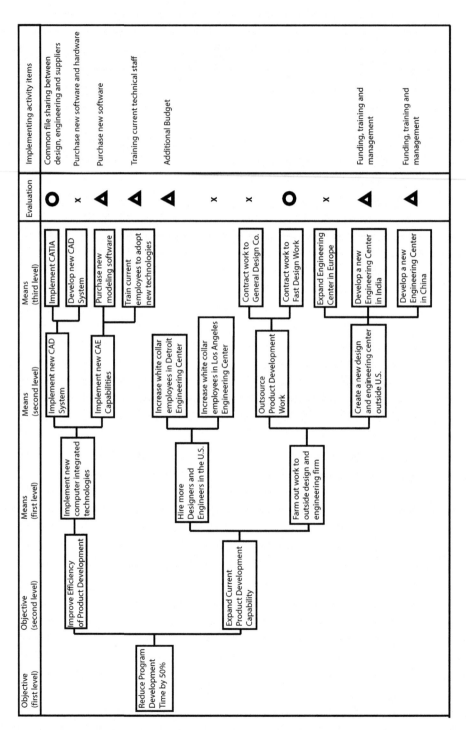

FIGURE 17.17 Systematic diagram for reducing program development time.

Traditional and New Quality Tools

10. Developing a new engineering center in India (categorized as uncertain because of the current unstable economic condition)
11. Developing a new engineering center in China (categorized as uncertain because of communication difficulties and training time).

The systematic diagram, thus, is a very effective tool in summarizing a lot of information and displaying all possible alternative solutions to meet the first-level objective. The systematic diagram also organizes all the possibilities into a hierarchy—from very general solutions to more specific solutions.

MATRIX DIAGRAM

Purpose

The purpose of the matrix diagram is to present the relationship between two sets of variables in a compact tabular format. The relationships are indicated in a qualitative or quantitative manner for each combination of variables represented by each cell of the matrix.

Description

A matrix diagram presents data in a tabular format organized in rows and columns. Each row and column are assigned to a variable, and each cell representing the intersection of the row and the column contains data about the relationship between the variables assigned to the row and the column. One set of variables can be assigned to the rows and another set to the columns. Thus, if a matrix contains m number of rows and n number of columns, mn numbers of cells present relationships between m number of variables in rows and n number of variables in columns. The variables assigned to a row, or a column can be from a different set or the same set of entities. For example, columns can be assigned to different products ($P_1, P_2, \ldots, P_n$ where $P_j = j$th product) and rows can be assigned to the type of users ($U_1, U_2, \ldots, U_m$; where $U_i = i$th user). The ijth cell will present data related to U_i user and P_j product (e.g., time required for ith user to operate jth product). The matrix can be symmetric (i.e., it will have the same number of rows and columns) if its rows and columns represent the same set of variables. For example, an $n \times n$ matrix containing n machines assigned to its columns and rows can indicate relationships between the ith machine to the jth machine in the ijth cell (e.g., distance between the ith and jth machine). The diagonal of the matrix where $i = j$ will not be useful as there is no relationship between the same machine.

The data contained in the matrix diagram thus provides information on the existence of relationships, levels of relationships (e.g., strength or correlation), and pattern of relationships (e.g., to determine cluster or group of issues or items that can be packaged or placed together or combined).

Example: Relationship between Vehicle Parameters and Vehicle Performance

Figure 17.18 presents a matrix diagram showing the relationship between the parameters of a vehicle and performance of the vehicle. The rows of the matrix represent five vehicle parameters ($X_1, X_2, \ldots, X_5$) and the columns represent vehicle performance

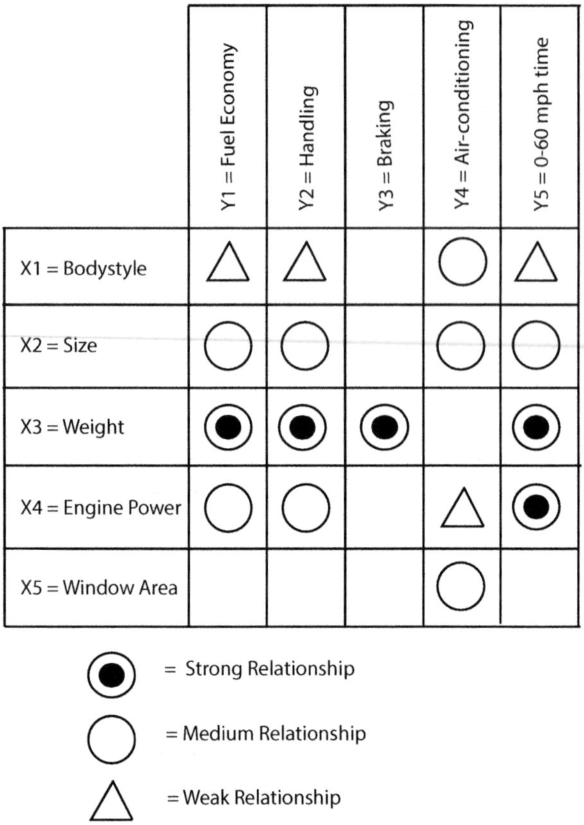

	Y1 = Fuel Economy	Y2 = Handling	Y3 = Braking	Y4 = Air-conditioning	Y5 = 0-60 mph time
X1 = Bodystyle	△	△		○	△
X2 = Size	○	○		○	○
X3 = Weight	●	●	●		●
X4 = Engine Power	○	○		△	●
X5 = Window Area				○	

● = Strong Relationship

○ = Medium Relationship

△ = Weak Relationship

FIGURE 17.18 Example of matrix diagram.

variables ($Y_1, Y_2, \ldots, Y_5$). The symbols in the matrix indicate the strength of relationships between the variables. For example, the weight of the vehicle (X_3) strongly affects its fuel economy (Y_1), handling (Y_2), braking (Y_3), and 0–60 mph acceleration time (Y_5). The vehicle body style (X_1) does not affect the vehicle-braking performance (Y_3).

Interface matrices described in Chapter 5 (see Figures 5.6 and 5.8) are also examples of matrix diagrams.

MATRIX DATA ANALYSIS

Purpose

The purpose of matrix data analysis is to analyze the data contained in the matrix to facilitate decision-making (e.g., to select important variables or to obtain prioritized list of variables used to define rows and/or columns of the matrix).

Description

The matrix data analysis requires that the data contained in the matrix is quantitative in nature. Numerical ratings, weights (e.g., importance weights) along with values of quantitative variables (i.e., continuous or discrete variables) can be transformed into evaluation measures that can be used to prioritize variables or issues to make decisions. Chapters 3 and 15 provided examples of several methods used for decision-making, which involved matrix data. The matrix data-based methods covered in Chapters 3 and 15 were as follows: (1) decision evaluation matrix and different principles used in evaluation (see Tables 3.2 through 3.4), (2) analytical hierarchical method (see Table 3.6), (3) total weighted score for product concepts (see Tables 3.7), (4) Pugh diagram (see Tables 15.2 and 15.3), and (5) relationship matrix and absolute importance ratings computation in the QFD (see Figure 15.4).

Examples of Matrix Data Analysis

As mentioned above, decision analysis, Pugh diagrams, and QFD covered in Chapters 3 and 15 are examples of matrix data analysis. Furthermore, any analysis of data contained in a tabular format in spreadsheets used for analysis of costs, budgets, comparative evaluations, and so forth are examples of matrix data analysis.

PROCESS DECISION PROGRAM CHART

Purpose

The purpose of the PDPC is to analyze a process involved in a program and develop countermeasures to eliminate problems that can occur during the process.

Description

Any product program generally involves a number of processes, and processes include series of activities or tasks to accomplish the objectives of the processes. During the execution of any process, unexpected problems can occur, and they can cause interruptions, rework, or delays. To minimize such problems, the teams involved in implementing and managing the process are asked to brainstorm and determine possible problems that can occur during each task in the process and come up with countermeasures to eliminate the problems. The PDPC provides a simplified chart to organize and document the information generated during the brainstorming sessions. The chart typically lists: (1) steps (or tasks) involved in the process, (2) problems that can occur (e.g., what if something goes wrong in each step), and (3) countermeasures that can be implemented to eliminate the problems or consequences (e.g., delays, additional costs) due to the problems.

The format of the PDPC varies between different implementers. Some PDPCs can include process flow charts including a sequence of major steps (or tasks) along with additional information on problems and countermeasures. Whereas others use tree-type charts diagrams where each branch of the tree represents the steps and subsequent parts of the branch provide a list of problems and countermeasures to eliminate the problems.

Example: PDPC for Reducing Problems in a Product Development Process

Figure 17.19 presents a PDPC for a product development process. The left-hand side of the figure presents a flow diagram showing major activities in the product development. The middle column presents possible problems in each major activity and the last column presents countermeasures that can be incorporated to eliminate or reduce the problems. The problems and the countermeasures were obtained during a brainstorming session conducted by the senior management to avoid delays in developing a new product program.

Arrow Diagrams

Purpose

The purpose of an arrow diagram is to show relationships, direction of flow (e.g., flow of materials, information, energy), and/or sequence (or order) of steps in a process.

Description

An arrow diagram is a flow diagram. The arrows are used to indicate the direction of flow or relationships between different systems or entities. Any diagram with arrows used to indicate relationship, flow, interaction, or interface can be considered as an arrow diagram. Thus, it could be a flow diagram, an interface diagram, a circuit diagram, a network diagram, or a project management representational diagram (such

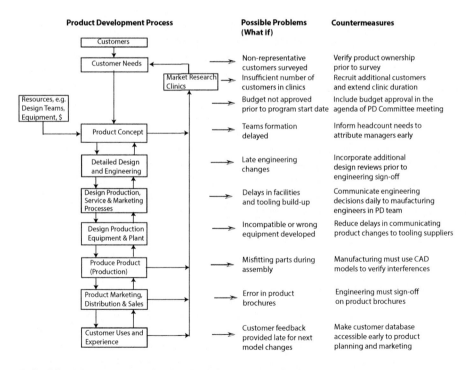

FIGURE 17.19 An example of a PDPC for a product development program.

as program evaluation review technique diagram, precedence diagramming method, or diagram used in the critical path method).

Examples

Some examples of the arrow diagrams can be found in Figures 17.17 (relations diagram), 1.1 (flow diagram), and 5.5 (interface diagram).

EXPERIMENT DESIGN

The experiment design is a major study area in statistics. When well-developed mathematical or computer models are not available to solve a problem at hand, engineers often resort to solve the problem by the experimental route. Thus, an experimental research program based on the right experiment design and sound statistical analysis techniques can lead to the development of an improved product much more efficiently than a trial-and-error approach to find the optimum combination of levels of independent variables that define the product characteristics.

The purpose of the experimental research is to allow the investigator to control the research situation (product or process design) so that causal relationships among the independent variables (that define the product or the process) and the dependent variable may be evaluated. An experiment is a series of controlled observations undertaken in an artificial situation with deliberate manipulation of some variables in order to answer one or more hypotheses. In an experiment, one or more variables (called independent variables) are manipulated and their effect on another variable (called dependent or response variable) is measured, while all other variables that may confound the relationship(s) are eliminated or controlled.

Steps involved in a product design-related experiment generally include the following:

1. Identify a problem (e.g., define the product to be designed).
2. Search for variables affecting the problem and variables that can be used to measure the performance of the product (e.g., use a C-E diagram, an FMEA, or design specifications selected from a QFD chart).
3. Determine response variables and independent variables with their levels to study.
4. Develop hypotheses to test.
5. Design a plan for the experiment.
6. Prepare test samples (or experimental units), experimental setup, and measurement procedure.
7. Conduct the experiment—collect data.
8. Analyze the data.
9. Recommend a solution (e.g., combination of independent variables and their levels that produce the best response).

An experiment design is a formal pattern or plan for collecting data (observations or responses). It includes the following:

1. Identification of independent variables (e.g., product parameters, factors) and their levels to be investigated
2. Selection of the response variable (e.g., product performance, percentage of customers satisfied) (Note: A complex product may need more than one response variable to evaluate each attribute)
3. Determining combinations of levels of independent variables to be tested and the number of replications to be tested
4. Procedure for measurement of the response variable
5. Decisions about the kind and number of experimental units
6. Carefully arranged scheme governing the order in which experimental trials are to be run.

The data obtained from the experiment are generally analyzed by using the following steps:

1. Plotting means and interactions (for multifactor experiments).
2. Conducting analysis of variance (ANOVA) to determine significant factors and their interactions (for multifactor experiments).
3. If some factors are statistically significant, then determine combination factor levels that provide the best response.
4. Analyze errors (residuals) to determine if assumptions related to the errors are not violated (e.g., the effect of order of experimental trials, learning, fatigue, and wear).

AN EXAMPLE: EXPERIMENT TO SELECT A DISPLAY WITH THE HIGHEST LUMINANCE

An engineer measured luminance (brightness) values (in foot-lambert [fL]) of six display samples (three samples of 1.0 mm thickness and three samples of 1.5 mm thickness) provided by each of the three display suppliers. The measured luminance values of the samples are provided in Table 17.4. Since the legibility of displays

TABLE 17.4
Luminance Levels of Display Screens for Combinations of Glass Thickness and Display Suppliers

Glass Thickness	Display Supplier		
	A	B	C
1.0 mm	280	300	290
	290	310	285
	285	295	290
1.5 mm	230	260	220
	235	240	225
	240	235	230

generally improves with an increase in display luminance, the objective of the engineer was to select the display with the highest luminance value.

From the values provided in Table 17.4, the following mean values of display luminance are computed:

a. Means for main effects plot:
   Mean luminance of 1.0 mm glass thickness = 291.667 fL
   Mean luminance of 1.5 mm glass thickness = 235.00 fL
   Mean luminance of supplier A displays = 260.00 fL
   Mean luminance of supplier B displays = 273.33 fL
   Mean luminance of supplier C displays = 256.667 fL
b. Means for interaction plot
   Mean luminance of 1.0 mm glass thickness displays of supplier A = 285.00 fL
   Mean luminance of 1.0 mm glass thickness displays of supplier B = 301.667 fL
   Mean luminance of 1.0 mm glass thickness displays of supplier C = 288.333 fL
   Mean luminance of 1.5 mm glass thickness displays of supplier A = 235.00 fL
   Mean luminance of 1.5 mm glass thickness displays of supplier B = 245.00 fL
   Mean luminance of 1.5 mm glass thickness displays of supplier C = 225.00 fL.

Figures 17.20 and 17.21 present the main effects and interactions plots from the above mean values. A visual examination of values of the means presented in Figures 17.20 and 17.21 shows that the highest mean luminance values are obtained for glass thickness of 1.0 mm and displays from supplier B. The ANOVA table presented in Table 17.5 shows that both main effects, namely due to, thickness (i.e., glass thickness) and supplier were significant at P values of 0.000 and 0.004, respectively. (Note: P value is the significance level. It is the probability at which the effect being tested can be rejected.) The effect due to interaction between the glass thickness and supplier (shown in Table 17.5 as Thickness × Supplier) was not significant

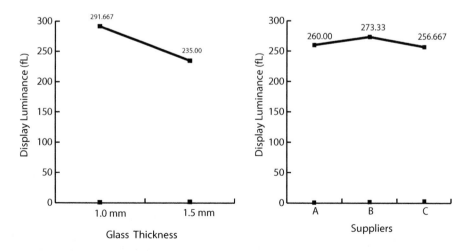

FIGURE 17.20 Main effect plots showing effects of glass thickness and suppliers.

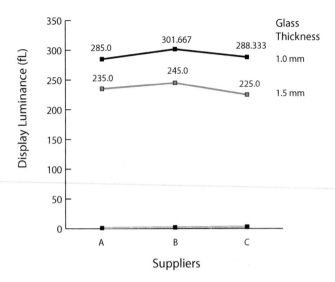

FIGURE 17.21 Interaction plot of glass thickness and supplier.

TABLE 17.5
Analysis of Variance Table

Source	Degrees of Freedom	Sum of Squares	Mean Squares	F	P
Thickness	1	14450.0	14450.0	273.8	0.000
Supplier	2	933.3	466.7	8.8	0.004
Thickness × Supplier	2	133.3	66.7	1.3	0.318
Error	12	633.3	52.8		
Total	17	16150.0			

as the P value of 0.318 was larger than the criterion value of 0.01 significance value (or 99% confidence probability) used to assess statistical support for the observed effects. Thus, 1.00 mm glass thickness and supplier B can be recommended for the displays. Additional information on the ANOVA technique can be found in the study by Kolarik (1995), Besterfield et al., (2003), NIST (2012), or other textbooks on experiment design.

MULTIVARIATE EXPERIMENT DESIGNS

Multivariate experiments have more than one independent variable. Many complex products have many independent variables and solving problems experimentally can become very expensive and time-consuming especially when the number of independent variables is large, and each independent variable (or factor) has many levels.

A treatment combination (TC) in an experiment is a combination of each level of each of the independent variables. The costs of preparing test samples and conducting a multifactor experiment are high because the product (samples) needs to be configured or made for each treatment combination to test and measure the values of its response variables. The number of TCs in a full-factorial design is equal to the total number of possible combinations of factor levels (e.g., in a 3-factor design with factors A, B, and C, the treatment combinations will be $a \times b \times c$, where a, b, and c are the levels of factors A, B, and C respectively). When the number of levels (l) of each factor are the same (e.g., $l = 2$ or 3), the number of treatment combinations (TC) will be equal to l^f, where f = number of factors. A full factorial (includes all possible TCs) design with replications conducted in a random manner is also called *F*actorial *A*rrangement of *T*reatments and *C*ompletely *R*andomized *D*esign. Thus, a three-factor experiment design with factors A, B, and C, each with two levels and two replications will have eight TCs and 16 trials (measurements or observations). And three-factor design with three levels of each factor and three replications will have 27 treatment combinations and 81 trials.

A complex product or its systems can have many independent variables and thus the number of treatment combinations to be tested in an experiment can be very large. Taguchi developed several concepts and approaches to design robust products and to reduce the cost of experimentation. Brief descriptions of the basic considerations in Taguchi's work are provided below.

TAGUCHI'S THREE-STEP PRODUCT DESIGN APPROACH

Taguchi introduced a three-step approach in product design to improve quality (Phadke,1989). The three steps involve the following:

1. *Concept design*: In this step, the product designer examines many different technologies and product configurations to meet the desired product requirements and selects the most suitable design concept.
2. *Parameter design*: In this step, a combination of the values of independent variables that define the product parameters is selected. The selection is based on experimentation by studying the product performance (response variable) under several combinations of levels of the product parameters. The results of the experiment in this step thus help in defining the product that produces the best (or maximum) performance.
3. *Tolerance design*: In this step, the tolerance levels on each of the parameters are selected to obtain the best product performance at minimum manufacturing costs.

The above approach thus involves selecting the best product concept, maximizing its performance through parameter selection, and minimizing its manufacturing costs by designing tolerances on the parameter values. In the above steps, Taguchi also incorporated two following important concepts: (1) product robustness and (2) quadratic costs.

TAGUCHI'S PRODUCT ROBUSTNESS AND QUADRATIC COSTS

The product "robustness" was defined in terms of a product design (i.e., a combination of levels of product parameters) that will produce the maximum performance that is also insensitive to variations in other uncontrollable variables (called the noise variables).

The concept of quadratic costs was also introduced in the experiment design. The concept of quadratic costs means that the costs related to use in service of the product increase proportional to the square of the deviation of the value of each product parameter from its respective target value. Thus, to minimize the total cost of the product, the values of product parameters should be very close to the target values. The concept of quadratic costs was used by Taguchi in the parameter design and also in the tolerance design steps.

In the parameter design, the variations in the response variable for each treatment combination are caused by the introduction of the noise factors (by defining an outer array) in the experiment design described in the next section. Whereas, in the tolerance design, the deviations from the target values are introduced due to the manufacturing variations in the process of realizing the product with the parameters.

TAGUCHI EXPERIMENTS

For the parameter design, Taguchi developed a unique procedure to select an experiment design and data analysis procedure to (a) reduce the total number of treatment combinations, (b) still use a large number of independent variables, (c) use his quadratic cost functions, and (d) include robustness concept (Phadke, 1989; Kolarik, 1995; Yang and El-Haik 2003; Besterfield et al., 2005).

Key features of Taguchi experiments are as follows:

1. *Use of two orthogonal arrays*: Two orthogonal arrays are used in the creation of its experiment design. (The orthogonal arrays allow the use of many factors and a relatively small number of TCs. However, these arrays only account for partial factorials, that is, they do not allow evaluation of each factor separately without confounding with other factors. An inner array is used to define TCs of control factors (independent variables) and another array, called the outer array, is used to create replications by using TCs of noise factors. (The use of noise factors and their levels selected in the outer array are considered to create a robust product solution.)
2. *Quadratic cost assumption*: The analysis uses "signal-to-noise ratio" (S/N ratio; also called eta, η) as the response variable of the experiment. The η is a transformation of the traditional response variable values (y_is) to account for the quadratic loss functions. Thus, η is used as the response variable instead of Y. The transformation of Y to η is defined such that the larger value of η is always better (and thus it should be maximized).
Computation of η values for each TC: the S/N ratio (η) is defined as transformation of n responses ($y_1, y_2, \ldots, y_n$) obtained from the outer array (or replications). Depending on the objective of the experiment (i.e., smaller, larger, or nominal value of the response variable[Y] is desired), the value of η for each TC is computed as follows:

   a. For smaller-the-better values of (y): $\eta = -10 \log_{10} \left\{ \dfrac{1}{n} \sum_{i=1}^{n} y_i^2 \right\}$

b. For nominal-the-best values of (y): $\eta = 10\log_{10}\left\{\dfrac{\mu^2}{\sigma^2}\right\}$

where μ and σ are the mean and standard deviation of responses (y_i, for $i = 1$ to n)

c. For larger-the-better values of (y): $\eta = -10\log_{10}\left\{\dfrac{1}{n}\sum_{i=1}^{n}\left(\dfrac{1}{yi^2}\right)\right\}$

3. *Use of confirmation runs*: Confirmation runs are made at the selected treatment combination that produces the highest value of η.

The data analysis procedure involves the following steps:

1. The data are analyzed by obtaining mean values of η at each level of each factor. Differences (deltas) in means (at each level) of each factor are computed and factors and their level with the highest mean value with at least half of the strongest difference (largest delta) are identified.
2. The above-identified combination of factors and their levels is selected as the best combination.
3. At least one or more additional trials at the selected combination are conducted as confirmation runs. If the computed value of η for the confirmation run is close (within an accepted percent difference) to the predicted value η, then the selected combination is considered to be confirmed.

The above-described features are, thus, Taguchi's attempt to combine his concepts of quadratic costs and robustness and to reduce TCs (by use of orthogonal arrays) and introduce simplicity in the data analysis. The arbitrariness used in combining the above concepts is however not accepted by many classical statisticians (Kolarik, 1995). For more detailed information on the above-described Taguchi concepts, experiment design, and data analysis, the reader is referred to the studies by Phadke (1989), Kolarik (1995), and Besterfield et al. (2003).

CONCLUDING REMARKS

This chapter described seven new tools and seven traditional tools used in the application of the TQM. The TQM tools along with techniques in the experiment design area and other product development tools (covered in Chapter 15), human factors engineering tools (covered in Chapter 18), and safety engineering tools (covered in Chapter 19) constitute a comprehensive set of tools used in creating quality products.

Illustration of a C-E diagram for an undesired event defined as "New Car's headlamps are misaimed." The causes of this event are placed into the following categories and placed on primary bones with labels such as operator related, aiming equipment related, plant management related, vehicle body related, and so forth. Secondary causes in each category are placed on the secondary bones corresponding to the category of the primary bones.

REFERENCES

Besterfield, D. H., Besterfield-Michna, C., Besterfield, G. H., and M. Besterfield-Scare. 2003. *Total Quality Management*. ISBN 0-13-099306-9. Third Edition. Upper Saddle River, NJ: Prentice Hall.

Ishikawa, K. 1985. *What Is Total Quality Control? The Japanese Way*. Translated by David I. Lu. Upper Saddle River, NJ: Prentice Hall.

Kolarik, W. J. 1995. *Creating Quality—Concepts, Systems, Strategies, and Tools*. New York: McGraw-Hill, Inc.

National Institute of Standards and Technology (NIST). 2012. *NIST/SEMATECH e-Handbook of Statistical Methods (Engineering Statistics Handbook)*. U.S. Department of Commerce. Website: http://www.itl.nist.gov/div898/handbook/ (accessed August 30, 2012).

Phadke, M. S. 1989. *Quality Engineering Using Robust Design*. Upper Saddle River, NJ: P T R Prentice-Hall, Inc.

Yang, K. and B. El-Haik. 2003. *Design for Six Sigma*. ISBN 0071412085. New York: McGraw-Hill.

18 Human Factors Engineering Tools

INTRODUCTION

Human factors engineers use a number of tools during product programs. The tools are used for the following purposes: (1) to obtain information about the characteristics, capabilities, and limitations of populations of users; (2) to apply the collected information to design products; and (3) to evaluate the products during different phases of the product programs. The goal of a human factors engineer is to ensure that the product is designed such that most individuals in its intended user population are able to use the product easily, comfortably, and safely.

The human factors methods described in this chapter are categorized as follows:

1. Databases on human characteristics and capabilities
2. Anthropometric and biomechanical human models
3. Checklists and scorecards
4. Task analysis
5. Human performance evaluation models
6. Laboratory, simulator, and field studies
7. Human performance measurement methods.

The application of the human factors tools requires the implementer to be knowledgeable about the human factors issues, principles, and research studies in the subject area. These tools cannot be easily mastered without sufficient background in the variables related to the users, the product, the users' tasks, and the product usage situations. The user's performance in using a product can be affected by many factors such as familiarity with the past models of similar products, adaptability to new situations, improved performance with practice, deliberately changing behavior to please or displease the evaluator (or experimenter), and so forth. Thus, many variables can affect the user's behavior, performance, and preferences. A trained human factors professional will generally take necessary precautions to ensure that biases are not introduced during the applications of human factors methods.

DATABASES ON HUMAN CHARACTERISTICS AND CAPABILITIES

A number of Human Factors Engineering handbooks, textbooks, and standards provide data on various human characteristics and capabilities (e.g., anthropometric, biomechanical, sensory, and information processing characteristics) for various populations (by gender, age groups, occupations, and national origin) (Garrett, 1971; Van Cott and Kinkade, 1972; Card et al., 1983; Sanders, 1983;

Jurgens et al., 1990; Woodson, 1992; Sanders and McCormick, 1993; Kroemer et al., 1994; Wickens et al., 1998; Konz and Johnson, 2004; Pheasant and Haslegrave, 2006; McDowell et al. 2008; SAE, 2009; Bhise, 2012). In addition, many research reports and journals related to human factors provide useful data from many studies.

These databases provide information on distributions and percentile values of various characteristics and capabilities that are needed to design products to fit most people in selected product user populations. (Note: The website of this book contains a data file on anthropometric dimensions of seven populations.) Such data are needed for designing products. For example, while designing an automotive product, the designer must ensure that a short female can reach and operate the pedals and see over the steering wheel and that a tall male can fit inside the cockpit with sufficient headroom, legroom, hip room, shoulder room, and elbow room. The designers also need to know the level of familiarity of the users with the operation of the controls and displays, the performance characteristics of the product, and the characteristics of the operational environment. Further, the decision-making and product-operating capabilities of the users must be known to the product designers.

ANTHROPOMETRIC AND BIOMECHANICAL HUMAN MODELS

A number of two- and three-dimensional (3D) anthropometric and biomechanical models are presented in the literature, and many are commercially available for design and evaluation purposes. These models can be configured to represent individual males and females in different percentile dimensions for different populations. Many of the models have built-in human motion, posture simulations, and biomechanical strength, as well as percentile force exertion prediction capabilities. Crash test dummies resembling human biomechanical characteristics are also used to evaluate the crashworthiness of vehicles in accident situations (Seiffert and Wech, 2003).

Integrated digital workplaces and digital manikins and visualization tools are available in several software applications. Computer-aided design (CAD) tools with manikin models (digital human models), such as Jack/Jill, SAFEWORK, RAMSIS, SAMMIE, and UM 3DSSP, are being currently used by different designers to assist in the product development process (Chaffin, 2001, 2007; Reed et al., 1999, 2003; Badler et al., 2005; Human Solutions, 2010). Many of these tools are being updated to incorporate additional capabilities.

Before using any of the models in the design process, the ergonomics engineer should conduct validation studies to determine if the population of the particular users of the product being designed can be accurately represented in terms of their dimensions, postures, motions, strength, and comfort. The postures assumed by the selected digital human model and their outputs should match closely with the postures and dimensions of real users under different actual usage situations.

HUMAN FACTORS CHECKLISTS AND SCORECARDS

Checklists and scorecards are commonly used by human factors experts to evaluate products. Checklists aid in evaluating applications of human factors guidelines, design considerations, and requirements, whereas scorecards help in summarizing the findings of human factors evaluations and enable the tracking of quantitative and comparative assessments of the product over time as the design progresses.

CHECKLIST

Human factors checklists typically include a series of questions related to meeting human factors guidelines. The product is usually evaluated by one or more human factors experts or the users of the product. Each evaluator would use the product and then answer each question. The answers to the questions (i.e., "yes" or "no" or a rating of how well the human factors guideline was met [10 = met very well; 1 = did not meet the guideline]) can be used to summarize the responses (e.g., percentage of human factors guidelines met and percentage of guidelines that met with a rating of 8 or above) in key areas such as locations of controls and displays, visibility of displays, legibility of displays, comprehension or interpretability of controls and displays, operation of controls, feedback from controls, and so on.

An Example: A Checklist for Evaluation of an Automotive Control

Table 18.1 presents a checklist created by the author to evaluate a control located in an automotive instrument panel (Bhise, 2012).

Scorecard

Human factors scorecards are created to provide feedback to design teams on the ergonomics characteristics of the product being designed. Ergonomics experts systematically develop scoring criteria and the evaluation procedure. The product design is analyzed by conducting evaluations based on objective analyses (e.g., performance evaluations, CAD analyses of reach and visibility, legibility predictions [Bhise, 2012]), as well as the expert's ratings for each criterion. A scorecard is prepared by summarizing the results of these evaluations. The scorecards are presented and discussed with the design, engineering, and program management teams during program review meetings.

The data gathered from the completion of checklists can be categorized for comparisons by types of users (e.g., based on familiarity/unfamiliarity with the product, male/female users, and young/old users) and scores can be developed by types of users (e.g., older user's scorecard, women user's scorecard, and unfamiliar user's scorecard).

An Example: Ergonomic Scorecard for Automotive Interior Evaluation

Figure 18.1 presents an ergonomic scorecard for controls and displays in the interior of an automotive product. The scorecard represents an ergonomics summary chart (called the "smiley faces" chart). The chart lists each control and display in different interior regions of the vehicle on the left-hand side of the table. The evaluation criteria are grouped into nine columns located in the middle of the table. The nine criteria groups are labeled as follows:

1. Visibility, obscurations, and reflections
2. Forward vision down angle

TABLE 18.1
Control Evaluation Checklist

	No.	Question	Yes	No
Findability	1	Can this control be easily found?		
	2	Is the control located in the expected region?		
	3	Is the control visible from the normal operating posture?		
	4	Can the control be seen without head and torso movements?		
	5	Is the control visible at night from the normal operating posture?		
Identification	6	Is the control logically placed and/or grouped to facilitate its identification?		
	7	Is the control properly labeled?		
	8	Is the label visible?		
	9	Can the label be read (legible?) from the normal operating posture?		
	10	Is the label illuminated at night?		
	11	Can the label be read (legible?) at night from normal operating posture?		
	12	Can the control be identified by touch?		
	13	Can the control be discriminated from other controls located close to it?		
Interpretability	14	Can the control be NOT confused with other controls or functions?		
	15	Can an unfamiliar operator guess the operation of the control?		
	16	Does the shape of the control convey/suggest activation directions?		
	17	Does the control work like most other controls of that control type?		
	18	Is the control grouped logically?		
	19	Is the control placed within a group of controls that control the same basic function?		
	20	Are there other controls within 2–3 inches that have a similar visual appearance or tactile feel?		
Control Location, Reach and Grasp	21	Is the control located within maximum comfortable reach distance?		
	22	Can the control be reached without excessive bending/turning of operator's wrist?		
	23	Is the target area of the control large enough to reach the control quickly?		
	24	Can the control be reached without complex/compound hand/foot motions?		
	25	Can the control be reached without torso lean?		
	26	Can the control be grasped comfortably without awkward finger/hand orientations?		
	27	Is there sufficient clearance while grasping the control?		
	28	Is there sufficient clearance for a person with long (15 mm) fingernails?		
	29	Is there sufficient clearance space for operator's hands/knuckles?		
	30	Is there sufficient clearance to grasp the control with winter gloves?		
	31	Is there sufficient foot clearance (if foot operated control)?		

(Continued)

Human Factors Engineering Tools

TABLE 18.1 (Continued)

	No.	Question	Yes	No
	32	Is the control located at "just about right" location?		
	33	Is the control NOT located too high?		
	34	Is the control NOT located too low?		
	35	Is the control located NOT too far?		
	36	Is the control located NOT too close to the driver?		
	37	Is the control located NOT too much to the left?		
	38	Is the control located NOT too much to the right?		
	39	Is the control oriented to facilitate its operation?		
	40	Is the control NOT combined or integrated with other controls?		
	41	Can the location of the control be NOT changed when the setting of any other control is changed?		
Operability	42	Can the control be operated quickly?		
	43	Can the control be operated blindly or with one short glance?		
	44	Is the operation of the control part of a sequence of control operations?		
	45	Can the control be operated without reading more than 2 words or labels?		
	46	Can the control be operated without looking at a display screen?		
	47	Can the control be operated easily without excessive force/torque/effort?		
	48	Does the control provide visual, tactile, or sound feedback on completion of the control action?		
	49	Does the control provide immediate feedback (without excessive time lag)?		
	50	Does the control move without excessive dead space, backlash, or lag?		
	51	Is sufficient clearance space provided for operating hand/foot as the control is moved through its operating movement?		
	52	Can the control be operated without regrasping?		
	53	Can the control be moved without excessive inertia or damping?		
	54	Does the control direction of motion meet the direction of motion stereotypes?		
	55	Can the control be operated without more than one simultaneous movement?		
	56	Is the direction of screen/display movement related to the control movement compatible?		
	57	Is the magnitude of displayed movement related to the control movement "about right"?		
	58	Can the control be activated easily with gloved hand?		
	59	Can the control be operated easily by a person with long fingernails?		
	60	Does the surface texture/feel of the control facilitate its operation?		
	61	Can the operation of the control be performed with little memory capacity (5 or fewer items)?		
	62	Are surfaces on the control rounded to reduce sharp corners and grasping discomfort during its operation?		
		Sum of "Yes" and "No" responses		

3. Grouping, association, and expected locations
4. Identification labeling
5. Graphics legibility and illumination
6. Understandability/interpretability
7. Maximum and minimum reach distance
8. Control area, clearance, and grasping
9. Control movements, efforts, and operability.

A five-point rating scale (with 5 = highest score [meets ergonomics guideline very well] and 1 = lowest score [does not meet the ergonomics guideline]) is used to evaluate the ergonomic guidelines in each of the aforementioned nine groups. The ratings are usually obtained from trained ergonomists by using inputs from the following: (1) measurements obtained from 3D-CAD model of the occupant package (e.g., reach distances and down angles), (2) ergonomics review by sitting in an interior buck (if available), and (3) results from applicable design tools and models (e.g., the Society of Automotive Engineers, Inc. [SAE], design practices [SAE, 2009; Bhise, 2012]). The ratings are graphically displayed by using a graphic scale of smiley faces for each of the aforementioned nine groups, for each item listed in each row (Figure 18.1).

The chart in Figure 18.1 provides an easy-to-view format that can be used to provide an overall ergonomics status of a vehicle interior and was found by the author to be a useful tool in various design and management review meetings. The objective of the ergonomics engineer is to convince the design team during design review meetings to remove as many black dots (rating = 1 to 2) and black doughnuts (rating = 3) from the charts and increase the number of smiley faces (rating = 4 to 5) by making necessary design changes.

TASK ANALYSIS

Task analysis is one of the basic tools used by ergonomists in investigating and designing tasks. It provides a formal comparison between the demands that each task places on the human operator and the capabilities the human operator possesses to respond to the demands. Task analysis can be conducted with or without a real product or a process. But it is easier if the real product or equipment is available and if the task can be performed by actual (representative) users under real usage situations to understand the subtasks.

The analysis involves breaking a task or an operation into smaller units (called subtasks) and analyzing subtask demands with respect to user capabilities. The subtasks are the smallest units of behavior that need to be differentiated to solve the problem at hand, for example, grasp a handle, read a display, and select a control setting. The user capabilities that are considered here are as follows: sensing, use of memory, information processing, and response execution (movements, reaches, accuracy, postures, forces, time constraints, etc.).

Task analysis can be conducted by using different formats. Table 18.2 presents a format presented by Drury (1983) that the author found to be useful during the automotive design process (Bhise, 2012). The left-hand side of Table 18.2 describes the

Human Factors Engineering Tools

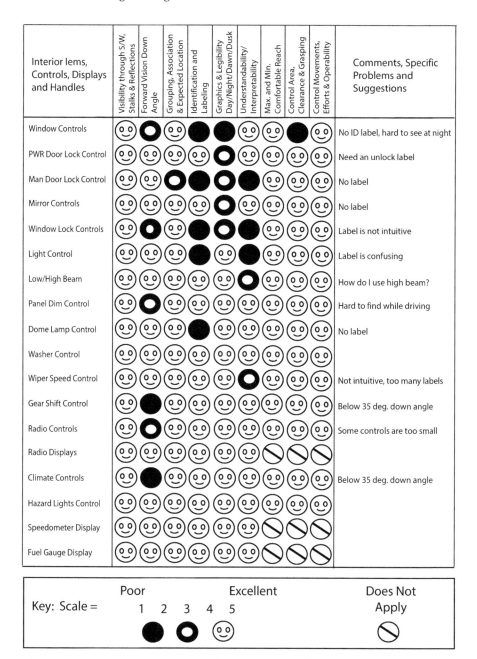

FIGURE 18.1 Ergonomics scorecard for automotive controls and displays evaluation.

subtasks involved in the task along with the purpose of each subtask. Thus, the description of each subtask makes the analyst (human factors specialist) think about the need for each subtask and, perhaps, even suggest a better way to do the task. The right-hand columns of the table forces the analyst to consider various human

TABLE 18.2
An Example of a Task Analysis for Opening the Liftgate and Removing a Jack and Wheel-Nut Wrench

Task Number	Task Description	Task Description — Task Purpose	Task Description — Scanning/Seeing	Task Analysis — Memory	Task Analysis — Interpolating	Task Analysis — Manipulating	Task Analysis — Possible Errors
1	Unlock liftgate	Allow access	Locate button on key fob	Fob/button location	Which button to press	Press key fob button	Cannot find key fob/button
2	Locate liftgate release	Prepare to open liftgate	Locate release	Release location	Location of release	Scan expected area	Release not found
3	Operate liftgate release	Unlatch liftgate	Confirm liftgate release	Release type	Release procedure	Squeeze release	Liftgate not unlatched
4	Pull up briefly	Start liftgate upward	Clearance for liftgate	Liftgate position	Liftgate travel arc	Pull-up	Pull too much/little
5	Stand back	Clearance for liftgate	Clearance for liftgate	Strut opening speed	Lift gate travel arc/rate	Stand back	Liftgate strikes operator
6	Locate jack compartment	Locate jack	Locate jack compartment	Jack location	Location of jack	Scan expected area	Compartment not found
7	Unlatch jack cover	Prepare to remove cover	Confirm latches opened	Latch type	Unlatch procedure	Unlatch jack cover	Jack cover not unlatched
8	Open jack cover	Open jack compartment	Cover clearance	Cover hinge type	Open procedure	Open jack cover	Jack cover not opened
9	Remove jack cover	Full jack access	Align tabs and slots	Tab/slot alignment	Orientation for removal	Remove jack cover	Cover not fully removed
10	Locate jack fastener	Prepare to access jack	Locate jack fastener	Fastener location	Location of fastener	Scan expected area	Fastener not located
11	Remove jack fastener	Access jack	Confirm removal	Fastener type	Amount of turns required	Unthread fastener	Fastener not fully removed
12	Pull jack upward	Remove from support clip	Clearance for jack & hands	Retention clip type	Force required for removal	Pull jack upward	Jack not freed from clip
13	Remove from compartment	Remove jack	Clearance for jack and panel	Orientation that allows removal	Optimal removal orientation	Orient and remove jack	Jack oriented incorrectly
14	Locate wrench fastener	Prepare to access wrench	Locate wrench fastener	Fastener location	Location of fastener	Scan expected area	Fastener not located
15	Remove wrench fastener	Access wrench	Confirm removal	Fastener type	Amount of turns required	Unthread fastener	Fastener not fully removed
16	Remove nut wrench	Remove nut wrench	Confirm removal	Direction of motion	Wrench removal procedure	Pull wrench off jack	Wrench not fully removed

functional capabilities, such as searching and scanning, retrieving information from the memory, interpolating (information processing), and manipulating (e.g., hand/finger movements), required in performing each subtask. The last column requires the analyst to think about possible errors that can occur in each subtask. This last column is the most important output that can be used to improve the task and/or improve the product to reduce errors, problems, and difficulties involved in performing the task.

If the task analysis is performed on a product that already exists (or for which a mock-up, prototype product, or simulation is available), then a number of user trials can be performed, and information can be gathered on how different users performed each subtask and the problems, difficulties, and errors experienced by the users (see Chapter 7 for methods of data gathering). The information can then be used in creating the task analysis table. Even if a product is not available, the task analysis can be performed on an early product concept by predicting the possible sequence of subtasks needed to use the product in performing each task (or product usage).

AN EXAMPLE: TASK ANALYSIS FOR OPENING A LIFTGATE AND REMOVING A JACK

The task analyzed in Table 18.2 is opening the rear liftgate of a sports utility vehicle and removing a jack and wheel-nut wrench (needed to change a tire). The last column of possible errors can provide insights that will assist in designing for ease in opening the liftgate and removing the jack and wheel-nut wrench.

HUMAN PERFORMANCE EVALUATION MODELS

Human factors researchers have developed a number of models to predict the performance of users under different product usage or work situations. The models are generally based on statistical analysis (e.g., regression models) of data gathered from experimental research studies. An available human performance model, if it is found to apply to a given product usage situation, can be applied to predict the performance of users under different product usage situations. The model predictions along with the results of additional experimental research can be used to narrow down a number of product concepts and designs during the early stages of the product development processes. For example, one of the oldest operator performance models to predict time requirements of factory jobs was "Methods-Time Measurement," called the MTM (Maynard et al., 1948). The model is based on breaking down a given task into a series of predefined micromotions. The time required to perform various micromotions (e.g., reach, move, grasp, release, turn, position, assemble, turn body, leg motions, eye motions, steps, side steps, etc.) are provided in tables. The applicable times from the tables can be added to predict the time to complete a given task. The times are for an experienced operator handling and assembling small parts. Other predetermined time prediction models are also used in the industry (Konz and Johnson, 2004).

A human information processing model for the prediction of human operator time requirements in information processing tasks was developed by Card et al. (1980, 1983). The basic approach involves the top–down, successive decomposition of a task. The analyst divides the task into logical steps. For each step, the analyst identifies the human and device task operators. This approach assumes error-free performance, well-learned tasks, and particular locations of controls.

Many other models are also available to predict human errors. Leiden et al. (2001) have provided a review of a number of cognitive models such as ACT-R, Air MIDAS, Core MIDAS, APEX, COGNET, D-OMAR, EPIC, GLEAN, SAMPLE, and Soar for prediction of human error. Many information processing and workload assessment models are also available in the human factors literature (covered in a later section of this chapter). Such models, after understanding the limitations on their applicability, can be used to evaluate user performance for a given product under a specified usage situation.

Bhise (2012) has described a number of models used for ergonomic evaluations in the automotive design process. The models include driver-positioning and occupant-packaging practices incorporated in SAE practices (SAE, 2009). Bhise (2012) also described a number of human vision models based on visual contrast thresholds and disability and discomfort glare prediction equations. The models have been used to evaluate automotive headlighting systems (Comprehensive Headlamp Environment Systems Simulation [CHESS] model) (Bhise et al., 1977, 1988; Bhise and Matle, 1989), legibility of displays (Bhise and Hammoudeh, 2004), and veiling glare effects from sunlight reflections into vehicle windshields (Bhise, 2007; Bhise and Sethumadhavan, 2008a,b).

Hankey et al. (2001) developed a model to predict the level of demand placed on the driver while using in-vehicle devices (e.g., an entertainment system or a navigation system). The model was called the IVIS DEMAnD model. Jackson and Bhise (2002) applied the model to evaluate relative sensitivities of vehicle parameters and other external factors such as traffic demand and driver age in evaluating figures of demand during various driving tasks.

The human factors models in general are based on modeling relationships between many independent variables and assumed characteristics of human operators. The relationships are usually developed based on experimental research. The models are thus approximations of human performance and should be used with caution. The results of models should be validated by performing studies using real subjects and products under actual usage situations. This issue of testing with real subjects is covered in the following section.

LABORATORY, SIMULATOR, AND FIELD STUDIES

Early product concepts, product prototypes, and final products (production versions) are generally evaluated by using real subjects who are asked to perform a number of tasks to verify and validate the product design. Product evaluations under actual usage situations involving field testing are generally preferred. However, due to several reasons such as high costs to build a working model of the product, costs in recruiting subjects, and time required to perform field tests, other research avenues such as laboratory tests, product simulations, use of early prototypes, and testing with simulators are commonly considered.

Five data collection methods employed in human factors studies include (1) observations of subjects performing given tasks with the product, (2) communication with the subjects to obtain ratings on selected product features or asking the subjects to describe problems and difficulties in using the product, (3) measurements of the performance of subjects and the product during product uses, (4) measurements of the physiological state of subjects while completing different usage tasks, and (5) obtaining subjective ratings to measure operator workload.

HUMAN PERFORMANCE MEASUREMENT METHODS

Human performance in the laboratory, field, and simulator studies can be measured by using a combination of methods of observation, communication, and experimentation. These methods are described in Chapter 7. Some examples of measurements in these evaluations can include the following:

- Observable human responses: we can observe a human operator's responses such as his or her visual information acquisition behavior through measurements of eye movements, head movements, eye glances, time spent in viewing different objects (e.g., displays), and control movements from measurements of movements of body parts such as hand and foot while operating various controls. We can also measure the physiological state of the operator by measuring variables such as the operator's heart rate, sweat rate, pupil diameter, and so forth.
- Operator's subjective responses: we can also develop a structured questionnaire and ask the operator a number of questions at different points in a test procedure (if an experimenter is present) or at the end of the procedure to understand the operator's problems, difficulties, confusions frustrations, and situational awareness issues. We can also ask the operator to provide ratings on his or her workload, comfort, ease in using different controls and displays, and so forth.
- State of product/equipment: we can also record the state of the product (or the equipment being used) by installing measuring instruments in the product to measure product outputs (as functions of time) such as speed, control positions and movements, distance traveled, vibrations, temperatures, and energy consumption.

RANGE OF HUMAN PERFORMANCE MEASURES

The range of human performance measures that can be used extend from the measurement of some early events, actions, or steps in a task to the measurement of final outcomes. For example, the range of events that can be measured in a target detection task while driving can involve measurements of locations and durations of eye fixations (eye search patterns), target detection response (correct detection or failure), response time to detect the target, detection distance, lane position variability while searching for the target, steering wheel deviations when searching for the target, erratic or evasive maneuvers as a result of late or no detection, and accident (if it occurs) resulting from non-detection of the target.

Behavioral measures can provide information on how, what, and when the user performed certain predefined steps. Measures such as total time spent, types and numbers of errors committed, and percentage of times the task was completed within the allocated time all provide information on how well a given task was completed. The physiological measures can provide information on the physiological state of the operator (i.e., how the human operator's body is responding while performing the task) by measuring variables such as heart rate, sweat rate, brain waves, and so forth. Subjective measures are very useful in which the subject can be asked to describe problems encountered during a task and provide ratings using scales developed to obtain impressions of the subject on task-related characteristics such as level of difficulty or ease, perception of spaciousness inside a cockpit, workload, and comfort.

Thus, in determining what measures to select for a given study it is important to ensure that the researcher can obtain useful and valid information. The skill in selecting performance measures depends on the researcher's knowledge of the research literature in the problem area, depth of the researcher's human factors research experience, data sensing and recording equipment availability, time and resources available, and the researcher's experience in statistical data analysis.

The selection of dependent variables, whether they are behavioral or performance-based, will depend on the problem and the operator's tasks associated with the issues in the problem. The behavioral variables are generally based on observations of the user's movements and actions related to the task being performed. The performance measures, on the other hand, measure how well the user performed the task. In addition to the measurements of operator performance, it is also useful to obtain synchronized data on the state of the product, physiological state of the operator, and video recordings of the product usage situation (e.g., the user's field of views to the front sides and the rear) so that the performance can be measured and categorized according to the situation and environmental characteristics.

Types and Categories of Human Performance Measures

Human performance measures can be categorized as follows:

Behavioral measures: measuring what behavior and what choices the user exhibits (i.e., what did he or she do) by recording his or her eye movements, body movements, sequences in performing different movements and tasks, and decisions made.

Physical measures: these measures are derived from data gathered by using physical instruments such as timers, pedometers, odometers, speedometers, accelerometers, dynamometers, movement sensors, goniometers, thermometers, and so on.

Subjective measures: based on judgments of the users (or of an observing experimenter), for example, ratings, preferences, thresholds of perception, detection, and equivalency judgments.

Physiological measures: based on measurements of physiological state of the user from changes in heart rate, sweat rate, oxygen intake, galvanic skin resistance, electrical activities in different skeletal muscles (electromyographs [EMGs]) or the heart (electrocardiograms [EKGs]), and so forth.

Accident-based safety performance measures: from data available in accident records such as the number of accidents of a given type (e.g., based on human error, equipment malfunction, and environmental condition), accident rate (e.g., number of accidents per distance traveled or hours worked), and accident severity (e.g., damage or injury level).

Monetary measures: measures based on costs such as trip costs, energy consumption, costs and benefits related to the usage, and so forth.

EXAMPLES OF BEHAVIORAL HUMAN PERFORMANCE MEASURES

Some commonly used observable behavioral measures of human performance used in studies reported in the literature are based on statistics such as mean, median, standard deviation, percentages, and percentiles of data collected on the following:

1. Task completion time
2. Errors made in performing a task (frequency of errors and error rates) (e.g., operated a wrong control, turned a control in the opposite direction, or forgot to use a display or a control)
3. Number of steps or tasks completed in a given time period
4. Total eye involvement time (spent in looking at a given location or a display)
5. Glance durations while viewing given objects (e.g., speedometer, radio, mirror, and sign)
6. Number of glances made on a device to complete a given task
7. Eye fixation durations while viewing a given object or a location
8. Number of eye fixations made on a given object or a location
9. Percentage of time eyes were diverted from a primary display
10. Blink (or eye closure) rate
11. Detection rates of targets or events
12. Detection distance (i.e., when a target was visible)
13. Identification distance (i.e., when a target was identified)
14. Hand involvement time (e.g., time spent by the driver's hand away from the steering wheel to perform a radio tuning task)
15. Reaction (or response) time to a given signal or a situation
16. Accident frequency and rates (e.g., number of accidents per exposure unit such as time, product usages, distance traveled, operational cycles, etc.).

In experimental evaluations, the dependent variables are generally related to, and affected by, characteristics of users (e.g., young vs. old, males vs. females, experienced vs. inexperienced, familiarity of the user, and country of origin or nationality of the user), characteristics of the product usage environment (e.g., day vs. night), and characteristics of the product design (e.g., analog vs. digital display, locations and types of controls, different controls and displays layouts, product features or design configurations, and similarity/familiarity with products made by other manufacturers).

Methods to Measure Human Operator Workload

In many product usage situations, it is important to determine the amount of workload experienced by the product user in relation to his or her capabilities. The workload of an operator under some stressful situations (e.g., when the operator has to attend to a number of inputs and make many control actions within a short period) can be very high in relation to the operator's capabilities. Examples of such situations are a fighter pilot in a combat situation, a nuclear power plant operator under a malfunctioning reactor situation, and a driver approaching a busy intersection and intending to make a quick left turn. The product designer must understand such high workload situations and design the product to keep the workload of most users under manageable levels.

Many approaches and methods are used by various researchers to determine an operator's workload. Some of the commonly used and useful methods are briefly described in the following subsections.

Operator Performance Measurements

Various operator performance measurement methods and measures can be used to determine the effect of operator workload on operator performance in performing different tasks. The changes in levels of performance measures obtained while the operator is simply using the product (i.e., the baseline, e.g., driving a car) versus when the operator is asked to perform tasks in addition to using the product (i.e., dual task, e.g., driving and dialing a cell phone) have been used to assess the effects of operator workload (Olsson and Burns, 2000).

Physiological Measurements

A number of physiological measurements have been used to determine the effects of workload on the human operator. These measures are based on the assumption that workload will affect bodily functions. They are based on measuring effects of arousal, excitement, stress/tenseness, thought processes, and use of body movements through muscle activations that are caused during the performance of tasks.

Some examples of physiological measurements are as follows: heart rate, respiration rate, brain's spontaneous electrical activity from electroencephalograms and evoked potential recordings, electrical activity of the heart from EKGs, electrical activity of muscles from EMGs, electrical activities of the eye muscles from electrooculograms, galvanic skin response, body and skin temperatures, sweat rate, pupil size, and eyeblink rate.

The variations in body functions due to extensive physical biomechanical workload can be measured by heart rate, respiration rate, oxygen intake, and EMGs. These measures related to muscular activities are relatively easier to measure and interpret compared to physiological variations due to changes in mental workload during information processing or driving. Brookhuis and Waard (2001) have reported studies that showed that the drivers' heart rate increased under higher stress and workload; for example, driving through traffic circles increased the heart rate compared to driving on straight roads and awaiting a traffic light change increased heart rate variability. Mehler et al. (2009) examined the sensitivity of heart rate, skin conductance, and respiration rate as measures of mental workload in a simulated driving environment using a sample of 121 young adults. Their results showed that the mental workload imposed by the "n-back task" (recalling a one-digit number from 0, 1, or 2 digits back when one-digit numbers are presented at a constant rate) increased the heart rate, skin conductance, and the respiration rate. Verwey and Zaidel (2000) found that the eyeblink rate was related to drowsiness, suggesting that the frequency of eye closures exceeding 1 second is an indicator of drowsiness.

The reliability of physiological measures in measuring a user's mental workload is poor because of large variations among individuals and many factors (e.g., anxiety and stress) that affect these measures. The physiological measures are rarely used during the product development process because of three reasons: (1) the links between the physiological measures and real-world performance are not clearly understood, (2) the physiological measures are difficult to obtain, and (3) the high expense and time associated with collecting large amounts of data and analyses requires more sophisticated techniques.

Subjective Assessments

Subjective ratings on the level of difficulty, stress, discomfort, mental workload, physical workload, and so forth, provided by subjects during performance of different tasks are commonly used as measures to assess the workload.

Three well-developed subjective workload measurement techniques used in this field are as follows: (1) the National Aeronautics and Space Administration (NASA) Task Load Index (TLX), (2) subjective workload assessment technique (SWAT), and (3) workload profile (WP) (Meshkati et al., 1992). These three methods are described in the following subsections.

National Aeronautics and Space Administration Task Load Index

The NASA TLX is a multidimensional rating procedure that derives an overall workload score based on a weighted average of ratings on six subscales (Hart and Staveland, 1988; Hitt II et al., 1999). These six subscales include (1) mental demand, (2) physical demand, (3) temporal demand, (4) own performance, (5) effort, and (6) frustration. The method has been used to assess workloads in various human–machine environments such as aircraft cockpits, workstations, process control environments, and various actual driving as well as simulated driving situations (Bhise and Bhardwaj, 2008).

The subscales can be specified by 5-point, 10-point, or 100-point interval scales. The standardized descriptors (questions) used for each subscale category and adjectives used to define their endpoints are presented in the study by Hart and Staveland (1988) and the NASA TLX manual (NASA, 1988) (Table 18.3). The ratings on individual subscales can be used as evaluation scores, or an overall workload score can be obtained from a weighted sum of the ratings on the six scales. The weightings of the subscales can be obtained after the subjects have performed all the tasks. A paired

TABLE 18.3
Descriptions of Six Scales Used in the Measurements of the NASA TLX

No.	Subscale (Workload Attribute)	Adjectives Used to Describe the Low and High End Points of the Scale	Questions Used to Describe the Scale Attribute
1	Mental demand	Low, high	How much mental and perceptual activity was required (e.g., thinking, deciding, calculating, remembering, looking, and searching)? Was the task easy or demanding, simple or complex, exacting or forgiving?
2	Physical demand	Low, high	How much physical activity was required (e.g., pushing, pulling, turning, controlling, and activating)? Was the task easy or demanding, slow or brisk, slack or strenuous, restful or laborious?
3	Temporal demand	Low, high	How much time pressure did you feel due to the rate or pace at which the tasks of task elements occurred? Was the pace slow and leisurely or rapid and frantic?
4	Performance	Good, poor	How successful do you think you were in accomplishing the goals of the task set by the experimenter (or yourself)? How satisfied were you with your performance in accomplishing these goals?
5	Effort	Low, high	How hard did you have to work (mentally and physically) to accomplish your level of performance?
6	Frustration level	Low, high	How insecure, discouraged, irritated, stressed, and annoyed versus secure, gratified, content, relaxed, and complacent did you feel during the task?

comparison method (e.g., Analytical Hierarchical method; see Chapter 7 or Thurstone's method [Bhise, 2012]) can be used to obtain the weightings based on the importance of the subscale categories associated with the tasks.

Subjective Workload Assessment Technique

SWAT involves asking the operator to rate his or her workload using three scales: (1) time load, (2) mental effort load, and (3) psychological stress load. Each scale has three levels, low, medium, and high. The descriptors used to define the three levels of each of the three scales are presented in Table 18.4 (Meshkati et al., 1992). The method uses conjoint measurements and scaling techniques to develop a single interval rating scale (Reid and Nygren, 1988).

Workload Profile

The WP method is based on the multiple resource model (Wickens, Gordon and Liu, 1998). It considers the following eight workload dimensions as attentional resources: (1) perceptual/central processing, (2) response selection and execution, (3) spatial processing, (4) verbal processing, (5) visual processing, (6) auditory processing, (7)

TABLE 18.4
Descriptors Used to Define Three Levels of Time Load, Mental Effort Load, and Psychological Stress Load Scales Used in SWAT

No.	Scale	Level	Descriptors
1	Time load	Low	Often have spare time. Interruptions or overlaps among activities occur infrequently or not at all.
		Medium	Occasionally have spare time. Interruptions or overlaps among activities occur infrequently.
		High	Almost never have spare time. Interruptions or overlaps among activities are very frequent, or occur all the time.
2	Mental effort load	Low	Very little conscious mental effort or concentration required. Activity is almost automatic, requiring little or no attention.
		Medium	Moderate conscious mental effort or concentration required. Complexity of activity is moderately high due to uncertainty, unpredictability, or unfamiliarity. Considerable attention required.
		High	Extensive mental effort and concentration are necessary. Very complex activity requiring total attention.
3	Psychological stress load	Low	Little confusion, risk, frustration, or anxiety exists and can be easily accommodated.
		Medium	Moderate stress due to confusion, frustration, or anxiety noticeably adds to workload. Significant compensation is required to maintain adequate performance.
		High	High to very intense stress due to confusion, frustration, or anxiety. High extreme determination and self-control required.

manual output, and (8) speech output. The subjects are asked to provide proportions for each of the eight workload dimensions used in each task (in a random order) after they have experienced all the tasks. Thus, each task is evaluated by providing eight ratings. The ratings will vary from 0 to 1 to represent the proportion of each attentional resource used in the task. Thus, a rating of "0" means that the task did not require the dimension (or resource) and "1" means that the task required maximum attention. The ratings on these eight dimensions of each task are later summed to obtain an overall workload rating for the task.

Subjective methods are commonly used in many product development programs because they are easier to obtain (require no instrumentation) and have high "face validity" as the "voice of the customer." The disadvantages of subjective methods are that the rater may find it difficult to understand many issues associated with comparing different products and situations, and the agreement between different raters may not be unanimous.

Secondary Task Performance Measurement

This approach uses an artificially added task called a "secondary task" while performing a primary task (of product usage) and assumes that an upper limit exists on the capacity of the user to gather and process information. The performance in the secondary task is used to measure the user's workload on the assumption that adding the secondary task on the top of the primary task will increase the user's total workload; if the secondary task is sufficiently difficult, the user will reach or exceed his or her overall capacity to perform both the tasks. By carefully controlling the user's priorities through instructions, the user will be asked to maintain performance in the primary task. Thus, the level of performance in the secondary task will indicate the amount of workload or capacity taken up by the primary task.

Many different secondary tasks have been used in the literature. Some examples of secondary tasks are peripheral detection tasks, arithmetic addition tasks, repetitive tapping tasks, time estimation, random number generation, choice reaction time tasks, critical tracking tasks, visual search tasks, and memory search tasks. For example, Olsson and Burns (2000) measured the reaction time to detect peripheral targets and hit rate (proportion of targets correctly detected) while driving a car under a baseline driving condition and in other conditions involving the baseline driving and performing radio station tuning and CD tasks (e.g., turn on the CD mode and select a given song/track). They found that when the drivers were asked to perform these additional radio and CD tasks, their peripheral detection performance, measured by both reaction time and hit rate, was degraded compared to their performance in the baseline driving condition.

The use of the secondary task as a method to measure user's workloads has a number of shortcomings. If the introduction of the secondary task modifies (or interferes) the user's primary task, then the user may be forced to change his or her strategy and, thus, the load imposed by the primary task may be distorted. The interference in the primary task is greater when the tasks share the same response resources than when the responses occupy different resources. Further, it is difficult for some users to maintain the same level of attention and priority in performing the primary task when the secondary task is introduced.

PRODUCT PSYCHOPHYSICS

Psychophysics is an area within psychology that deals with the relationship between physical characteristics of stimuli and psychological perceptions, that is, how a physical event is perceived by a person. The physical event associated with a product such as a movement of a control or occurrence of a signal (e.g., lighting up of a warning light on an instrument panel) is perceived by its user. The relationship between a movement of a control (e.g., change in deflection of the gas pedal) and the amount of product output (e.g., sound of the engine) is perceived by the user as he or she moves the control. These relationships between the amounts of physical changes in the product (or its characteristics) and the level of expected change in the feedback received by the users are examples of psychophysical problems. Proper applications of psychophysical principles will produce just the right levels of relationships between the physical characteristics of the product and the user's perception. Ergonomics engineers can help in solving such psychophysical problems.

During usage of any physical product, a user receives a number of sensory inputs. The user senses the inputs and processes the information to make judgments. Psychophysics helps us to analyze the relationships between physical characteristics of a stimulus (e.g., visual detail) and psychological response (i.e., how a person perceives and reacts to the stimulus). Some typical psychophysical problems are as follows: (1) detection of a signal (or a visual target or a change in the pointer position in a gauge), (2) just noticeable difference between two stimuli (e.g., the smallest difference in brightness or color between two tail lamps that can be recognized by an observer), (3) equality (when the two signals are perceptually equal in magnitude, e.g., determining if two stop lamps on the back of a car are equally bright), (4) magnitude estimation (e.g., estimating the speed of a vehicle or distance between two vehicles), (5) interval estimation (e.g., determining the time [interval] required to reach a particular location with a given vehicle speed), (6) production of the magnitude of a stimulus (e.g., pushing the brake pedal hard enough to produce the required stopping distance), and (7) rating on a scale (e.g., providing a rating on the ease or difficulty using a 10-point scale, where 1 = very difficult and 10 = very easy).

It should be noted that any one of the aforementioned psychophysical problems involves information processing and decision-making. Thus, the application of psychophysics to achieve the right amount and type of relationship between physical and psychological perceptions will help in improving the functionality and customer satisfaction of the product.

CONCLUDING REMARKS

The Human Factors Engineering tools covered in this chapter help design teams to understand how an evaluated product concept or product will be perceived by its users in terms of ease of use, comfort, convenience, and performing required tasks with minimum human error. It is important to know that if the product can be used with minimum mental and physical effort, its users will prefer it over other similar products sold in the market. Constant feedback from the human factors engineers through the applications of ergonomics tools is essential to make sure that the work planned according to the systems engineering management plan will produce an efficient product that is not only within planned timings and budget but also liked by its users.

REFERENCES

Badler, N., Allbeck, J., Lee, S.-J., Rabbitz, R., Broderick, T. and K. Mulkern. 2005. New Behavioral Paradigms for Virtual Human Models. *SAE Transactions Journal of Passenger Cars—Electronic and Electrical Systems*. Paper 2005-01-2689. Presented at the 2005 SAE Digital Human Modeling Conference, Iowa City, Iowa.

Bhise, V. D. 2007. Effects of Veiling Glare on Automotive Displays. *Proceedings of the Society of Information Display Vehicle and Photons Symposium*, Dearborn, Michigan.

Bhise, V. D. 2012. *Ergonomics in the Automotive Design Process*. Boca Raton, FL: CRC Press.

Bhise, V. D. and S. Bhardwaj. 2008. Comparison of Driver Behavior and Performance in Two Driving Simulators. SAE Paper No. 2008-01-0562. Paper presented at the SAE World Congress, Detroit, Michigan.

Bhise, V. D., Farber, E. I., Saunby, C. S., Walnus, J. B. and G. M. Troell. 1977. Modeling Vision with Headlights in a Systems Context, pp 54. SAE Paper No. 770238. Paper presented at the 1977 SAE International Automotive Engineering Congress, Detroit, Michigan.

Bhise, V. D. and R. Hammoudeh. 2004. A PC Based Model for Prediction of Visibility and Legibility for a Human Factors Engineer's Tool Box. *Proceedings of the Human Factors and Ergonomics Society 48th Annual Meeting*, September 2004; 48(7): 1074–1076. New Orleans, Louisiana.

Bhise, V. D. and C. C. Matle. 1989. Effects of Headlamp Aim and Aiming Variability on Visual Performance in Night Driving. *Transportation Research Record*, 1247: 46–55. Transportation Research Board, Washington, DC.

Bhise, V. D., Matle, C. C. and E. I. Farber. 1988. Predicting Effects of Driver Age on Visual Performance in Night Driving. SAE Paper No. 881755 (also No. 890873). Paper presented at the 1988 SAE Passenger Car Meeting, Dearborn, Michigan.

Bhise, V. and S. Sethumadhavan. 2008a. Effect of Windshield Glare on Driver Visibility. Transportation Research Record (TRR). *Journal of the Transportation Research Board*, 2056: 1–8.

Bhise, V. and S. Sethumadhavan. 2008b. Predicting Effects of Veiling Glare Caused by Instrument Panel Reflections in the Windshields. SAE Paper No. 2008-01-0666. *International Journal of Passenger Cars—Electronics Electrical Systems*, 1(1): 275–281. Society of Automotive Engineers, Inc., Warrendale, Pennsylvania.

Brookhuis, K. A. and D. Waard. 2001. Assessment of Driver's Workload: Performance and Subjective and Physiological Indexes. In *Stress, Workload and Fatigue* (Eds.) Hancock, P. A. and P. A. Desmond (pp. 321–333). Mahwah, NJ: Lawrence Erlbaum Associates, Inc.

Card, S. K., Morgan, T. P. and A. Newell. 1980. The Keystroke-Level Model for User Performance Time with Interactive Systems. *Communications of the ACM*, 23(7): 396–410.

Card, S. K., Morgan, T. P. and A. Newell. 1983. *The Psychology of Human-Computer Interaction*. Hillsdale, NJ: Lawrence Erlbaum Associates.

Chaffin, D. B. 2001. *Digital Human Modeling for Vehicle and Workplace Design*. SAE International. ISBN: 978-0-7680-0687-2, New York: Association of Computer Machinery.

Chaffin, D. B. 2007. Human Motion Simulation for Vehicle and Workplace Design: Research Articles. *Human Factors in Ergonomics & Manufacturing*, 17(5): 475–484.

Drury, C. 1983. Task Analysis Methods in Industry. *Applied Ergonomics*, 14: 19–28.

Garrett, J. W. 1971. The Adult Human Hand: Some Anthropometric and Biomechanical Considerations. *Human Factors*, 13(2): 117–131.

Hankey, J. M., Dingus, T. A., Hanowski, R. J., Wierwille, W. W. and C. Andrews. 2001. *In-Vehicle Information Systems Behavioral Model and Design Support: Final Report.* Report No. FHWA-RD-00-135 sponsored by the Turner-Fairbank Highway Research Center of the Federal Highway Administration, Virginia Tech Transportation Institute, Blacksburg, Virginia.

Hart, S. G. and L. E. Staveland. 1988. Development of NASA-TLX (Task Load Index): Results of Empirical and Theoretical Research. In *Human Mental Workload* (Eds.) Hancock, P. A. and N. Meshkati, 139–183. Amsterdam, the Netherlands: North Holland.

Hitt II, J. M., Kring, J. P., Daskarolis, E., Morris, C. and M. Mouloua. 1999. Assessing mental workload with subjective measures: An analytical review of the NASA-TLX index since its inception. *Proceedings of the 43rd Annual Meeting of the Human Factors and Ergonomics Society*, Santa Monica, California.

Human Solutions. 2010. RAMSIS model applications. Website: http://www.human-solutions.com/automotive/index_en.php (accessed May 3, 2013).

Jackson, D. and V. D. Bhise. 2002. An Evaluation of the IVIS-DEMAnD Driver Attention Model. SAE Paper No. 2002-01-0092. Paper presented at the SAE International Congress in Detroit, Michigan.

Jurgens, H., Aune, I. and U. Pieper. 1990. *International Data on Anthropometry*. Geneva, Switzerland: International Labour Organization.

Konz, S. and S. Johnson. 2004. *Work Design-Industrial Ergonomics*. Sixth Edition. Scottsdale, AZ: Holcomb Hathaway.

Kroemer, K. H. E., Kroemer, H. B. and K. E. Kroemer-Elbert. 1994. *Ergonomics—How to Design for Ease and Efficiency*. Englewood Cliffs, NJ: Prentice Hall.

Leiden, K., Laughery, K. R., Keller, J., French, J., Warwick, W. and S. D. Wood, 2001. *A Review of Human Performance Models for the Prediction of Human Error*. Boulder, CO: Micro Analysis & Design, Inc.

Maynard, H., Stegemerten, G. and J. Schwab. 1948. *Methods-Time Measurement*. New York: McGraw-Hill.

McDowell, M. A., Fryor, C. D., Ogden, C. L. and K. M. Flegal. 2008. *Anthropometric Reference Data for Children and Adults: United States 2003–2006*. National Health Statistics Reports, Vol. 10. Website: http://www.cdc.gov/nchs/data/nhsr/nhsr010.pdf (accessed may 3, 2013).

Mehler, B. R., Reimer, B., Coughlin, J. F. and J. A. Dusek. 2009. Impact of Incremental Increases in Cognitive Workload on Physiological Arousal and Performance in Young Adult Drivers. *Transportation Research Record, Journal of the Transportation Research Board*, 2138: 6–12.

Meshkati, N., Hancock, P. and M. Rahimi. 1992. Techniques in Mental Workload Assessment. In *Evaluation of Human Work, A Practical Ergonomics Methodology* (Eds.) Wilson, J. and E. Corlett, 605–627. London, England: Taylor & Francis.

NASA circa. 1988. NASA Task Load Index (v.10)- Paper and Pencil Package. Human Performance Research Group, NASA Ames Research Center, Moffett Field, California. Website: http://humansystems.arc.nasa.gov/groups/TLX/downloads/TLX_pappen_manual.pdf (accessed July 2, 2012).

Olsson, S. and P. C. Burns. 2000. *Measuring Driver Visual Distraction with a Peripheral Detection Task*. Linköping, Sweden: Department of Education and Psychology, Linköping University.

Pheasant, S. and C. M. Haslegrave. 2006. *Bodyspace: Anthropometry, Ergonomics and the Design of Work*. Third Edition. London, England: CRC Press, Taylor & Francis Group.

Reed, M. P., Parkinson, M. B. and D. B. Chaffin. 2003. A New Approach to Modeling Driver Reach. Technical Paper 2003-01-0587. *SAE Transactions: Journal of Passenger Cars—Mechanical Systems*, 112: 709–718.

Reed, M. P., Roe, R. W. and L. W. Schneider. 1999. Design and Development of the ASPECT Manikin. Technical Paper 990963. *SAE Transactions: Journal of Passenger Cars*, 108. (Also available as SAE Paper no. 1999-01-0963 in website: http://mreed.umtri.umich.edu/mreed/pubs/Reed_1999-01-0963.pdf; accessed: October 8, 2022).

Reid, G. B. and T. E. Nygren. 1988. The Subjective Workload Assessment Technique: A Scaling Procedure for Measuring Mental Workload. In *Human Mental Workload* (Eds.) Hancock, P. A. and N. Meshkati, 139–183. Amsterdam, the Netherlands: North Holland.

Sanders, M. S. 1983. U.S. *Truck Driver Anthropometric and Truck Work Space Data Survey*. Report No. CRG/TR-83/002. Canyon Research Group, Inc., West Lake Village, California.

Sanders, M. S. and E. J. McCormick. 1993. *Human Factors in Engineering and Design*. Seventh Edition. New York: McGraw-Hill.

Seiffert, U. and L. Wech. 2003. *Automotive Safety Handbook*. Warrendale, PA: SAE International.

Society of Automotive Engineers, Inc. 2009. *SAE Handbook*. Warrendale, PA: Society of Automotive Engineers, Inc.

Van Cott, H. P. and R. G. Kinkade (Eds.). 1972. *Human Engineering Guide to Equipment Design*. New York: McGraw-Hill, Inc./U.S. Government Printing Press.

Verwey, W. B. and D. M. Zaidel. 2000. Predicting Drowsiness Accidents from Personal Attributes, Eyeblinks, and Ongoing Driving Behaviour. *Personality and Individual Differences*, 28(1): 123–142.

Wickens, C., Gordon, S. E. and Y. Liu. 1998. *An Introduction to Human Factors Engineering*. Upper Saddle River, NJ: Pearson Prentice Hall (Addison Wesley Longman, Inc.).

Woodson, W. E. 1992. *Human Engineering Design Handbook*. Second Edition. New York: McGraw-Hill Book Co.

19 Safety Engineering Tools

INTRODUCTION

Safety engineers use a variety of tools to solve various safety problems ranging from identifying hazards inherent in a product to analyzing accidents occurring during the product use and monitoring safety performance during the product life cycle. This chapter presents many of the widely used tools. The chapter begins with various hazard identification and risk-reduction tools such as hazard analysis, failure modes and effects analysis (FMEA), and fault tree analysis (FTA). These tools help in reducing potential occurrences of accidents during product uses. After accidents occur, they are investigated, and the collected data are entered into accident databases. The data in the accident databases can be sorted, tabulated, and analyzed to understand occurrence rates and causal factors of certain types of accidents. The causes are further studied to determine countermeasures to reduce future accidents. Cost–benefit analyses help in selecting one or more countermeasures. The effectiveness of any of the countermeasures can be analyzed by maintaining control charts on occurrences of accidents and determining whether the accident rates are reduced.

This chapter also presents some approaches for improving product reliability by studying the effects of series, parallel, and hybrid product configurations. Additional material is also included in the reliability engineer's role in product development and tasks in relation to the systems engineering process.

HAZARD IDENTIFICATION AND RISK REDUCTION TOOLS

HAZARD ANALYSIS

A hazard can be defined as an unsafe situation that can cause injuries and losses if allowed to remain without the removal of underlying causes of its occurrence. The occurrence probability of an unsafe situation can be generally estimated from historic information or by interviewing experts who have experienced or studied similar situations. Hazard analysis can be defined as a systematic method used by safety engineers to study the causation of an unsafe situation by identifying its characteristics such as unsafe conditions, unsafe acts or actions committed by people, or combinations of unsafe acts and conditions and determining their probabilities of occurrence. Many safety researchers such as Heinrich et al. (1980), Tarrants (1980), and Brown (1976) have suggested different formats for documentation of hazard analysis with differing levels of details.

GENERAL HAZARD ANALYSIS

Brown (1976) suggested a simple hazard analysis involving a general hazard analysis card. This card can be filled out by anyone involved or familiar with a process or a situation that can involve a hazard. The person filling out the card is required to provide the following information: (1) hazard description, (2) location of the hazard in the workplace, (3) severity level (nuisance, marginal, critical, or catastrophic), (4) probability of accident (unlikely, probable, considerable, or imminent), and (5) possible costs due to an accident resulting from the hazard (prohibitive, extreme, significant, or nominal). The cards are then reviewed by a safety professional who categorizes each card into the following actions: defer, need more analysis, or immediate action needed.

Hazard analysis thus involves an overall look at the system under consideration to identify the safety problems that require more detailed analysis. The sources of inputs for general hazard analysis include the following: (1) obvious hazards due to the nature of the process (e.g., likelihood of a fire in processes involving flammable fluids and of electrocution hazards with electrical systems involving high voltages), (2) hard recordkeeping data required by law or organizational policies (e.g., accident reports and injury records, violation of safety practices), and (3) other investigations that stress the cooperation of the line and staff personnel.

DETAILED HAZARD ANALYSIS

A detailed hazard analysis generally involves breaking down a task or a process into a series of steps (or subtasks) that may contain one or more hazards. A detailed hazard analysis is documented by creating a matrix. The steps are listed as rows and the hazards are listed as columns of the matrix. The matrix is usually created by a safety professional. The cells of the matrix where a particular hazard element is present in each step are identified by entering "X" marks. The safety professionals evaluate each X-marked cell to determine one or more possible countermeasures that could be applied to avoid the hazard. The effect of each countermeasure on all possible hazards in all rows (steps) are rated in terms of the following: (1) R_i = the hazard can be reduced by the countermeasure to a degree indicated by the subscript i, (2) E = the hazard in the step can be eliminated, (3) I = the hazard would increase by the countermeasure, or (4) X = the hazard still exists in the step. The aforementioned ratings are placed in the corresponding cells to understand the effect of each countermeasure. The analysis is usually performed by applying combinations of different countermeasures to select the best set of countermeasures. In addition, a cost–benefit analysis can be performed on each alternative (or countermeasure) to determine the most cost-effective alternative that produces the highest benefits-to-costs ratio. The benefits can be defined as the reduction in costs due to the elimination or reduction of accidents. The costs are defined by adding all costs involved in the implementation of the countermeasures.

METHODS SAFETY ANALYSIS

A methods safety analysis is also a hazard analysis technique that involves breaking down a task or a process into a series of steps. Application of this technique involves a safety professional to analyze each step of the process (usually by discussing each step with the team members involved in the process) to identify possible hazards in

Safety Engineering Tools 381

each step and the causes of the hazards and propose improvements in the steps to eliminate the hazards. The improvements are incorporated in the process, and a new improved method is developed. Thereafter, the team members are trained to follow the newly developed process. This technique is referred to as methods safety analysis because it is similar to the traditional methods analysis (or operations analysis) technique used by industrial engineers. In methods analysis, an industrial engineer analyzes the method involved in performing a task or an operation by systematically breaking down each step (or work motion) and then improving each step by asking questions such as can the step be eliminated, combined, or made more efficient by reducing unnecessary motions. In methods safety analysis, the emphasis is on improving safety. Thus, methods safety analysis is primarily concerned with investigating hazards involved in the methods of an operation. The fundamentals of methods safety analysis are the same as those of methods analysis. The steps in methods safety analysis are as follows: (1) breaking down the job or operation into its elemental tasks, (1) listing them in proper order, (3) examining them critically to find opportunities to eliminate the hazards, and (4) developing a new safer and more efficient method.

The advantages of methods safety analysis are as follows:

1. It maps out all the details of an operation so that they can be studied and restudied.
2. It is quick, simple, factual, and objective.
3. It permits the comparison of current and proposed methods.
4. It presents a picture of the effect on production by the safety improvements.
5. It aids management in reviewing the benefits.
6. It permits the engineer to improve productivity.
7. It facilitates analysis of the safety potential of an operation before an accident occurs, that is, it is proactive.
8. It assists in thorough investigations of methods or operations involved in the accident where the causes are obscured.

CHECKLISTS TO UNCOVER HAZARDS

Determining causes for hazards and developing countermeasures require specialized safety professionals with experience and knowledge in hazard recognition and accident causation theories (see Chapter 12). The hazard recognition abilities can be enhanced by the creation of hazards checklists. For example, Hammer (1980) provided a comprehensive set of checklists to analyze product designs. Hammer's checklists included the following categories of hazards: (1) acceleration (e.g., does the product contain any spring-loaded or cantilevered object or device that could be affected by acceleration or deceleration?), (2) chemical reaction (e.g., is there any material present that will react with the oxygen in the air to produce a toxic, corrosive, or flammable material or one that will ignite spontaneously?), (3) electrical shock (e.g., are the voltages and amperages high enough to cause arcing or sparking, which can ignite a flammable gas or combustible material?), (4) explosives and explosions, (5) flammability and fires, (6) heat and temperatures, (7) mechanical items (e.g., sharp points and edges), (8) pressure (e.g., high-pressure lines and vessels), (9) radiation, (10) toxic materials, (11) vibration and noise, and (12) other miscellaneous hazards. Product designers should construct specialized checklists containing issues related to hazards inherent in their product and use the checklists to identify all possible hazards.

RISK ANALYSIS

Risk analysis is a methodology used to compute the level of risk in a given product, process, or situation. The level of risk is generally expressed in terms of a risk priority number called the RPN. The computation of an RPN is similar to the approach used in FMEA (see the following section and Chapter 15), which consists of the multiplication of three rating values, namely, severity rating (S), occurrence rating (O), and hazard detection rating (D). Thus, RPN = $S \times O \times D$. Many models of RPN with differing scales are used in the industry and also by government agencies for risk analysis.

Figure 19.1 presents an example of a nomograph (also called a nomogram) used to illustrate risk assessment. Three situations S1, S2, and S3 are shown by joining points on the three scales through a tie line between the first two scales. The first situation S1 involves a situation with moderate potential for injury, highly probable hazard occurrence, and improbable hazard recognition resulting in extremely high risk. The second situation S2 involves minor potential for injury, probable hazard occurrence, and highly improbable hazard recognition resulting in remote to virtually nonexistent risk. The third situation S3 involves minor potential of injury, highly probable hazard occurrence, and improbable hazard recognition resulting in extremely low risk.

The problems with the RPN computation methodology are as follows: (1) the selection of values of the ratings is based on subjective judgments of the analyst, (2) lack of objective data to determine values of the ratings, and (3) assumptions are made about consumer or user behavior. Thus, RPN-based risk analysis models are not precise. But they can be used as guides along with other sources of information such as recommendations from multiple experts and discussions between decision makers and experts.

SYSTEMS SAFETY ANALYSIS TOOLS

FAILURE MODES AND EFFECTS ANALYSIS

The failure modes and effects analysis (FMEA) is a commonly recognized tool in the industry. Its applications to study a process (P for process) or a product design (D for design) are commonly referred to as PFEMA and DFEMA, respectively. FMEA is a proactive and qualitative tool used by quality, safety, and product/process engineers to improve the reliability of products or processes (by reducing or eliminating failures, thus improving quality, i.e., customer satisfaction). It seeks to identify (1) possible failure modes and mechanisms, (2) the effects or consequences that the failures may have on performance of the product or process, (3) methods of detecting the

Safety Engineering Tools

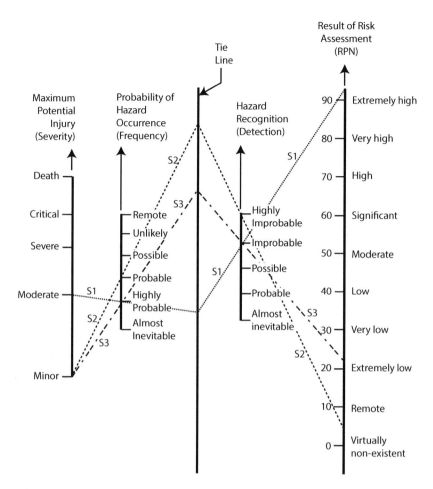

FIGURE 19.1 Example of a nomograph for risk assessment. (Redrawn from: Risk & Policy Limited. 2006. "Establishing a Comparative Inventory of Approaches and Methods Used by Enforcement Authorities for the Assessment of the Safety of Consumer Products Covered by Directive 2001/95/EC on General Product Safety and Identification of Best Practices." Final Report, February 2006 prepared for DG SANCO, European Commission.)

identified failure modes, and (4) possible means for prevention. It is very effective when performed early in product/process development and by experienced multi-functional teams.

FMEA encourages the systematic evaluation of a product or process at the specified levels of system complexity (defined by RPN). It postulates single-point failures, identification of possible failure mechanisms, examination of associated effects, the likelihood of occurrence, and preventive measures. It also creates systematic documentation (in a tabular format) of potential product or process nonperformance. The FMEA technique is described in detail in Chapter 15. An example of FMEA is provided in Chapter 15, Table 15.7.

Products designed using DFEMA principles should have higher quality and reliability than those developed using traditional design methods (e.g., design reviews). DFEMA also ensures that the transition from the design phase to the production phase is as smooth and rapid as possible.

Fault Tree Analysis

Purpose

The purpose of a fault tree is to fully describe the occurrence of an event placed on top of the fault tree diagram. The event is called the "head event." Another purpose of the fault tree is to determine the probability of occurrence of the head event.

Description

In a fault tree, all possible combinations of events that can cause the head event to occur are described as branches underneath the head event. The relationships between all events can be described in terms of a series of Boolean algebraic equations. The equations can be used to analyze occurrences of any of the events in the tree. The Boolean expressions can be used to determine the probability of occurrence of any of the events by using the probabilities of terminating events under each of the branches.

Fault tree is a logical tree diagram showing how an event shown at the top of the tree (head event) can occur. It describes all possible events that can lead to the occurrence of the head event. A fault tree has only one head event. It is generally an "undesired" event. Various events that can lead to an event are shown by using logical operators called "gates." "AND" and "OR" gates (and other gates) are used to show how an event described on the top of the gate can occur due to the occurrences of events shown below the gate.

Application of Boolean Algebra

Boolean algebra can be used to provide the logical description of a fault tree and its branches.

It can also help in performing computational analyses for large, complex fault trees using computers. A given Boolean expression defining the same head (output) event can be rewritten in various equivalent expressions for (1) restricting the tree, (2) equation simplifications of the tree, and (3) determination of mutually exclusive branches or events. The probability of a head event can be computed from its Boolean expression defining the event. The fault trees of product failures can also be constructed in configurations involving different product design alternatives or countermeasures, and comparisons of quantitative evaluations of the alternative configurations of fault trees can be made to reduce the occurrences of undesired events and to determine the effectiveness of the countermeasures. The basics of Boolean algebra are described below.

Let us assume that $X, Y, Z, A, B, C, D, \ldots$ are Boolean algebra variables (i.e., logical variables). Each Boolean variable defines an event. Variable X can be defined to convey that the event X exists, that is, X is true. Then $\bar{X}$ (i.e., X bar) denotes that the event X does not exist, or X is not true, that is, X is false.

Safety Engineering Tools

The primary logical operators are as follows: (1) + (plus), which denotes OR gate situation, and (2) · (dot), which denotes AND gate situation.

Therefore, Boolean equations can be written as follows:

$$X + X = X$$
$$\bar{X} + X = 1$$
$$X \cdot X = X$$
$$X \cdot \bar{X} = 0$$

where 0 = null (no event exists) and 1 = all events coexist (i.e., the universe).

Figure 19.2 illustrates AND, OR, and inhibit gates. All the events above, below, and within the gates are defined by Boolean variables. Figure 19.2 thus illustrates the following:

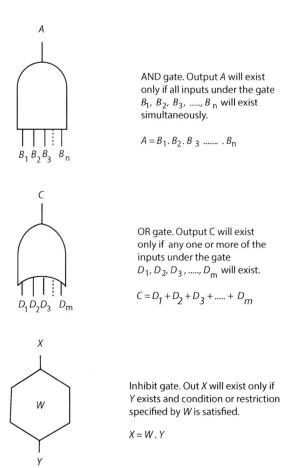

AND gate. Output A will exist only if all inputs under the gate $B_1, B_2, B_3, ..., B_n$ will exist simultaneously.

$$A = B_1 \cdot B_2 \cdot B_3 .... \cdot B_n$$

OR gate. Output C will exist only if any one or more of the inputs under the gate $D_1, D_2, D_3, ..., D_m$ will exist.

$$C = D_1 + D_2 + D_3 + ... + D_m$$

Inhibit gate. Out X will exist only if Y exists and condition or restriction specified by W is satisfied.

$$X = W \cdot Y$$

FIGURE 19.2 AND, OR, and inhibit gates used in fault trees.

1. The AND gate illustrates that $A = B_1 \cdot B_2 \cdot B_3 \cdots B_n$.
2. The OR gate illustrates that $C = D_1 + D_2 + D_3 + \cdots + D_m$.
3. The inhibit gate illustrates that $X = W \cdot Y$.

The events denoted by Boolean variables are shown in the fault tree diagram by using different shaped boxes. The notations used for different types of events are shown in Figure 19.3. The description of each event is written inside the box in words or in terms of its Boolean variable (such as G, F, E, and H in Figure 19.3). The output of any event box is on the top side, and the input is on the bottom side.

AND Gate

Assume that events A and B are required to be present simultaneously to generate event X. Then, the Boolean (logical) expression can be written as follows: $X = A \cdot B$.

The probability of $X = P(X)$ can be defined as follows:

$P(X) = P(A) \cdot P(B \mid A)$ (if A and B are not independent events)

$$= P(B) \cdot P(A \mid B)$$

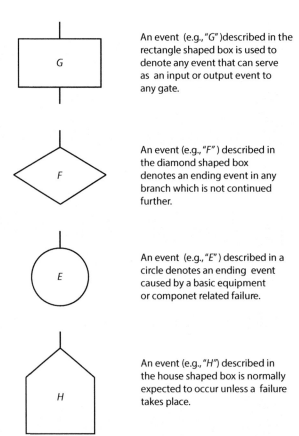

An event (e.g., "G") described in the rectangle shaped box is used to denote any event that can serve as an input or output event to any gate.

An event (e.g., "F") described in the diamond shaped box denotes an ending event in any branch which is not continued further.

An event (e.g., "E") described in a circle denotes an ending event caused by a basic equipment or componet related failure.

An event (e.g., "H") described in the house shaped box is normally expected to occur unless a failure takes place.

FIGURE 19.3 Types of events shown in different shaped boxes in fault trees.

Otherwise,
$P(X) = P(A) \cdot P(B)$ (if A and B are independent events)

OR Gate

Assume that an event A or B is required to generate the event Y. Then, the Boolean (logical) expression can be written as follows: $Y = A + B$.

The probability of $Y = P(Y)$ can be defined as follows:

$P(Y) = P(A) + P(B) - P(AB)$ (if A and B are not mutually exclusive events)
$P(Y) = 1 - P(\bar{A}) \cdot P(\bar{B})$
Otherwise,
$P(Y) = P(A) + P(B)$ (if A and B are mutually exclusive events)

If $P(AB) = 0$, events A and B cannot coexist. Thus, A and B are mutually exclusive. Therefore, $P(A + B) = P(A) + P(B)$. This mutually exclusive reasoning can be applied and generalized for any n mutually exclusive events (A_1 to A_n), and the probability of occurrence of any one or more of the events (A_1 to A_n) can be written as follows:

$$P(A_1 + A_2 + A_3 + \cdots + A_n) = \sum_{i=1}^{n} P(A_i)$$

The examples in the following subsections will help illustrate the independent and mutually exclusive events.

An Example: Two-Engine Aircraft

Let us assume that a twin-engine airplane has two identical engines. The probability that an engine will work in the flight is 0.99. Compute the probabilities that at least one engine will work in the flight and that both the engines will work in the flight. Assume that the two engines are independent (i.e., operation of one engine does not affect the other engine).

The first part of the problem can be solved by considering an OR gate where at least engine A or B will work. The second problem is solved by considering an AND gate (see Figure 19.4).

Fault Tree Development Rules

Brown (1976) has provided three rules for fault tree development (discussed in the following subsections). These rules allow the systematic development of a fault tree by using OR gates first (at the top and every subsequent event) and organizing conditional events to reduce errors in computations of the probabilities.

Rule 1: Fault Tree Development Rule

For any event that requires further development, analyze all possible OR input events first prior to the analysis for AND inputs. This will hold for the head event or any subsequent event that needs further development.

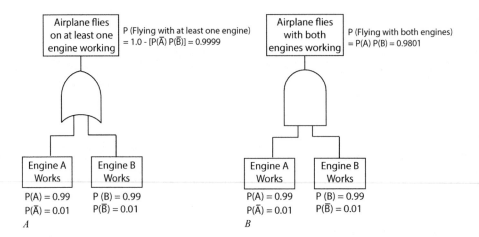

FIGURE 19.4 Illustration of the application of OR and gates.

This rule is based on data collection completeness (think about all reasons for the occurrence of each event first) and probability considerations leading to an event. An OR gate should be used just below the head event.

Rule 2: OR-Gate Event Rules

The input events to an OR gate must be defined such that together they constitute all possible ways that the output event can occur. In addition, each event must include the occurrence of the output event.

This rule is based on the inclusion of all possible causes for completeness. Otherwise, the probability estimate will be underestimated. All events placed under an OR gate must collectively exhaust all possible events leading to the event above the OR gate. Thus, use an event (in a diamond-shaped box; Figure 19.3, the second event from top) to include events occurring due to "all other" reasons (allows for future inclusions or expansions). (An example of this situation is shown in Figure 19.5; see event G.) If all the events leading to an OR gate are mutually exclusive (i.e., the events do not overlap), then their probabilities can be simply added.

Rule 3: AND-Gate Event Rule

The input events to an AND gate must be defined such that the second event is conditioned on the first, the third event is conditioned on the first and the second, and so on, and the last is conditioned on all others. In addition, at least one of the events must include the occurrence of the output event.

This rule is for taking into account non-independent events under the AND gate. It also helps to reduce uncertainty related to the independence of events under the AND gate and increasing the accuracy of calculating the occurrence probability of the output event of the AND gate.

Fault Tree Example: Printer Fails to Print

Figure 19.5 illustrates a fault tree drawn for the following head event: printer fails to print a report. The head event is labeled as A. The head event can occur due to any of

Safety Engineering Tools

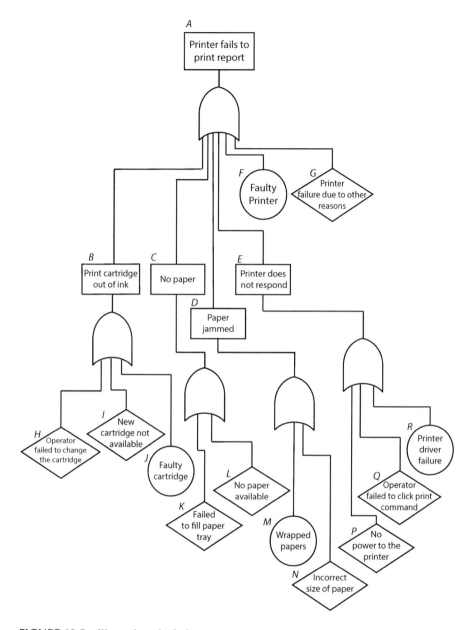

FIGURE 19.5 Illustration of a fault tree.

the events *B*, *C*, *D*, *E*, *F*, and *G* shown below the OR gate under the head event (see rule 1). Note that event *G* is defined to satisfy rule 2.

Thus, using Boolean algebra, *A* can be defined as follows:

$$A = B + C + D + E + F + G$$

Similarly, other events can be defined as follows:

$$B = H + I + J$$

$$C = L + K$$

$$D = M + N$$

$$E = P + Q + R$$

The probability of occurrence of the head event can be computed by using the following equations:

$$P(A) = 1.0 - [P(\bar{B}) \cdot P(\bar{C}) \cdot P(\bar{D}) \cdot P(\bar{E}) \cdot P(\bar{F}) \cdot P/(\bar{G})]$$
$$P(B) = 1.0 - [P(\bar{H}) \cdot P(\bar{I}) \cdot P(\bar{J})]$$
$$P(C) = 1.0 - [P(\bar{L}) \cdot P(\bar{K})]$$
$$P(D) = 1.0 - [P(\bar{M}) \cdot P(\bar{N})]$$
$$P(E) = 1.0 - [P(\bar{P}) \cdot P(\bar{Q}) \cdot P(\bar{R})]$$

The probability of the head event can be computed by using the probabilities of the ending events H, I, J, K, L, M, N, P, Q, R, and G. It is assumed that the numeric values of the probabilities of each of the ending events are given (from estimates of experts or assumptions).

Advantages of Fault Tree Analysis

The advantages of using FTA are as follows: (1) it allows better understanding of the system through causal relationships of events, (2) it helps in communicating issues among design team members, (3) it provides probability estimates for occurrences of all events for better decision making, and (4) it points out critical paths (branches in the fault tree with high occurrence probabilities of undesired events) and improvements in branches (or events in the branches) that can reduce the probability of occurrence of the undesired event.

ACCIDENT DATA ANALYSIS TOOLS

PURPOSE OF ACCIDENT DATA COLLECTION

The purpose of accident data collection is to get all the relevant facts about one or more accidents such as how, when, and what happened, and who caused it to understand the factors and causes associated with the accidents. The ultimate goal is to reduce such accidents by developing accident prevention programs and to monitor the effectiveness of implemented countermeasures for evaluating the state of safety. The collected accident data can also be compared with the accident data from past time intervals and the accident data available from other organizations or situations.

Flow of Accident Data Collection

Accident data originate from the accident reports that are generally prepared by the supervisor, the police officer, the medical staff, or the persons involved in the accident. The accident reports are collected and entered in summary sheets and databases by data entry personnel. A completed accident report form contains information such as date and time of the accident, persons involved, location of the accident, description of injuries, equipment involved, description of the loss, the situation and the environment present at the time of the accident, accounts of how the accident occurred, diagram/map with the locations of various objects, and movements of the equipment/vehicles. Some accidents are investigated at a greater detail by accident investigators or multidisciplinary teams (involving researchers, engineers, medical personnel, lawyers, and other experts) and the data are entered into databases. Their report is called the "accident-analysis report."

Accident researchers and statisticians analyze the data by sorting, summarizing in tabular and other formats (e.g., pie charts and tree diagrams), and conducting statistical analyses to determine significant effects or differences due to certain variables. For example, the accidents can be sorted by conditions (e.g., day vs. night), situations (e.g., normal vs. emergency), before versus after the implementation of countermeasures, equipment used (e.g., make, model, and brand of equipment involved in the accidents), operator characteristics (e.g., young vs. old), and so on.

Accident Data Reporting Thresholds

All accidents have some type of reporting threshold, that is, when an accident report must be filed. The thresholds are typically based on the level of severity of the injury or property damage over a certain amount. Unfortunately, the thresholds vary between different organizations. For example, police-recordable accidents are defined based on the occurrence of a personal injury or property damage over $100–$200. The workplace-recordable accidents defined by the Occupational Safety and Health Administration are more severe than just first-aid injuries and involve loss of time or a doctor's judgment. Thus, what is reported can vary based on the accident location and the judgment of the person treating the injured or filing the report.

Accident Investigations

The primary reason for investigating an accident is not to identify a scapegoat but to determine the causes of the accident by gathering factual information about the details that led to the accident.

Generally, an accident report (supervisor or police reported) does not answer the why question. (It answers who, what, where, and when related to the accident.) Thus, the purpose of accident investigation is to understand why the accident occurred and how it occurred so that future accidents of this type can be prevented.

Ideally, accidents should be investigated as soon as all emergency procedures have been accomplished. The accidents can be investigated by the supervisor, an investigative team, or an outside specialist to collect facts that led to the causation of the accident. During the investigation, the accident scene is isolated to maintain the conditions that existed at the time of the accident and all the evidence (e.g., photographs and/or video of the accident scene and interviews of witnesses) are gathered. Later, simulations may be created to understand and determine the series of events in the accident and, finally, a report is prepared.

ACCIDENT DATA SOURCES AND USERS

Accident reports can be categorized by their sources. Typical sources of accident reports are as follows: private organizations (e.g., companies, manufacturers, service providers, insurance companies, and health-care providers), public institutions (e.g., universities), law enforcement organizations (e.g., police in city, county, or state), state accident databases, and databases created by federal agencies.

The U.S. Consumer Products Safety Commission (CPSC) maintains a national-level accident database called the National Electronic Injury Surveillance System (NEISS). It represents a national probability sample of hospitals in the United States and its territories. Patient information is collected from each NEISS hospital for every emergency visit involving an injury associated with consumer products (CPSC, 2012). The National Highway Traffic Safety Administration (NHTSA) (U.S. Department of Transportation) created the Fatality Analysis Reporting System (FARS) in 1975 (NHTSA, 2012). This accident data system was conceived, designed, and developed by the National Center for Statistics and Analysis to assist the traffic safety community in identifying traffic safety problems and evaluating both motor vehicle safety standards and highway safety initiatives.

Accident data files store data on a number of parameters obtained from the accident reports and accident investigations. Typically, the parameters include the following: accident case number, location of the accident (e.g. city, county, or state), date and time of the accident, characteristics of each injured person (e.g., age, gender, driver's license number, and address), nature of injury (e.g., severity level and body parts injured), details of equipment involved in the accident (e.g., vehicles, model number, manufacturer, and vehicle identification number), safety equipment used (e.g. seat belts, hard hats, and guards), narrative of how the accident occurred, nature of property damage, and so forth.

The primary purposes of accident data analyses are as follows: (1) to estimate the magnitude of the accident problem, frequencies, and rates by accident types or categories, and (2) to understand causes to develop prevention countermeasures to reduce accidents. The accident data are used by many decision makers such as company management personnel (e.g., to determine the state of safety in their products, operations, costs, and countermeasures), government officials (e.g., for standards development, recalls, and public notices), safety researchers (e.g., to analyze, evaluate, and propose countermeasures), lawyers and legal professionals (e.g., to establish evidence, link accidents to design, or manufacturing defects of products, errors, and negligence), and insurance company researchers (e.g., to determine incident rates, severity, costs, and insurance premiums).

SAFETY PERFORMANCE MONITORING, EVALUATION, AND CONTROL

INTERVIEW AND OBSERVATIONAL TECHNIQUES FOR NON-ACCIDENT MEASUREMENT OF SAFETY PERFORMANCE

In many situations requiring the measurement of safety performance, non-accident events (e.g., unsafe acts, unsafe conditions, and near accidents) can be measured instead of waiting for accidents to occur. Critical incident technique (CIT) and behavioral sampling are two commonly used techniques in the non-accident-based safety performance measurement area. CIT involves interviewing a preselected group of individuals to recall details about predefined critical incidents (e.g., potential or actual accidents). Behavioral sampling involves the traditional "occurrence sampling" or "work sampling" application by observing and recording unsafe acts and unsafe conditions in a workplace. The collected information is used to monitor safety performance.

CRITICAL INCIDENT TECHNIQUE

The critical incident technique (CIT) is a method of identifying errors and unsafe conditions that contribute to both potential and actual injurious accidents within a given population by means of a stratified random sample of participant-observer selected within the population. Quantitative information (e.g., frequencies and proportion of unsafe behaviors) obtained from the questions asked during the interviews can be used to measure safety performance (Tarrants, 1980).

One of the oldest and famous applications of CIT was by Fitts and Jones (1961a,b) to study pilot errors to help design better aircraft controls and displays. Soon after World War II, the air force launched a systematic study of errors made by pilots in situations where accidents and near accidents occurred. The pilots were asked to recall incidents in which they almost lost an airplane or witnessed a copilot make an error in reading aircraft displays or operating controls. From the analyses of the data gathered from these critical incidents, Fitts and Jones found that practically all the pilots, regardless of experience or skill, reported making errors in using cockpit controls and instruments. They also concluded that it should be possible to eliminate or reduce most of these pilot errors by designing equipment in accordance with human requirements. Similarly, driver errors in using displays and controls can be reduced if they are designed in accordance with the human engineering criteria.

The steps involved in conducting CIT are as follows:

1. Define the objectives of the study, for example, to solve a safety problem by identifying the causes of a certain type of accidents and/or measuring the level of safety (frequency or rates of unsafe occurrences) in a given situation.
2. Determine sample characteristics of participant-observer (their locations, representation).

3. Develop questions to ask and obtain (1) descriptions of situations related to the errors or unsafe conditions, (2) descriptions of the errors and unsafe conditions, (3) wordings of probe questions to assist in recalling incidents related to each particular type of error and unsafe condition.
4. Carefully select a representative sample of participant-observer based on valid stratification criteria (e.g., stratification by shift, locations of product uses).
5. Conduct preliminary interviews (to facilitate recall) with each participant-observer (about 24 hours before the data collection interview) to inform them about the purpose of the study, explain the type of information desired, and explain the procedure.
6. Conduct interview with each participant-respondent separately for about 45 minutes. Record the interview for subsequent data retrieval.
7. Replay the audio recording and extract details of each incident (when, where, how it occurred, who, and possible cause), classify the mentioned unsafe acts and conditions, and keep records on their counts (frequency) and data.
8. Prepare a table of frequencies of unsafe acts and conditions.
9. Conduct statistical analyses: (1) to determine the distribution of unsafe acts and conditions (e.g., by types, occurrence time, and locations), (2) to compare with past data, and (3) to maintain control charts of frequency (C or U charts) and rates or proportion of unsafe acts.
10. Document and present conclusions to management.

During interviews (in step 6), the interviewer first starts with open-ended questions. Some examples of open-ended questions are as follows: (1) Would you describe as fully as you can the last time you saw an unsafe human error or unsafe condition in your department? and (2) Would you please describe as fully as you can other unsafe human errors or conditions in your department during the past 2 years? Next, the interviewer probes systematically for other incidents that they may have forgotten. Examples of questions used here are as follows: (1) Have you ever seen anyone cleaning a machine or removing a part while the machine was in motion? (2) Have you ever seen anyone speeding or any other improper handling of a power truck? (3) Have you ever seen anyone "beating" or "cheating" a machine guard?

BEHAVIORAL SAMPLING

Behavioral sampling is really the occurrence sampling (also known as work sampling) technique commonly used by industrial engineers to obtain information on the operational status (e.g., working or not working) of human operators or machines/equipment in workplaces (Konz and Johnson, 2004). Occurrence sampling is a process of observing at discrete random points (times) to obtain the estimate of occurrence proportions of certain preselected events. A trained observer observes a process at the preselected random observation times and classifies each observation into the preselected categories.

Safety Engineering Tools

The steps involved in conducting a behavior sampling study are as follows:

1. Define the objectives of the study, for example, to determine the frequency and proportion or percentage of time spent by operators in working in an unsafe manner and/or under unsafe conditions.
2. Set the time period of the study.
3. Define observable elements, for example, safe and unsafe acts and conditions or situations.
4. Determine preliminary estimates of the proportion of time spent in safe and unsafe elements (usually based on about 100–200 samples).
5. Determine the required number of observations.
   The required number of observations (n) can be computed as follows:

   $$n = \frac{z^2 p(1-p)}{(A^2)}$$

   where
   p = Initial (or preliminary) estimate of the proportion of time spent in unsafe conditions (in decimal, e.g., $p = 0.05$ for 5%)
   z = Number of standard deviations for the confidence level desired ($z = 1.96$ for 95% confidence level)
   $A = sp$ = absolute accuracy desired on the value of proportion of unsafe conditions
   s = relative accuracy desired (in decimal, e.g., 0.1 for 10%)

   The number of observations (n) obtained from the aforementioned formula represents the number of observations required to obtain an estimate of p within the accuracy of $\pm A$ for the given confidence probability (defined by the input value of z).
6. Establish observation intervals and observation times to make n observations.
7. Design an observation record sheet (e.g., tally sheets with predetermined randomized times and observation routes).
8. Conduct orientation of persons undergoing the study (e.g., keep people informed to get buy-in and train observers).
9. Conduct observations, classify elements, and record them.
10. Evaluate results (determine the proportion of time spent in different types of observable elements).
11. Maintain control charts on proportion of time spent in unsafe conditions (P-chart) or number of unsafe acts and unsafe conditions (C-chart) observed in a selected observation time interval.

To ensure that the data gathered during the observations are unbiased and representative of the situation in the observation area, many precautions must be exercised due to problems such as (1) the operators may change their behavior when they see an observer in their work area; (2) observer biases due to reasons such as lack of training and falsification; (3) inability of the observers to observe (e.g., due to obstructed or hidden workplaces or unlighted observation areas); (4) observation plan fails to include stratification based on influencing factors such as time of the day, type of work or department, and type of operations; and (5) certain behaviors may occur regularly and happen to take place (or not take place) in synch with the observations.

CONTROL CHARTS

Control charts can be maintained based on observations in behavior sampling studies or accident data gathered from available databases. Control charts provide visual as well as statistical information to monitor the observed processes over time. Some examples of the types of controls charts that can be used are as follows: (1) C-charts for accident frequencies (or unsafe behavior), (2) P-charts for the proportion of accidents of a certain type (or type of observed unsafe behavior), and (3) X-bar charts for lost workdays (accident severity). Control charts can be used when sufficient data are available. Generally, only large organizations can maintain and use charts. Additional information on control charts is provided in Chapter 17 and Appendix 8.

BEFORE VERSUS AFTER STUDIES

Safety performance needs to be measured before the implementation of a safety countermeasure and after the countermeasure is implemented to compare the data and determine if a statistically significant difference exists between before and after situations. If the statistical analysis confirms that a significant difference in safety performance exists, then the magnitude of the difference in safety performance can also be estimated. Based on the measured value of the difference, additional actions can be planned. The before versus after study must be carefully planned to ensure that the information gathered is valid and that sufficient sample sizes are obtained for proper statistical inference.

COST–BENEFIT ANALYSIS

The costs associated with the implementation of different countermeasures and the possible benefits that can be realized must be estimated to determine ratios of benefits-to-costs for different countermeasures (alternatives). The ratios can be used to determine the most cost-effective alternative (see the study by Brown [1976] for more information). Chapters 3, 20, and 24 provide additional information on decision-making and examples on cost–benefit analyses. The effect of the selected alternatives also needs to be evaluated by considering possible trade-offs between different product attributes affected by the selection of the countermeasure with the highest benefits-to-costs ratio.

RELIABILITY ANALYSES

DEFINITIONS OF RELIABILITY AND MAINTAINABILITY

Reliability can be defined as the probability that a system, subsystem, or component will perform successfully for a specified amount of time or usage or for a given mission, when operated under specified conditions.

Safety Engineering Tools

The reliability of a product A can be defined as the ratio of the number of successful occurrences (or operations) to the total number of attempts or trials made to use the product under the same specified conditions. The reliability of A is defined as follows:

$$R(A) = \left[\frac{\text{(Number of successes)}}{\text{(Total number of attempts)}} \right]$$

The reliability engineers assigned to the product programs are also asked to perform maintainability analyses. Maintainability can be defined as the probability that a failed system, subsystem, or component will be restored or repaired to a specified condition within a period when the maintenance is performed in accordance with the prescribed procedures.

RELIABILITY OF A SERIES SYSTEM

When a number of systems (subsystems or components) are arranged in a series, that is, they work in series with the output of a preceding system becoming the input to the next system in the series, the reliability of the entire system will depend on the reliability of each of the component systems (called components here).

Figure 19.6 shows a series system with $A_1, A_2, A_3, \ldots, A_n$ as its components.

The Boolean expression for the entire system to work (T) can be written as follows:

$$T = A_1 \cdot A_2 \cdot A_3 \cdots A_n$$

where $A_j = j$th component works.

Thus, the operation of a series system acts as an AND gate where all the inputs under the gate must occur simultaneously to provide the output.

Assuming that all the components $A_1, A_2, A_3, \ldots, A_n$ are independent (i.e., they do not depend on each other for their operation), the reliability of the entire system can be computed as follows:

$$R(T) = \prod_{i=1}^{n} R(A_i)$$

Using the aforementioned formula, Table 19.1 presents system reliability as a function of number of independent components (n) in series and reliability for each component. The second row of the table shows that when a component's reliability is 0.95, the system with 100 components in series will have a reliability of 0.005920

FIGURE 19.6 A series system.

TABLE 19.1
Reliability of a System as a Function of Component Reliability and Number of Series Components in the System

Sr. No.	Component Reliability = $R(A_i)$	Number of Components (n) in Series						Reference for Reliability Based on a CTS Variable
		10	20	50	100	1,000	10,000	
1	0.9	0.3486784401	0.1215766546	0.0051537752	2.6561398888E-05	1.7478712517E-46	0.000000000E+00	
2	0.95	0.5987369392	0.3584859224	0.0769449753	0.005920529	5.29182E-23	1.7221E-223	Within $\pm 2\sigma$ limits of normal distribution
3	0.9545	0.6277104045	0.3940203519	0.0974533194	0.009497149	5.96943E-21	5.7455E-203	
4	0.9973	0.9733256992	0.9473629167	0.8735564043	0.763100792	0.066960554	1.81213E-12	Within $\pm 3\sigma$ limits of normal distribution
5	0.999937	0.9993701786	0.9987407538	0.9968548571	0.993719606	0.93894161	0.532581231	Within $\pm 4\sigma$ limits of normal distribution
6	0.9999966	0.9999660005	0.9999320022	0.9998300142	0.999660057	0.996605768	0.966571449	Within $\pm 6\sigma$ limits and $\pm 1.5\sigma$ shift in mean of normal distribution
7	0.99999943	0.9999943000	0.9999886001	0.9999715004	0.999943002	0.999430162	0.994316213	Within $\pm 5\sigma$ limits of normal distribution
8	0.999999998	0.9999999800	0.9999999600	0.9999999000	0.9999998	0.999998	0.99998	Within $\pm 6\sigma$ limits of normal distribution

(i.e., 0.95^{100}). But the sixth row of the table shows that when a component has a reliability of 0.9999966, the reliability of the 100 components in the series system will be 0.99966.

This situation (in the sixth row of Table 19.1) is equivalent to 3.4 defects per million opportunities considered in the definition of Six-Sigma methodology. Here, the component production measured by using a variable, that is, "critical to customer satisfaction" (CTS), has its customer-derived tolerance (i.e., upper specification limit minus lower specification limit) equal to 12σ ($\pm 6\sigma$), and the mean of the variable varies between $\pm 1.5\sigma$ from the midpoint of the tolerance. (See the last column of reference in Table 19.1. σ = standard deviation of the component's variable, that is, CTS. The variable is assumed to be normally distributed.) (Kolarik, 1995; Yang and El-Haik, 2003).

The data in Table 19.1 show the importance of achieving high levels of component reliability to achieve high levels of series system reliability as the number of components in the series increases (which happens in complex products).

RELIABILITY OF A PARALLEL SYSTEM

A parallel system has all its components arranged in parallel, as shown in Figure 19.7. Here, the failure of any one of the components will not affect the failure of the entire system. The entire system can only fail if all of the components of the system fail simultaneously. Incorporation of parallel branches thus adds backup components to a product, and this in effect can improve its reliability considerably. However, such an addition of parallel components will increase costs, weight, and packaging space of the product.

Thus, assuming that all the component systems are independent, for the entire parallel system (T) to work at least one of its component systems must work. The following Boolean expression describes a parallel system (T) with its parallel components $A_1, A_2, A_3, \ldots, A_n$ (see Figure 19.7).

$$T = A_1 + A_2 + A_3 + \cdots + A_n$$

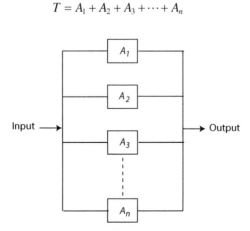

FIGURE 19.7 A parallel system.

Thus, the operation of a parallel system acts as an OR gate where at least one of the inputs under the gate must occur to provide output. The reliability of the parallel system can be computed as follows:

$$R(T) = 1.0 - \prod_{i=1}^{n}[1 - R(A_i)]$$

Table 19.2 presents system reliability values for different values of the number of components (n) in parallel for different values of reliability for each component using the aforementioned formula. These suggest that if a four-engine aircraft can continue flying on only one engine, then each engine with only 0.9 reliability can maintain the overall airplane reliability at the level of 0.9999 (see row 3 of Table 19.2 for $n = 4$). The system reliability values in Table 19.2 thus suggest that very high levels of system reliability can be obtained by incorporating only a few components in parallel.

RELIABILITY OF HYBRID SYSTEMS

Hybrid systems take advantage of the increasing reliability by incorporating parallel branches for critical components or systems. In a further effort to keep the cost down, only selected components whose reliability cannot be improved are duplicated (or laddered) in parallel branches. Figure 19.8 shows four different arrangements using components ($A, B, C, \ldots, G$) in series. System (a) shows a series system with components $A, B, C, \ldots, G$. System (b) shows several parallel branches for each of its components to increase the reliability of each component. System (c) shows several parallel branches of the entire series system. And system (d) shows only a few critical components (namely, A and C) with parallel branches. Thus, depending on the reliability of individual components and the desired overall reliability of the entire system, alternate configurations can be developed and evaluated to determine the most reliable system for the given cost, weight, and space objectives of the product program.

DESIGNING FOR RELIABILITY

A complex product will have a number of systems, and for the whole product to work successfully all systems of the product must work. Thus, the simplest and the first-level reliability analysis will consider a reliability model with all systems arranged in series (see Figure 19.6). Thus, failure of any one of the systems will be regarded as failure of the product. Similarly, the series model can also be used to evaluate the reliability of any system by considering that all of its subsystems are arranged in series. The series model can also be applied for the functioning of any subsystem by assuming that all components in a subsystem are arranged in series.

Let us assume that a product contains n number of systems ($i = 1$ to n). The ith system can be assumed to have m_i number of subsystems. And each subsystem

TABLE 19.2
Reliability of a System as a Function of Component Reliability and Number of Parallel Components in the System

Sr. No.	Component Reliability $= R(A_i)$	Number of Components (n) in Parallel						Reference for Reliability Based on a CTS Variable
		2	3	4	5	6	7	
1	0.7	0.9100000000	0.9730000000	0.9919000000	0.9975700000	0.9992710000	0.9997813000	
2	0.8	0.9600000000	0.9920000000	0.9984000000	0.9996800000	0.9999360000	0.9999872000	
3	0.9	0.9900000000	0.9990000000	0.9999000000	0.9999900000	0.9999990000	0.9999999000	
4	0.95	0.9975000000	0.9998750000	0.9999937500	0.9999996875	0.9999999844	0.9999999992	
5	0.9545	0.9979297500	0.9999058036	0.9999957141	0.9999998050	0.9999999911	0.9999999996	Within $\pm 2\sigma$ limits of normal distribution
6	0.9973	0.9999927100	0.9999999803	0.9999999999	1.0000000000	1.0000000000	1.0000000000	Within $\pm 3\sigma$ limits of normal distribution
7	0.999937	0.9999999960	1.0000000000	1.0000000000	1.0000000000	1.0000000000	1.0000000000	Within $\pm 4\sigma$ limits of normal distribution
8	0.9999966	1.0000000000	1.0000000000	1.0000000000	1.0000000000	1.0000000000	1.0000000000	Within $\pm 6\sigma$ limits and $\pm 1.5\sigma$ shift in mean of normal distribution
9	0.99999943	1.0000000000	1.0000000000	1.0000000000	1.0000000000	1.0000000000	1.0000000000	Within $\pm 5\sigma$ limits of normal distribution
10	0.999999998	1.0000000000	1.0000000000	1.0000000000	1.0000000000	1.0000000000	1.0000000000	Within $\pm 6\sigma$ limits of normal distribution

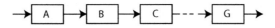

(a) A series system

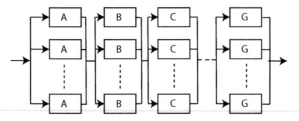

(b) A hybrid system with redundant parallel components

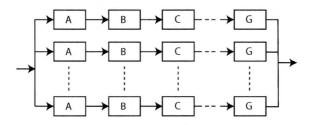

(c) A hybrid system with redundant parallel branches

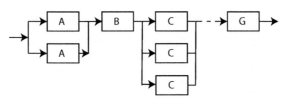

(d) A hybrid system with only critical components (A and C) in parallel

FIGURE 19.8 Different arrangements of systems. (a) A series system. (b) A hybrid system with redundant parallel components. (c) A hybrid system with redundant parallel branches. (d) A hybrid system with only critical components (A and C) in parallel.

(defined as jth subsystem in ith system) can be assumed to have l_{ij} number of components. Then, the reliability of the whole product (R_p) can be computed as follows:

$$R_p = \prod_{i=1}^{n} \prod_{j=1}^{mi} \prod_{k=1}^{lij} R_{ijk}$$

where R_{ijk} = reliability of kth component in jth subsystem of ith system.

Safety Engineering Tools

FIGURE 19.9 Reliability block diagram of the product in Chapter 2, Figure 2.1.

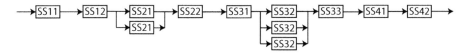

FIGURE 19.10 Illustration of the backup subsystems added in parallel to the product to improve product reliability.

Thus, for the product shown in the decomposition illustration in Chapter 2, Figure 2.1, the values of the variables in the aforementioned expression would be as follows: $n = 4$, $m_1 = 2, m_2 = 2, m_3 = 3, m_4 = 2, l_{11} = 5, l_{12} = 3, l_{23} = 4, l_{31} = 2, l_{32} = 5, l_{33} = 4, l_{41} = 3$, and $l_{42} = 5$. Assuming that the reliability of each component is equal to 0.999, the reliability of the product (with 36 components) will be $(0.999)^{36} = 0.96462$.

The reliability block diagram (showing only the subsystems) of the product in Figure 2.1 is shown in Figure 19.9.

If a component cannot be made with a given level of reliability, then one alternative would be to include one or more additional subsystems in parallel with the subsystem containing the problem component. For example, Figure 19.10 shows that subsystem SS21 has one backup (or redundant) system and subsystem SS32 has two additional backup systems added to the product as parallel systems to improve product reliability. Such an approach would obviously increase the cost of the product and also increase other product characteristics such as weight and volume (or packaging space) requirements.

APPROACHES FOR RELIABILITY IMPROVEMENTS

Reliability is one of the key attributes of product quality. Thus, it should be considered at the earliest stages of product conceptualization, functional allocation, and design synthesis. There is no unique approach to achieve the highest level of reliability. However, many of the following steps should be included during the product development process:

1. Collect historical data on past failures of similar systems to understand the ranges of reliability achieved in past systems.
2. Understand all possible failure modes of existing and proposed systems (including human errors and software failures).
3. Understand the benefits and costs associated with the use of new technologies for reliability improvements (e.g., newer filament-free technologies for light sources such as light-emitting diodes (LED) that can substantially increase the life of a light source compared to old tungsten-filament lamps; however, the newer sources generally cost much more).

4. Set reliability requirements on the product and cascade the reliability requirements to systems, subsystems, and component levels.
5. Consider available alternatives such as purchased components or systems from various suppliers (e.g., supplier capability), material selection for components (e.g., material properties such as tensile and compressive strengths, hardness, impact value [toughness], fatigue life, and creep), stress–strength analysis, redundancies, technologies, simplification, derating (using a component significantly below its rated value), and trade-offs associated with the allocation of functions to humans versus machines (or computers) (Ebeling, 1997).
6. Use redundancy (i.e., incorporate parallel components or branches as discussed earlier).
7. Conduct comparative analyses (using combinations of tools covered in Part II of this book) iteratively by using different configurations and estimates of reliability (e.g., optimistic, nominal, or pessimistic) for various elements of the product (systems, subsystems, and components) by considering trade-offs between product performance, reliability, costs, and other attributes of the product.
8. Use Six-Sigma approaches (i.e., DMAIC [Define, Measure, Analyze, Improve, and Control] or IDOV [Identify, Design, Optimize, and Verify] approaches) (see Chapter 14, Tables 14.2 and 14.3) with combinations of tools such as cause–effect diagrams, FMEA, FTA, and reliability models with block diagramming.

A Reliability Engineer's Tasks

The reliability engineers assigned to a product development team generally perform the following tasks during different product phases (Ebeling, 1997):

Conceptual phase: (1) develop reliability and maintainability goals and specifications, (2) gather reliability data from existing products and literature and conduct competitive analysis, and (3) allocate reliability requirements to various systems and apportion life cycle costs.

Detailed design and engineering phase: (1) conduct reliability predictions and testing, (2) conduct trade-off analyses and suggest design changes to meet reliability requirements, and (3) develop and communicate plans for reliability and maintainability to various teams within product development.

Production phase: conduct demonstrations of reliability and maintainability by conducting various verification tests (e.g., environmental and stress testing and implement quality assurance program by conducting inspections and acceptance testing).

Operational phase: (1) establish preventive maintenance programs and (2) determine spare parts plan and maintenance resource levels.

CONCLUDING REMARKS

Many commonly used safety engineering and reliability tools are presented in this chapter. Some tools such as hazard analysis, FMEA, and FTA are useful during product conceptualization, the allocation of functions to various systems and subsystems, and evaluating product configurations and designs. CIT and accident data analyses can provide an understanding on the types of safety issues and failure modes of past products. Learning from past products in terms of their failures and accident involvement can provide useful insights into creating safer new products. Data from reliability testing during product verification stages can provide estimates of the reliability of various components, subsystems, and systems in a product.

The product design teams need to use combinations of all possible tools to gather information to learn from past products and their failures and apply as many possible techniques to improve safety, performance, and reliability. Tools covered in other chapters in Part III of this book will further help in improving other product characteristics related to quality and human factors and also help in improving the product development process.

REFERENCES

Brown, D. B. 1976. *Systems Analysis and Design for Safety*. Upper Saddle River, NJ: Prentice-Hall, Inc.

Consumer Products Safety Commission (CPSC). 2012. *The National Electronic Injury Surveillance System (NEISS)*. Website: http://www.cpsc.gov/library/neiss.html (accessed March 9, 2012).

Ebeling, C. E. 1997. *An Introduction to Reliability and Maintainability Engineering*. New York: McGraw-Hill.

Fitts, P. M. and R. E. Jones. 1961a. Analysis of Factors Contributing to 460 "Pilot Errors" Experiences in Operating Aircraft Controls. Memorandum Report TSEAA-694-12. Aero Medical Laboratory, Air Materiel Command, Wright-Patterson Air Force Base, Dayton, Ohio, July 1947. In *Selected Papers on Human Factors in the Design and Use of Control Systems* (Ed.) Wallace Sanaiko H. New York: Dover Publications, Inc.

Fitts, P. M. and R. E. Jones. 1961b. Psychological Aspects of Instrument Display I: Analysis of Factors Contributing to 270 "Pilot Errors" Experiences in Reading and Interpreting Aircraft Instruments. Memorandum Report TSEAA-694-12A. Aero Medical Laboratory, Air Materiel Command, Wright-Patterson Air Force Base, Dayton, Ohio, July 1947. In *Selected Papers on Human Factors in the Design and Use of Control Systems* (Ed.) Wallace Sanaiko H. New York: Dover Publications, Inc.

Hammer, W. 1980. *Product Safety Management and Engineering*. Upper Saddle River, NJ: Prentice-Hall, Inc.

Heinrich, H. W., Petersen, D. and N. Roos. 1980. *Industrial Accident Prevention*. Fifth Edition. New York: McGraw-Hill.

Kolarik, W.J. 1995. *Creating Quality—Concepts, Systems, Strategies, and Tools*. New York: McGraw-Hill, Inc.

Konz, S. and S. Johnson. 2004. *Work Design—Industrial Ergonomics*. Sixth Edition. Scottsdale, AZ: Holcomb Hathaway, Publishers.

National Highway Traffic Safety Administration (NHTSA). 2012. *The Fatality Analysis Reporting System (FARS)*. Website: http://www-fars.nhtsa.dot.gov/Main/index.aspx (accessed May 3, 2013).

Tarrants, W. E. 1980. *The Measurement of Safety Performance*. New York: Garland STPM Press.

Yang, K. and B. El-Haik. 2003. *Design for Six Sigma*. ISBN 0071412085. New York: McGraw-Hill.

20 Cost–Benefit Analysis

INTRODUCTION

Cost–benefit analysis is a useful method for decision-making. The analysis requires understanding and use of costs and benefits associated with each alternative involved in the problem being solved. The benefits generated from the use of a system should be higher than the costs associated in operating and maintaining the system. This chapter provides information on how to create a cost–benefit analysis and presents an application of the method to determine the acceptability of a solar photovoltaic power source for residential purposes.

COST–BENEFIT ANALYSIS: WHAT IS IT?

A cost–benefit analysis is a process by which a decision maker can analyze available alternatives and make decisions related to selection of an alternative involving a system or a product. To conduct a cost–benefit analysis, a decision model is first developed by identifying alternatives and outcomes, and by determining benefits and costs of each combination of alternatives and outcomes. The decision is made based on one or more decision principles and maximum values of evaluation measures obtained by (a) subtracting the costs from benefits, and/or (b) computing ratios of benefits-to-costs (see decision matrix covered in Chapter 3).

WHY USE COST–BENEFIT ANALYSIS?

Cost–benefit analysis is an objective method of decision-making. However, to determine the alternatives, outcomes, costs, and benefits (associated with each combination of alternatives and outcomes), the cost–benefit analysis relies on the abilities of the analyst and/or the team involved in its formulation and estimation of the values of the variables included in the analysis.

Organizations rely on cost–benefit analysis to support decision-making because it provides an objective and evidence-based view of the issues being evaluated—without the influences of opinions, politics, or biases of decision makers and other individuals within and outside the organization who may directly or indirectly be affected by the decision. By providing an unclouded view of the consequences of a decision, the cost–benefit analysis is an invaluable tool in developing business strategy, evaluating a proposal, or making resource allocation or purchase decisions. In business, government, finance, and even the nonprofit world, the cost–benefit analysis offers unique and valuable insights. Some examples of its applications are:

a. Comparing project proposals
b. Deciding whether to pursue a proposed project

c. Evaluating alternate locations for a new product manufacturing plant
d. Weighing different investment opportunities
e. Measuring social benefits of proposed changes in regulations
f. Appraising the desirability of suggested policy alternatives
g. Assessing change initiatives
h. Determining effects of economic or political change on stakeholders and participants.

STEPS INVOLVED IN COST–BENEFIT ANALYSIS

While there is no "standard" format for performing a cost–benefit analysis, there are certain core elements that will be present across almost all such analyses. The five basic steps to performing a cost–benefit analysis include:

1. Identify alternatives, outcomes, and probabilities of the outcomes
2. Identify costs and benefits so they can be categorized by type (e.g., fixed costs, variable costs, and safety costs)
3. Calculate costs and benefits for each combination of alternative and outcome over the assumed life of the project or initiative at the beginning of the project planning time using the "present value" method
4. Select principles to be used in evaluating the alternatives (see Chapter 3)
5. Calculate evaluation measures for all alternatives and make recommendations. The most used evaluation measures are (a) the net present value (i.e., the sum of the present value of benefits over the life of the project minus the sum of the present value of costs over the life of the project, and (b) the benefit-to-cost ratio (i.e., the ratio of the sum of the present value of benefits over the life of the project to the sum of the present value of costs over the life of the project). The evaluation measures are computed for each combination of alternatives and outcomes, and the alternative that has the highest value of the evaluation measure is usually selected (or considered further for company management concurrence).

As with any process, it is important to work through all the steps thoroughly and not give in to the temptation to cut corners or base assumptions on opinions or guesses. It is important to ensure that the analysis is as comprehensive as possible (i.e., it covers all costs and benefits incurred over the entire life cycle of the project). Cost–benefit analysis does not require any specific tool to display and calculate costs, benefits, and evaluation measures. However, tabular formats are generally used to display data and spreadsheets are commonly used to perform calculations.

SOME EXAMPLES OF PROBLEMS FOR APPLICATION OF COST–BENEFIT ANALYSIS

Any decision-making problem in a product or system development and operation area can be analyzed by applying the cost–benefit analysis. Some examples of

Cost–Benefit Analysis

problems in the energy systems area that involve computations of costs and benefits associated with different alternatives considered in problem solving are described below:

1. <u>Select a technology for building a new electric power plant</u>: Alternatives considered here are different technologies (e.g., natural gas fired plant, nuclear power plant, geothermal power plant, wind turbines, and so forth) used to produce electricity. The costs involved are plant construction and equipment installation costs (capital costs), operating and maintenance costs, insurance costs, safety, and accident costs. The benefits are typically revenue generated, rebates and tax credits, and economic impact on the community where the plant will be located. (See Chapter 24 for a detailed example covering this problem).
2. <u>Select a technology for carbon capture and sequestration for fossil fuel power plants</u>: Alternatives here will be different methods used to capture and sequestrate the carbon and the percentage of emitted carbon that will be captured. The costs will include construction and installation of the carbon capturing and sequestration equipment, operating and maintenance of the specialized equipment, safety, and accident costs. The benefits will include reduction in pollution-related costs (e.g., social cost of carbon based on reduction of metric tons of carbon dioxide over a given period and discount rate).
3. <u>Develop a plan for improvement in existing electric grid</u>: Here the alternatives will be different approaches and specific levels of improvements that will be considered in improving the grid (e.g., adding more energy sources and powerlines, incorporating energy demand measurement and electricity distribution control center, storage of excess energy). The costs will include installation costs of additional equipment, and additional operating and maintenance and safety and accident costs. The benefits will be gains due to improved grid efficiency, reduced power outages, and increased power usage.
4. <u>Develop future requirements to control pollution from various industries</u>: The alternatives that are considered here are different methods that can be used in controlling pollution and the level of pollution or percent reduction in pollution that can be achieved by each alternative (e.g., pollution controls on factories producing paints and plastics, steel producing plants and retiring coal-fueled power plants). The costs will be those associated with various changes that need to be made in different industries affected by the new requirements. The gains will be based on reduction in social and health-related costs and additional economic benefits gained (e.g., more people moving into the area with reduced pollution levels and additional jobs created) due to the implementation of the proposed requirements.
5. <u>Develop future requirements on transportation and distribution of fuel and/or energy</u>: The alternatives to be considered here are the types of requirements and their applicability to different fuel and power

distribution systems. The costs will include capital costs in incorporating the changes, operating and maintenance costs associated with the changes and changes in other costs (e.g., safety costs). The benefits will include revenues from the additional sale of fuel/energy and avoided costs in fuel and power distribution due to the changes and economic impact on the local community due to the changes.
6. <u>Develop future requirements on fuel economy and emissions from motor vehicles</u>: The alternatives will be the type of changes to be made to future motor vehicles (e.g., weight reduction, changes in powertrains, and aerodynamic improvements). The costs will include costs to develop and incorporate new changes in future vehicles; and benefits will include reduction in fuel costs and pollution-related costs (e.g., medical costs, climate change-related costs).

COST–BENEFIT ANALYSIS OF RESIDENTIAL SOLAR PANELS: AN EXAMPLE

Problem

The problem considered in this section is to conduct two cost–benefit analyses for installing 12 kW residential photovoltaic (PV) solar energy systems to generate electric energy in two average size homes, one in Detroit, MI and the other in Phoenix, AZ. The analysis procedure used here was similar to one described by the Solar Foundation (2012).

Two spreadsheets (one for each city) were prepared by using assumptions and data from the websites given below. The data were used to compute benefit-to-cost ratios for homes in the two cities.

Assumptions

1. Life of Equipment: 25 years
2. Financing Period: 25 years
3. Loan Interest (or discount rate): 5% per year.

Websites

1. Solar Photovoltaic System Costs:
   U.S. Solar Photovoltaic System Cost Benchmark: Q1 2017 (Fu, et al., 2017; Barbose and Darghouth, 2019).
2. Energy Consumption:
   http://www.eia.gov/consumption/residential/reports/2009/state_briefs/
   http://www.eia.gov/consumption/residential/reports/2009/state_briefs/pdf/mi.pdf
   http://www.eia.gov/consumption/residential/reports/2009/state_briefs/pdf/az.pdf
3. PV Watts:
   http://rredc.nrel.gov/solar/calculators/PVWATTS/version1/

Cost–Benefit Analysis

4. Financial Incentives for Solar PV:
   http://programs.dsireusa.org/system/program?state=MI
   http://programs.dsireusa.org/system/program?state=AZ

Costs of Going Solar

1. The average weighted installed cost of solar for a residential non-utility solar energy system was assumed to be $3.30/Watt (Barbose and Darghouth, 2019).
2. Equipment costs are those associated with purchasing the hardware necessary for installing a solar energy system. For a rooftop photovoltaic (PV) system, hardware components include the PV modules, solar power inverters, mounting and racking hardware, meters, disconnect devices, and system wiring.
3. Securing interconnection approvals and system inspection costs
4. Project financing costs
5. Sales taxes on the purchase of equipment and services
6. Operation and maintenance costs (O&M costs) (about $6–27/kW): System cleaning, replacement of broken panels, and inverter replacement.
7. Cost of supplemental insurance (0.25–0.50% of installed costs)
8. Power Back-up System Costs: Solar plants can only provide power during the daytime and the power output will depend upon intensity of sun illumination. Thus, a backup power supply (e.g., from the utility company) or a battery would be needed to provide power when the solar system cannot provide the required level of electric power.

Benefits of Going Solar

1. Avoided energy costs: These are long-term energy cost savings (avoid paying retail costs of energy consumed and other fixed costs per month for service) due to the use of electricity generated by the solar system.
2. Value of excess solar generated energy sold to the utility company
3. Benefits from reduced pollution: These are reduced costs due to less pollution generated by utility company-supplied power. These can be in the form of rebates provided by the utility company to the solar electricity producers (e.g., solar renewable energy credits [SREC]).
4. Incentives provided by federal, state, and local governments: Cash rebates and grants, federal investment tax credit, and low-interest loans.
5. Utility company provided incentives.

Solar Energy Output

1. Capacity Rating is a measure of the size of a solar energy system, typically measured in watts (W) or kilowatts (kW).
2. The output of the solar panels is reduced due to losses in DC to AC conversion. Derate factor is used to account for energy losses in conversion.

3. Peak Sun Hours is a measure of the number of hours per day that solar irradiance (the amount of solar radiation falling on a particular area) is at it is maximum (i.e., 1,000 W/m^2).
4. The annual energy output of the solar plant can be calculated by using the following formula:

$$\text{Annual output} = DCR \times DF \times PSH \times 365 \text{ kwh/yr} \quad (20.1)$$

Where,
DCR = DC Rating of the solar plant (kW)
DF = Derate factor
PSH = Peak sun hours per day (kWh/m^2/day).

Cost–Benefit Analysis and Calculations

In this study, two cost–benefit analyses were conducted for installing 12 kW residential PV solar energy systems to generate energy in two average size homes, one in Detroit, MI and the other in Phoenix, AZ (Schwager, Tate and Kim, 2020). The spreadsheets for the two analyses are presented in Tables 20.1 and 20.2. The life of the equipment and finance period evaluated was over a 25-year period, and the loan interest (or discount rate) assumed was 5% per year. This study compared the present value of the costs, present value of the benefits, the net present value of the benefits (benefits minus the costs) and the ratio of present value of benefits to present value of costs over 25 years for each PV solar system.

The costs that were considered included (a) the PV solar system installation cost, (b) the operation and maintenance cost, and (c) the cost of insurance for a PV system. The benefits considered included (a) the avoided electrical utility cost, (b) SREC benefits, (c) tax credits, and (d) net metering payback from the utility company.

Installed Costs

The cost of installing a 12kW solar panel system was determined for Michigan and for Arizona. While the cost of installing the solar panel system has decreased over time, according to the 2019 annual "Tracking the Sun" report from Berkeley National Labs (Barbose and Darghouth, 2019), the median cost to install a solar panel system in Arizona was $3.30/W in 2018. Prices in the state of Michigan were not included in the report. Therefore, the national median cost for installation, $3.30/W (Barbose and Darghouth, 2019) was used for Michigan.

The installed cost assumed the customer financed the entirety of the system over 25 years, with a 5% interest loan, compounded annually, where the interest was spread out over the life of the loan and included in payments. The equal payment capital recovery equation below was used.

$$A = P[(i(1+i)^n)/((1+i)^n - 1)] \quad (20.2)$$

TABLE 20.1
Spreadsheet for PV Solar Residential System in Detroit, MI

MI

	Costs					Benefits									
Year	Installed Costs ($)	O&M Costs ($)	Insurance ($)	Total Costs ($)	Present Value of Costs ($)	Total Electrical Energy Usage (kwh)	Energy Generated by PV (kWh)	Energy Usage Avoided (kwh)	Average Detroit Electrical Cost ($/kwh)	Avoided Electricity Costs ($)	Excess Generated (kwh)	SREC Revenue ($)	Tax Credit ($)	Total Benefits ($)	Present Value of Benefits ($)	Net Present Value = Present Value of Benefits Minus Present Value of Total Costs ($)	Net Cumulative Present Value
1	2810	240	672	3722	3722	8000	15514	8000	0.12	2327	7514	0	10296	12623	12623	8901	8901
2	2810	240	672	3722	3545	8080	15436	8080	0.12	2339	7356	0	0	2339	2227	-1318	7583
3	2810	240	672	3722	3376	8161	15359	8161	0.12	2350	7198	0	0	2350	2132	-1244	6339
4	2810	240	672	3722	3215	8242	15282	8242	0.12	2362	7040	0	0	2362	2040	-1175	5164
5	2810	240	672	3722	3062	8325	15206	8325	0.12	2374	6881	0	0	2374	1953	-1109	4055
6	2810	240	672	3722	2916	8408	15130	8408	0.13	2385	6722	0	0	2385	1869	-1047	3007
7	2810	240	672	3722	2777	8492	15054	8492	0.13	2397	6562	0	0	2397	1789	-989	2018
8	2810	240	672	3722	2645	8577	14979	8577	0.13	2409	6402	0	0	2409	1712	-933	1085
9	2810	240	672	3722	2519	8663	14904	8663	0.13	2421	6241	0	0	2421	1639	-881	205
10	2810	240	672	3722	2399	8749	14830	8749	0.13	2433	6080	0	0	2433	1568	-831	-627
11	2810	240	672	3722	2285	8837	14755	8837	0.13	2445	5919	0	0	2445	1501	-784	-1411
12	2810	240	672	3722	2176	8925	14682	8925	0.13	2457	5756	0	0	2457	1437	-740	-2150
13	2810	240	672	3722	2073	9015	14608	9015	0.14	2469	5594	0	0	2469	1375	-698	-2848
14	2810	240	672	3722	1974	9105	14535	9105	0.14	2481	5431	0	0	2481	1316	-658	-3506
15	2810	240	672	3722	1880	9196	14463	9196	0.14	2494	5267	0	0	2494	1259	-620	-4126
16	2810	240	672	3722	1790	9288	14390	9288	0.14	2506	5103	0	0	2506	1205	-585	-4711
17	2810	240	672	3722	1705	9381	14318	9381	0.14	2518	4938	0	0	2518	1154	-551	-5263
18	2810	240	672	3722	1624	9474	14247	9474	0.14	2531	4772	0	0	2531	1104	-520	-5782
19	2810	240	672	3722	1547	9569	14175	9569	0.14	2543	4606	0	0	2543	1057	-490	-6272
20	2810	240	672	3722	1473	9665	14105	9665	0.14	2556	4440	0	0	2556	1011	-461	-6734
21	2810	240	672	3722	1403	9762	14034	9762	0.15	2569	4273	0	0	2569	968	-435	-7168
22	2810	240	672	3722	1336	9859	13964	9859	0.15	2581	4105	0	0	2581	927	-409	-7578
23	2810	240	672	3722	1272	9958	13894	9958	0.15	2594	3936	0	0	2594	887	-386	-7963
24	2810	240	672	3722	1212	10057	13825	10057	0.15	2607	3767	0	0	2607	849	-363	-8326
25	2810	240	672	3722	1154	10158	13756	10158	0.15	2620	3598	0	0	2620	812	-342	-8668
Total ->	70243	6000	16809	93052	55082	225946	365446			61768		0		72064	46413	-8668	
Benefits-to-Costs Ratio =	0.843																

TABLE 20.2
Spreadsheet for PV Solar Residential System in Phoenix, AZ

AZ

	Costs					Benefits								Net Present	
Year	Installed Costs ($)	O&M Costs ($)	Insurance Costs ($)	Total Costs ($)	Present Value of Costs ($)	Total Electrical Energy Usage (kwh)	Energy Generated by PV (kWh)	Energy Usage Avoided (kwh)	Average Detroit Electrical Cost ($/kwh)	Avoided Electricity Costs ($)	Excess Generated (kwh)	SREC Revenue ($)	Net Metering Credits ($)	Tax Credit ($)	Total Benefits ($)	Present Value of Benefits ($)	Net Present Value = Benefits Minus Present Value of Total Costs ($)	Cumulative Net Present Value ($)
---	---	---	---	---	---	---	---	---	---	---	---	---	---	---	---	---	---	---
1	2810	240	269	3319	3319	14000	21989	14000	0.15	2100	7989	0	770	11296	14166	14166	10847	10847
2	2810	240	269	3319	3161	14140	21879	14140	0.15	2142	7739	0	746	0	2888	2751	-410	10437
3	2810	240	269	3319	3010	14281	21770	14281	0.15	2185	7489	0	722	0	2907	2637	-374	10063
4	2810	240	269	3319	2867	14424	21661	14424	0.15	2229	7237	0	698	0	2927	2528	-339	9725
5	2810	240	269	3319	2731	14568	21553	14568	0.16	2274	6984	0	673	0	2947	2425	-306	9419
6	2810	240	269	3319	2601	14714	21445	14714	0.16	2320	6731	0	649	0	2969	2326	-275	9144
7	2810	240	269	3319	2477	14861	21338	14861	0.16	2366	6477	0	624	0	2991	2232	-245	8899
8	2810	240	269	3319	2359	15010	21231	15010	0.16	2414	6221	0	600	0	3014	2142	-217	8682
9	2810	240	269	3319	2246	15160	21125	15160	0.16	2462	5965	0	575	0	3037	2056	-191	8492
10	2810	240	269	3319	2139	15312	21019	15312	0.16	2512	5708	0	550	0	3062	1974	-166	8326
11	2810	240	269	3319	2038	15465	20914	15465	0.17	2562	5450	0	525	0	3088	1896	-142	8184
12	2810	240	269	3319	1941	15619	20810	15619	0.17	2614	5190	0	500	0	3114	1821	-120	8064
13	2810	240	269	3319	1848	15776	20706	15776	0.17	2666	4930	0	475	0	3142	1749	-99	7966
14	2810	240	269	3319	1760	15933	20602	15933	0.17	2720	4669	0	450	0	3170	1681	-79	7887
15	2810	240	269	3319	1676	16093	20499	16093	0.17	2775	4406	0	425	0	3199	1616	-60	7826
16	2810	240	269	3319	1596	16254	20397	16254	0.17	2830	4143	0	399	0	3230	1554	-43	7784
17	2810	240	269	3319	1520	16416	20295	16416	0.18	2887	3879	0	374	0	3261	1494	-26	7757
18	2810	240	269	3319	1448	16580	20193	16580	0.18	2945	3613	0	348	0	3294	1437	-11	7746
19	2810	240	269	3319	1379	16746	20092	16746	0.18	3005	3346	0	323	0	3327	1383	3	7749
20	2810	240	269	3319	1313	16914	19992	16914	0.18	3065	3078	0	297	0	3362	1330	17	7766
21	2810	240	269	3319	1251	17083	19892	17083	0.18	3127	2809	0	271	0	3397	1280	30	7796
22	2810	240	269	3319	1191	17253	19792	17253	0.18	3189	2539	0	245	0	3434	1233	41	7837
23	2810	240	269	3319	1135	17426	19693	17426	0.19	3254	2267	0	219	0	3472	1187	52	7890
24	2810	240	269	3319	1081	17600	19595	17600	0.19	3319	1995	0	192	0	3511	1143	63	7952
25	2810	240	269	3319	1029	17776	19497	17776	0.19	3386	1721	0	166	0	3552	1101	72	8024
Total ->	70243	6000	6732	82975	49117					67350		0	11816		90462	57141	8024	

Benefits-to-Costs Ratio = 1.163

Cost–Benefit Analysis

Where, A is the yearly payment, P is the principal (total amount financed), i is the annual interest rate, and n is the number of annual payments

For both systems (installed in Arizona and Michigan), the total cost of installation, including interest, was $70,242.93 at a yearly payment of $2,809.72.

Operation and Maintenance Cost

According to the U.S Department of Energy (Walker, 2020), the yearly operation and maintenance cost of a residential PV system was about $20/kW/year. These costs include replacing solar power inverters, cleaning the panels, and other potential maintenance costs. This is one of the costs that increase with larger PV systems.

Insurance

In both Arizona and Michigan there is not a separate insurance to cover solar panels for your home. However, you can still have your solar panels covered by your homeowner's insurance by increasing the coverage for your home (Nationwide, 2020).

The insurance cost in Phoenix, Arizona is .0068 of the value of the home. Thus, multiplication of 0.0068 by the total cost of the solar system of $ 39,699 (i.e., 12kW × 3.3 $/kW) provides an estimate of $ 269/year cost to insure the solar system. Michigan charges an insurance cost of .017 times $ 39,600 (the added value from the solar system). Thus, the cost to insure the solar system in Michigan will be $ 672 per year.

Present Value of Cost

For each of the 25 years, the sum of the yearly installment cost, O&M cost, and insurance cost was computed. Then the present value of the cost was calculated for each year with the equation below.

$$P = F/(1+i)^n \qquad (20.3)$$

where P is the present value, F is the future value of the payment made in n^{th} annual payment with discount rate (i).

Using the above equation and summing the present values of cost for 25 years for each PV system, it was found that the present value of cost for the PV system in Detroit was $55,081 and the present value of the cost for the PV system in Phoenix was $49,116.65 (see Tables 20.1 and 20.2, respectively).

Avoided Electric Utility Cost

The average electrical utility usage for the average residential house was taken at the state level for both Detroit and Phoenix. The average residential yearly consumption for Michigan was 8000 kWh/year and for Arizona, 14000 kWh/year (EIA, 2020a, 2020b and 2020c). The lower electrical consumption in Michigan can largely be explained by the colder climate and the resultant lower air conditioning usage.

Using the equation of annual output (see equation 20.1 above), a derate factor of 0.77 and the peak sun hours average from PV Watts (NREL, 2020), 4.6 kWh/m²/day for Detroit and 6.52 kWh/m²/day for Phoenix, the energy generated by PV system in Detroit was 15,514 kWh/year and the energy generated by PV system in Phoenix was

21,989 kWh year. PV Watts gave similar numbers of 15,742 kWh/year for Detroit and 20,482 kWh/year for Phoenix with inputs of a 12 kW system that was roof-mounted with a 20-degree tilt and 14.08% efficiency loss. PV Watts (NREL, 2020) values were within 2% for Detroit and 7% for Phoenix of the above estimates. A 1% efficiency loss each following year was included in the calculations to account for equipment degradation.

The avoided yearly electricity cost was computed by determining the energy supplied by the PV generation up to the yearly electrical consumption. The 12 kWh systems in both states generated more than the consumption, so the avoided electrical usage generation was less than the PV generation for each year. Then the avoided electrical usage was multiplied by the electrical cost to obtain the avoided electrical utility cost. The average electricity cost used for Detroit was $0.15/kWh and for Phoenix was $0.12/kWh. It was also assumed that the electrical usage and electricity costs would increase by 1% a year.

SREC, Net Metering, and Tax Credit Revenue

Revenue generated by the solar power system is highly dependent on the state and even local utility regulations. Revenue generation opportunities made up the difference in why the Michigan system is a negative value proposition and the Arizona system is a positive one. For this analysis, three sources of revenue from the solar power system were considered, namely, Solar Renewable Energy Credits (SREC) revenue, Net Metering Credits, and Tax Credits.

SREC are credits that the owners of the solar energy generators can sell to utilities in states where there is a requirement that power companies produce a certain portion of their energy output using solar power. If a utility company does not own the solar infrastructure themselves, they can pay solar energy generators (owners of the solar generators) for credit for producing that energy. While neither Michigan nor Arizona have such requirements for utilities, Michigan solar producers could sell credits to Ohio, which did have a solar carve-out, until 2019. The passage of HB6 eliminated the solar energy requirement, and with it, the market for Ohio SRECs (Callender et al., 2019). For the analysis, the amount for SREC credits was zero for all 25 years (DSIRE, 2020).

A federal tax credit of 26% of the total cost to install the solar power system is available to homeowners for systems installed in 2020 (IRS, 2020). Taxpayers have the option of claiming the credit all in one year or splitting the amount between two years. As it is more financially advantageous for this analysis, the tax credit was claimed in the first year for both the Michigan and Arizona analyses. In addition to the federal tax credit, Arizona residents are eligible for a tax credit of $1000, or 25% of the cost of installation, whichever is less (Arizona State Government, 2019). For this system, $1000 is the lesser value, and this value was used. This $1000 was also claimed entirely in the first year of installation, for a total value of tax credits in the first year of $11,296.00 for Arizonans, and $10,296.00 for Michiganders.

Net metering, which is the ability to sell excess energy generated back to utilities, is available in Michigan and Arizona. In Arizona, residents receive $0.0964 per kWh sent to the grid (Wichner, 2018). This credit is disbursed monthly in the form of a bill credit, and at the end of the year can be cashed out (TEP, 2018). Since the system

Cost–Benefit Analysis

produces more energy than the house consumes, net metering credits produced a total of $11,816 over 25 years, assuming the homeowner elected to cash out bill credits at the end of each billing year. While net metering is available in Michigan, there is no provision to allow a homeowner to cash out excess solar bill credits (DTE, 2020). For this reason, net metering revenue was not included in the calculation of Michigan benefits, since a bill credit is not a realizable gain. Ideally, to take advantage of net metering in Michigan, the solar power system would be smaller than required to always meet all energy demands to allow a homeowner to use the bill credits to offset the cost of energy from the utility.

Net Present Value

The present value of the benefits was calculated using the same equation (Equation 20.3) used for the present value of costs. Then the net present value for each year was calculated by subtracting the present value costs from the present value of benefits. The net cumulative present value (over the 25 years) of a PV system in Detroit was −$8668, and the net cumulative present value of a PV system in Phoenix was $8024 (see Tables 20.1 and 20.2).

The benefits-to-costs ratios obtained for the two analyses show that the value of the ratio for Detroit, MI was 0.843 (less than 1.0); whereas the value of the ratio was 1.163 for Phoenix, AZ. The curves for the cumulative present value shown in Figure 20.1 also show that implementation of a solar PV system for an average home in Phoenix will be profitable but it will not be profitable for an average home in Detroit.

Conclusions of the Cost–Benefit Analyses

Based on the data presented in Tables 20.1 and 20.2, the net cumulative present value over 25 years for the Detroit PV system was a loss of $8668 and for the Phoenix PV system was a net benefit of $8024.

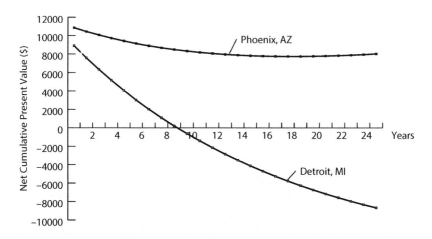

FIGURE 20.1 Net Cumulative Present Value of Solar PV Systems in Detroit, MI and Phoenix, AZ Homes.

Many different factors impact the overall cost–benefit analysis. In particular, the environment at the location affects the consumption and the amount of energy that can be generated. A home in cooler city uses less air conditioning and thus less electricity, and a home located in higher peak sun hours city allows for more electricity generation by the PV system. In general, more electrical usage with more peak sun hours allows for more electricity utility cost avoidance, and this was one of the factors that lead to the Phoenix systems showing more profitability than the Detroit system.

It was also found that a 12kW system was too large for the average residential house. PV systems are more expensive with the larger DC rating, and the excess electrical generation does not give that much advantage. Additional calculations run for an 8kW system showed that with scaling the system down to meet the consumption of electricity, the loss over 25 years for the Detroit system decreased by $2800 and the profit by the Phoenix system increased by $8500. Phoenix gives out a net metering credit for excess electrical generation, but it was still financially better over the 25 years to downsize the PV system.

The analyses also showed that factors such as the cost of insurance, government incentives, and net metering affect the evaluation measures. Michigan had higher insurance costs, while Arizona had larger government incentives. And Arizona's net metering credit made generating excess electricity profitably.

Overall, it is important to consider all the factors when looking into if a PV system will have a positive net cumulative present value. Adding more factors to such an analysis will make the analysis more comprehensive and representative of the actual situation. Further, the assumptions based on the past data may not apply well over future years because the financial incentives and costs of the solar system will change in the future. Such analysis should also be iterated by changing the kW power output capability of the PV system to determine the optimum capability of the PV system to achieve maximum long-term profitability. These calculations showed that 8 kW systems were more profitable for the average household better than a 12 kW system for both the locations analyzed. Finally, not all PV systems will necessarily be profitable, so homeowners must consider how much they are willing to pay to lower emissions.

EXERCISING COST–BENEFIT MODEL FOR SENSITIVITY ANALYSIS

After the creation of the cost–benefit model (i.e., the spreadsheet) such as those shown in Tables 20.1 and 20.2, they can be exercised (i.e., iterated) by using different values of input variables to determine the sensitivity of changes in the input variables to overall conclusions. Sensitivity analyses thus conducted can provide a better understanding of the risks associated in determining the best course of action.

RISKS AND UNCERTAINTIES IN COST–BENEFIT ANALYSIS

Despite its usefulness, cost–benefit analysis has several associated risks and uncertainties that are important to note. These risks and uncertainties can result from incomplete and erroneous formulation of the analysis, inappropriate use of evaluation measures, and/or criteria to reach conclusions. Much of the risk involved with cost–benefit analysis can be related to deliberate manipulations of the data leading

Cost–Benefit Analysis

to the selection of a wrong or inappropriate alternative. Stakeholders or interested parties may try to influence results by over- or understating costs and benefits. In some cases, supporters of a project may insert a personal or organizational bias into the analysis.

On the data side, there can be a tendency to rely too much on data compiled from previous projects. This may inadvertently yield results that do not directly apply to the situation being considered. Since data leveraged from an earlier analysis may not directly apply to the circumstances at hand, this may yield results that are not consistent with the requirements of the situation being considered. Using heuristics to assess the dollar value of intangibles may provide quick, "ballpark-type" information, but it can also result in errors that produce an inaccurate picture of costs that can invalidate findings.

In addressing risk, it is helpful to utilize probability theory (i.e., assign probabilities to cost and benefit estimates and to outcomes) to identify and examine key patterns that can influence the selection of an alternative.

UNCERTAINTIES

There are several considerations that can influence the results of any cost–benefit analysis, and while they would not apply in every situation, it is important to keep them in mind as these analyses are conducted and exercised. These considerations are:

1. Inaccuracies in cost and benefit information can diminish the credibility of findings.
2. Revenue and costs are moving targets, and thus, selecting values of their parameters can be very challenging.
3. Income level can influence a customer's or decision maker's ability or willingness to make decisions.
4. Some benefits cannot be directly reflected in dollar amounts because of the subjective considerations and judgments associated with the values of many variables.
5. Many costs and benefits are not just linear functions of input or output variables. Thus, non-linearities must be carefully studied before simply linear-scaling the values of evaluation measures.

CONTROVERSIAL ASPECTS

The most controversial aspects of cost–benefit analysis are costs and benefits that are associated with the subjective nature of their values that lead to intangibles. Many variables related to costs and benefits are difficult to quantify, such as human life, pain, and sufferings due to adverse health effects of environment or injuries due to accidents, and other considerations such as level of comfort and convenience, brand equity, customer loyalty, and so forth.

With respect to intangibles, using the cost–benefit analysis process to drive more critical thinking around all aspects of value can provide beneficial outcomes.

Cost–benefit analysis assumes that a monetary value can be placed on all the costs and benefits of a program, including tangible and intangible returns. A major advantage of cost–benefit analysis lies in forcing people to explicitly and systematically consider the various factors which should influence strategic choice.

CONCLUDING REMARKS

During the development of any cost–benefit analysis, care must be exercised to ensure that uncertainties in (a) the prediction of scenarios, (b) model development, and (c) data must be considered. Uncertainty in scenarios can occur due to errors in selecting alternatives, outcomes, and their probabilities. The model could be conceptually imperfect due to omission or inappropriate consideration of many variables. And the data used to estimate costs and benefits may not be valid and contain errors. Model validation is very important and necessary in many cases and involves comparison of model predictions with other existing or future observations. Confidence in model results can be significantly increased by model calibration, verification, and validation exercises.

REFERENCES

Arizona State Government. 2019. Arizona Form 310: Credit for Solar Energy Devices. Website: https://azdor.gov/forms/tax-credits-forms/credit-solar-energy-credit (Accessed: January 22, 2020).

Barbose, G. and N. Darghouth, *Tracking the Sun: Pricing and Design Trends for Distributed Photovoltaic Systems in the United States.* Lawrence Berkeley National Laboratory, October 2019. Website: https://emp.lbl.gov/tracking-the-sun (Accessed: July 27, 2020).

Callender, Wilkin, et al. 2019. As Passed by Senate: Am. Sub.H. B. No. 6. *The Ohio Legislature, 133 General Assembly*, 2019. Website: https://www.legislature.ohio.gov/legislation/legislation-documents?id=GA133-HB-6 (Accessed: January 22, 2020).

DSIRE. 2020. *Database of State Incentives for Renewables & Efficiency.* NC State University. Websites: http://programs.dsireusa.org/system/program?state=AZ and http://programs.dsireusa.org/system/program?state=MI (Accessed: January 22, 2020).

DTE Energy (DTE). 2020. Distributed Generation FAQs. Website: https://www.newlook.dteenergy.com/wps/wcm/connect/8dc7d28e-5621-41ba-a874-5dddc8c56703/DistributedGenerationFAQ.pdf?MOD=AJPERES (Accessed: January 20, 2020).

EIA, 2020a. Household Energy Use in Arizona. Website: http://www.eia.gov/consumption/residential/reports/2009/state_briefs/pdf/az.pdf (Accessed January 20, 2020).

EIA, 2020b. Household Energy Use in Michigan. Website: http://www.eia.gov/consumption/residential/reports/2009/state_briefs/pdf/mi.pdf (Accessed January 20, 2020).

EIA, 2020c. Residential Energy consumption Survey (RECS) -Analysis & Projections. Website: http://www.eia.gov/consumption/residential/reports/2009/state_briefs/ (Accessed January 20, 2020).

Fu, Ran, Feldman, David, Margolis, Robert, Woodhouse, Mike and Kristen Ardani. 2017. U.S. Solar Photovoltaic System Cost Benchmark: Q1 2017. National Renewable Energy, Technical Report NREL/TP-6A20-68925, September 2017.

Internal Revenue Service (IRS), 2020. Instructions for IRS form 3468. Department of the Treasury, Internal Revenue Service, 2019. Website: https://www.irs.gov/pub/irs-pdf/i3468.pdf; (Accessed: January 22, 2020).

Nationwide Mutual Insurance Company (Nationwide). 2020. Are solar panels covered by home insurance? (Website: https://www.nationwide.com/lc/resources/home/articles/solar-panel-insurance; accessed: January 20, 2020).

National Renewable Energy laboratory (NREL). 2020. PV Watts Calculator. Website: https://pvwatts.nrel.gov/ (Accessed January 20, 2020).

Schwager, M. Tate, G. and J. Kim, 2020. Pp-4 Project Report. Prepared for ESE 504 Course, University of Michigan-Dearborn, April 2020.

The Solar Foundation. 2012. Solar Accounting: Measuring the Costs and Benefits of Going Solar. Sponsored by the U.S. Department of Energy under Award Number DE-EE0003525. August 23, 2012. Website: https://farm-energy.extension.org/wp-content/uploads/2019/04/DOE-Solar-Foundation.pdf (Accessed: July 26, 2020.)

Tucson Electric Power Company (TEP). 2018. Rider 4: Net Metering for Certain Partial Requirements Service, *Tucson Electric Power Company*, 21 Sep 2018. Website: https://www.tep.com/wp-content/uploads/2018/10/704__tep_rider.pdf (Accessed: January 22, 2020).

Walker, Andy. 2020. PV O&M Cost Model and Cost Reduction. NREL, US Department of Energy. NREL/PR-7A40-68023 217AD. Website: www.nrel.gov/docs/fy17osti/68023.pdf (Accessed: October 4, 2020).

Wichner, D. 2018. New Tucson Electric Power Customers to Get Lower Credits for Excess Electricity. *Arizona Daily Star*, 12 September 2018. Website: https://tucson.com/news/local/new-tucson-electric-power-solar-customers-to-get-lower-credits/article_c1c16729-8333-57e1-9c39-3a03cdd4b3c2.html (Accessed: January 22, 2020).

21 Life Cycle Analyses

INTRODUCTION

One of the major considerations in implementing systems engineering is to have life cycle consideration in developing systems or products. This means that during the development of a system or a product the design team must consider and evaluate the entire life cycle of the system (or product) from its concept development to its disposal (after retirement), i.e., often called the "cradle-to-grave" considerations or life cycle analyses (LCA). Typically, the present value of costs incurred during the entire product program from the product concept generation to the retirement of the product is added, and the sum is subtracted from the present value of the total revenues generated by the product sales (number of units sold during the life cycle times the selling price). The resulting profit (revenues minus the costs) is used to evaluate the product program.

In the concept development phase, many alternate product concepts are developed. Leading product concepts are further evaluated by estimating their life cycle costs and revenues. And the best alternative that provides the highest benefit-to-cost ratio (i.e., revenue-to-costs ratio) is selected (see Chapter 20 on cost–benefit analysis). This life cycle analysis involving costs and revenue is called the life cycle costing analysis (LCCA). Whereas the life cycle analysis of other used, emitted, or polluting materials such as carbon, carbon dioxides, methane, or water consumed are considered in life cycle analyses (LCAs).

The life cycle costing analysis (LCCA) is most used in industries. However, due to the increasing importance of sustainability, LCA is also used to determine the effects due to environmental pollution, climate change, and emphasis on improving efficiencies in consumption of resources such as energy and water. The LCA has been used to evaluate alternatives in problems involving design for environment and design for sustainability (see Chapter 13). Here alternate product or system designs can be evaluated over the entire life cycle of the product being designed by computing evaluation measures based on environmental impact (e.g., pollution such as carbon emitted [carbon footprint], equivalent CO_2 emitted, ozone depleted, and water used) or sustainability measures (e.g., energy consumed or saved, materials recycled, components reused, and so forth).

For example, electric vehicles (EVs) are considered by many as the "zero" emission vehicles as compared to the traditional internal combustion engine (ICE) equipped vehicles. However, when a LCA of carbon produced by including in raw materials extraction, processing, manufacturing, operation, disposal, and recycling of all the energy and materials used in and for these vehicles, one quickly realizes that the creation of EVs also produces a lot of carbon.

The design for environment (DFE) and design for sustainability (DFS) are covered in Chapter 13.

WHAT IS PRODUCT LIFE CYCLE?

All products undergo the phases of development, growth, and decline. A clear understanding and knowledge of the product life cycle and its stages will help to act in the right direction to prolong the lifetime of the product. Every manufactured product goes through the following main stages of its life cycle:

   a. Research to assess customer needs and product concept generation
   b. Engineering design of the product
   c. Development of manufacturing equipment and facilities for the producing the product
   d. Manufacturing and assembly operations for the product
   e. Distribution and sales of the product (during which the product sales experience growth, maturity—achieving peak sales levels and later decline in sales as the product design gets old and obsolete)
   f. Product retirement (i.e., production is stopped, plant is retooled for the next product or model, or plant is disposed-off, retired products, and plant equipment and materials are disassembled, recycled, or scrapped).

LIFE CYCLE ANALYSIS

Life cycle analysis (LCA) is a general term that has many names such as life cycle assessment, cradle-to-grave analysis, and eco-balance. The LCA is a tool that can be used to evaluate the potential environmental impacts of a product, system, material, or process (e.g., the process of generating electric energy for electric vehicles) considering all stages of its life from raw material extraction through materials processing, manufacture, distribution, use, repair and maintenance, and disposal or recycling.

The LCA methodology evaluates the energy requirements, environmental impacts, and life cycle costs of a process or product by quantifying all the material inputs and outputs and their respective energy and environmental effects. The analysis begins at the "cradle" stage with the raw materials, continues through all of the manufacturing/processing steps (including pre- and post-processing) and follows the product through its useful life span up to its final disposal and/or recycling ("grave" stage). LCA applies to the full life cycle.

LCA can be utilized to analyze the advantages and disadvantages of a technology (ISO, 2006). LCA has been used in most industries, either for process improvements or determining how certain processes compare to others. The assessment is not limited to any one industry but can be applied to manufacturing, business growth decisions, research, and development, as well as benchmarking and a number of other areas to determine future pathways for growth.

Objectives of LCA

The primary objectives of the LCA are (a) to compare the full range of environmental effects assignable to products, systems, and services by quantifying all inputs and outputs of material flows and assessing how these material flows affect the

Life Cycle Analyses

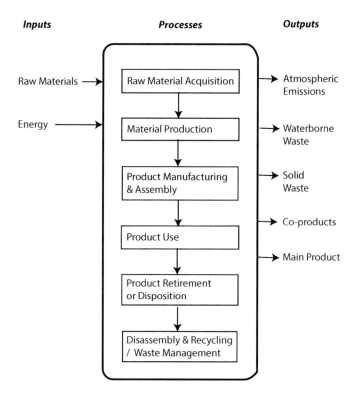

FIGURE 21.1 Inputs, outputs, and processes involved during the life cycle of a product.

environment, and (b) to use this information to improve processes (or develop new processes), support policy and provide a sound basis for informed decisions.

Figure 21.1 illustrates the inputs, outputs, and processes involved during the life cycle of a product. The processes begin with acquisition of the raw materials and energy required to create the product and all other associated materials (e.g., metals, concrete, plastics, wood, coolants, fuel, lubricants, paints required to create, install, and operate the product). The materials are used during the manufacturing and assembly processes to produce the product, install it (if needed in a facility), and use (or operate) it during its life. Later at the end of the life, the product is pulled of the service and retired. The retired product is disassembled and materials are recycled for future use or scrapped. The plant producing the product is also retired, dismantled, and the plant site is restored to its original natural state.

The left-hand side of the figure shows that different raw materials and energy sources are required to run the above processes. The right-hand side of the figure shows outputs of different processes such as the main product, co-products, waste, and emissions. The LCA requires that quantities of all the inputs and outputs of the processes must be considered and the information must be used to make decisions related to issues such as selecting a product, a system or a process, or a policy (e.g., an organization or a government policy).

Since a complex product involves many systems and each system includes many components, the above block diagram is also complex. Different raw materials and manufacturing methods are needed for each component and the energy needs for processing of materials, inputs, manufacturing processes, and their outputs (types and amounts) will be also different.

LCA Impact Categories

The LCA impact categories typically include resource consumption, energy use, water use, landfill space use (hazardous and non-hazardous), global warming, ozone depletion, photochemical smog, acidification, air quality (particulates), water eutrophication (hypertrophication or enrichment of a water body with nutrients), water quality, human health toxicity (occupational and public), ecotoxicity (aquatic and terrestrial), and aesthetics (e.g., appearance and odor).

The environmental impacts that may be covered by LCA include climate change, human respiratory health decrement from particulates, human health decrement from toxics, human health decrement from carcinogens, acidification, eutrophication, ecosystem toxicity, ozone depletion, smog formation, habitat alteration, biodiversity decrease, resource depletion, water consumption, land use, and/or land use change. The impact levels are calculated through the use of measures such as the Gross Energy Requirement (GER), Global Warming Potential (GWP_{100}), Ozone Depletion Potential (ODP), Acidification Potential (AP), Photochemical Ozone Creation Potential (POCP), Photochemical Oxidation, Eutrophication, and Human Toxicity.

The exact steps involved in the life cycle impact assessment depend upon the product or the system being analyzed. However, each component of the product needs to be evaluated in terms of processing steps that are performed from raw material extraction to assembly of the component to operation and disposal of the product and plant producing the product. During the assessment of each step, the amount of emissions and/or waste products are calculated and their overall impact is summed for each category of impact assessment.

The formulations and models for conducting LCA and LCCA also vary depending upon the product or processes being analyzed. Many software applications are also commercially available for impact assessments. The software applications are designed for different types of users and designed for different types of LCA applications. The main differentiations of LCA software are in the database and in the methodology adopted. The most used LCA packages are SimaPro (SimaPro, 2022), Easewaste (Bhandar et al., 2010), Umberto (Umberto, 2020), and GaBi (Sephera, 2022).

Additional examples of inputs (resources consumed) and outputs (including emissions) are:

a. Fuel (e.g., oil, coal, natural gas, wood) burnt, water consumed, other chemicals used, and emissions (e.g., carbon, ozone, methane) and waste products (e.g., ash, heat loss) generated during the production of raw materials (steel, plastics, rubber, glass, etc.) and components in a vehicle

b. Paint, electricity, and compressed air used and emissions generated during painting and assembly operations
c. Crude oil, natural gas, and electricity consumed and emissions from electricity generation plants (or in gasoline refining)
d. Emissions during vehicle usage (equivalent CO_2 for mixture of gases)
e. Emissions during disassembly, recycling, and disposal of automotive interiors
f. Toxicity/contaminations in water or air.

CARBON FOOTPRINT

The carbon footprint is the amount of carbon dioxide released into the atmosphere because of the activities of a particular individual, organization, or community. A carbon footprint is the total amount of greenhouse gases (including carbon dioxide and methane) that are generated by our actions. The average carbon footprint for a person in the United States is 16 tons, one of the highest rates in the world. Globally, the average carbon footprint is closer to 4 tons.

Carbon footprint: CO2 equivalent: Carbon Footprint accounts for the full inventory of all greenhouse gas emissions released throughout the product life cycle and focuses on one metric in CO_2-eq. units. The Greenhouse Gas Equivalencies calculator provided on EPA's website (EPA, 2022) is a useful tool. It allows conversion of emissions or energy data to the equivalent amount of carbon dioxide (CO_2) emissions from using that amount. Life cycle analysis processes and procedures can be used to calculate carbon footprint. Companies determining the carbon footprint of their products can also benefit from the assistance of external LCA practitioners who can conduct carbon footprint calculations based on ISO 14067:2018 standards which specify the principles, requirements, and guidelines for the quantification and reporting of the carbon footprint of a product and also in compliance with ISO 14040/44 standards for life cycle assessment (ISO, 2022).

Ulrich and Eppinger (2016) provide excellent lists of goals and guidelines for design for environment. Table 21.1 provides some of the goals and guidelines for design for environment projects.

FOUR PHASES OF LIFE CYCLE ASSESSMENT

ISO standards 14040 and 14044 define the following four phases of LCA.

1. **Goal and Scope Definition.** This phase includes the objectives of the study, the functional unit, the system boundaries, the data needed, the assumptions, and the limits that must be defined. Particularly, the functional unit is the reference unit used to normalize all the inputs and outputs to compare them with each other.
2. **Life Cycle Inventory.** This phase refers to the analysis of the material and energy flows and the study of the working system. The data are collected

TABLE 21.1
DFE Goal and Guidelines During Life Cycle of a Product Program

Life Cycle Phase	Design for Environment (DFE) Goals and Guidelines
Raw Materials Acquisition	Select plentiful, renewable raw materials
	Eliminate toxic and hazardous raw materials
	Reduce raw material transportation costs
	Reduce raw material storage space.
Material Production	Reduce use of raw materials
	Eliminate toxic materials
	Reduce and discard waste
	Reduce production
Production: Manufacturing and Assembly	Select processes with high energy efficiency
	Eliminate unnecessary manufacturing/assembly steps. Combine steps. Reduce the number of components.
	Reduce waste/scrappage of materials
	Reduce component handling, transportation, and assembly time
	Specify materials that do not require surface coatings/paintings.
Distribution	Plan the most energy-efficient shipping method
	Reduce emissions during transport
	Eliminate toxic and dangerous packaging materials
	Eliminate packaging
	Reuse packaging materials
	Specify light-weight materials
Uses/Operations	Eliminate emissions and reduce energy consumption during use
	Extend useful product life
	Reduce refueling/recharging time
Recovery (Disassemble/Recycling) and Disposal	Facilitate product disassembly to separate materials
	Enable recovery and remanufacturing of components
	Facilitate material recycling
	Reduce waste volume for incineration and landfill disposal

for the entire life cycle of all components of the system being analyzed. The critical aspect of this phase is the quality of inputs, which must be verified and validated in order to guarantee the data reliability and correct use. During this phase, a conversion of the available data to appropriate indicators takes place. The indicators are given per functional unit used.

3. **Impact Assessment.** This phase includes the assessment of the potential impacts associated with the identified forms of resource use and environmental emissions. The impact assessment methods, which are used in LCA can be divided into two categories: those that focus on the amounts of resources used per unit of materials (upstream) and those which estimate the emissions of the system (downstream waste).

4. **Interpretation.** In this phase the analyst scrutinizes the results and discusses them, giving as much precise information as possible to the decision makers. This phase may also highlight some problems in the LCA development which may need a more detailed approach. For example, it can be decided to improve the quality level of some data collected from the

literature, because they describe a process that significantly influences environmental pressure, and therefore a more elevated accuracy of them may guarantee less variability in the results. This mechanism of the LCA assures the improvement of results.

LIFE CYCLE COST ANALYSIS

Life cycle cost analysis (LCCA) is a tool to determine the most cost-effective option among different alternatives to build/purchase, own, operate, maintain and, finally, dispose off an object/product or process. The LCCA thus considers all costs associated with a product, system, or service being analyzed "from cradle-to-grave." All the costs are usually discounted and totaled to a present-day value known as net present value (NPV).

Thus, for example, in conducting a LCCA for a power plant project, in addition to the initial construction cost, LCCA will account for all the user costs (e.g., all direct and indirect costs such as land costs, financing costs, materials purchase costs, operating and maintenance costs, labor costs, insurance costs, and taxes), and agency costs related to future activities, including future periodic maintenance and rehabilitation.

Objectives of LCCA

The objectives of LCCA are (a) help management understand the cost consequences of a project, (b) identify areas of possible cost reductions, and (c) evaluating alternatives.

SOME EXAMPLES OF LCA AND LCCA APPLICATIONS

Examples of LCA

Emissions from Automotive Products

Kukreja (2018) conducted a life cycle analysis (LCA) to comparatively analyze two vehicle models of similar size of each type (ICEV and EV) currently used in the City's fleet. Ford Focus was chosen for the ICEV (internal combustion engine vehicle) and Mitsubishi i-MiEV for the EV (electric vehicle), both with a vehicle life of 150,000km. Carbon emissions and energy consumption were analyzed for each phase from cradle-to-grave for both vehicles: raw material production, vehicle manufacture, transportation, operation, and decommissioning. The analysis showed that the electric vehicle has notably lower carbon emissions and lower energy consumption per kilometer. After considering all phases, the Ford Focus emits 392.4 gCO_2-eq/km and Mitsubishi i-MiEV emits 203.0 gCO_2-eq/km over the vehicle life. Corresponding energy consumption is 4.2 MJ/km for Ford Focus and 2.0 MJ/km for Mitsubishi i-MiEV.

The assumption throughout the NASEM (2021) report is that vehicle electrification improves fuel economy (e.g., in hybrid electric vehicles [HEVs] and plug-in hybrid electric vehicles [PHEVs]), or eliminates the use of petroleum-based fuels (e.g., BEVs). If full fuel cycle emissions per mile are considered, the assumptions are more complex and depend upon the upstream emissions of the charging electricity source. When and where electricity is generated with low carbon sources, emissions per mile are significantly reduced relative to an internal combustion engine vehicle (ICEV). However, when and where electric systems depend upon high emitting generation facilities, the emission benefits are reduced. In 2025–2035, it is anticipated that the U.S. grid will continue to work toward net-zero emissions, which will drive a decrease in total emissions for electrified vehicles.

EXAMPLES OF LCCA

Chapter 24 provides case studies of cost–benefits studies provided by using this approach. It describes two studies in which the outputs of JEDI models were used.

JEDI Models: The Jobs and Economic Development Impact (JEDI) models are user-friendly tools that estimate the economic impacts of constructing and operating power generation and biofuel plants at the local and state levels (EIA, 2021). Using JEDI, you can analyze the energy impacts of wind, biofuels, concentrating solar power, geothermal, marine and hydrokinetic power, coal, and natural gas power plants. The benefits assessed by these models include revenue generated from sale of electricity, revenue from additional jobs created due to the new power plant during construction, as well as during operation and maintenance. The costs primarily include power plant development and construction costs, financing costs, and operating and maintenance costs.

Cost–Benefit Analysis of Photovoltaic Solar Panels

Chapter 20 describes a cost–benefit study conducted to compare the life cycle analysis of solar panel-based home power generators installed in two cities, namely, Detroit, MI and Phoenix, AZ.

Levelized Cost of Technologies

EIA calculates two following measures, levelized cost of electricity (LCOE) and levelized cost of avoided electricity (LACE) that, when used together, largely explain the economic competitiveness of electricity generating technologies (EIA, 2020).

Levelized Cost of Electricity:

> The levelized cost of electricity (LCOE) is a LCCA application to determine the cost of electricity over the life cycle of an electricity generating plant. LCOE represents the installed capital costs and ongoing operating costs of a power plant, converted to a level stream of payments over the plant's assumed financial lifetime. Installed capital costs include construction costs, financing costs, tax credits, and other plant-related subsidies or taxes. Ongoing costs include the cost of generating fuel (for power plants that consume fuel), expected maintenance costs, and other related taxes or subsidies based on the operation of the plant.

Life Cycle Analyses

The levelized cost of energy (LCOE), or levelized cost of electricity, is a measure of the average net present cost of electricity generation for a generating plant over its lifetime. The LCOE is calculated as the ratio between all the discounted costs over the lifetime of an electricity generating plant divided by a discounted sum of the actual energy amounts delivered. It is measured in $/MWh. The LCOE is used to compare different methods of electricity generation on a consistent basis.

The LCOE represents the average revenue per unit of electricity generated that would be required to recover the costs of building and operating a generating plant during an assumed financial life and duty cycle. Inputs to LCOE are chosen by the estimator. They can include the cost of capital, fuel costs, fixed and variable operations and maintenance costs, financing costs, and an assumed utilization rate.

Levelized cost of electricity (LCOE) represents the average revenue per unit of electricity generated that would be required to recover the costs of building and operating a generating plant during an assumed financial life and duty cycle. LCOE is often cited as a convenient summary measure of the overall competitiveness of different generating technologies.

Key inputs to calculating LCOE include capital costs, fuel costs, fixed and variable operations and maintenance (O&M) costs, financing costs, and an assumed utilization rate for each plant type. The importance of each of these factors varies across technologies.

The LCOE is calculated as follows:

$$\text{LCOE} = \frac{\text{Sum of costs over lifetime}}{\text{Sum of electrical energy produced over lifetime}}$$

Where

Sum of costs over the lifetime = $\sum (I_t + M_t + F_t)/(1+r)^t$
Sum of electrical energy produced over the lifetime = $\sum E_t/(1+r)^t$

I_t = investment expenditures in the year t
M_t = operations and maintenance expenditures in the year t
F_t = fuel expenditures in the year t
E_t = electrical energy generated in the year t
r = discount rate
n = expected lifetime of system or power station
$\sum$ = Sum over from $t=1$ to n years

Figure 21.2 presents levelized cost of electricity for different electricity generating technologies. (EIA, 2020).

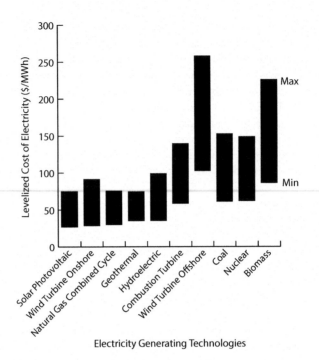

FIGURE 21.2 Levelized cost of electricity for different electricity generating technologies. [Data projected for 2025 from: Annual Energy Outlook, EIA, 2020].

Levelized Avoided Cost of Electricity:

The levelized avoided cost of electricity (LACE) represents a power plant's value to the grid. A generator's avoided cost reflects the costs that would be incurred to provide the electricity displaced by a new generation project (e.g., a solar generator produced power sold back to the utility company) as an estimate of the revenue available to the plant. As with LCOE, these revenues are converted to a level stream of payments over the plant's assumed financial lifetime.

CONCLUDING REMARKS

In conducting the LCCA and LCA, it is important that reliable and accurate estimates are obtained for all variables and used in the analyses. Otherwise, the results obtained from the analysis will not be accurate and useful. First, it is important that the modeler of the analysis understands the product or system being analyzed in terms of how it operates and used, how various components are created, materials used, processes used for producing the components, assembly processes used in lower level systems and how the lower level systems are assembled to create the whole product or system. The analyst needs to know how the whole product or the system is used during its operation in terms of product usage, i.e., usage cycles or demands in its outputs, inputs, operating conditions, and so forth, to estimate inputs, outputs, and associated costs of operations. Thus, it is recommended that these life cycle analyses should be conducted by using multidisciplinary teams with a background in technologies used in all the life cycle phases (e.g., materials, manufacture, operational, and retirement and recycling) of the products or systems being analyzed.

REFERENCES

Bhandar, Gurbakhash Singh, Christensen, Michael and Zwicky Hauschild. 2022. EASEWASTE-life cycle modeling capabilities for waste management technologies. International Journal of Life Cycle Assessment, Vol 15, Issue 4, 2010, pp. 403–416. (Website: https://orbit.dtu.dk/en/publications/easewaste-life-cycle-modeling-capabilities-for-waste-management-t/; Accessed: June 24, 2022).

EIA. 2020. Annual Energy Outlook 2020 with projections to 2050. U.S. Energy Information Administarion, Washington, DC., January 2020.

EIA. 2021. JEDI: Jobs & Economic Development Impact Models. (Website: https://www.nrel.gov/analysis/jedi/; Accessed: December 27, 2021).

EPA. 2022. Greenhouse Gas Equivalencies Calculator. Website: https://www.epa.gov/energy/greenhouse-gas-equivalencies-calculator (Accessed: May 6, 2022).

Kukreja, Balpreet. 2018. Life Cycle Analysis of Electric Vehicles: Quantifying the Impact Report prepared for: Adrian Cheng, Equipment Services, City of Vancouver, August 2018.

ISO. (2006). *Environmental management —life cycle assessment —principles and framework International Organization for Standardization.*

National Academies of Sciences, Engineering, and Medicine (NASEM). 2021. *Assessment of Technologies for Improving Light-Duty Vehicle Fuel Economy 2025-2035.* Washington, DC: National Academies Press. 2021 (Website: http://nap.edu/26092; Accessed: December 28, 2021).

Sephera. 2022. GaBi Software Suite - #1 LCA Software Website: https://gabi.sphera.com/america/software/ (Accessed: May 6, 2022).

SimaPro. 2022. LCA software for informed change-makers. Website: https://simapro.com/ (Accessed: May 6, 2022).

Ulrich, Karl T. and Steven D. Eppinger. 2016. Product Design and Development. ISBN 978-0-07-802906-6. Publisher: McGraw-Hill, Sixth Edition, 2016.

Umberto. 2020. Umberto LCA+ : LCA Software for Life Cycle Assessment. Website: https://www.ifu.com/umberto/lca-software/ (Accessed: May 6, 2022).

Part IV

Applications, Case Studies, and Integration

22 Applications of Systems Engineering Tools
A Case Study on an Automotive Powertrain System

INTRODUCTION

This chapter provides a case study involving applications of systems engineering tools to study subsystems, interfaces, requirements, and trade-offs associated in the design of an automotive powertrain system. The study involved familiarizing with a powertrain system of a late model production vehicle and creating the following: (1) a system decomposition tree, (2) an interface diagram, (3) an interface matrix, (4) an attribute requirements cascade, and (5) a list of trade-off considerations with other vehicle attributes. The project presented in this chapter is described in Appendix 3. The project was part of assignments in the author's automotive systems engineering course described in Chapter 25.

AUTOMOTIVE POWERTRAIN PROJECT

PROJECT OBJECTIVES

The objectives of the project presented in this chapter were the following:

1. To understand automotive systems, subsystems, and their interfaces
2. To cascade the system requirements into subsystem and component requirements
3. To understand coordination in system design tasks between different design and engineering activities and issues associated with the trade-offs in packaging and assembly.

PROJECT STEPS

The project was performed by completing the following steps:

1. Select a vehicle system of a current production vehicle.
2. Study the selected system, identify all subsystems of the system, and develop a decomposition tree for the selected system.
3. Develop an interface diagram showing interfaces (links) between the subsystems of the selected system and other systems in the vehicle.

4. Develop an interface matrix including all of the subsystems (of the selected system) and other vehicle systems. Specify interface type in each cell in the matrix by using the following letter codes: P = physical, S = spatial/packaging space, F = functional, E = energy transfer, M = material flow, I = information flow, and 0 = none.
5. State at least three important engineering requirements on the selected system.
6. Cascade system requirements (mentioned in item 5) to each of the subsystems. Using a tabular format, list the cascaded requirements on each subsystem.
7. Describe at least three major trade-offs that should be considered in designing the system to ensure that the system will fit and work with other systems in the vehicle.

SYSTEMS, SUBSYSTEMS, AND SUB-SUBSYSTEMS

The powertrain system of a late model, front-wheel drive, small (C-segment) passenger car was selected for this project (Prado, 2012). The system involved a 1.4 L four-cylinder double-overhead cam (DOHC) turbo-charged gasoline engine coupled to a six-speed automatic transmission. It delivered 140 HP and achieved 42 mpg fuel consumption in highway driving. The powertrain system was decomposed into the following three major subsystems: (1) engine, (2) transmission, and (3) drivetrain. Prado's original description of the decomposition of the three subsystems was rewritten as shown in the following subsections.

ENGINE SUB-SUBSYSTEM

The sub-subsystems of the engine subsystems are as follows:

Engine block and cylinder head: this sub-subsystem holds all the components of the engine. Basically, it works as a frame for the moving parts and the support devices.
Power conversion: all the moving parts that generate mechanical power are part of this sub-subsystem, such as pistons, piston rings, piston pins, connecting rods, bearings, and crankshaft.
Valvetrain: in combination with the intake and exhaust systems, this sub-subsystem allows the engine to breathe. The main components are valves, springs, retainers, sleeves, camshafts, sprockets, and timing belts.
Intake, exhaust, and turbo system: these three sub-subsystems provide an air and fuel mixture to the engine and remove exhaust gases. The intake sub-subsystem has the throttle body, intake pipes, and location for sensors. The exhaust is made up of an exhaust manifold, pipes, a catalytic converter, the location for sensors, and mufflers. The turbocharger is merged between intake and exhaust, combining the kinetic energy of the exhaust gases and converting this sub-subsystem to boost the engine in the intake side.
Fuel system: this sub-subsystem manages fuel delivery in the cylinders through a complex network of components such as a fuel tank, fuel lines, a fuel pump, fuel injectors, sensors, and a fuel pressure valve.

Cooling system: this sub-subsystem's function is to keep the engine running within the best operating temperature range regardless of the climate. Its components are the radiator, the coolant pump, coolant lines, hoses, sensors, and actuators (such as a thermostat).

Accessories: they accomplish other functions in the engine to keep it running (e.g., an oil pump and lubrication) or providing other needs onboard a vehicle such as supplying energy for the climate control system (heating, ventilating, and air-conditioning systems), steering system, electrical system, and so forth.

TRANSMISSION SUB-SUBSYSTEMS

The sub-subsystems of the transmission subsystem are as follows:

Casing: it holds all the moving parts such as bearings, shafts, and gears. It provides protection and isolation for all the components located inside. It also provides for lubrication of the moving parts.

Power conversion: it contains all the moving parts that transmit or change the mechanical power and are part of this group, such as transmission pistons, shafts, clutches, gears, and so forth.

Differential: it is a special kind of gear arrangement that is located at the final drive and makes it possible to split mechanical power among the left and the right powered front wheels in the proportion needed for the vehicle during all vehicle maneuvers. Its components are conical gears, carriers, special bearings, shims, and so forth.

Valve-body system: it is a labyrinth structure with a set of control valves that regulates the pressure and the flow rate of the oil that circulates inside the transmission. It is usually attached to or a part of the main casing.

Lubrication system: it manages the flow of oil for lubrication of the moving parts. For automatic transmission, it works in combination with the valve-body system.

DRIVETRAIN SUB-SUBSYSTEMS

The sub-subsystems of the drivetrain subsystem are as follows:

Driveshafts: they couple the differential outputs to the front wheels. They are made of steel and feature constant-velocity joints at their pivot points.

Wheels and tires: they consist of wheels with mounted tires and wheel hubs.

FASTENERS

There are many different types of fasteners (bolts, nuts, pins, clips, etc.) that provide mechanical joints between the subsystems and sub-subsystems described in the earlier subsections.

Decomposition Tree for the Powertrain System

Based on the description of subsystems and sub-subsytems given in preceding subsections, a decomposition tree for the powertrain system was created. The decomposition tree is presented in Figure 22.1. It shows the major subsystems, sub-subsystems, and components of the powertrain system in a hierarchical arrangement.

Interfaces

The vehicle powertrain system also interfaces with the following vehicle systems: body system, chassis system, fuel system, electrical system, and climate control system. Figure 22.2 presents an interface diagram of the powertrain system. The interface diagram shows interfaces between the powertrain system's three subsystems and their sub-subsystems and other vehicle systems by means of arrows. The type of each interface is indicated by a letter code placed next to each arrow. The letter codes are P = physical, S = spatial-packaging space, F = functional, E = energy transfer, M = material flow, and I = information flow.

Table 22.1 presents an interface matrix of the powertrain system. It shows all the sub-subsystems of the powertrain system and other major vehicle systems in its columns and rows. Each interface is represented by an individual cell of the matrix. The letters inside each cell indicate the type of interfaces associated with the systems defined by the row and column of the cell. The letter code used here is the same as that used in the interface diagram in Figure 22.2. The blank cells show the absence of interfaces between the rows and columns corresponding to the cells. As expected, the engine block and cylinder head subsystem, the transmission casing, and the powertrain conversion subsystems (for the engine and the transmission) have the most interfaces. Thus, powertrain engineers involved in the design of the aforementioned subsystems have to spend a lot of time in understanding all the interfaces and their requirements to ensure that all the sub-subsystems operate to provide the necessary subsystem performance and they function well with other systems in the vehicle. Since the entire powertrain system is installed in the vehicle using the vehicle body and vehicle chassis system, the body and chassis engineers have to work very closely with the powertrain engineers to ensure that all subsystems of the powertrain system can be packaged and installed to provide their respective functions.

Requirements of the Powertrain System
The powertrain system must be designed to meet requirements of the following three major attributes: (1) performance, (2) fuel economy, and (3) costs. Brief descriptions of the three attributes are provided in the following subsection.

Attributes of the Powertrain System
Fuel Economy
The vehicle must meet the U.S. Environmental Protection Agency (EPA) and the National Highway Traffic Safety Administration fuel economy and emission standards. To verify that the vehicle meets the requirements, it must undergo a series of EPA fuel economy and emissions verification tests including city and highway duty cycles and idling. The powertrain can also be tested on a dynamometer and the

Applications of Systems Engineering Tools

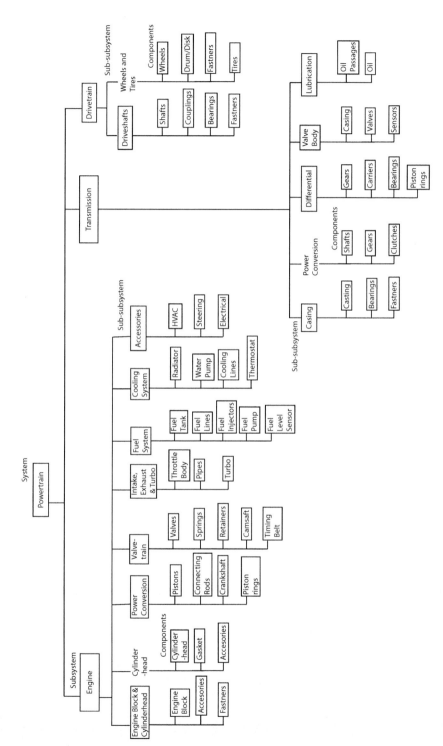

FIGURE 22.1 Decomposition tree of the automotive powertrain system.

442 Designing Complex Products with Systems Engineering Processes

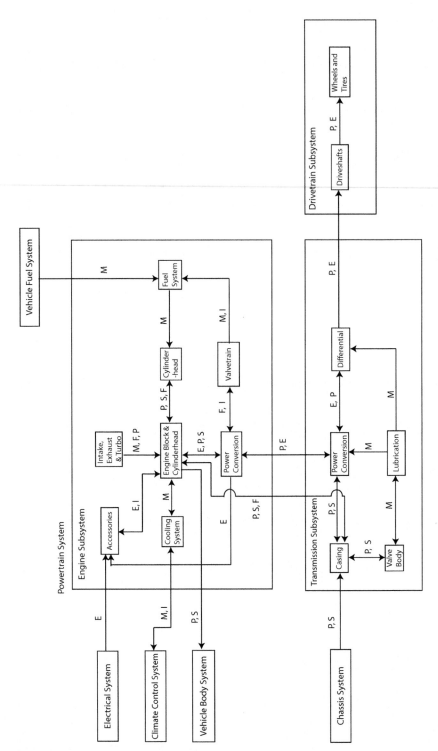

FIGURE 22.2 Interface diagram showing interfaces between the subsystems of the powertrain system and other vehicle systems.

TABLE 22.1
Interface Matrix Showing Interfaces between the Subsystems of the Powertrain System and Other Vehicle Systems

	1	2	3	4	5	6	7	8	9	10	11	12	13	14	15	16	17	18	19	20
	Engine Block and Cylinder-head	Intake, Exhaust, & Turbo	Power Conversion (Engine)	Cylinder-head	Cooling System	Valvetrain (Engine)	Fuel System (Engine)	Accessories (Engine)	Trans. Casing (trans.)	Power Conversion (trans.)	Valve Body (trans.)	Lubri-cation (trans.)	Differ-ential	Drive-shafts	Wheels and Tires	Vehicle Body System	Vehicle Chassis System	Climate Control System	Vehicle Electrical System	Vehicle Fuel System
Engine Block and Cylinderhead				P, S, F				E, I	P, S, F	E,P,S						P, S				
Intake, Exhaust, & Turbo	M, F, P																			
Power Conversion (Engine)	E, P, S					F, I		E		P, E										
Cylinderhead	P, S, F			M																
Cooling System	M				M, I													M, I		
Valvetrain (Engine)			F, I	M			M, I													
Fuel System (Engine)	M																			
Accessories (Engine)	E, I																	E		
Trans. Casing	P,S,F							I, E	P, S	P, S						I	S		I	
Power Conversion (trans.)	E, P,S		P, E						P, S		P, S		E, P							
Valve Body (trans.)										M	M									
Lubrication (trans.)											M	M								
Differential										E, P			M	P, E		S				
Driveshafts															P, E	S				
Wheels and Tires														S		S	S			
Vehicle Body System									P, S						S	P,S	P,S	P,S	P,S	
Vehicle Chassis															S	S	S			
Climate Control System					M, I														I	
Vehicle Electrical System								E								P		E, I		E, I
Vehicle Fuel System							M									P	S		E, I	

required EPA tests can be simulated by varying combinations of wheel speed and torque in a test cell.

Performance

Several methods can be used to test powertrain performance. For example, the vehicle can be tested using a dynamometer, road tests, long-term durability tests, cold and hot environmental tests under different maneuvers, and so forth. A professional driver can be used to conduct the acceleration–time test from 0 to 60 mph on a test track. The powertrain can be tested in a dynamometer at combinations of wheel speed and torque over different simulated vehicle maneuvers.

Costs

The vehicle cost can be based on the manufacturer-suggested retail price (MSRP) and benchmarking prices of similar automotive products. The powertrain costs can be estimated based on an acceptable percentage of the vehicle MSRP minus the sum of dealer and marketing costs and the desired company profit. The powertrain cost can be also estimated from the manufacturing and/or supplier costs of the powertrain system. Generally, developing a cost target of the powertrain from a competitive benchmarking exercise followed by reductions in manufacturing costs to meet target costs is considered to be a reasonable approach.

Cascading Vehicle Attribute Requirements to Powertrain Requirements

Table 22.2 presents a tabular chart created to understand the role of vehicle attributes and their sub-attributes in the development of powertrain system design requirements. The table presents all the vehicle attributes in the first column. The sub-attributes of each of the vehicle attributes are presented in the second column of the table. And the third column presents the powertrain design requirements developed to support each of the vehicle sub-attributes shown in the second column. The last three columns briefly describe key requirements on the three powertrain subsystems. These subsystem requirements are cascaded (or resulted) from the vehicle attributes and their sub-attributes listed in the first two columns of this table. The tabular format is very useful in realizing that the powertrain system needs to meet certain requirements of each of the vehicle sub-attributes that have a relationship with it. The tabular format thus shows traceability of requirements from the product level to the subsystems level.

Trade-Offs in Powertrain Development

There are several trade-offs between the three major attributes of the powertrain system. Higher vehicle performance requires a more powerful engine and higher capacity transmission and drivetrain, which increase the powertrain costs. Achieving a higher performance requires a more powerful powertrain system, which results in challenges in packaging its larger and heavier hardware. This also negatively impacts achieving a better fuel economy. Similarly, greater levels of fuel economy (i.e., higher miles per gallon) are also linked to higher costs due to the extra hardware

TABLE 22.2
Vehicle Attributes Affecting Powertrain Requirements

Vehicle Attribute	Sub-attribute	Powertrain Design Requirements	Engine Subsystem Requirements	Transmission Subsystem Requirements	Drivetrain Subsystem Requirements
Package	Seating package (driver and passengers). Entry/exit Luggage/cargo package Fields of view/visibility				
	Powertrain package	The powertrain must fit within the vehicle space defined by overall vehicle dimensions, passenger space, and suspension and steering system space. The powertrain for a front-wheel drive vehicle must fit within the engine compartment.	Must fit within the engine compartment.	Must fit within the engine compartment.	Must fit within the engine compartment.
	Suspensions and tires package				Must share front fender space with wheels, tires, suspension, and steering system.
	Other mech package		Must share space with bumpers, grill, forward lighting system, and energy-absorbing front structure.		
Controls, displays, and ergonomics	locations—layout	Display powertrain-related functions (e.g., tachometer, fuel gauge, and fuel economy).	Communicate engine speed and fuel consumption data.	Display gear/shift position.	
	Hand and foot reach		Service points within reach distances from outside the vehicle.	Service points within reach distances from outside the vehicle.	

(continued)

TABLE 22.2 (Continued)

Vehicle Attribute	Sub-attribute	Powertrain Design Requirements	Engine Subsystem Requirements	Transmission Subsystem Requirements	Drivetrain Subsystem Requirements
	Visibility and obscurations Operability		Visible engine service labels. Visible hood controls. Hood-opening controls and hood prop-up system ergonomics.	Visible engine transmission labels.	
Safety	Front impact	Body components in the frontal region and powertrain packaging must be considered simultaneously.	Mechanical properties (e.g., weight, center of gravity) must be considered in impact performance.	Mechanical properties (e.g., weight, center of gravity) must be considered in impact performance.	Mechanical properties (e.g., weight, center of gravity) must be considered in impact performance.
	Side impact Rear impact Roof crush Sensors, belts, and air bags Other safety features				
Styling/appearance	Exterior—shape, proportions, etc.	The powertrain must accommodate within the hard points (e.g., the cowl point) and front bumper to firewall length and overall vehicle width.	Must fit below cowl point.	Must fit within the engine compartment.	Must fit within the overall vehicle width (with tires, wheels, and suspensions).
	Interior—I/P, console, trim, etc. Luggage/cargo/storage Underhood appearance				

Category	Subcategory	Description
	Color/texture mastering	Consider an engine cover to improve underhood appearance. Specify colors and textures of visible surfaces.
	Craftsmanship	Color code and label all engine service points. Locate transmission fluid-level dipstick within reach and view.
Thermal and aerodynamics	Aerodynamics	Grill openings, hood size, and shape affect aerodynamics. Minimize drag because of engine compartment and front fenders.
	Thermal management	The engine cooling and exhaust heat protection must be considered. Provide airflow for radiator cooling.
	Water management	Engine electrical systems must be shielded from rainwater or placed in water-tight compartments. Shield engine electrical and electronics from water damage. Shield transmission electrical and electronics from water damage.
Performance and drivability	Performance feel	These are key attributes during powertrain development. Matching of the three subsystems is most important to provide needed engine speeds to wheel speed and required torque-level combinations during up and down shifting. Must meet long-term durability requirements of 100,000 mi. for all powertrain and chassis components.
	Fuel economy	
	Long-range capabilities	
	Drivability	
	Manual shifting	
Vehicle dynamics	Ride	Powertrain parameters (e.g., weight and center of gravity) affect vehicle dynamics. Must consider the entire powertrain system weight, weight distribution, and subsystem locations for vehicle dynamics. Must meet long-term durability requirements of 100,000 mi. for all powertrain and chassis components.
	Steering and handling	
	Braking	

(Continued)

TABLE 22.2 (Continued)

Vehicle Attribute	Sub-attribute	Powertrain Design Requirements	Engine Subsystem Requirements	Transmission Subsystem Requirements	Drivetrain Subsystem Requirements
Noise, vibrations, and harshness (NVH)	Drivability Manual shifting	Important for customer perception of vehicle image and quality.	Must meet NVH requirements.	Must meet NVH requirements.	Must meet NVH requirements.
	Wind noise				
	Electrical/mechanical	Electrical connections to alternator, starter motor, powertrain sensors, CPUs. Sensors and actuators.			
	Brake NVH				
	Squeaks and rattles				
	Pass by noise	Exhaust noise must meet acceptable sound requirements.			
Interior climate comfort	Heater performance A/C performance	A/C and heater performance is dependent on powertrain system characteristics.	Consider engine power loss because of the water pump and A/C system.		
Security	Water ingestion Vehicle theft Contents/component theft Personal security				
Emissions	Tailpipe emissions	These requirements must be integrated during powertrain development.	Must meet emissions requirements.		
	Vapor emissions Onboard diagnostics		Engine CPU (microprocessor serving as a central processing unit) data transfer to diagnostic and display system.		

Applications of Systems Engineering Tools

Category	Sub-item	Powertrain			
Weight	Body				
	Chassis				
	Powertrain	Must consider powertrain weight.	Meet target weight for engine.	Meet target weight for transmission.	Meet target weight for drivetrain.
	Climate control		Meet coolant transfer flow to heater core.		
	Electrical				
Cost	Cost to the customer	Powertrain system costs affect vehicle cost and corporate financial picture.	Meet cost targets for engine.	Meet cost targets for transmission.	Meet cost targets for drivetrain.
	Cost to the company				
Customer life cycle	Purchase and service experience	Powertrain system will affect all sub-attributes related to customer life cycle.	Consider service costs. Increase oil-change mileage.	Long life transmission fluid.	Service-free lubrication.
	Operating experience				
	Life stage changes				
	System upgrading				
	Disposal/recyclability				
Product/process compatibility	Powertrain system design will be affected by decisions on all the product/process compatibility issues. Commonality Carryover Complexity Tooling/plant life cycle	Reusability	Consider modular designs and component sharing to reduce costs.		

(e.g., turbocharger), costly light-weight and/or high-strength materials (e.g., aluminum, magnesium, and titanium), more sophisticated engine and transmission (e.g., turbo-boost, low-friction bearings, and eight-speed transmissions), low drag aerodynamic body shape, engine start–stop feature, and well-integrated vehicle systems (e.g., to reduce weight and increase crashworthiness capabilities) needed.

CONCLUDING REMARKS

Vehicle design involves managing many customer needs. It is important to highlight that cascading system requirements from the product level to lower levels of entities is essential to the development of a product that the customer really wants. This analysis shows that many vehicle attributes affect powertrain design. Very often these requirements produce opposing effects and the trade-offs between the requirements require careful selection of technologies and optimization of design parameters. The three attributes, fuel economy, performance, and costs, rank high in importance to the customers. And the trade-offs between them are complex as they require infusion of many new technologies that, in turn, require substantial amounts of additional research, design iterations, and testing.

REFERENCE

Prado, M. 2012. *Project P-2: Vehicle Systems Analyses: Requirements, Interfaces, Trade-offs and Verification.* Unpublished report prepared for AE 500 Class (Instructor: Prof. V. Bhise), University of Michigan-Dearborn, MI.

23 Case Studies and Integration

INTRODUCTION

This chapter provides nine case studies to illustrate product development issues and applications of several techniques covered in the preceding chapters. The case studies are intended to help the reader in understanding how relatively simple tools can be applied to gain insights into product design issues and to solve problems related to the complex products.

The first case study involves applications of product decomposition, product attributes, and use of matrix data to determine the relative importance of systems in a motorcycle. The second case study illustrates a benchmarking study of automotive steering wheels and evaluations of the steering wheels in a driving simulator. The third case study shows how a Pugh diagram can be used to improve an automotive design concept by comparing it with existing competitors and the manufacturer's current product. The fourth study is about a product development success story where a design team visited users of pneumatic grinders and developed their new grinder that won a lot of awards. The fifth case study involves a review of a unique product design and assembly features of "Smart"—a two-passenger microcar. The sixth case study involved observations made from watching a 5-hour video on the development of Boeing 777 commercial airliner. The seventh case study is on some unique product features of Boeing 787 Dreamliner. The eighth case study illustrates a flexible process for assembling a family of complex laptop computer products with a large product variety. Finally, the ninth case study involves translation of customer needs of an electric car into its engineering specifications.

CASE STUDY 1: MOTORCYCLE SYSTEMS

OBJECTIVES

1. To illustrate product decomposition, product attributes, and use of matrix data to determine the relative importance of systems in a motorcycle (a complex product)
2. To determine the most important systems within a motorcycle from the viewpoint of customer satisfaction.

PROJECT DESCRIPTION

One of the key techniques in managing complex product development is to decompose the product into a number of manageable systems so that the systems can be designed in coordination with their interfacing systems. Therefore, this section uses a motorcycle as an example of a complex product and illustrates its decomposition into a number of systems.

The systems in the motorcycle are illustrated in Figure 23.1. Each of the systems can be further decomposed into its subsystems and components of each of the subsystems, and a decomposition tree of the motorcycle can be constructed.

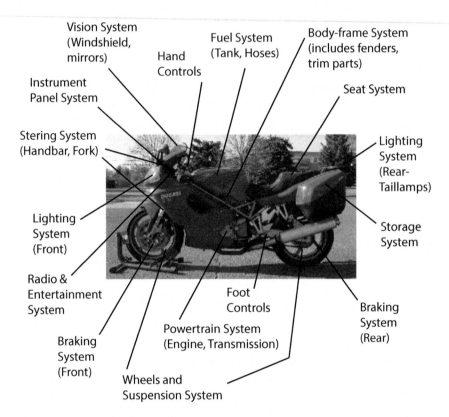

FIGURE 23.1 Illustration of the decomposition of a motorcycle into a number of systems.

Case Studies and Integration 453

MOTORCYCLE ATTRIBUTES TO SYSTEMS RELATIONSHIPS

The next important step is to identify attributes of the complex product by communicating with its customers. Table 23.1 shows six attributes of the motorcycle in its left-hand column obtained from interviews with a number of motorcycle owners to understand their needs.

The table presents a matrix of the strength of relationships between the attributes and the systems in the motorcycle illustrated in Figure 23.1. The strengths of relationships are indicated by using a 10-point rating scale, where 10 = very strong relationship, 1 = very weak relationship, and a blank or no scale value illustrates no relationship. The strengths of relationships were developed by the engineers involved in benchmarking the motorcycles (see Table 15.1 and Figures 15.1 and 15.2).

The importance ratings of each of the attributes are presented in the last column of Table 23.1. The importance ratings were obtained by asking the customers to rate the importance of each attribute by using a 10-point importance rating scale, where 10 = most important and 1 = least important.

The absolute importance ratings of each of the systems were obtained by summing the products (multiplications) of the strength of the relationship of each attribute to the system and importance rating of each attribute. The relative importance ratings are expressed in percentages of the sum of all absolute importance ratings. (Note that the sum in Table 23.1 is 2749. The sum is calculated by using the same formula described in Chapter 15 for calculating the absolute importance ratings in the quality function deployment [QFD].) The most important systems of the motorcycle (that affect the product attributes) were the body-frame system and hand controls (both received 11% [highest] relative importance ratings). The steering system, powertrain system, and foot controls (with 10% relative importance ratings) were the next important systems.

Table 23.2 presents an interface matrix of the motorcycle systems. The matrix shows that the systems with the most interfaces (ordered from highest to next lower) are the body-frame system, hand controls, foot controls, and instrument panel system. Engineers assigned to design each system must coordinate their designs with all the interfacing systems shown in the matrix.

The motorcycle case study thus illustrates the use of customer-based product attributes, decomposition of the product into manageable number of systems, and understanding the role of the important systems and interfaces between the systems—all of which are the early steps in the systems engineering implementation aimed toward designing the "right" product.

TABLE 23.1
Attributes to Systems Relationship Matrix

Attributes	Body-frame System	Steering System	Wheels and Suspension System	Barking System	Seat System	Powertrain System	Fuel System	Electrical System	Lighting System	Vision System	Instrument Panel System	Hand Controls	Foot Controls	Storage System	Radio and Entertainment System	Importance Rating (10 = Most Important; 1 = Least Important)
Power and performance						9	5	5								10
Maneuvering, handling, and braking	9	9	9	9		9						9	9			10
Comfort and convenience			5		9					6	6	5			3	7
Safety	9	9		9	3	3			9	9	5	9	9	9		9
Styling and appearance	9	5	5	3	3		5		3	3	5	3	3		3	6
Costs	9	9	9	5	5	9	5	5	5	3	5	3	3	1	1	8
Absolute importance ratings	297	273	227	211	121	279	120	90	139	165	157	298	263	62	47	2,749
Relative importance ratings	11	10	8	8	4	10	4	3	5	6	6	11	10	2	2	100

Case Studies and Integration

TABLE 23.2
Interface Matrix of Motorcycle Systems

	Body-frame System	Steering System	Wheels and Suspension System	Barking System	Seat System	Powertrain System	Fuel System	Electrical System	Lighting System	Vision System	Instrument Panel System	Hand Controls	Foot Controls	Storage System	Radio and Entertainment System
Body-frame system		P, F	P, F		P, C	P, F	P	P	P	P	P	P	P	P	
Steering system	P, F		F									P			
Wheels and suspension system	P, F	F		P, F		F									
Barking system	P	P	P, F					F	F			F	F		
Seat system	P, C									C		C, S	C, S		
Powertrain system	P, F	P	F					F			C	F	F		
Fuel system	P							F			F				
Electrical system	P			F		F	F		F		F	F	F		F
Lighting system	P	F		F		F		F			F	F	F		
Vision system	P				C				F		C				
Instrument panel system	P				C	F	F	F	F	C		F	F		F
Hand controls	P	P, F		F	C, S	F		F	F						
Foot controls	P			F	C, S	F		F	F			C			C
Storage system	P														
Radio and entertainment system	P							F			C				

Note: P = Physical attachment; F = Functional relation; C = Comfort related; S = Safety related.

CASE STUDY 2: BENCHMARKING AND EVALUATION OF STEERING WHEELS

Objectives

The objective of this case study is to illustrate how existing product designs can be benchmarked and evaluated.

Project Description

The project was undertaken at a request of an automotive steering wheel design engineer. The steering wheel engineer was asked by his product engineering vice president to drive a particular European luxury car and design a similar steering wheel for their new vehicle that was being designed for a luxury market segment. The engineer was not sure if the European vehicle had the best steering wheel for his automotive product. Thus, he undertook a benchmarking study by gathering a large number of steering wheels from different automotive manufacturers. He also assembled a team of designers and engineers to study the design and manufacturing issues with the steering wheels. The benchmarking study enabled the team to understand different design variables of the steering wheels. The team selected a few steering wheels with different combinations of the design variables. The selected steering wheels were used in an evaluation experiment in a driving simulator. The benchmarking and the evaluation studies are described below.

Benchmarking Study

The steering wheels of a large number of current production vehicles and some steering wheels sold in the aftermarket (for nonproduction or custom-designed vehicles) were reviewed by the design team consisting of steering wheel engineers and interior designers. They also collected a large number of steering wheels from different United States, European, and Asian automotive manufacturers. Figure 23.2 presents photographs of some of the steering wheels. The following characteristics of the steering wheels were measured and a database involving the following variables was created.

1. Number of spokes (3 or 4)
2. Radial locations of the spokes
3. Rim cross section (shapes and dimensions)
4. Rim material (leather, plastic, wood, simulated wood, or other)
5. Rim color (dark, light, or wood-like)
6. Rim surface hardness (or softness)
7. Rim surface smoothness (or roughness)
8. Rim texture
9. Spoke width at the rim
10. Spoke thickness at the rim
11. Spoke width at the hub
12. Spoke thickness at the hub

Case Studies and Integration 457

FIGURE 23.2 Illustrations of steering wheel designs included in the study.

13. Stitch pattern (if the rim was leather-wrapped)
14. Hardness or softness of stitches (stitch feel)
15. Stitched thread diameter
16. Stitch thickness (stitch protrusion from the surrounding surface).

The analyses of the benchmarking database showed that all the steering wheels of the luxury vehicles had 3 or 4 spokes, the rim diameters ranged between 28 and 35 mm, the rims were leather-wrapped or partially covered with laminated wood or wood-like materials, and most leather-wrapped wheels had stitches along the inside edge of the rim. The hardness ratings of the steering wheel rims ranged between "somewhat soft" and "very hard." The rim hardness values were also measured by using a hardness tester.

EVALUATION IN A DRIVING SIMULATOR

The design team selected 12 steering wheels from the benchmarking study for customer evaluation. After extensive discussions, the team agreed that the selected steering wheels should not be evaluated in a static nondriving situation. A test vehicle with a special steering column could be used for evaluations under actual driving situations. But the time required to change the steering wheels and the total time required for the drive evaluations of all the steering wheels was found to be impractical. Thus, the team decided to conduct the study in a driving simulator where the entire test could be conducted within less than 1.5 hours for each test subject. The steering column in the driving simulator was modified to change steering wheels very quickly (within 15 seconds). Eighty subjects participated in the simulator study, where each subject drove the simulator with each steering wheel in a random order for a few minutes. During the simulator driving, an experimenter seated in the front passenger seat asked a number of questions to the driver. Two pictures of the simulator setup are shown in Figure 23.3.

The instructions to the test drivers and some questions asked by the interviewers are shown in Figure 23.4. For example, in question #1, the interviewer asked, "Is the steering wheel rim comfortable or uncomfortable?" If the subject answered that it was "comfortable," then the interviewer asked if it was "somewhat comfortable" or "very comfortable." If the subject could not make up his/her mind on whether it was comfortable or uncomfortable, then the middle category of response, namely, "neither uncomfortable nor comfortable" was selected.

After all subjects were tested, the data were summarized by computing the percentages of responses in each category of the five-point scales for each question. The steering wheels with highest percentages of desired answers (e.g., "liked very much" and "liked"; "very comfortable" and "comfortable") were further analyzed to determine characteristics (e.g., rim thickness, slipperiness, and softness) of the highly preferred steering wheels.

This case study, thus, illustrates how a design team can obtain information from benchmarking many competitive products and by conducting quick evaluations using the customers to help select products with preferred characteristics.

Case Studies and Integration 459

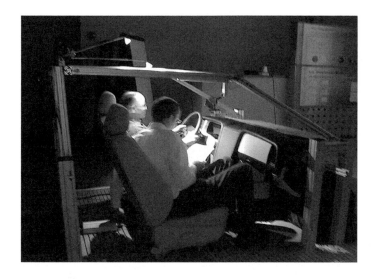

FIGURE 23.3 Driving simulator used for evaluations of the steering wheels.

CASE STUDY 3: PUGH ANALYSIS OF AN AUTOMOTIVE CONCEPT

OBJECTIVE

The objective of this case study was to illustrate the usefulness of a Pugh diagram in selecting and improving a product concept.

Problem: New Product Concept

A Pugh diagram provided in Table 23.3 was completed by an automotive product development team during an early concept generation stage. Based on the data

Instructions: Hold your hands on the top of the steering wheel and answer the following questions:

Question No.	Desription of the Question	Response Scale (Select one)				
1	Is the steering wheel rim	Too Uncomfortable	Somewhat Uncomfortable	Neither Uncomfortable nor comfortable	Somewhat Comfortable	Too Comfortable
2	Is the steering wheel rim	Too Thin	Somewhat Thin	About right	Somewhat thick	Too Thick
3	Is the steering wheel rim	Too Slippery	Somewhat Slippery	Neither Slippery nor Sticky	Somewhat Slicky	Too Sticky
4	Is the touch feel of the rim surface	Too Dry	Somewhat Dry	Neither Dry nor Wet	Somewhat Wet	Too Wet

Instructions: Now hold the steering wheel by grasping the top spokes such that the each thumb hooks on the spoke near the rim and answer the follwing questions:

Question No.	Desription of the Question	Response Scale (Select one)				
5	How is the touch feel of the top side of the spokes	Like Very Much	Like Somewhat	Neither Like nor Dislike	Dislike Somewhat	Dislike Very Much
6	Is the topside of the spokes	Too Soft	Somewhat Soft	Neither Soft nor Hard	Somewhat Hard	Too Hard
7	Is the topside of the spokes	Too Smooth	Somewhat Smooth	Neither Smooth nor Rough	Somewhat Rough	Too Rough

FIGURE 23.4 An illustration of a questionnaire used during the simulated driving.

provided in the Pugh diagram, what conclusions can you draw? What advice can you give to the team to improve the acceptability of their new product? Provide reasons to support your advice.

ANALYSIS OF THE PROBLEM

The Pugh diagram presented in Table 23.3 compares the new product concept (called the "New Product" in the third column of the table) along with three competitor products (called competitor #1, competitor #2, and competitor #3) with the "current product" as the "datum" (shown in the last column of the table). The comparisons were made on 13 customer-based product attributes presented in the second column of the table. The notations used to display the results of the comparisons of the products (based on a given attribute) were as follows: + = compared product was better than the current product, 0 = compared product was the same as the current product, and − = compared product was worse than the current product. Additional information on the Pugh diagram is provided in Chapter 15.

The bottom three rows of the table present sums of all pluses, minuses, and net score ("sum of pluses" minus "the sum of minuses" for each column). Comparing the products based on the net score, the new product is only 1 point ahead of their current product (datum) and it does not perform better than competitors #1 and #2. Competitor #2 is better than the new product in the following four attributes: (1) customer life cycle, (2) cost, (3) electrical and electronics, and (4) weight. Competitor #1 is better than the new product in the following six attributes: (1) customer life cycle, (2) cost, (3) emissions, (4) electrical and electronics, (5) weight, and (6) noise, vibrations, and

TABLE 23.3
Pugh Diagram Showing Comparative Evaluation of a New Product Concept

Sr. No.	Customer-Based Product Attribute	New Product	Competitor #1	Competitor #2	Competitor #3	Current Product
1	Customer Life cycle	0	+	+	0	
2	Cost	−	0	0	0	
3	Styling and Appearance	+	−	+	0	
4	Package and Ergonomics	0	0	0	0	
5	Performance and Economy	+	+	+	0	
6	Safety/Security	0	0	0	0	
7	Vehicle Dynamics	+	0	+	−	
8	Interior Comfort Environment	0	0	0	−	
9	Emissions	0	+	0	0	
10	Electrical and Electronics	0	+	+	0	
11	Weight	−	0	0	0	
12	Noise, Vibrations, and Harshness	0	+	0	+	
13	Aerodynamics and Thermal Management	0	0	0	0	
	Sum of Pluses	3	5	5	1	
	Sum of Minuses	2	1	0	2	
	Net Score	1	4	5	−1	

harshness. However, the new product is better than competitor #1 in the following two attributes: (1) styling and appearance and (2) vehicle dynamics.

Therefore, the new product must be minimally improved in the following attributes: (1) customer life cycle, (2) cost, and (3) weight. These changes would reduce the sum of minuses from 2 to 0 and would increase the sum of pluses from 3 to 4. Thus, the net score of the new product can be brought up to 4 from 1, which will make the new product on par with competitor #1. In the electrical and electronics attribute, the new product will be as good as their current product but will not be better than competitors #1 and #2. The team was comfortable with the above decisions as they felt that improving electrical and electronics would have added too much cost to the new product.

The Pugh diagram thus allows the decision makers to make many comparisons and look for opportunities to make further changes in their new product concept until the design team feels that they had incorporated sufficient improvements to make the product attractive to their customers.

CASE STUDY 4: CYCLONE GRINDER DEVELOPMENT

OBJECTIVE

The objective of this project was to review an article published in a book by Peters (1992) on the development of Ingersoll Rand's (IR) new Cyclone grinder (see Figure 23.5). The project was revolutionary (at the time when it was undertaken) as it could substantially reduce the product development time and created a totally new and improved grinder design to a level that it won many awards and was loved by its end users.

The specific objectives in reviewing the article were as follows:

1. List customer needs of the grinder (as a customer/user would mention in his/her words)

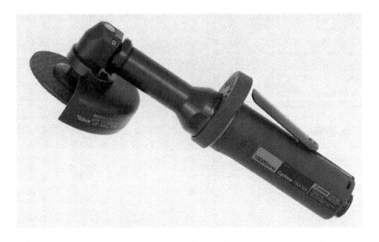

FIGURE 23.5 Ingersoll Rand pneumatic grinder.

Case Studies and Integration

2. Generate a list of functional requirements of the grinder (as an engineer would specify)
3. Describe the decomposition of the grinder (systems and components within each system)
4. Develop project activities and milestones chart showing all the key activities involved in the development of the Cyclone grinder
5. Describe key concepts/ideas that contributed to the successful design of the Cyclone grinder
6. Describe risks associated in the grinder development program This project is also described in Appendix 1 as objective #1.

PROJECT DESCRIPTION

The project was revolutionary (when it was undertaken in the late 1980s). It reduced the product development cycle by about half to only 1 year, using a small but co-located and dedicated team of professionals from different disciplines (involving personnel from engineering, industrial design, manufacturing, project management, marketing, sales, distributors, suppliers, and customers) and completely redesigned the product by listening to its customers and improved its cost, robustness, safety, and user comfort.

CUSTOMER REQUIREMENTS FOR THE GRINDER

Note: This list of requirements is stated from the views of the customer/user.

1. The grinder must be comfortable to use. [Customer comment: "After eight hours of nearly straight use it would be nice if my hands weren't numb and stinging at the end of the day."]
2. The grinder must be reliable. [Customer comment: "Getting my job done in a timely fashion is my highest priority. This is not possible if my tools are constantly breaking down."]
3. It must be fast and easy to change disks. [Customer comment: "Heavy use causes the grinding disks needing changes several times a day. The faster the disks can be changed, the more efficiently I can complete my job."]
4. The grinder must be powerful. [Customer comment: "Performing my job duties can get frustrating if the grinder stalls under normal use."]
5. The grinder must be safe. [Customer comment: "Power tools like this are inherently dangerous and any safety measures that can be incorporated into the tool give me peace of mind."]

FUNCTIONAL REQUIREMENTS FOR THE GRINDER

Note: This list of requirements is stated from the views of an engineer.

1. Composite material used for the grinder body must be durable. [Engineer comment: "Customer's might be worried about the switch from a metal to a plastic housing. Durability of the composite housing must be on par with metal."]

2. Grinder must be powerful and not stall at any speed. [Engineer comment: "Grinder must be highly durable against stalling under both normal and abusive use conditions."]
3. Grinder must be cost-effective. [Engineer comment: "The new Grinder cannot cost more to produce than previous models."]
4. Grinder must be safe, that is it should have a hand guard. [Engineer comment: "A hand guard needs to be used in order to protect the user's fingers."]
5. Disks must be easy to discriminate and attach. [Engineer comment: "Color coding the grinders and disks will help the user to correctly match the products."]
6. Grinder must be competitively appealing. [Engineer comment: "The aesthetics of the products must be at least on-par with those of competitors."]
7. Grinder must be easy to manufacture. [Engineer comment: "Ease of manufacturing cuts cost and increases quality."]

Systems and Components of the Grinder

The three major systems of the grinder and their components are presented below.

1. *Head section* (business end): spindle/collet, collet nut, grinding wheel, wheel flange, flange spacer, flange nut, wheel guard, guard lock washer, guard screws, guard label, extension (optional), angle head (optional)
2. *Body section* (motor housing and user interface area): composite shell molded over a steel liner, safety flange and flange clamp, locking or non-locking throttle lever, throttle lever pin, throttle valve plunger, "drop-in" motor and vanes, front and rear rotor bearings, front end plate and spacer
3. *Inlet and exhaust sections*: inlet assembly, ball valve, inlet screen, inlet seal, exhaust hose adapter, exhaust hose, exhaust hose retainer, flow ring.

Grinder Development Project Schedule

Figure 23.6 presents the project activities and milestone chart. It shows all major activities and project phases. The milestones are major events (e.g., management concurrence or approval, beginning of a next activity). The milestones are indicated by black-filled triangles.

Key Concepts for Successful Cyclone Grinder Design

Numerous concepts/ideas led to the success of the Cyclone grinder design. Some of these included the following:

1. *Motivation*: The motivation behind this project was one of the important factors. The demand for the grinder was enormous at that time. This air-power hand tool was used for material removal and fine finishing in aircraft engine turbine blades, auto engine blocks, and even in bar stool legs and in several other applications. Customers such as Boeing and Caterpillar bought these grinders by the thousands. Thus, the motivation for being a market leader in this industry was high for IR.

Case Studies and Integration 465

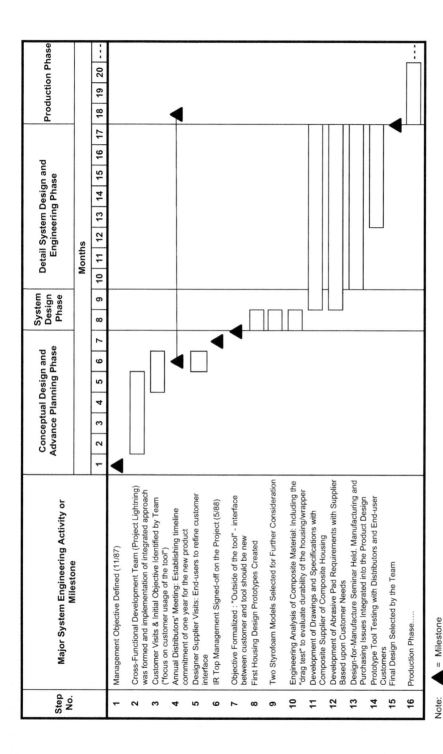

FIGURE 23.6 Project activities and milestones chart.

2. *Determination for improvement*: In 1987, IR's air-powered grinder was ranked third in the industry. It had been reconceived four times in the preceding dozen years and with each round, the innovation was less creative, more expensive, and more time-consuming than the round before. With this kind of history, the determination of improving the process was extraordinary for IR's managers. They were inclined to think of a better way to improve their new product development process.
3. *Project planning*: Presence of a plan before the start of the project helped in guiding and managing the entire project throughout the term.
4. *Customer focus*: Traditionally, engineering would have kicked-off the new product development process by working on designing a "better" grinder without knowing how it would be liked by its customers. IR's decision to visit the customer sites and see the actual product in action was the most important pillar of this project. The engineers at IR were able to take the feedback from the customers and design the product that was ideal and safer for the customer, as the customer wanted. This was a big selling point for this product.
5. *Systems Engineering*: The systems engineering approach for this product helped in adding creativity and maintaining a low cost and implementing time management to the project. Thus, implementing the SE approach was a key contributor in achieving huge success in the project.
6. *Concurrent engineering approach*: In the past, the IR team had used the traditional sequential process. This time the concurrent engineering approach was used. Every major player in the project was involved right from the beginning. This way everyone was able to contribute and provide their feedback right from the beginning. And thus, the chances of going through several redesigning iterations were significantly reduced.
7. *Team building events*: The small events such as barbecue or dinner at the Guthrie Inn did not seem like too much but it really helped the team to bond and work together in harmony.
8. *Industrial design*: Getting an external industrial design firm involved in the project was another addition to the pillars of the success of the project. The firm came up with 20–30 different design concepts, delivered on time, and produced Styrofoam models that helped the IR managers visualize how the product would look like. This entire process helped in promoting the urgency and importance of the project and the product among the design team.
9. *Supplier's involvement*: The suppliers that provided the major components for the new grinder were made full-blown members of the development team. This way the IR team had plenty of time to go back and forth between the suppliers and add their creativity and experience to their product design. Several important decisions were made during this process, which turned out to be major selling points for this product.
10. *Workers involvement*: Several of the front-line workers had engineering-breakthrough ideas in their mind. Involving workers and getting their feedback provided a big help for the project. This also helped in boosting morale

among the workers and provided them with motivation for the breakthrough design.
11. *Team involvement*: Program manager's decision to not decide early between the routine and revolutionary design was a big psychic boost for the members in team. A decision on which design to go with was not made till the entire team and key outsiders were onboard to commit to one design.
12. *Validation/verification*: IR's decision about validating the material by conducting the "drag test" (tying the grinder bodies on the back bumper of a car and dragging them in the parking lot for hours to demonstrate the scratch-resistant property of the new composite bodies), and to perform the "halt test" (to illustrate high torque capability to avoid stalling) while demonstrating the product to distributors and customers for a large number of samples was another contribution to the success of the project. This verification/validation increased the confidence while demonstrating the product to the customers and helped support the new product design.

RISK MANAGEMENT

The IR management decisions played a decisive role in the eventual success of the project. In the end, risks in several categories were effectively managed. The risks could be classified into the following four categories:

1. Programmatic risk
   a. Minimal involvement of the upper management (to allow a team to take the lead in product development activities)
   b. Lack of concept buy-in from the distributors or the upper management during the early stages added to uncertainty in program approval
   c. Insufficient support from the chosen supplier partners
2. Technical risk
   a. Challenge of creating a revolutionary design
   b. Assignment of nontechnical team members to technical tasks
3. Cost risk
   a. Expenses for team building exercises
   b. Single supplier sourcing
   c. Travel to end users' work and distributor sites
   d. Large overall team size
   e. Maintaining multiple design concepts (until final concept selection)
   f. Creating a high number of prototype products (for end user evaluation trials)
4. Schedule risk
   a. Aggressive program target timing
   b. Fixed distributor conference dates.

Although the risks were split into the above four categories, they were highly inter-related. For example, the decision to create a completely new design could result in several additional costs and scheduling risks if many developmental and customer acceptance problems were not fully understood or properly resolved.

KEY OBSERVATIONS

The eventual success of the Cyclone grinder cannot be linked to a single source. Several key concepts during the development contributed to its success. Creating a cross-functional team and a simultaneous engineering approach helped to reduce costly product design iterations by having all disciplines in the process represented at each design step. Allowing the entire diverse team to determine the product targets through extensive interactions with actual end users at the start of the project also aided in design decisions—as each team member could understand the trade-offs needed to satisfy the concerns of the end users. Clear definitions of the product targets from the end user involvement coupled with the design-for-manufacturing approach further aided in drastically reducing the part count. This in turn reduced the number of product failures and overall costs for IR and its customers. Both the key suppliers were involved in the program right from the early stage. This led to leveraging of the suppliers' technical expertise in creating unique innovative features in the grinder. Finally, successful communications with the distributors and upper management, as well as team-building exercises, helped in reducing the product development time. Without all these factors, the grinder may have not achieved its industry leading status and won many awards.

CASE STUDY 5: SMART CAR DESIGN AND PRODUCTION

OBJECTIVES

The objectives of this project were to understand the customer, product design, and business issues associated with the product development process of smart car. More specific objectives were to prepare the following:

1. List of smart car's customer needs
2. Provide a comparison of product specifications and available features of smart car with two other vehicles made by different vehicle manufacturers that can compete with it in the U.S. market
3. Describe important product design/development issues in creating the smart car including safety-related changes
4. Describe key business issues in involving suppliers at the assembly location.

The information about the smart vehicle was obtained from a number of sources (Siekman, 2002; Brooke, 2008; smart car, 2012; also see Appendix 1).

PROJECT INTRODUCTION

This case study involves smart car product design (see Figure 23.7), assembly, and business issues. The Micro Compact Car (MCC) Smart GmbH began with

Case Studies and Integration

FIGURE 23.7 Smart car.

Mercedes-Benz and Swatch in 1994. The idea of this vehicle was heavily debated and much skepticism arose around the success of this vehicle, especially in the U.S. market. smart car is still getting much attention even today with the release of a four-passenger version. It was a gamble and a different outlook on the design and manufacturing of an automobile; a gamble that proved to be a success story and also with many questions.

This case study outlines some of the key consumer needs as well as compares with its two closest competitors in the U.S. market, namely, Honda Fit and Ford Fiesta. The Chevy Spark is a future concept vehicle that would more closely resemble the size than the two benchmarked vehicles as well as the Fiat 500 that recently debuted in the United States; however, currently, there is no exact benchmark in this microsized car segment in the United States.

This section will also include important design, development, and manufacturing issues specific to Smart, as well as the key business issues facing this type of automotive philosophy. This leads to the discussion of the differences and similarities of the Smart design and assembly compared to a more traditional process followed in the United States.

SMART CAR'S CUSTOMER NEEDS

The key needs of smart car customers were as follows:

1. High fuel economy rating (low fuel consumption; greater than 40 mpg city driving)
2. Environmentally friendly

3. Comfortably accommodate two people of all sizes including tall people
4. Highly maneuverable in densely populated urban areas
5. Easy to park in small spaces
6. Safe vehicle (meets applicable safety regulations)
7. Stylish exterior
8. Affordably priced (price range $10,000–$16,000)
9. Low maintenance costs
10. Light weight
11. Ample storage space
12. Must have a few options to select from, mainly different levels of comfort and infotainment
13. Easy to enter and exit the vehicle.

The smart car was originally conceived as a societal and environmental statement to address a congestion issue in large European cities by someone outside the auto industry. This microcar was creating a new segment of the vehicle that was focused on urban commuters where driving and parking in very congested spaces is very important. Although some of the original concepts from the founder of Swatch have not materialized, the car produced by Daimler AG today does focus on many of the same customer needs from the original concept, but some have changed to meet the actual customer needs. The key customer needs with special appealing considerations are described below:

- *Safety*—although it is a small vehicle, it has to be safe. smart has responded to initial concerns, as well as the entry into the U.S. market, by continuing to make safety improvements to meet both intrusion and crash pulse requirements. The 2010 model achieved NHTSA (National Highway Traffic Safety Administration) 5 stars for driver side, 4 stars for driver front, and 3 stars overall.
- *Options*—although the original concept for smart was to manufacture a very simple vehicle available in four colors and no options from the assembly plant, before the car even launched in Europe prospective owners wanted more colors and more options. Today in the United States, the vehicle is available in seven colors, three models, and five option packages.
- *Nimble handling/performance in congested environment*—part of the original concept was that the overall size of a microcar would make it very nimble in an urban environment. This continues to be a key selling point and reason for purchase of the vehicle.
- *Ease of parking in congested environment*—similar to nimble handling and performance, the microcar has a significant advantage over larger-sized vehicles in trying to find a parking spot in congested cities. Again, this continues to be a key selling point and a reason for people to purchase the vehicle. (The length of this vehicle [2692 mm (106″)] is about the same as the width of a traditional U.S. large vehicle [2286 mm (90″) with mirrors], thus it is not uncommon in Europe to see them parked straight in against the curb where other cars are parallel parked.)

- *Fuel economy/environment*—one key motivator of the original concept was for the vehicle was to be "green," so the design intent was for the vehicle to be small, simple, and light, which also provides for high fuel economy and continues to meet the needs of the actual customer.
- *Practical/sensible/affordable*—the original altruistic model for Smart looked at the number of empty seats in vehicles commuting to work and looked to create a simple and affordable mode of transportation that met the need. Reading customer reviews online, there are many people who love this car for what it is, a small commuter car. They are very passionate about their statements, so these people understand and agree with the original Swatch concept.

BENCHMARKING OF THE SMART CAR

Table 23.4 presents the comparison of various dimensions and features of smart car with two other small cars in the U.S. market. Comparisons of exterior dimensions, weight, and engine data of the Smart with the other two benchmarked vehicles show that the smart car is considerably smaller than the two vehicles. smart is designed to carry two

TABLE 23.4
Comparison of smart Car with Two Other Small Cars in the U.S. Market

	Vehicle		
	2011 Smart Passion Coupe	2011 Honda Fit	2011 Ford Fiesta
Body Type	Coupe	Hatchback	Sedan
Seating capacity	2	5	5
Base price	$14,690	$15,100	$13,320
Exterior dimensions	L: 2,695 mm (106.1") W: 1,559 mm (61.38") H: 1,542 mm (60.71")	L: 4,105 mm (161.6") W: 1,524 mm (60.0") H: 1,694 mm (66.7")	L: 4,409 mm (173.6") W: 1,697 mm (66.8") H: 1,473 mm (58.0")
Wheelbase	1,811 mm (71.3")	2,408 mm (94.8")	2,489 mm (98.0")
Interior volume	1.29 m^3 (45.4 ft^3)	2.57 m^3 (90.8 ft^3)	2.41 m^3 (85.1 ft^3)
Weight	820.1 kg (1,808 lbs)	1,129.0 kg (2,489 lbs)	1,169.4 kg (2,578 lbs)
Engine	3 Cylinder, 1.0 Liter Premium gasoline	4 cylinder, 1.5 Liter Standard gasoline	4 cylinder, 1.6 Liter Standard gasoline
Transmission	5 Speed automated manual	5 Speed manual	5 Speed manual
Suspension	Front: MacPherson Strut Rear: DeDion axle with coil springs and shock absorbers on rear	Front: MacPherson Strut Rear: Torsion-Beam rear suspension	Front: MacPherson Strut Rear: Torsion-Beam rear suspension
Fuel economy	33 city/41 highway (mpg)	27 city/33 highway (mpg)	29 city/38 highway (mpg)
Performance	70 hp at 5,800 rpm 68 lb ft torque at 4,500 rpm	117 hp at 6,600 rpm 106 lb ft torque at 4,800 rpm	120 hp at 6,350 rpm 112 lb ft at 5,000 rpm
Internal cargo space	0.22 m^3 (7.8 ft^3)	0.58 m^3 (20.6 ft^3)	0.36 m^3 (12.8 ft^3)

passengers as compared to the other two that can carry five passengers. However, the base price of smart is comparable with that of the two benchmarked vehicles.

Key Product Design Development Issues

1. Use of plastic panels to reduce weight led to the uncertainty of crashworthiness of the vehicle and occupant safety.
2. Designing the shell frame—"Safety shell" and enhancing the strength of the structure was a challenge.
3. Packaging the engine in the rear.
4. Visible parts of the steel body were powder coated and painted plastic panels were used. This eliminated volatile emissions in the painting process.
5. Features such as air conditioning, radio, and automatic shifting were integrated into modules for packaging within the compact space.
6. Due to the immaturity of available hybrid technology at that point of time, the design team had to tweak the existing gasoline powered engines to improve their performance.
7. Avoiding the influence of Mercedes Benz and maintaining the program goal until the launch of the vehicle.

Key Business and Supply Chain Issues

1. Supplier selection and ensuring that they had the latest technology and availability of worksite to reduce wastage of material and time.
2. Higher scrap rate due to inexperienced suppliers and new processes.
3. Five labor unions had formed and demanded for reduced work hours and were upset because of pay difference between the supplier companies and OEM all working in the same premises.
4. Though all the suppliers and the OEM were working under the same roof, each had its management. This led to disagreement on various policy issues.
5. Suppliers had to move into the main assembly plant that required large initial investments for the setup and complexity of operations.
6. The incorporation of variants was unplanned and required many modifications to the assembly line and mutual agreement with the suppliers.
7. Policies such as "Pay on Build" introduced to cut down the costs and improve quality were harsh on the suppliers.
8. MCC was not producing the initial volumes promised to the suppliers and there was no compensation paid in return.

CASE STUDY 6: PROBLEMS DURING BOEING 777 DEVELOPMENT

Objective

The objective of this project was to understand issues involved in a complex product development by observing videos of the Boeing 777 product development.

Case Studies and Integration

PROJECT DESCRIPTION AND UNCOVERED PROBLEMS

The graduate students from the author's Automotive Systems Engineering Course (AE 500: Automobile—An Integrated System) were asked to view 5 hours of video tapes created by PBS entitled "21st Century Jet—The Building of the 777" (PBS, 1995). They were asked to describe problems presented in these Boeing 777 development videos that have issues similar to the problems encountered during the automotive product development process (Boeing Company, 2011).

Brief descriptions of uncovered problems are provided below.

1. Interfacing large number of disciplines and departments
2. Including customer's personnel (e.g., United Airlines flight crews, and maintenance personnel) in the same teams of the Boeing employees involved in the product development
3. Communications and coordination in tasks with large number of people in many teams, including suppliers from other countries
4. Multitude of issues and trade-offs in meeting regulatory standards and extensive testing to demonstrate compliance
5. Problems encountered in implementing new technologies (e.g., fly-by-wire, new materials)
6. Cabin configuration for maximizing number of seats and occupant comfort (e.g., trade-offs between people space and equipment space)
7. Implementation of new design/build process and discussions in team meetings on a regular basis
8. Extensive systems and product testing (verification and validation tests) before the design was completed (e.g., force required to open frozen aircraft doors, cabin pressure and seal leakage tests, wing load tests, weld strength tests, engine tests under various loads, brake testing, and flight tests over various extreme environmental conditions and durations)
9. Implementation of system to track defects (e.g., dents, damage, mis-drilled holes, and interferences) during the production process
10. Handling late product design changes and their effects on part/subassembly deliveries (e.g., changes made to the parts while original design was being assembled by a supplier)
11. Managing delays caused by late deliveries by contractors (Sharing of computer-aided design [CAD] databases between Boeing and its contactors helped in reducing the delays.)

This Boeing 777 video is an excellent educational tool to understand the complexities in development of the giant airplane. It covers issues in all phases of the product program from product concept development, team building, personnel management, product design, design reviews, computer modeling, engineering, verification testing (of fuselage, doors, wings, engines, brakes, etc.), prototype testing, manufacturing, and flight testing to selling of the first airplane to the United Airlines.

CASE STUDY 7: BOEING 787 DREAMLINER DESIGN AND PRODUCTION

Objective

The objective of this project was to develop a list of new product features introduced in the Boeing 787 Dreamliner and to understand some problems encountered during its product development process.

Project and Product Description

This project was accomplished by conducting a literature review of information available in various magazines, web sites, and during a plant visit to the Boeing 787 assembly facility in Everett, Washington (Boeing Commercial Airplanes, 2011).

The first Boeing787 was delivered to Japan's All Nippon Airways on September 26, 2011—about 36 months late. Why did such a major implementer of the Systems Engineering process fail so much? The reasons for the delays included the following: (1) early airplane versions were overweight (e.g., more time needed to redesign with different materials), (2) about 70% of the parts were outsourced to suppliers all around the world, many suppliers had difficulties in producing needed parts early (e.g., fastener shortage), (3) machinists' strikes in 2008, (4) an electrical fire during a test run, and (5) challenges in implementation of the new material, carbon fiber, for its exterior structural components (e.g., fuselage [see Figures 23.8 and 23.9], wings, nose and tail cone, and tail pieces).

FIGURE 23.8 Boeing 787 carbon fiber fuselage section.

Case Studies and Integration

FIGURE 23.9 Boeing 787 fuselage close-up view showing carbon fiber stiffeners and about 12-mm-thick outer skin.

Some unique features of the airplane are described below:

1. Larger windows (65% larger area than on previous Boeing commercial planes) with dimensions of 27 × 47 cm (10.7″ × 18.4″ with electric dimmers rather than pull-down shades. (The larger windows were possible due to new stronger carbon fiber material.)
2. Light-emitting diode illuminated smooth mood lighting inside the passenger compartment. It is also claimed to contribute to the perception of a higher ceiling and more spaciousness from the passenger eye level. Cabin interior width is approximately 547 cm (18 ft) at the armrest height.
3. Quieter inside and outside (e.g., lower take-off noise and use of engine nacelles with noise-reducing serrated edges).
4. The internal cabin pressure increased to the equivalent of 1800 m (6000 ft) altitude instead of the 2400 m (8000 ft) on conventional aircrafts. Thus, increased oxygen at higher pressure will make the passengers feel more comfortable and less tired.
5. Higher humidity in the passenger cabin is possible because of the use of composites, which do not corrode. Cabin air is provided by electrically driven compressors using no engine-bleed air. An advanced cabin air-conditioning system provides better air quality due to: (1) Ozone removal from outside air, (2) removal of bacteria, viruses, and fungi using high-efficiency

particulate air filters, and (3) removal of odors, irritants, and gaseous contaminants by a gaseous filtration system.
6. Larger overhead storage bins.
7. Interior designed to better accommodate persons with mobility disabilities (e.g., a 142 cm [56″] by 145 cm [57″] convertible lavatory includes a movable center wall that allows two separate lavatories to become one large, wheelchair-accessible facility).
8. Greater fuel economy (about 20% less fuel consumption than earlier commercial jetliners) due to greater use of light-weight materials composites/carbon-fiber material instead of aluminum for its shell, raked wingtips, and more fuel efficient engines from GE and Rolls-Royce.
9. Heavy use of computers in product design (CAD and CAE) and airplane operation (about 14 million lines of code, which is about six to seven times the lines of code used in previous Boeing 777 airplane).
10. Reduced customization options for the airlines (to reduce build complexity and costs).
11. Global subcontractors required to do more assembly themselves and deliver completed subassemblies to Boeing for final assembly. This approach resulted in a leaner and simpler assembly line and lower inventory with preinstalled systems reducing final assembly time by about 3 days.
12. Cruising speed 903 km/h (561 mph) and range 15,000 to 15,700 km (8,000–8,500 miles), enough to cover the Los Angeles to Bangkok or New York City to Hong Kong routes.

Features (1) to (7) above were aimed toward pleasing passengers and making them feel more comfortable. Whereas features (8) to (12) helped provide better business results to the airlines (also customers).

Airlines have high hopes for the plane. About 821 airplanes have already been sold (as of October 2011), putting pressure on Boeing to deliver. The plane was announced in 2003 and listed for between $185 and $218 million each, though discounts were often given. Boeing has spent about $17–23 billion over the past decade in developing the airplane.

Production Issues

The airplane is estimated to contain over a million parts. To simplify the production and reduce costs, Boeing outsources about 70% of the production to over 50 major suppliers all around the world (from countries such as Japan, Korea, the United Kingdom, Italy, France, Germany, Sweden, Australia, and Canada, and from over 15 states within the United States). Like the Boeing 777, 3D CAD modeling system was used to maintain databases on all systems, subsystems, components, and assemblies. The CAD databases were available to all suppliers. Many simulation models were used to design and test various systems of the airplane. Boeing modified a large airplane called the Dreamlifter to transport large pieces (e.g., fuselage, wings, tail section) from supplier production facilities to their assembly plant in Everett, WA. Communications with the suppliers and

Case Studies and Integration 477

building the supplier capability, however, was a challenge as many of the initial supplier-produced parts would not fit or were incomplete. Boeing had to spend considerable manpower to fix the supplier problems and improve the production capabilities of the suppliers.

The technology of building carbon-fiber parts for the airplane structures was not new. It was used before for other air force planes. However, designing larger components such as the wings and the fuselage sections required costly design of production equipment and testing to ensure the required level of strength, durability, and safety. Actual flight testing of the entire airplane was extensive—over 1000 flights of six early-built 787 airplanes over 20 months since December 15, 2009.

CASE STUDY 8: FLEXIBLE ASSEMBLY LINE FOR LAPTOP COMPUTERS

Objectives

The objective of this project was to illustrate the manufacturing of a family of complex products with a large product variety.

Background

Chapters 1 and 2 provided illustration of the systems and components of the laptop computer. The laptop computer is not only a complex product with a number of systems and components, but it can be produced by a large number of manufacturers in a variety of configurations or models. The economies of scale have led to the formation of large high-volume suppliers in countries such as China and Korea that have developed assembly facilities with the flexibility to produce a large variety of laptop computers.

Laptop computer assembly lines can be designed by using a number of computer-integrated manufacturing techniques. The line can be designed to be "flexible," that is, it can produce a wide variety of laptop computer models for different brands and models with different product families. Some of the variables that can create different models are listed below:

1. Screen type (resolution, technology, and size [13″, 15″, and 17″ (measured diagonally), and also with different aspect [length-to-width] ratios)
2. Processor (chips made by different manufacturers using different configurations and capacities)
3. Read-only-memory capacity
4. Hard disk data storage capacity
5. DVD/CD disk drive-type and read/write capabilities
6. Battery type and capacity
7. Exterior case type (brand-related features [e.g., style, shape, logo, material, surface finish, and color])
8. Wireless card type and characteristics
9. Graphics card characteristics

10. Manual input system characteristics (keyboard, touchpad/track button/ mouse, and microphone)
11. Video camera characteristics
12. Audio card and system characteristics
13. Number and types of ports for connections to universal serial bus (USB) devices, projectors, screens, cameras, cell phones, and so forth.

Assembly Line Configuration

The assembly line can be designed to assemble a large variety of laptop computers with different combinations of above characteristics for different brands of computers by using flexible manufacturing equipment, such as (1) product identification tag and identification code recognition system (e.g., bar codes, radio frequency identification), (2) numerical control machines, (3) robots, (4) material transportation systems [conveyors], (5) automated material storage and retrieval systems, and (6) central computer system that coordinates all plant floor equipment according to a master production schedule of different models and brands of laptop products along with many human operators working at different workstations and tending different automated and semi-automated machines. Groover (2008) describes many concepts, considerations, and techniques associated with the above-mentioned computer-integrated and flexible manufacturing areas.

Figure 23.10 presents a flow diagram of a hypothetical assembly plant of the laptop assembler.

The figure shows the following: (1) assembly workstations S1 to S18 arranged in a "U"-shaped workcell, (2) components and preassembled systems (shown as entities E1 to E19) of the laptop are delivered by an *a*utomated *g*uided *v*ehicle (AGV) from an *a*utomated *s*torage and *r*etrieval *s*ystem (ASRS, shown on the left lower side) to the workstations, (3) the assembly and delivered entities are transferred to workstations by moving conveyors, (4) the assembly work (involving loading/unloading of entities, insertion of parts, soldering, etc.) at each workstation performed by robots or human operators, and (5) all plant floor equipment controlled by a central computer system that schedules quantities and models of different laptop products to be assembled.

The assembly begins with the selection of a printed circuit board (E1) at the first workstation (S1). A bar code label containing coded information on the laptop model is glued to the printed circuit board. The bar code is scanned at entry location of every workstation and required types of entities (corresponding to the model) are delivered and assembled at each workstation. The last few workstations involve software loading, testing, and verification of operation of different systems as well as the whole computer and packing each computer in a box along with CDs containing owner's manual and backups of operating system and software applications (ordered by the customer). The packed boxes of computers are sent to an ASRS (shown on the lower right side) for subsequent palletizing (by destinations) and shipping to distributors or customers.

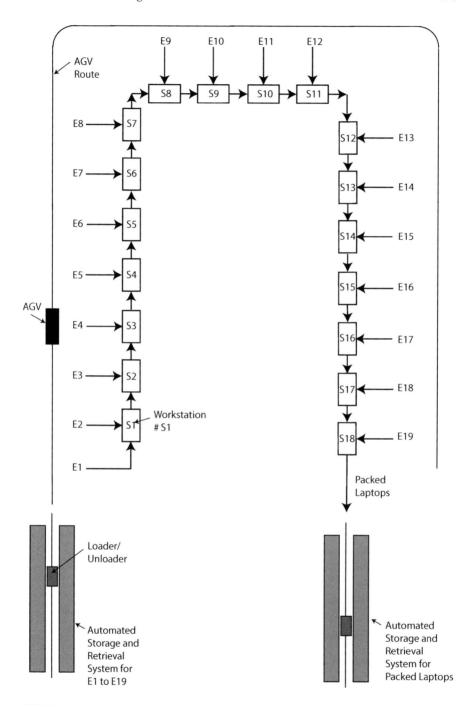

FIGURE 23.10 Plan view of assembly workstations, material transportation, and storage systems for a laptop computer.

The key learning points in this assembly case study are as follows:

1. Realizing that complex products such as the laptop computers are generally designed to allow production of many design variations (i.e., models, product families) to satisfy needs of different customers.
2. Use of common platforms and sharing of carryover (existing) components and systems in complex products allows reduction of the number of tasks, time, costs, and risks in development of new complex products.
3. The design variations in complex products are usually created by development of common platforms (or a chassis with common major systems) and other modular systems to offer different options (e.g., processors in computers, engines in cars and planes) to the customers.
4. Use of computer-assisted technologies such as CAD (computer-aided design) and CAM (computer-aided manufacturing) (programmable CNC [computer numerical control] machines, conveyors, robots, AGVs, and ASRSs) and concepts such as group technologies and flexible manufacturing allow incorporation of flexibilities in product design and manufacturing.
5. Flexible manufacturing techniques allow incorporation of changes in the products (both existing and future products) and still retain the advantages of economy of scale (high total product volume by considering product variations) in production and assembly operations.

CASE STUDY 9: SPECIFICATIONS FOR AN ELECTRIC CAR

OBJECTIVE

The objective of this project was to illustrate the use of the matrix data analysis technique to develop specifications of a complex product.

PROJECT BACKGROUND

A research project was undertaken to develop specifications for a future electric car. The matrix data analysis technique was used to translate the customer needs of an electric vehicle into its specifications. The matrix data analysis technique is used in the QFD to translate the customer requirements into the functional specifications of a product (see Chapters 15 and 17 for details on the QFD and the matrix data analysis, respectively) (Besterfield et al., 2003).

APPLICATION OF THE MATRIX DATA ANALYSIS

The following customer requirements were identified by interviews with potential drivers of future electric cars:

1. Long travel range on one-time charge
2. Less time to fully recharge
3. Electric energy efficiency

4. Hassle-free charging from home or office
 5. Sound feedback
 6. Safety system to prevent short circuits, current leakage, and fires
 7. Vehicle parameters monitoring
 8. Simple, easy to use, and informative interfaces
 9. Constant display for information (e.g., range display, energy use display [kwh/km or kwh/mile], cost display [cents/km or cents/mile], and battery charge gauge [percentage of battery capacity available])
 10. Memory storage space/capacity for trip data, songs, phonebook, and so forth
 11. Availability of trip data and analysis of vehicle information from a remotely located PC
 12. Handheld device to carry vehicle information and help in trip planning.

The areas for engineering specifications included in the analysis were as follows:

 1. Type of battery
 2. Charge of the battery
 3. Buffer stock of charge
 4. Torque of the motor
 5. Power of the motor
 6. Driver–vehicle interface
 7. House–utility company interface
 8. Utility company–vehicle interface
 9. Vehicle–house interface
 10. Data connectivity
 11. Power connectivity
 12. Outside safety (i.e., safety of people outside the vehicle due to electrical hazards and sound level to hear an approaching electric vehicle)
 13. Inside safety (safety of people inside the vehicle due to electrical hazards).

The matrix data analysis chart presented in Table 23.5 was completed by the researchers by determining the importance ratings of the customer requirements (second column from left) and the strength of relationships between the customer requirements and the functional specifications (the matrix of 12 rows and 13 columns) based on the information obtained from the driver interview data and literature surveys. The importance of each customer need was rated by using a 10-point scale, where 10 = most important and 1 = least important (see the column with the heading "Importance Rating" in Table 23.5). Each cell of the relationship matrix provides a number to illustrate the strength of the relationship between the customer need and the functional specification defining the cell. The weights of 9, 3, and 1 were used to define strong, medium, and low relationships, respectively.

The bottom two rows of the chart show the absolute importance ratings and relative importance ratings of the specifications (see Chapter 15 for calculations of the ratings in the QFD section). The relative importance weight row at the bottom of the matrix data chart shows that the three most important engineering requirements

TABLE 23.5
Matrix Data Analysis Chart for Electric Car

Row Number	Importance Rating	Description of Customer Need	1: Type of battery	2: Charge of the battery	3: Buffer stock of charge	4: Torque of the motor	5: Power of the motor	6: Driver-vehicle interface	7: House-utility company interface	8: Utility company to vehicle interface	9: Vehicle-house interface	10: Data connectivity	11: Power connectivity	12: Outside safety	13: Inside Safety
1	10	Long travel range on one-time charge	9	9	9	3	3	9	1	3	1	1	1	1	1
2	9	Less time to fully recharge	9	9	9	1	1	1	9	9	9	1	3	1	1
3	7	Electric energy efficiency	9	1	1	9	9	9	9	3	1	3	9	1	1
4	9	Hassle-free charging from home or office	9	1	1	1	1	9	3	1	9	1	9	9	9
5	8	Sound feedback	1	1	1	9	9	9	1	1	1	1	1	9	1
6	9	Safety system to prevent short circuits, leakage, fires, etc.	3	3	1	1	1	9	1	1	1	3	9	9	9
7	7	Vehicle parameters monitoring	3	9	9	9	9	9	1	1	9	9	9	9	9
8	9	Simple easy to use and informative interfaces	1	1	1	1	1	9	1	1	9	1	1	9	9
9	8	Constant display for information (e.g., kwh/mile)	9	9	9	1	1	9	1	1	9	9	9	1	1
10	6	Memory storage space for trip data, songs, etc.	1	1	1	1	1	9	1	3	3	1	3	3	3
11	6	Availability of trip data and analysis from a remote PC	1	1	1	1	1	9	1	3	3	9	1	1	1
12	7	Handheld device to carry vehicle information and trip planning	1	3	1	1	1	9	3	9	9	9	1	9	9
		Absolute importance ratings	471	399	367	291	291	783	255	281	511	351	445	499	435
		Relative importance ratings	8.8	7.4	6.8	5.4	5.4	14.6	4.7	5.2	9.5	6.5	8.3	9.3	8.1

(specifications) are as follows: (1) driver–vehicle interface (14.6% relative importance rating), (2) vehicle–house interface (9.5% relative importance rating), and (3) outside safety (9.3% relative importance rating).

CONCLUDING REMARKS

The case studies included in this chapter were purposely simplified. The goal was to stress on the applications of several important techniques and many issues covered in this book. The case studies help emphasize many points such as: (1) complex products have many customer needs that can be translated into product attributes, (2) product attributes can be translated into functional specifications, (3) complex product have many systems, (4) attribute requirements need to be cascaded to various subsystem requirements, (5) trade-offs between requirements must be carefully evaluated, (6) matrix data analysis tool helps in organizing and prioritizing functional specifications, (7) complex products can be designed to create many variations, (8) multifunctional collocated teams are useful in designing complex products, (9) computer-assisted technologies can play a major role in designing and producing complex products, and (10) complex products will most likely be assembled by using components or subassemblies produced by many different suppliers all around the world requiring constant communication between many suppliers and the assemblers.

REFERENCES

Besterfield, D. H., Besterfield-Michna, C., Besterfield, G. H. and M. Besterfield-Scare. 2003. *Total Quality Management*. Third Edition. ISBN 0-13-099306-9. Upper Saddle River, NJ: Prentice Hall.

Boeing Commercial Airplanes (a business unit of The Boeing Company). 2011. Website: http://www.boeing.com/commercial/787family/dev_team.html (accessed September 26, 2011).

Boeing Company. 2011. Boeing 777 Assembly LineVideo. Website:http://www.boeing.com/Features/2010/05/bca_moving_line_05_24_10.htm (accessed September 26, 2011).

Brooke, L. 2008. Little car, big job. *Automotive Engineering (aei-online.org)*, February, 2008: 76–77.

Groover, M. P. 2008. *Automation, Production Systems, and Computer-Integrated Manufacturing*. Third Edition. ISBN 0-13-239321-2. Upper Saddle River, NJ: Prentice Hall.

Peters, T. J. 1992. Ingersoll Rand: Barbecues, Drag Tests, Medieval Warriors; and Slowing Down to Speed Things Up (Chapter 6 on Cyclone grinder development story) in *Liberation Management* (pp. 72–85). New York: Alfred A. Knopf, Inc.

Public Broadcasting Service (PBS). 1995. *21st Century Jet—The Building of the 777*. Producers: Karl Sabbagh, David Davis and Peggy Case. PBS Home Video (5 hours). Produced by Skyscraper Products for KCTS Seattle and Channel 4 London.

Siekman, P. 2002. The Smart Car is Looking More So. *Fortune*, April 15, 2002: 310[I]–310[P].

Smart car. 2012. Website: http://www.smartusa.com/models/electric-drive/overview.aspx (accessed December 12, 2012).

24 Case Studies in Cost–Benefit Analysis

INTRODUCTION

This chapter presents five case studies of cost–benefit analyses conducted on large projects that extend over many years. The first cost–benefit analysis was for a product development situation where an automotive company needs to decide where, i.e., in which country should a new vehicle be developed and built. The decision is complex because it involves consideration of wages to be paid to technical and hourly employees in different countries and increased vehicle shipping costs over many years of production. Cases studies 2 and 3 provide insights into costs and revenues from electric power generation plants that use different technologies (e.g., wind turbines vs. natural gas combined cycle power plant) over the long periods of plant construction and power generation. Case study 4 illustrates a large cost–benefit analysis conducted by the Environmental Protection Agency (EPA) and the National Highway Transportation Safety Administration (NHTSA) to justify more stringent emission and corporate average fuel economy requirements on 2017 to 2025 model-year light vehicles in the United States based on the costs of implementation of new fuel saving and emission reduction technologies in future car and light truck products and the benefits that will be incurred due to fuel savings and reduction in greenhouses gases. Case study 5 illustrates a cost–benefit analysis conducted to support a decision to implement robots for improving the assembly costs and output of differential carrier assemblies in automotive products.

CASE STUDY 1: COST–BENEFIT ANALYSIS OF AUTOMOTIVE PRODUCT DEVELOPMENT PROGRAMS

A financial model developed to estimate costs and benefits to develop, produce, and sell an automotive product described in Bhise (2017) was exercised here to evaluate the following three alternatives:

Alternative 1 (A1): Domestic Production—Design and produce a new vehicle in the U.S. for sale in the U.S.
Alternative 2 (A2) Production in a Low-Wage Country—Design and produce a new vehicle in a Low-Wage Country and sell in the U.S. market.
Alternative 3 (A3): Domestic Production with Light-Weight Materials—Design and produce a new vehicle in the U.S. with light-weight materials and sell in the U.S. market.

Three outcomes were considered as follows:

Outcome 1 (O1): Vehicle production output at base (current) levels
Outcome 2 (O2): Vehicle production output at 110% of base levels
Outcome 3 (O3): Vehicle production output at 90% of base levels

The financial analysis of the vehicle program was performed by using an excel spreadsheet-based model which computed various product development, manufacturing, and sales and marketing costs and revenue generated from the product sales based on a number of inputs. The rows of the spreadsheet represented time in months from −40 months (when the vehicle program is kicked-off) to +60 months from Job#1 (Note: Job#1 is when the first production vehicle rolls out of its assembly plant).

Table 24.1 provides the spreadsheet for the base condition involving Alternative 1 and Outcome 1. The columns of the spreadsheet included variables such as the number of technical people, number of manufacturing and assembly hours required, number of vehicles produced, costs of supplier-produced entities (components, subsystems, and systems), sales and marketing costs, and present values of total costs, revenues, and revenues minus total costs. The outputs of the model were used to generate curves of present values (at −40 months) of total costs, revenues, and revenue minus total cost as functions of the time from −40 months to +60 months from Job #1.

Assumptions:
1. Technical personnel (e.g., engineer) salary: $8000/month in the U.S.; $2500/month in a low-wage country (e.g., Mexico)
2. Hourly employee (e.g., production worker): $30/hour in the U.S.; $5/hour in a low-wage country
3. Marketing and vehicle transportation cost: $2000 per vehicle from a U.S. plant location; $2750 per vehicle from a low-wage country. Thus, using light-weight entities incur $750 additional cost per vehicle.
4. Cost of parts and materials purchased from suppliers and overheads: $8000/vehicle; $8750 vehicle with light-weight materials
5. Discount rate: 3%

The columns of the spreadsheet (Table 24.1) provided are labeled as A to T. The description of the variables provided in each column is as follows:

A: Time in months from Job#1
B: Product development headcount (number of technical people working on the vehicle program—includes all team members from various disciplines such as engineers, designers, product planners, marketing researchers, financial analyst, and technicians.
C: Product development manpower costs—Total salary and benefits costs of the employees involved in product development. Assumed to be $8000/month per employee for domestic (U.S.) and $2500/month per employee for a low-wage country.

TABLE 24.1
Financial Calculations Spreadsheet for an Automotive Program

A	B	C	D	E	F	G	H	I	J	K	L	M	N	O	P	Q	R	S	T
Months from Job #1	Product Development Headcount	Prod. Dev. Manpower Costs	Services and Supplies Costs	Facilities and Tooling Costs	Product Development Costs Subtotal	Present Value of Development Costs Subtotal	Mfg. Headcount	Mfg. Manpower Costs	Number of Vehicles Produced	Parts, Materials, & overhead Costs	Manufacturing Costs Subtotal	Sales and Marketing Costs	Total Costs	Revenue from Vehicle Sales	Present Value of Total Cost	Present Value of Revenue from Vehicle Sales	Present Value of Cumulative Total Cost	Present Value of Cumulative Total Revenue	Present Value Cash Flow
-40	50	$400,000	$140,000		$540,000	$540,000							$540,000		$540,000	$0	$540,000	$0	-$540,000
-39	100	$800,000	$280,000		$1,080,000	$1,077,307							$1,080,000		$1,077,307	$0	$1,617,307	$0	-$1,617,307
-38	200	$1,600,000	$560,000		$2,160,000	$2,149,240							$2,160,000		$2,149,240	$0	$3,766,547	$0	-$3,766,547
-37	500	$4,000,000	$1,400,000		$5,400,000	$5,359,702							$5,400,000		$5,359,702	$0	$9,126,249	$0	-$9,126,249
-36	800	$6,400,000	$2,240,000		$8,640,000	$8,554,137							$8,640,000		$8,554,137	$0	$17,680,386	$0	-$17,680,386
-35	1000	$8,000,000	$2,800,000		$10,800,000	$10,666,007							$10,800,000		$10,666,007	$0	$28,346,393	$0	-$28,346,393
-34	1000	$8,000,000	$2,800,000		$10,800,000	$10,639,408							$10,800,000		$10,639,408	$0	$38,985,801	$0	-$38,985,801
-33	1200	$9,600,000	$3,360,000		$12,960,000	$12,735,451							$12,960,000		$12,735,451	$0	$51,721,252	$0	-$51,721,252
-32	1200	$9,600,000	$3,360,000		$12,960,000	$12,703,692							$12,960,000		$12,703,692	$0	$64,424,944	$0	-$64,424,944
-31	1200	$9,600,000	$3,360,000		$12,960,000	$12,672,012							$12,960,000		$12,672,012	$0	$77,096,956	$0	-$77,096,956
-30	1200	$9,600,000	$3,360,000		$12,960,000	$12,640,411							$12,960,000		$12,640,411	$0	$89,737,366	$0	-$89,737,366
-29	1200	$9,600,000	$3,360,000		$12,960,000	$12,608,889							$12,960,000		$12,608,889	$0	$102,346,255	$0	-$102,346,255
-28	1200	$9,600,000	$3,360,000		$12,960,000	$12,577,445							$12,960,000		$12,577,445	$0	$114,923,700	$0	-$114,923,700
-27	1200	$9,600,000	$3,360,000		$12,960,000	$12,546,080							$12,960,000		$12,546,080	$0	$127,469,780	$0	-$127,469,780
-26	1200	$9,600,000	$3,360,000		$12,960,000	$12,514,793							$12,960,000		$12,514,793	$0	$139,984,573	$0	-$139,984,573
-25	1200	$9,600,000	$3,360,000	$40,000,000	$52,960,000	$51,013,164							$52,960,000		$51,013,164	$0	$190,997,736	$0	-$190,997,736
-24	1200	$9,600,000	$3,360,000	$40,000,000	$52,960,000	$50,885,949							$52,960,000		$50,885,949	$0	$241,883,685	$0	-$241,883,685
-23	1200	$9,600,000	$3,360,000	$40,000,000	$52,960,000	$50,759,051							$52,960,000		$50,759,051	$0	$292,642,736	$0	-$292,642,736
-22	1200	$9,600,000	$3,360,000	$40,000,000	$52,960,000	$50,632,470							$52,960,000		$50,632,470	$0	$343,275,206	$0	-$343,275,206
-21	1200	$9,600,000	$3,360,000	$40,000,000	$52,960,000	$50,506,204							$52,960,000		$50,506,204	$0	$393,781,410	$0	-$393,781,410
-20	1200	$9,600,000	$3,360,000	$40,000,000	$52,960,000	$50,380,254							$52,960,000		$50,380,254	$0	$444,161,664	$0	-$444,161,664
-19	1200	$9,600,000	$3,360,000	$40,000,000	$52,960,000	$50,254,617							$52,960,000		$50,254,617	$0	$494,416,281	$0	-$494,416,281
-18	1200	$9,600,000	$3,360,000	$40,000,000	$52,960,000	$50,129,294							$52,960,000		$50,129,294	$0	$544,545,575	$0	-$544,545,575
-17	1200	$9,600,000	$3,360,000	$40,000,000	$52,960,000	$50,004,283							$52,960,000		$50,004,283	$0	$594,549,858	$0	-$594,549,858

(Continued)

TABLE 24.1 (Continued)

A	B	C	D	E	F	G	H	I	J	K	L	M	N	O	P	Q	R	S	T
Months from Job#1	Product Development Headcount	Prod. Dev. Manpower Costs	Services and Supplies Costs	Facilities and Tooling Costs	Product Development Costs Subtotal	Present Value of Development Costs Subtotal	Mfg. Headcount	Mfg. Manpower Costs	Number of Vehicles Produced	Parts, Materials, & overhead Costs	Manufacturing Costs Subtotal	Sales and Marketing Costs	Total Costs	Revenue from Vehicle Sales	Present Value of Total Cost	Present Value of Revenue from Vehicle Sales	Present Value of Cumulative Total Cost	Present Value of Cumulative Total Revenue	Present Value Cash Flow
-16	1200	$9,600,000	$3,360,000	$40,000,000	$52,960,000	$49,879,584							$52,960,000		$49,879,584	$0	$644,429,443	$0	-$644,429,443
-15	1200	$9,600,000	$3,360,000	$40,000,000	$52,960,000	$49,755,196							$52,960,000		$49,755,196	$0	$694,184,639	$0	-$694,184,639
-14	1200	$9,600,000	$3,360,000	$40,000,000	$52,960,000	$49,631,119							$52,960,000		$49,631,119	$0	$743,815,757	$0	-$743,815,757
-13	1200	$9,600,000	$3,360,000	$40,000,000	$52,960,000	$49,507,350							$52,960,000		$49,507,350	$0	$793,323,108	$0	-$793,323,108
-12	1200	$9,600,000	$3,360,000	$40,000,000	$52,960,000	$49,383,890							$52,960,000		$49,383,890	$0	$842,706,998	$0	-$842,706,998
-11	1200	$9,600,000	$3,360,000	$40,000,000	$52,960,000	$49,260,739							$52,960,000		$49,260,739	$0	$891,967,737	$0	-$891,967,737
-10	1200	$9,600,000	$3,360,000	$40,000,000	$52,960,000	$49,137,894							$52,960,000		$49,137,894	$0	$941,105,630	$0	-$941,105,630
-9	1200	$9,600,000	$3,360,000	$40,000,000	$52,960,000	$49,015,355							$52,960,000		$49,015,355	$0	$990,120,986	$0	-$990,120,986
-8	1200	$9,600,000	$3,360,000	$40,000,000	$52,960,000	$48,893,123							$52,960,000		$48,893,123	$0	$1,039,014,109	$0	-$1,039,014,109
-7	1200	$9,600,000	$3,360,000	$40,000,000	$52,960,000	$48,771,195							$52,960,000		$48,771,195	$0	$1,087,785,303	$0	-$1,087,785,303
-6	1000	$8,000,000	$2,800,000	$40,000,000	$50,800,000	$46,665,374							$50,800,000		$46,665,374	$0	$1,134,450,677	$0	-$1,134,450,677
-5	1000	$8,000,000	$2,800,000	$40,000,000	$50,800,000	$46,549,001							$50,800,000		$46,549,001	$0	$1,180,999,678	$0	-$1,180,999,678
-4	800	$6,400,000	$2,240,000	$40,000,000	$48,640,000	$44,458,606							$48,640,000		$44,458,606	$0	$1,225,458,284	$0	-$1,225,458,284
-3	600	$4,800,000	$1,680,000	$40,000,000	$46,480,000	$42,378,347	500	$6,000,000	1,000	$8,000,000	$14,000,000	$2,000,000	$62,480,000	$28,000,000	$56,966,418	$17,682,426	$1,282,424,702	$17,682,426	-$1,264,742,276
-2	400	$3,200,000	$1,120,000	$40,000,000	$44,320,000	$40,308,187	600	$7,200,000	4,000	$32,000,000	$83,520,000	$8,000,000	$135,840,000	$112,000,000	$123,543,865	$69,856,498	$1,405,968,567	$87,538,924	-$1,318,429,643
-1	300	$2,400,000	$840,000		$3,240,000	$2,939,369	800	$9,600,000	8,000	$64,000,000	$76,840,000	$16,000,000	$96,080,000	$224,000,000	$87,164,998	$137,988,144	$1,493,133,565	$225,527,069	-$1,267,606,496
0	200	$1,600,000	$560,000		$2,160,000	$1,954,693	1000	$12,000,000	10,000	$80,000,000	$94,160,000	$20,000,000	$116,320,000	$280,000,000	$105,263,824	$170,355,734	$1,598,397,389	$395,882,803	-$1,202,514,586
1					$0	$0	1000	$12,000,000	24,000	$192,000,000	$204,000,000	$48,000,000	$252,000,000	$672,000,000	$227,478,789	$403,806,184	$1,825,876,178	$799,688,987	-$1,026,187,191
2					$0	$0	1000	$12,000,000	24,000	$192,000,000	$204,000,000	$48,000,000	$252,000,000	$672,000,000	$226,911,510	$398,820,923	$2,052,787,688	$1,198,509,909	-$854,277,779
3					$0	$0	1000	$12,000,000	24,000	$192,000,000	$204,000,000	$48,000,000	$252,000,000	$672,000,000	$226,345,646	$393,897,207	$2,279,133,334	$1,592,407,117	-$686,726,218
4					$0	$0	1000	$12,000,000	24,000	$192,000,000	$204,000,000	$48,000,000	$252,000,000	$672,000,000	$225,781,193	$389,034,279	$2,504,914,528	$1,981,441,396	-$523,473,132
5					$0	$0	1000	$12,000,000	24,000	$192,000,000	$204,000,000	$48,000,000	$252,000,000	$672,000,000	$225,218,148	$384,231,387	$2,730,132,675	$2,365,672,783	-$364,459,893
6					$0	$0	1000	$12,000,000	24,000	$192,000,000	$204,000,000	$48,000,000	$252,000,000	$672,000,000	$224,656,507	$379,487,789	$2,954,789,182	$2,745,160,572	-$209,628,610
7					$0	$0	1000	$12,000,000	25,000	$200,000,000	$212,000,000	$50,000,000	$262,000,000	$700,000,000	$232,988,975	$390,419,536	$3,187,778,157	$3,135,580,108	-$52,198,049
8					$0	$0	1000	$12,000,000	26,000	$208,000,000	$220,000,000	$52,000,000	$272,000,000	$728,000,000	$241,278,488	$401,023,524	$3,429,056,645	$3,536,603,632	$107,546,987

9		$0	$0	1000	$12,000,000	27,000	$216,000,000	$228,000,000	$54,000,000	$282,000,000	$756,000,000	$249,525,207	$411,306,178	$3,678,581,852	$3,947,909,810	$269,327,958
10		$0	$0	1000	$12,000,000	28,000	$224,000,000	$236,000,000	$56,000,000	$292,000,000	$784,000,000	$257,729,296	$421,273,818	$3,936,311,148	$4,369,183,628	$432,872,480
11		$0	$0	1000	$12,000,000	29,000	$232,000,000	$244,000,000	$58,000,000	$302,000,000	$812,000,000	$265,890,914	$430,932,653	$4,202,202,062	$4,800,116,281	$597,914,219
12		$0	$0	1000	$12,000,000	30,000	$240,000,000	$252,000,000	$60,000,000	$312,000,000	$840,000,000	$274,010,223	$440,288,790	$4,476,212,285	$5,240,405,070	$764,192,785
13		$0	$0	1000	$12,000,000	30,000	$240,000,000	$252,000,000	$60,000,000	$312,000,000	$840,000,000	$273,326,906	$434,853,126	$4,749,539,191	$5,675,258,196	$925,719,005
14		$0	$0	1000	$12,000,000	30,000	$240,000,000	$252,000,000	$60,000,000	$312,000,000	$840,000,000	$272,645,293	$429,484,569	$5,022,184,484	$6,104,742,765	$1,082,558,280
15		$0	$0	1000	$12,000,000	30,000	$240,000,000	$252,000,000	$60,000,000	$312,000,000	$840,000,000	$271,965,379	$424,182,290	$5,294,149,864	$6,528,925,055	$1,234,775,191
16		$0	$0	1000	$12,000,000	30,000	$240,000,000	$252,000,000	$60,000,000	$312,000,000	$840,000,000	$271,287,162	$418,945,472	$5,565,437,025	$6,947,870,526	$1,382,433,501
17		$0	$0	1000	$12,000,000	30,000	$240,000,000	$252,000,000	$60,000,000	$312,000,000	$840,000,000	$270,610,635	$413,773,305	$5,836,047,660	$7,361,643,831	$1,525,596,171
18		$0	$0	1000	$12,000,000	30,000	$240,000,000	$252,000,000	$60,000,000	$312,000,000	$840,000,000	$269,935,796	$408,664,993	$6,105,983,456	$7,770,308,824	$1,664,325,368
19		$0	$0	1000	$12,000,000	30,000	$240,000,000	$252,000,000	$60,000,000	$312,000,000	$840,000,000	$269,262,639	$403,619,746	$6,375,246,095	$8,173,928,570	$1,798,682,475
20		$0	$0	1000	$12,000,000	30,000	$240,000,000	$252,000,000	$60,000,000	$312,000,000	$840,000,000	$268,591,161	$398,636,786	$6,643,837,256	$8,572,565,356	$1,928,728,101
21		$0	$0	1000	$12,000,000	30,000	$240,000,000	$252,000,000	$60,000,000	$312,000,000	$840,000,000	$267,921,358	$393,715,344	$6,911,758,613	$8,966,280,701	$2,054,522,087
22		$0	$0	1000	$12,000,000	30,000	$240,000,000	$252,000,000	$60,000,000	$312,000,000	$840,000,000	$267,253,225	$388,854,661	$7,179,011,838	$9,355,135,362	$2,176,123,524
23		$0	$0	1000	$12,000,000	30,000	$240,000,000	$252,000,000	$60,000,000	$312,000,000	$840,000,000	$266,586,758	$384,053,986	$7,445,598,595	$9,739,189,348	$2,293,590,753
24		$0	$0	1000	$12,000,000	30,000	$240,000,000	$252,000,000	$60,000,000	$312,000,000	$840,000,000	$265,921,953	$379,312,579	$7,711,520,548	$10,118,501,927	$2,406,981,379
25				1000	$12,000,000	30,000	$240,000,000	$252,000,000	$60,000,000	$312,000,000	$840,000,000	$265,258,806	$374,629,708	$7,976,779,354	$10,493,131,635	$2,516,352,281
26				1000	$12,000,000	30,000	$240,000,000	$252,000,000	$60,000,000	$312,000,000	$840,000,000	$264,597,312	$370,004,650	$8,241,376,666	$10,863,136,284	$2,621,759,618
27				1000	$12,000,000	30,000	$240,000,000	$252,000,000	$60,000,000	$312,000,000	$840,000,000	$263,937,469	$365,436,691	$8,505,314,135	$11,228,572,975	$2,723,258,840
28				1000	$12,000,000	30,000	$240,000,000	$252,000,000	$60,000,000	$312,000,000	$840,000,000	$263,279,271	$360,925,127	$8,768,593,406	$11,589,498,102	$2,820,904,696
29				1000	$12,000,000	30,000	$240,000,000	$252,000,000	$60,000,000	$312,000,000	$840,000,000	$262,622,714	$356,469,261	$9,031,216,120	$11,945,967,363	$2,914,751,243
30				1000	$12,000,000	30,000	$240,000,000	$252,000,000	$60,000,000	$312,000,000	$840,000,000	$261,967,794	$352,068,406	$9,293,183,914	$12,298,035,769	$3,004,851,855
31				1000	$12,000,000	30,000	$240,000,000	$252,000,000	$60,000,000	$312,000,000	$840,000,000	$261,314,508	$347,721,882	$9,554,498,422	$12,645,757,652	$3,091,259,229
32				1000	$12,000,000	30,000	$240,000,000	$252,000,000	$60,000,000	$312,000,000	$840,000,000	$260,662,851	$343,429,020	$9,815,161,273	$12,989,186,671	$3,174,025,398
33				1000	$12,000,000	30,000	$240,000,000	$252,000,000	$60,000,000	$312,000,000	$840,000,000	$260,012,819	$339,189,155	$10,075,174,092	$13,328,375,827	$3,253,201,734
34				1000	$12,000,000	30,000	$240,000,000	$252,000,000	$60,000,000	$312,000,000	$840,000,000	$259,364,408	$335,001,635	$10,334,538,500	$13,663,377,461	$3,328,838,961
35				1000	$12,000,000	30,000	$240,000,000	$252,000,000	$60,000,000	$312,000,000	$840,000,000	$258,717,614	$330,865,812	$10,593,256,114	$13,994,243,274	$3,400,987,160
36				1000	$12,000,000	30,000	$240,000,000	$252,000,000	$60,000,000	$312,000,000	$840,000,000	$258,072,433	$326,781,049	$10,851,328,547	$14,321,024,323	$3,469,695,776
37				1000	$12,000,000	29,000	$232,000,000	$244,000,000	$58,000,000	$302,000,000	$812,000,000	$249,177,936	$311,988,491	$11,100,506,482	$14,633,012,814	$3,532,506,332
38				1000	$12,000,000	28,000	$224,000,000	$236,000,000	$56,000,000	$292,000,000	$784,000,000	$240,326,195	$297,511,375	$11,340,832,677	$14,930,524,189	$3,589,691,512
39				1000	$12,000,000	27,000	$216,000,000	$228,000,000	$54,000,000	$282,000,000	$756,000,000	$231,517,053	$283,344,167	$11,572,349,731	$15,213,868,356	$3,641,518,626
40				1000	$12,000,000	26,000	$208,000,000	$220,000,000	$52,000,000	$272,000,000	$728,000,000	$222,750,353	$269,481,421	$11,795,100,083	$15,483,349,777	$3,688,249,693

(Continued)

TABLE 24.1 (Continued)

A	B	C	D	E	F	G	H	I	J	K	L	M	N	O	P	Q	R	S	T
Months from Job #1	Product Development Headcount	Prod. Dev. Manpower Costs	Services and Supplies Costs	Facilities and Tooling Costs	Product Development Costs Subtotal	Present Value of Development Costs Subtotal	Mfg. Headcount	Mfg. Manpower Costs	Number of Vehicles Produced	Parts, Materials, & overhead Costs	Manufacturing Costs Subtotal	Sales and Marketing Costs	Total Costs	Revenue from Vehicle Sales	Present Value of Total Cost	Present Value of Revenue from Vehicle Sales	Present Value of Cumulative Total Cost	Present Value of Cumulative Total Revenue	Present Value Cash Flow
41							1000	$12,000,000	25,000	$200,000,000	$212,000,000	$50,000,000	$262,000,000	$700,000,000	$214,025,937	$255,917,779	$12,009,126,020	$15,739,267,555	$3,730,141,535
42							1000	$12,000,000	24,000	$192,000,000	$204,000,000	$48,000,000	$252,000,000	$672,000,000	$205,343,649	$242,647,968	$12,214,469,669	$15,981,915,523	$3,767,445,854
43							1000	$12,000,000	24,000	$192,000,000	$204,000,000	$48,000,000	$252,000,000	$672,000,000	$204,831,570	$239,652,314	$12,419,301,239	$16,221,567,837	$3,802,266,598
44							1000	$12,000,000	24,000	$192,000,000	$204,000,000	$48,000,000	$252,000,000	$672,000,000	$204,320,768	$236,693,643	$12,623,622,007	$16,458,261,480	$3,834,639,473
45							1000	$12,000,000	24,000	$192,000,000	$204,000,000	$48,000,000	$252,000,000	$672,000,000	$203,811,240	$233,771,500	$12,827,433,247	$16,692,032,980	$3,864,599,733
46							1000	$12,000,000	24,000	$192,000,000	$204,000,000	$48,000,000	$252,000,000	$672,000,000	$203,302,982	$230,885,432	$13,030,736,229	$16,922,918,412	$3,892,182,182
47							1000	$12,000,000	24,000	$192,000,000	$204,000,000	$48,000,000	$252,000,000	$672,000,000	$202,795,992	$228,034,994	$13,233,532,221	$17,150,953,406	$3,917,421,184
48							1000	$12,000,000	24,000	$192,000,000	$204,000,000	$48,000,000	$252,000,000	$672,000,000	$202,290,267	$225,219,747	$13,435,822,488	$17,376,173,153	$3,940,350,665
49							1000	$12,000,000	24,000	$192,000,000	$204,000,000	$48,000,000	$252,000,000	$672,000,000	$201,785,802	$222,439,257	$13,637,608,290	$17,598,612,410	$3,961,004,119
50							1000	$12,000,000	24,000	$192,000,000	$204,000,000	$48,000,000	$252,000,000	$672,000,000	$201,282,596	$219,693,093	$13,838,890,886	$17,818,305,503	$3,979,414,617
51							1000	$12,000,000	24,000	$192,000,000	$204,000,000	$48,000,000	$252,000,000	$672,000,000	$200,780,644	$216,980,833	$14,039,671,530	$18,035,286,335	$3,995,614,805
52							1000	$12,000,000	24,000	$192,000,000	$204,000,000	$48,000,000	$252,000,000	$672,000,000	$200,279,944	$214,302,057	$14,239,951,475	$18,249,588,392	$4,009,636,918
53							1000	$12,000,000	24,000	$192,000,000	$204,000,000	$48,000,000	$252,000,000	$672,000,000	$199,780,493	$211,656,352	$14,439,731,968	$18,461,244,745	$4,021,512,777
54							1000	$12,000,000	24,000	$192,000,000	$204,000,000	$48,000,000	$252,000,000	$672,000,000	$199,282,287	$209,043,311	$14,639,014,255	$18,670,288,056	$4,031,273,801
55							1000	$12,000,000	24,000	$192,000,000	$204,000,000	$48,000,000	$252,000,000	$672,000,000	$198,785,324	$206,462,529	$14,837,799,579	$18,876,750,585	$4,038,951,006
56							1000	$12,000,000	24,000	$192,000,000	$204,000,000	$48,000,000	$252,000,000	$672,000,000	$198,289,600	$203,913,609	$15,036,089,179	$19,080,664,195	$4,044,575,016
57							1000	$12,000,000	24,000	$192,000,000	$204,000,000	$48,000,000	$252,000,000	$672,000,000	$197,795,112	$201,396,157	$15,233,884,291	$19,282,060,352	$4,048,176,061
58							1000	$12,000,000	24,000	$192,000,000	$204,000,000	$48,000,000	$252,000,000	$672,000,000	$197,301,858	$198,909,785	$15,431,186,149	$19,480,970,137	$4,049,783,988
59							1000	$12,000,000	24,000	$192,000,000	$204,000,000	$48,000,000	$252,000,000	$672,000,000	$196,809,833	$196,454,109	$15,627,995,982	$19,677,424,246	$4,049,428,264
60							1000	$12,000,000	24,000	$192,000,000	$204,000,000	$48,000,000	$252,000,000	$672,000,000	$196,319,035	$194,028,749.4	$15,824,315,017	$19,871,452,995	$4,047,137,978

Discount rate for present value calculations = 3

Selling price ($) per vehicle = 28000

Prod. Dec engr 8000/month
Prod worker 30/hr
Marketing 2000/veh

Benefit/cost Ratio 1.26
Benefits−costs $4,047,137,978

Case Studies in Cost–Benefit Analysis

D: Services and supplied costs—monthly costs of purchasing materials and services from other companies (suppliers, contractors, and consultants)

E: Facilities and tooling costs—monthly costs incurred in building facilities (manufacturing and assembly plants and equipment modifications such as designing, building and installing workstations, tools, fixtures, and equipment [test and production])

F: Product development costs subtotal (sum of costs in columns C, D, and E)

G: Present value of development costs subtotal (i.e., value at −40 months [assumed to be the "present" time] by considering discount rate. Using the formula $P = F[1/(1 + i)^n]$, where P = present value, F = future value and i annual discount rate).

H: Manufacturing headcount—number of hourly employees in manufacturing and assembly plants of the manufacturer

I: Manufacturing manpower costs—total number of hours spent in each month by hourly employees multiplied by the average hourly rate (including benefits)

J: Number of vehicles built—number of vehicles built (assembled) and released by the vehicle assembly plant in each month.

K: Parts, materials, and overhead costs—total cost of entities (components, sub-assemblies and assemblies) purchased from suppliers, raw materials purchased, and overheads (e.g., maintenance, utilities, insurance) for all vehicles built in each month

L: Manufacturing costs subtotal—sum of amounts in columns I and K.

M: Sales and marketing costs—selling and marketing costs per vehicle incurred by the vehicle manufacturer (includes the cost of advertising, incentives paid to the customers, and dealer commissions) multiplied by the number of vehicles produced in the month (in column J)

N: Total costs—sum of F, L, and M

O: Revenue from vehicle sales—total amount received from vehicle sales (selling price x number of vehicle produced)

P: Present value of total costs—present value at −40 months of the total cost (in column N

Q: Present value of revenue from vehicle sales—present value at −40 months of total revenue (in column O).

R: Present value of cumulative total costs—sum of all present values of total costs up to and including the month shown in column A.

S: Present value of cumulative total revenue—sum of all present values of total revenues up to and including the month shown in column A.

T: Present value cashflow—Net profit or loss, i.e., value in column S minus value in column R.

Results from the cost–benefit analyses conducted for the combinations of alternatives and outcomes are shown in Tables 24.2 and 24.3. The benefit-to-cost ratio was computed as the ratio of the present value of the cumulative revenue (value in the last row of column S) divided by the present value of cumulative total costs (value in the last row of column R) (see Table 24.2). The present value of net profit (or loss) was

TABLE 24.2
Cost-to-Benefit Ratios for Combinations of Alternatives and Outcomes

		Benefit-to-costs Ratio			Average Value (Based on Laplace Principle)
	Outcomes -->	O1	O2	O3	
Alternative	Description	Product Output at 100%	Product Output at 110%	Product Output at 90%	
A1	Domestic production	1.26	1.27	1.24	1.26
A2	Production in Low-Wage Country and shipping products to the U.S.	1.24	1.25	1.23	1.24
A3	Domestic production with light-weight components	1.18	1.19	1.16	1.18

TABLE 24.3
Cashflow at 60 Months after Job#1 for Combinations of Alternatives and Outcomes

		Present Value of Cashflow at 60 months after Job#1 (in billlions)			Average Value (Based on Laplace Principle)
	Outcomes -->	O1	O2	O3	
Alternative	Description	Product Output at 100%	Product Output at 110%	Product Output at 90%	
A1	Domestic production	$4.05	$4.65	$3.44	$4.05
A2	Production in Low-Wage Country and shipping products to the U.S.	$3.83	$4.33	$3.33	$3.83
A3	Domestic production with light-weight components	$3.00	$3.51	$2.51	$3.01

computed by subtracting the present value of cumulative total costs (value in the last row of column R) present value of the cumulative revenue (value in the last row of column S) (see Table 24.3).

Table 24.2 shows that the highest value of benefit-to-cost ratio was 1.27 and it was achieved under alternative 1 (domestic production) and outcome 2 (110% of base output). Producing vehicles in a low-wage country was not found to provide higher

Case Studies in Cost–Benefit Analysis

benefit-to-cost ratios as compared to the domestic production due to higher shipping costs. Using light-weight materials (alternative 3) also did not provide higher benefit-to-cost ratios over alternative 1 due to higher costs of materials.

Table 24.3 provides values of cashflow (net present value of revenue minus costs after 60 months from Job#1). It shows that alternative 1 was the best among the three alternatives and outcome 2 provided the highest cashflow.

The advantages of the model described above are that it can be exercised under different set of input variables (to create different alternatives and outcomes) by financial planners to determine the locations of the product design and manufacturing facilities.

CASE STUDY 2: JEDI MODEL APPLICATIONS FOR COMPARISON OF COSTS AND ECONOMIC BENEFITS OF WIND TURBINE POWER PLANT WITH NATURAL GAS POWER PLANT

JEDI (Jobs and Economic Development Impact) models developed by the National Renewable Energy Laboratory (NREL) (NREL, 2022a and 2022b) were used here to compare the costs and benefits of constructing and operating 200 MW wind turbines vs. natural gas-combined cycle power plant. The JEDI models are spreadsheet-based and work using the Microsoft Excel application.

The models were downloaded from the NREL website. The basic inputs and outputs of the models for the wind turbine and natural gas combined cycle power plant are provided in Tables 24.4 and 24.5, respectively.

TABLE 24.4
Wind Turbine JEDI Models Inputs and Outputs

Project Description

Project Name	Wind Plant
Project Location (State)	MICHIGAN
Project Location (Region)	#N/A
Total Project Size (MW)	200
Number of Projects (included in Total Project Size)	1
Number of Turbines	80

Turbine Information

Turbine Rating (MW)	2.5
Hub Height (m)	0
Rotor Diameter (m)	0
Turbine Spacing (times rotor diameter)	0
Row Spacing (times rotor diameter)	0

Construction Information

Year of Construction	2022
Money Value (Dollar Year)	2022
Total Project Construction Time (months)	0
Rate of Deliveries (turbines per week)	0

(Continued)

TABLE 24.4 (Continued)

Project Description

Detailed Wind Farm Project Data Costs	MICHIGAN
Construction Costs	Costs
Equipment	
Turbines	$120,600,000
Blades	$37,600,000
Towers	$43,800,000
Transportation	$20,278,776
Equipment Total	$222,278,776
Balance of Plant	
Materials	
Construction (concrete rebar, equip, roads and site prep)	$18,560,960
Transformer	$0
Electrical (drop cable, wire,)	$5,682,993
HV line extension	$1,563,881
Materials Subtotal	$25,807,834
Labor	
Foundation	$4,239,948
Erection	$7,790,048
Electrical	$1,773,965
Management/supervision	$2,047,903
Misc.	$1,295,777
Labor Subtotal	$17,147,641
Development/Other Costs	
HV Sub/Interconnection	
Materials	$4,638,519
Labor	$1,987,937
Engineering	$2,376,670
Legal Services	$716,476
Land Easements	$0
Site Certificate	$437,230
Development/Other Subtotal	$10,156,831
Balance of Plant Total	$53,112,305
Sales Tax (Materials & Equipment Purchases)	$13,668,470
Total Project Costs	$289,059,552
Wind Plant Annual Operating and Maintenance Costs	Costs
Labor Costs	
Personnel	
Field Salaries	$480,751
Administrative	$58,962
Management	$97,274
Labor/Personnel Subtotal	$636,987
Materials and Services	
Vehicles	$233,196
Misc. Services	$90,946
Fees, Permits, Licenses	$45,473
Misc. Materials	$181,893
Insurance	$1,748,967
Fuel (motor vehicle gasoline)	$90,946
Tools and Misc. Supplies	$591,151
Spare Parts Inventory	$5,180,441
Materials and Services Subtotal	$8,163,013

(Continued)

Case Studies in Cost–Benefit Analysis

TABLE 24.4 (Continued)

Project Description

Sales Tax (Materials & Equipment Purchases)	$357,535
Other Taxes/Payments	$0
Total (with Sales Tax and Other Taxes/Payments)	$9,157,535
Debt Payment (average annual)	$16,600,075
Equity Payment—Individuals	$0
Equity Payment—Corporate	$36,855,084
Property Taxes	$2,474,350
Land Lease	$1,680,600
Total Annual Operating and Maintenance Costs	$66,767,643

TABLE 24.5
Natural Gas Power Plant JEDI Models Inputs and Outputs

Natural Gas Power Plant Project Description

Project Location	MICHIGAN
Year Construction Starts	2022
Project Size—Nameplate Capacity (MW)	200
Capacity Factor (Percentage)	65%
Heat Rate (Btu per kWh)	7,000
Construction Period (Months)	36
Plant Construction Cost ($/KW)	$1,250
Cost of Fuel ($/mmbtu)	$0.00
Produced Locally (Percent)	-50%
Fixed Operations and Maintenance Cost ($/kW)	$8.25
Variable Operations and Maintenance Cost ($/MWh)	$2.90
Money Value (Dollar Year)	2022

Annual Operating and Maintenance Costs	Cost
Fixed Costs	
Labor	$690,030
Materials	$99,997
Services	$859,973
Fixed Subtotal	$1,650,000
Variable Costs	
Water	$170,820
Catalysts & Chemicals	$341,640
Variable Subtotal	$3,302,520
Fuel Cost	$35,553,336
Sales Tax (Materials & Equipment Purchases)	$36,747
Other Taxes/Payments	$0
Total (with Sales Tax and Other Taxes/Payments)	$40,542,603
Financing (avg ann debt payment)	$13,600,000
Equity Payment—Individuals (avg ann payment)	$0
Equity Payment—Corporations (avg ann payment)	$0
Property Tax	$0
Land Lease	$0
Total (with financing)	$54,142,603
Total (without debt, equity, taxes, lease payments)	$40,505,856

TABLE 24.6
Comparison of Operating Characteristics of Wind Turbine Power Plant with Natural Gas Power Plant

Variable	Wind Turbine Power Plant	Natural Gas Power Plant
Capacity Factor	0.40	0.65
Total energy produced (W/yr)	700,800,000,000	1,138,800,000,000
Total revenue ($) at $0.12/kWh	$84,096,000	$136,656,000
Total annual O & M Costs ($/yr)	$66,767,643	$54,142,603
Profit ($/yr)	$17,328,357	$82,513,397
Benefit-to-cost ratio	1.26	2.52
	(this wind model) W10.30.2020	Old.jedi.ng-model NG4.17.17

Table 24.6 provides a comparison of the annual revenue and annual operating costs of the two power plants. It was assumed that each power plant runs 24 hours at a constant output at the specified capacity factor and the electricity generated was sold at $0.12/kWh. The annual operating and maintenance costs (O and M costs) computed by the JEDI models include plant financing, fuel, and other plant operating and maintenance costs. Benefits-to-cost ratios for the two plants were simply estimated by dividing the annual revenue by annual O and M costs for each plant. Table 24.6 shows that the benefit-to-cost ratios for the wind turbine and the natural gas power plant were 1.26 and 2.52, respectively. The natural gas power plant was much more economical to operate because of its higher capacity factor and lower O and M costs.

CASE STUDY 3: EVALUATION OF FIVE ELECTRIC POWER GENERATION ALTERNATIVES

A similar and more detailed cost–benefit analysis was conducted to evaluate five different alternatives to supply an additional 500 MW of electricity distribution capacity to customers of a power company located in Michigan (Bhise, 2022). It was assumed that the power company had decided to study the following five alternatives:

A1: Build new land-based wind turbines to generate the additional capacity
A2: Build a natural gas-fueled power plant to generate the additional capacity
A3: Build a geothermal plant to generate the additional capacity
A4: Build a concentrating solar plant to generate the additional capacity
A5: Do not build a new plant and purchase the needed energy from other utility companies.

Case Studies in Cost–Benefit Analysis

The company had also decided to study the problem by considering the following four outcomes:

O1: The economy will grow at current 1% annual rate of increase in electricity demand.
O2: The economy will accelerate to 2% annual rate of increase in electricity demand.
O3: The economy will be stagnant with no increase in electricity demand.
O4: The economy will get worse and decelerate at 2% annual rate of electricity demand.

The following input conditions were assumed:

1. Probability of occurrences of outcomes O1, O2, O3, and O4 were 0.25, 0.50, 0.20, and 0.05, respectively.
2. The capacity factors of wind turbines (A1), natural gas-fueled plant (A2), geothermal plant (A3), and concentrating solar plant (A4) were assumed to be 35%, 87%, 92%, and 25%, respectively.
3. The utility company can finance 100% of the project at 3.0% interest over the next 33 years (the first 3 years using a construction loan with annual interest-only payments, and 30 years of annual payments for principal and interest during the plant operation).
4. In the first year of the plant operation, the company will begin selling 400MW at $0.13/kWh rate and the rate will increase at 1% per year after the first year of operation.
5. Additional energy, if needed, can be purchased from other utility companies at a prenegotiated fixed rate of $0.12/kWh over the 30 years of new plant operation (4th to 33rd year after plant construction).
6. The utility company can gain a benefit (or rebate from the state government) equal to 5% of the economic impact generated during the construction and operation of the plant. The economic benefit will be estimated by using the Jobs and Economic Development Impact (JEDI) models developed by the National Renewable Energy Laboratory (NREL, 2020a).
7. The power plant will operate 24 hours per day and over all 365 days per year.
8. All power plants will be assumed to be built in a midwestern state of the U.S.
9. The 3% discount rate will be used for present value calculations.
10. Annual safety costs were assumed to be as follows:
    a. Wind turbines at 2% of its capital
    b. Natural gas plant at 0.5% of its capital cost
    c. Geothermal plant at 1.75% of its capital cost
    d. Solar plant at 0.5% of its capital.

TABLE 24.7
Decision Matrix of Benefit-to-Cost Ratios for Five Power Generation Alternatives and Four Outcomes

	Outcomes & Occurrence Probabilities				
Occurrence Probability →	0.25	0.5	0.2	0.05	
Alternatives ↓	O1	O2	O3	O4	Expected Value
A1: Wind	1.77	1.92	1.56	1.22	1.77
A2: Natural Gas	1.94	2.10	1.77	1.44	1.96
A3: Geothermal	0.87	0.94	0.77	0.60	0.87
A4: Concentrating Solar	0.76	0.82	0.68	0.54	0.76
A5: Purchasing Agreement	1.24	1.25	1.23	1.21	1.24

METHODOLOGY

A decision matrix with five alternatives and four outcomes was set up to conduct a cost–benefit analysis (see Chapter 6 for the decision matrix). The values of the evaluation measures were the net present values of benefits minus costs and benefit-to-cost ratios computed over the total 33 years (first 3 years of plant construction time and the next 30 years of electricity generation) for each combination of alternatives and outcomes.

Detailed spreadsheets for each combination of alternatives and outcomes are provided by Bhise (2022). The benefits-to-cost ratios obtained from the spreadsheet based on cumulative present values are presented in Table 24.7. The expected values of the benefit-to-cost ratio show that the natural gas power plant is the most economical and concentrating solar power plant is the least economical.

CASE STUDY 4: NHTSA/EPA COST–BENEFIT ANALYSIS: INCREASES IN VEHICLE COSTS, FUEL SAVINGS, AND AVOIDED POLLUTION

On October 15, 2012, EPA and NHTSA published a coordinated National Program to improve fuel economy and reduce greenhouse gas emissions of light-duty vehicles for model years 2017–2025. In the joint notice of the two government agencies, the EPA announced final greenhouse gas (GHG) emissions standards for model years 2017–2025. And NHTSA announced the final Corporate Average Fuel Economy (CAFE) for MYs 2017–2021 and augural standards for MYs 2022–2025. These standards apply to passenger cars, light-duty trucks, and medium-duty passenger vehicles (i.e., sport utility vehicles, cross-over utility vehicles, and light trucks) (EPA and NHTSA, 2012).

The National Program is estimated to save approximately 4 billion barrels of oil and to reduce GHG emissions by the equivalent of approximately 2 billion metric tons over the lifetimes of those light-duty vehicles produced in MYs 2017–2025. EPA standards are projected to require on an average industry fleet-wide basis, 163 grams/mile of carbon dioxide (CO_2) in the model year 2025, which is equivalent

Case Studies in Cost–Benefit Analysis

to 54.5 mpg if this level were achieved solely through improvements in fuel efficiency.

The first phase, MYs 2017–2021, is projected to require, on an average industry fleet-wide basis, a range from 40.3 to 41.0 mpg in MY 2021. NHTSA projected that on an average industry fleet-wide basis, a range from 48.7–49.7 mpg in the model year 2025. The agencies project that fuel savings will far outweigh higher vehicle costs, and that the net benefits to society of the MYs 2017–2025 National Program will be in the range of $326 billion to $451 billion (by considering 7% and 3% discount rates, respectively) over the lifetimes of those light-duty vehicles sold in MYs 2017–2025. The agencies estimate that technologies used to meet the standards will add, on average, about $1,800 to the cost of a new light-duty vehicle in MY 2025, consumers who drive their MY2025 vehicle for its entire lifetime will save, on average, $5,700 to $7,400 (by considering 7% and 3% discount rates, respectively) in fuel, for a net lifetime savings of $3,400 to $5,000.

INCREASE IN VEHICLE PRICE VS. FUEL SAVINGS

a. Despite estimated increases in average vehicle prices of between $183 to $287 per vehicle in MY 2017 to between $1,461 and $1,616 per vehicle in MY 2025, NHTSA estimates that discounted fuel savings over the vehicles' lifetimes will be sufficient to offset initial costs.

b. Even discounted at 7%, lifetime fuel savings are estimated to be more than 2.5 times the incremental price increase induced by manufacturers' compliance with the standards.

The EPA and NHTSA (2012) notice of the final rule provided detailed descriptions of the analyses conducted to support the final rule. The considerations and variables used are summarized below.

1. Costs of implementation of new technologies (e.g., turbo-boost, hybrid and electric powertrains, low friction bearings, low rolling resistance tires, aerodynamic changes in the vehicle bodies, light-weight materials, and so forth) by the vehicle manufacturers in their fleet of 2017–2025 MY vehicles projected to be sold in the U.S. These costs were obtained by NHTSA from 20 automobile manufacturers that sold their products in the U.S.
2. Savings from reduced fuel consumption by the vehicle manufacturers' fleet of 2017–2025 MY vehicles to be sold in the U.S.
3. Projected reduction in health-related costs due to reduction in pollution (greenhouse gases) emitted by these 2017–2025 MY vehicles.
4. The final rule also allowed for flexibilities to vehicle manufacturers by credit averaging, banking, and trading of accumulated credits. A manufacturer will generate credits if its car and/or truck fleet achieves a fleet average CO_2/CAFE level better than its car and/or truck standards. Conversely, a manufacturer will incur a debit/shortfall if its fleet average CO_2/CAFE level does not meet the standard when all credits are considered.

5. Incentives for Electric Vehicles (EV), Plug-in Hybrid Electric Vehicles (PHEV), Fuel Cell Vehicles (FCV), and Compressed Natural Gas (CNG) Vehicles (through multipliers) were also established to promote the use of alternate fuel vehicles. This multiplier approach means that each EV/PHEV/FCV/CNG vehicle would count as more than one vehicle in the manufacturer's compliance calculation.
6. The concept of the social cost of carbon was applied to estimate the social cost of carbon. The social cost of carbon ($SC-CO_2$) is a measure, in dollars, of the long-term damage done by one metric ton of carbon dioxide (CO_2) emissions in a year. EPA and other federal agencies use estimates of the social cost of carbon ($SC-CO_2$) to value the climate impacts of rulemaking (EPA, 2020a and 2020b). This dollar figure also represents the value of damages avoided for a small emission reduction (i.e., the benefit of a CO_2 reduction).

The $SC-CO_2$ is meant to be a comprehensive estimate of climate change damages and includes changes in net agricultural productivity, human health, property damages from increased flood risk, and changes in energy system costs, such as reduced costs for heating and increased costs for air conditioning. However, given current modeling and data limitations, it does not include all important damages. The IPCC (Intergovernmental Panel on Climate Change) Fifth Assessment report observed that $SC-CO_2$ estimates omit various impacts that would likely increase damages.

One of the most important factors influencing $SC-CO_2$ estimates is the discount rate. A large portion of climate change damages are expected to occur many decades into the future and the present value of those damages (the value at present of damages that occur in the future) is highly dependent on the discount rate.

Present value effect: To understand the effect that the discount rate has on present value calculations, consider the following example. Let us say that you have been promised that after 50 years you will receive $1 billion. In "present value" terms, that sum of money is worth $291 million today with a 2.5% discount rate. In other words, if you invested $291 million today at 2.5% and let it compound, it would be worth $1 billion in 50 years. A higher discount rate of 3% would decrease the present value today to $228 million, and the present value would be even lower—$87 million with a 5% rate. This effect is even more pronounced when looking at the present value of damages further out in time.

The values of the social cost of CO_2 for 2015 to 2050 for each metric ton of CO_2 increased (benefit if one metric ton of CO_2 is reduced) by assuming different discount rates (5% average, 3% average, 2.5% average discount rate, and 95th percentile value [as a worse case at 3% discount rate]) are provided in EPA (2020a).

From the overall values of the costs and benefits provided in EPA and NHTSA (2012), Table 24.8 was constructed to obtain benefit-to-cost ratios. The benefit-to-cost ratios of both the 3% and 7% discount rate calculations were well above 1.0 indicating that the regulation will be economically justifiable.

TABLE 24.8
Benefit-to-Cost Ratios at 3% and 7% Discount Rates for Automotive Fuel Economy and Emissions Requirements Mandated by NHTA and EPA in 2012

Period over Which the Present Values Were Computed	Total Costs* (Billions)	Fuel Savings (Billions)	Other Benefits** (Billions)	Net Benefits (Billions)	Benefits-to-Cost Ratio
2017-2025 MY Lifetime at 3% Discount Rate	$150.00	$475.00	$126.00	$601.00	4.01
2017-2025 MY Lifetime at 7% Discount Rate	$144.00	$364.00	$106.00	$470.00	3.26

*Vehicle program costs (costs to incorporate new technological changes in vehicles)
**Other benefits include reduced costs related to health, social cost of carbon, accidents, noise, increased drivng, less-frequent refueling

CASE STUDY 5: MANUFACTURING AND ASSEMBLY LINE: ROBOTIC ASSEMBLY OF AN AUTOMOTIVE DIFFERENTIAL GEAR CARRIER

The problem illustrated in the following pages of this chapter was originally created by the author for his Automotive Assembly Systems course at the University of Michigan-Dearborn for students to learn applications of design for assembly techniques developed by Boothroyd and Dewhurst (2011) to estimate and reduce the cost of assembly. The predicted assembly time and robot costs were used to estimate the benefit-to-cost ratio to justify a robotic assembly of an automotive differential gear carrier.

Problem:
A chassis assembly manager of an automotive company wanted to automate the task of assembling ring gear to the differential carrier and two tapered bearings at the left and the right sides to hold the side gears and the axle shafts. Figure 24.1 below shows a picture of the differential carrier with the ring gear and a tapered bearing on the right side. (Note that the left side tapered bearing is hidden under the ring gear.)

A simplified cross-sectional view of the differential assembly with the differential carrier along with other surrounding components in the assembly is shown in Figure 24.2. The end views of the ring gear, the differential carrier, the shim, and the tapered roller bearing to be assembled are provided in Figure 24.3.

Currently, the differential carrier assembly is performed in three in-line workstations. The components inside the differential carrier such as the side gears, pinion gears with the cross-pin shaft are manually assembled in the first assembly workstation. (Note that these components are not shown in Figures 24.2 and 24.3).

Then in the second assembly workstation, the two shims and tapered bearings are attached manually at the two ends of the carrier. The tapered bearings require about 20 lbs of force to slide (i.e., to mount the roller bearings by using a rubber hammer) on the ends of the carrier. The bearings are thus force-fit and are tiring to the assembler. Further, the assembler needs to reorient (turn) the differential carrier by

FIGURE 24.1 Picture of the left side of differential carrier with the ring gear attached with 16 ring gear bolts and a tapered roller bearing.

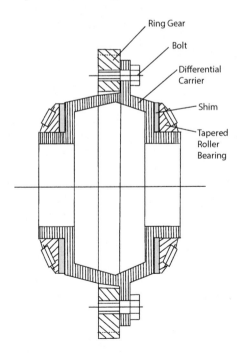

FIGURE 24.2 Cross-section of differential carrier.

Case Studies in Cost–Benefit Analysis

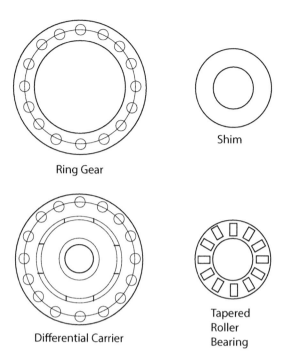

FIGURE 24.3 Four components of the differential carrier assembly (shown in plan view).

180 degrees to mount the second tapered bearing after the first tapered bearing is mounted in the vertical direction.

In the third workstation, the ring gear is placed face down (as shown in Figure 24.1) and the carrier assembly is placed on the top of the ring gear. The holes on the carrier assembly are aligned manually with the threaded holes in the ring gear. Then the 16 bolts are manually inserted and tightened by using a motorized nut-driver with a built-in torque limiter.

The completed carrier assembly is placed on a conveyer on the right side of the operator and the assembly is taken to the next workstation where it is mounted inside the outer casing of the differential.

The manager would like to design a robotic assembly line of workstations to replace the second and third workstations described above. The assignment thus is to develop an efficient assembly method and workstation layout for the robotic assembly of the shims, the tapered bearings, and the ring gear. The differential assemblies need to be delivered to the chassis assembly line at the rate of 60 units/hour.

BASELINE MANUAL ASSEMBLY

Boothroyd et al. (2011) manual assembly method was first used to obtain baseline time estimates to manually assemble the five types of components in the differential carrier assembly using the current assembly method described above. The five types of components are: (1) differential carrier, (2) ring gear, (3) 16 bolts, (4) 2 shims, and

TABLE 24.9
Manual Assembly Codes and Times for Differential Carrier Assembly Using Boothroyd and Dewhurst Manual Assembly Method

Component Name	Number of Components	Manual Handling Code	Handling Time per Component, s	Manual Insertion Code	Insertion Time per Component, s	Total Operation Time, s	Description
Carrier	1	91	3.00	06	5.50	8.50	Add carrier
Shim	2	02	1.88	00	1.50	6.76	Add shim
Roller Bearing	2	14	2.55	06	5.50	16.10	Add roller bearing
Press Operation	2			90	4.00	8.00	Opeartion
Ring Gear	1	90	2.00	06	5.50	7.50	Add ring gear
Reorient Assembly	2			98	9.00	18.00	Operation
Bolt	16	14	2.55	38	6.00	136.80	Add bolt
Drive Bolt	16			92	5.00	80.00	Operation
					Total-->	281.66	

(5) 2 roller bearings. Table 24.9 shows the manual assembly data obtained from the use of Boothroyd et al. manual assembly tables presented in Boothroyd et al. (2011) for assembling the five components. Here the carrier shim is first inserted over the lefthand side of the carrier, and the tapered roller bearing is press-fitted over it. The carrier assembly is then reoriented over the ring gear. The second carrier shim is then inserted over the righthand side of the carrier, and the tapered roller bearing is press-fitted over it. And finally, the 16 bolts are installed to attach the ring gear to the differential carrier.

This shows that the manual assembly for the 4 components, 16 bolts including reorientation of the assembly, and to press-fit the roller gearings would take 281.66 s, requiring 5 manual assembly stations and operators to achieve the same output. This provides the baseline supporting data to compare with the robotic assembly line data described below.

ROBOTIC ASSEMBLY METHOD

An assembly line considered here required six robots, a press (to slide tapered roller bearings), and seven conveyors to bring the following components: (a) ring gear, (b) differential carrier, (c) 16 bolts through bolt feeder, (d) left side shim, (e) left side tapered roller bearing, (f) right side shim, (g) right side tapered roller bearing. Figure 24.4 presents a plan view of the assembly line layout.

The proposed assembly system is a multi-station solution that starts with the ring gear being placed onto a fixture on the conveyor system (see the bottom side of

Case Studies in Cost–Benefit Analysis

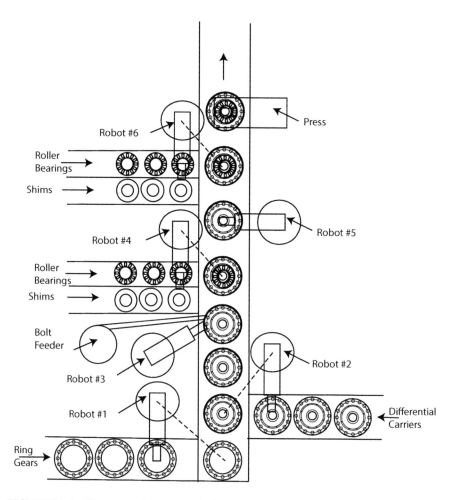

FIGURE 24.4 Plan view of the assembly line with six robots and one press.

Figure 24.4). The main assembly line moves from bottom to top in a vertical direction in Figure 24.2. Robot #1 places the ring gear in a fixture mounted in the main assembly line. Robot #2 places a differential carrier on top of the ring gear. The fixture helps align the holes of the differential carrier with the holes in the ring gear. The bolt feeder (located just to the left of robot #3) orients bolts and feeds them to robot #3 for inserting and fastening the ring gear to the differential carrier in a sequential order. The carrier and the ring gear mounted in the fixture indexes as the bolts are torqued by robot#3.

Fixture

The differential carrier is placed on the fixture by robot #2 after leaving the first station where the ring gear is placed by robot #1. This fixture allows the differential carrier to be placed on top, and ride on the conveyor system. The fixture is what is tracked by the robots so that the workstations are aware of where the components are oriented on the assembly line. Another key feature of this fixture is that the bottom is open and allows ease of access by the robots to do their job. This is most important for the press operation that simultaneously presses the top and bottom of the differential carrier to firmly install the bearings. Operating the press on both ends at the same time allows the cycle time to be much faster and minimizes needless operations.

Bolt Feeding

The feeder for the 16 bolts has a vibratory track that suspends the bolts by their head in a vertical orientation. The bolts slide down until they fall into the eight spots on the rotary feeder. This arrangement suspends and orients the bolts above the holes. Robot #3 (with eight sockets suspended above the rotary feeder) comes down to magnetically grab the eight bolts, lifts and rotates them (to be placed above the differential carrier on the line), and then torques them down. This process is repeated for the other eight bolts required to completely secure the ring gear.

Robot #4 picks up one shim and a roller bearing and places them on the top of the pinion shaft on the left side of the differential carrier. The roller bearing is also pushed in by robot #4 by about 3 mm to hold the tapered bearing on the pinion shaft.

This above process is repeated once the assembly is flipped by robot #5 so the second shim and bearing can be installed by robot #6.

After the shim and bearings are installed, the press at the next workstation compresses these together from the top and bottom to completely secure the bearings/shims to the carrier.

Shim and Bearing Feeding

The shim and bearing are fed to the station on vibratory feeders that present the components flat. The feeder is designed such that the component is always presented in the same orientation by a notch that prevents parts from arriving at the robot in any other orientation. The end of the feeder track is designed with a cutout at the end of it, so the component is able to be gripped from the top and bottom without the feeder track getting in the way. This gripper has two arms so that it can simultaneously grab both the bearing and the shim at once, as they are presented right next to each other. This saves cycle time since the gripper only needs to make one rotation between the feeders and the conveyor belt per part. The gripper then drops off the components, shim first then the bearing, onto the differential carrier.

Grippers

The gripper used on robots #1 and #2 to pick up and maneuver the ring gear and differential carrier are both very similar. They are three-finger single-gripper designs that retain the parts with friction to easily place them on the assembly line fixture and maintain orientation when releasing the parts.

The gripper for the ring gear bolts of robot #3 picks up the eight bolts via an electrically magnetic-induced bolt-holding mechanism. This allows for precise locating and orienting of the bolts to secure them onto the ring gear/carrier assembly.

To grasp the shim and bearings that flow down the feeder track and arrive at the workstation, a double gripper robot with eight flat horizontal fingers is utilized to grasp the shim and bearing simultaneously from the top and bottom of the parts. Four

Case Studies in Cost–Benefit Analysis

fingers per part are used to provide more stability which allows the robot to clamp down and handle the parts to keep them in place during rotation to the conveyor. The gripper is also able to freely move along its horizontal axis to have the ability to first place the shim, slide over, and then place the bearing on top of that.

ROW AND COLUMN CODES FOR COMPONENTS

The Boothroyd and Dewhurst (2011) method for single station single arm robotic assembly requires codes for selected the row and column for each component handled by the robot [Refer to Figure 5.21 in Boothroyd and Dewhurst (2011)]. Applying the Boothroyd and Dewhurst (2011) method for robotic assembly the row and column codes were first selected for each component and then the values of variables AR (relative robot cost), TP (relative effective basis operation time), AG (relative addition gripper or tool cost), and TG (relative time penalty for gripper or tool change) were read from the cell corresponding to the selected row and column. The codes and the values of the four parameters for the components are given below.

Ring Gear

   a. Row code = 2; Part added but not finally secured, involving motion along or about more than one axis.
   b. Column code = 5; Part requires a change to a special gripper, no holding down, not easy to align.
   c. Therefore, AR = 1.5, TP = 1.9, AG = 1.5, TG = 2.1.

Place Differential Carrier

   a. Row code = 2; Part added but not finally secured, involving motion along or about more than one axis.
   b. Column code = 5; Part requires change to a special gripper, no holding down, not easy to align.
   c. Therefore, AR = 1.5, TP = 1.9, AG = 1.5, TG = 2.1.

Torque Bolts

   a. Row code = 3; Part added and secured immediately, using motion along or about the vertical axis.
   b. Column code = 7; Force or torque levels within robot capability, part requires a change to the special gripper, screw fastening or nut running, not easy to align.
   c. Therefore, AR = 1, TP = 0.08, AG = 1.5, TG = 0.7.

Shims and Tapered Bearings

   a. Row code = 5; Part added and secured immediately involving motion along or about more than one axis.

b. Column code = 5; Force of torque levels within robot capability. Part requires a change to a special gripper. Snap or push fit. Not easy to align.
c. Therefore, AR = 1.5, TP = 1.9, AG = 0.5, TG = 0.

Reorient Operation—Rotate Differential in Fixture

a. Row code = 8; Separate securing operation or manipulation, or reorientation, or addition of non-solid, using motion along or about the vertical axis.
b. Column code = 7; Operation requires a change to special gripper or tool, reorient or unload assembly.
c. Therefore, AR = 1.5, TP = 1.8, AG = 0.5, TG = 0.

Press Operation—Boothroyd High Speed Automated Assembly Operation [from Figures 5.3 to 5.9 of Boothroyd and Dewhurst (2011)]. Note: A is the length of the longest side of the component and C is the shortest length of the component (enclosing a rectangular prism). E is orienting efficiency and C_R is relative feeder cost.

a. First digit code = 8; Non-rotational—Cubic A/B ≤ 3, A/C ≤ 4
b. Second digit code = 4; Part has no symmetry, orientation defined by one main feature.
c. Third digit code = 5; Through grooves parallel to Z axis and >0.1B
d. Forth digit code = 8; Parts are very small or large but are nonabrasive
e. Fifth digit code = 9; Large Parts, non-rotational, A/B ≤ 3, A/C ≤ 4
f. Therefore, E = 0.1, $C_R = 1.5 + 9 = 10.5$

Table 24.10 provides a completed worksheet prepared from the application of the Boothroyd and Dewhurst robotic assembly method. Table 24.11 provides the completed worksheet for the automatic assembly method used for the press operation.

ASSUMPTIONS REGARDING PARTS AND ASSEMBLY CONSIDERATIONS

Several costs and time values were taken from Boothroyd and Dewhurst (2011) for preparing the above two tables. The standard values (before applying multipliers for relative costs) were (a) robot cost was assumed to be $60,000, (b) gripper cost was assumed to be $5,000, (c) standard robot operation time was assumed to be 3 seconds. Total robot costs were $ 420,000, total gripper costs were $27,500 and the press cost was $30,000. (Total equipment cost $477,500).

Further, it was assumed the updated process was for a high-volume application, and therefore speed was prioritized.

Because it was unsure what would be considered easy or not easy to align for a robot in this scenario, everything was assumed as not easy to align.

This report assumed indexing 1/16 turn for bolting operation was covered in the Boothroyd operation chosen. The arm that secures bolts can secure eight bolts at once. In this case, the robot would secure every other bolt, then turn around 22.5 degrees to account for 1/16th of a turn and secure the remaining 8 bolts.

Case Studies in Cost–Benefit Analysis 509

TABLE 24.10
Completed Worksheet for Robotic Assembly Using Boothroyd and Dewhurst (2011) Method

Robot #	Component	Description of Step	Repeat Count	Code	AR	TP	AG	TG	Robot Cost	Gripper Cost	Robot Time (s)
1	Ring Gear	Place Ring Gear	1	25	1.5	1.9	1.5	2.1	$90,000	$7,500	9.9
2	Differential Carrier	Place Differential	1	25	1.5	1.9	1.5	2.1	$90,000	$7,500	9.9
3	Bolts	Torque Bolts	2	37	1	0.08	1.5	0.7	$60,000	$7,500	3.28
4, 6	Shim and Tapered Roller Bearing	Place Bearing + Shim	2	55	1.5	1.9	0.5	0	$90,000	$2,500	11.4
5	Differential Assembly	Reorient Differential in Fixture	1	87	1.5	1.8	0.5	0	$90,000	$2,500	5.4
Press	Differential Assembly	Press Operation	1	Automated Assembly Required							5.2
										Total Time	45.08

TABLE 24.11
Completed Worksheet for Automatic Workhead Assembly Using Boothroyd and Dewhurst (2011) Method

Name	Repeat Count	Feed Code	Orient Efficiency, E	Relative Feeder Cost, Cr (cents)	Max Feed/min Fm	Feeding Cost, Cf (cents)	Insertion Code	Relative Workhead Cost, Wc (cents)	Insertion Cost, Ci (cents)	Total Cost, Ct (cents)	Min Parts
Press Operation	1	[84589]	0.1	10.5	11.54	1.64	[00]	1.6	0.0036	1.6416	1

To fasten a bolt in the multi-arm workstation model, the part requires a change to a special gripper. In the workstation design chosen, the bolt-torquing robot head is only used for that one operation, so the part does not require a change to a special gripper. As this was the closest option for the operation, the code 37 was chosen despite not needing a gripper change.

The press operation in this project was determined to be an automated assembly operation rather than a robotic operation. As a result, several assumptions were made including the cost of using the standard workhead for 1 second of 0.06 cents, cost of press ($ 30,000), and the length of the differential which was assumed to be 13 inches.

It was assumed that the shim and bearing could be fed into the fixture which tracked with the differential while suspended simultaneously. This assumption reduced assembly time considerably because it removed a reorient operation and allowed the robot arm that grasped the shim and bearing to travel to the shim feeder and the bearing feeder, then the fixture and differential, removing the need to travel between the fixture and differential to deposit each part, then to the next feeder.

COST–BENEFIT ANALYSIS

Table 24.12 presents the costs and benefits of the manual assembly vs. the robotic assembly. The total benefit of robotic assembly per year was equal to $ 201,169 (i.e., $365,297 of labor cost saved from the manual assembly, and $108,128 cost of principal and interest for robotic assembly line equipment and $56,000 added from robotic assembly labor (for 1 line operator over 2000 hrs/yr). Benefit-to-cost ratio was 1.23 (i.e., $201,169 divided by $ 164,128). The benefit minus the costs for the robotic assembly was $37,041 (net annual saving) as compared to $365, 297 of assembly

TABLE 24.12
Cost Comparisons for Manual vs. Robotic Assembly of Differential Carrier

Variable	Manual Assembly	Robotic Assembly
Time per differential assembly (sec)	281.66	45.08
Allowance (additional time) (%)	15	10
Additional Equipment cost (robots, grippers and press) ($)		$477,500
Annual cost of financing (principal and inteest) at 6% interest rate over 10 years ($)		$108,128
Annual production volume (number of differential carrier needed/year)	145000	145000
Plant hours available per year (hr/yr)	2000	2000
No. of differential carrier assemblies per 2000 hours	22228	145196
Labor cost ($/hr)	$28	$28
Total assembly labor costs ($)	$365,297	$56,000
Total labor plus equipment costs ($/yr)	$365,297	$164,128
Benefits (savings per year) ($)	$0	$201,169
Benefit-to-cost ratio	0.00	1.23
Benefit minus cost ($)	−$365,297	$37,041

labor costs for the manual assembly. (Note that the present value method was not used here because the sum of discount and inflation rate was assumed to be 0%.)

LIFE CYCLE COST ANALYSIS

Life cycle cost analysis approach involves considering all costs and benefits to be computed over the entire life cycle) of the product program (from product conceptualization until its termination with disposal). The above four case studies have only included major design, construction, and operational phases of the systems such as automotive product development and production in case study 1 and power plant construction and operation in case studies 2 and 3. Additional costs at the end of the life usually include costs of plant termination, restoration of the plant site, or modification of the plants and equipment for incorporating next-generation of products and processes, and other costs related to recycling, scrapping, and disposal of components and equipment.

CONCLUDING REMARKS

The five case studies presented in this chapter illustrate that present values of all revenues and costs incurred over the entire life cycle of each alternative must be included in the cost-benefit analysis to get a complete understanding of the financial picture before a decision can be made. The use of spreadsheets also provides the decision maker a better understanding of relative values of different types of costs and how they vary as functions of time and how the present value approach provides the effect of discount factor on the entire life cycle of the projects that extend over many years.

REFERENCES

Bhise, V. D. 2017. Financial Analysis in Automotive Programs, Chapter 19 in *Automotive Product Development: A Systems Engineering Implementation*. ISBN: 978-1-4987-0681-0. Publisher: CRC Press, Boca Raton, FL: CRC Press, 2017.

Bhise, V. D. 2022. *Decision-Making in Energy Systems*. ISBN: 978-0-367-62015-8. Boca Raton, FL: CRC Press.

Boothroyd, G., Dewhurst, P., and W. A. Knight. 2011. *Product Design for Manufacturing and Assembly*. 3rd Edition, Boca Raton, FL: CRC Taylor & Francis Group.

EPA. 2020a. Social Cost of Carbon. (Website: https://19january2017snapshot.epa.gov/climatechange/social-cost-carbon_.html; accessed: October 3, 2020).

EPA. 2020b. Greenhouse Gas Equivalencies Calculator. (Website: https://www.epa.gov/energy/greenhouse-gas-equivalencies-calculator; accessed July 21, 2020).

EPA and NHTSA. 2012. 2017 and Later Model Year Light-Duty Vehicle Greenhouse Gas Emissions and Corporate Average Fuel Economy Standards. Published by the Environmental Protection Agency (EPA) and the National Highway Traffic Safety Administration (NHTSA, DOT). Federal Register / Vol. 77, No. 199, October 15, 2012. Pages 62623–63200.

NREL. 2022a. JEDI Wind Models. *JEDI Land Based Wind Model Beta rel.* W10.30.20. Website: https://www.nrel.gov/analysis/jedi/wind.html (Accessed: March 25, 2022).

NREL. 2022b. JEDI Natural Gas Model, Release number: W10.30.2020. Website; https://www.nrel.gov/analysis/jedi/natural-gas.html (Accessed: March 25, 2022).

25 Challenges and Future Issues in Systems Engineering

INTRODUCTION

In spite of the many design and analysis tools available in the field and many presented in this book, the demand for better tools is growing more acute because of the increasing demand for more complex products and systems. One primary reason for the increasing demand for complex products or systems is the increase in the integration of digital technologies, software, and human factors in hardware. Digital technologies with ever-increasing computing power and miniaturization of microprocessors and the advances in fields such as sensors, actuators, wireless communications, global positioning technologies, and Internet are creating new opportunities to develop complex products. Many of the new products offer programmable and reconfigurable features that offer further flexibility and diversity in their applications. This chapter also discusses challenges and future research needs for improved and faster applications of systems engineering. Some information is also included on a related and important issue of how to teach systems engineering and its concepts to our future students.

CHALLENGES IN SYSTEMS ENGINEERING

The challenges facing systems engineering are many (Boehm, 2005, 2006; Kamrani and Azimi, 2011). Some of the issues related to the challenges are presented as follows:

1. Customers are getting more sophisticated and as a result are expecting products with more features, service, upgradability, connectivity, and integrated functioning with other products or systems.
2. Program management is facing increasing pressures to reduce costs and timings of most program activities. The pressures on program managers are due to reasons such as scarcity of capital, increased competition, and customer expectations of "doing it right the first time." The company management must understand the collaborative working of systems engineers with other disciplines and must continually promote communication between departments to manage systems engineering activities (see Chapter 2).
3. The software development task in many complex products (e.g., aircraft) is much more complex and time-consuming than their hardware development. This requires increased resources in the integration of software engineering and systems engineering. The growth of software is also much faster and

larger than hardware development (e.g., Augustine's law, which states that software grows by an order of magnitude every 10 years) (Nidiffer, 2007).
4. The increased digital communication speed (e.g., terabits per second) allows for greater and higher levels of communication (e.g., voice, video, and net meetings facilitating virtual presence) worldwide quickly in highly structured database environments. Thus, cyberspace and physical space are increasingly intertwined.
5. Demands for complex systems (or products) and higher-level "system of systems" configurations are increasingly becoming more common. This requires applications of systems engineering concepts and techniques early in the development of systems.
6. As the systems are becoming more complex, there is an increasing demand from human factors engineers to reduce systems complexity in future systems. This demand is also driven by the trends in making products more compact (in part by miniaturizing microprocessors), adding more features, and providing higher flexibility and programmability in product functions.
7. Greater emphasis on the involvement of multifunctional teams (including human factors engineering) has led to the development of more coordinated product designs and creative user interfaces.
8. Systems engineering is recognized as an important discipline. Increased customer requests for systems engineering support in the earlier part of the product life cycle have increased demands for systems engineers.
9. Globalization in product design has become more common due to greater connectivity between designers, producers, and suppliers; economies of scale; and the capability to outsource and to reduce costs and timings (Friedman, 2005).
10. Severe constrains on supplier capabilities due to Covid-19 outbreaks and Ukraine war are now affecting the availability of supplier-based product development and production outputs.

NEED FOR TOOLS IN COMPLEX PRODUCT DEVELOPMENT

The need to create more integrated tools to undertake systems engineering functions has also increased. However, the market for generalized and more integrated tools is still small because of the enormous costs in developing such software tools.

Some areas of needed software capabilities are as follows: (1) relational database management with capability to manage components, systems, functions, requirements, interfaces, and traceability, (2) graphical editing and browsing, (3) expert system capability with reasoning and knowledge base; (4) system decomposition and block diagramming capabilities, (5) modeling, simulation, and prototyping, (6) distributed networks with import/export capability to interact with many currently available computer-aided design (CAD), computer-aided engineering (CAE), and computer-aided manufacturing (CAM) applications (new data exchange standards and file compatibility and transparency will greatly reduce these data exchange problems), (7) document generators, and (8) free and open source software with different toolkits for flexibility in software development and cost reduction. Model-based systems engineering (MBSE, see Chapter 15) is being increasingly used to expedite early

Challenges and Future Issues in Systems Engineering

work in systems engineering implementation but still not yet considered as a mainstream application due to unique needs with different product development projects.

TOOLS TO MANAGE MULTIFUNCTIONAL AND MULTIPLE REQUIREMENTS

Tools to enable management of many product attribute requirements and interactions between many systems and multidisciplinary considerations and trade-offs are also seriously lacking.

The reasons for the problem are as follows:

1. Many of the issues are design dependent (i.e., the problems depend on the solution that in turn depends on the selected design approach and type of technology considered (e.g., the problems in the operation of light-emitting diode lamps are entirely different from those involving tungsten filament lamps; new generation of computer chips require different manufacturing processes and equipment)
2. The data on interactions and trade-offs between different variables are not explicitly available, and thus are not modeled for computerized tool development.
3. High cost of tool development and limited applicability of the tools across different product industries (e.g., an automated packaging tool for engine packaging [or fuel tank packaging] in the automotive industry will not be useful for packaging engines in the aircraft industry). Furthermore, many organizations have their own management processes and the level of depth and details considered are very different between different organizations.

COORDINATION OF GLOBAL DESIGN TEAMS

When lower-level entities within complex products are designed by different product development teams located in different geographical locations (cities, countries, and time zones), coordination and communication between teams becomes more challenging. The team members need to understand cultural and economic differences between different countries but agree on common goals and requirements of the product. Interfacing systems designed by different teams in different countries are especially challenging because they need to have constant communication to understand detailed design issues of the interfaces while realizing the product-level goals. Use of common design tools, databases, and constant communication methods such as design team meetings and management reviews are very important. In addition, use of common project management tools, immediate access to the latest approved designs to all, tight control over change management systems, transparencies, and traceability of all product decisions would help the teams stay on a single course.

COMMONALITY

Sharing of common entities (i.e., components, subsystems, and systems) presently available and used in many models of the products can reduce time and costs in developing any new unique shared entities. One major disadvantage of entity sharing is that it does not allow incorporation of new improvements (e.g., due to advances in new technologies) as the changes may affect all other existing products. Furthermore, the customers may realize that the new product is not any better than the previous or other versions of the products that share these common entities.

MODULARITY

When a product is designed in a modular configuration, many of the modules can be designed with common interfaces so that the modules can be easily interfaced and still provide different functionality by exchanging one module with another with different capabilities (e.g., battery packs with different battery types but with common connecting interfaces). Use of modular configurations can reduce development time and can provide the ability to create greater product variety. The modularization can also help in reducing service and maintenance times as one module can be simply swapped with another of the same or a different capacity according to usage needs and schedules.

CAD AND CAE INTEGRATION

The use of common data files to store product designs for CAD as well as CAE and CAM are needed. Such commonization can allow for conducting specialized engineering analyses (e.g., structural analysis, fluid flow analyses, and thermal analyses) and manufacturing tasks (e.g., to program a tool cutter path to produce a part) by sharing the same data files. Computerization is generally costly, and resource-limited, but many engineering analysis software systems are now available. Advances in parametric models and analyses allow designers to make quick changes in product configurations by merely changing values of input parameters. Of course, humans with specialized knowledge are always needed to work with such integrated computer-assisted technologies.

ERGONOMIC NEEDS IN DESIGNING PRODUCTS

With the increase in complexity in future products and systems, human factors engineering will need to play a key role in simplifying features and user interfaces so that the products will be easier to learn and use. Many advances in new technologies can be used to accomplish ergonomics goals. Some examples of possible features are (1) providing flexibility through the use of programmable or reconfigurable displays and controls, (2) use of remote controls that can be activated through wireless and internet technologies, (3) smart displays that provide processed information on current state of the product and recommended actions to maintain required functional state, (4) voice controls to reduce manual workloads, (5) providing user aids with artificial intelligence capabilities, (6) automated diagnostics capabilities during malfunctions, (7) automatic takeover of the product functions and safe shutdown capability in case of emergency, and so on. Users should also be able to personalize their interfacing configurations and settings so that needed functions can be performed quickly and at lower error rates. Thus, future systems cannot only be convenient, comfortable, and safe but also reduce operator workload under normal usage and emergency situations and thus make the products less stressful and more enjoyable to use.

FUTURE TECHNOLOGICAL CHALLENGES

Many new and emerging technologies should be considered during the development of a new technology implementation plan in the very early stages of product planning. The cost of development of new technologies, development time, and the probability of successful implementation of the new technologies are always the major issues. Depending on the product and its proposed functional characteristics, many new technologies can be considered from new materials development and new hardware configurations to digital data communications, advances in microprocessors, their configurations, computers, software, and so on. Some issues related to the implementation of new technologies are (1) trade-offs between costs and timings, (2) trade-offs between reliability and complexity, (3) fully automated functionality versus human intervention and takeover of controls during emergency or malfunctions, and (4) operator preferences related to new technology features.

BRIGHT FUTURE FOR SYSTEMS ENGINEERS

Due to the many future advances discussed in the preceding pages, the demand for systems engineers is expected to be very strong with challenging opportunities for career growth. Many systems engineering programs taught in universities have specialized courses that offer the students not only opportunities to understand the problem-solving approaches and available tools but also opportunities to work in multidisciplinary team environments, work alongside industry specialists, and apply the concepts and tools to develop integrated solutions. The students are thus better prepared to use the latest tools and technologies to turn complex ideas into reality.

CHARACTERISTICS OF A GOOD SYSTEMS ENGINEER

Some important and desired characteristics of systems engineers are as follows:

1. Must be a strong team player and must have team management skills
2. Must have abilities to understand the "big picture" and also must be able to work with many details (e.g., issues, considerations, and requirements)
3. Must have the ability to understand and communicate with technical personnel in multiple disciplines and must possess diverse technical skills
4. Must have the ability to integrate a number of requirements and understand trade-offs
5. Must have formal training in systems engineering processes and techniques (e.g., covered in this book) and some work experience in product programs
6. Must be a good communicator
7. Must understand program management and have the ability to work on time and within budgetary constraints

8. Must be willing to take on responsibilities and take risks
9. Must be comfortable with uncertainty and constantly changing design concepts and work environments
10. Must be adaptable and willing to learn new issues.

TEACHING SYSTEMS ENGINEERING

Teaching systems engineering is challenging as designing complex products requires integrated considerations from many disciplines simultaneously. Integration requires team effort and understanding trade-offs and prioritizing requirements. The author has been offering an automotive systems engineering course at the University of Michigan-Dearborn campus Dearborn, Michigan, for many years. The course is titled "Automobile—An Integrated System." The objectives of this course are to cover the following topics in an integrated manner:

1. Systems engineering approach and its implementation
2. Product development processes
3. Automobile and its systems
4. Development of vehicle specifications
5. Tools and methods used in automotive product development
6. Multidisciplinary nature of decision-making and problems facing the auto industry
7. Integration of methods such as design for manufacture, design for assembly, design for environment, sustainability, and recycling in automotive product development and production systems
8. Concepts of total quality management, "creating quality," and "variability reduction."

The semester-long (14-week) course includes lectures and project work. The following topics are covered in the lectures:

1. Introduction, vehicle design process, and systems engineering
2. Quality, benchmarking, Pugh diagrams, and quality function deployment (QFD)
3. Vehicle systems review
4. Product planning and project costs
5. Attributes, requirements, vehicle systems, interfaces, and systems engineering process
6. Decision models, costs, trade-offs, timings, and cost–benefit analysis
7. Business plan development
8. Vehicle design packaging and trade-offs
9. Production systems and vehicle assembly
10. Design trends and new technologies
11. Regulations, standards, and vehicle evaluations
12. Program planning and management
13. Development of Systems Engineering Management Plan (SEMP).

Challenges and Future Issues in Systems Engineering 519

The lecture material was supplemented with case studies and examples of integration issues. Several videos on Boeing 777 product development and automotive assembly were also shown in the class (see Chapter 23, Case Study 6). The students were required to complete seven class projects in teams of two to three students. The handouts of the seven projects are presented in Appendices 1 through 7. The handouts provided the objectives of each project, tasks to perform in each step, and requirements on the class presentations and project reports. The seven projects provided the students with opportunities to discuss and perform various tasks related to vehicle design, development, business planning, and the applications of many of the systems engineering tools covered in Part III of this book. A new format of four projects with three to five students in each team has been used over the past few years to manage the in-class presentation time due to increased student enrollment (30-50 graduate students per term and the course is now offered in both fall and winter terms). The textbook used for the course is by Bhise (2017).

OBJECTIVES OF THE PROJECTS

The overall objectives of the projects were as follows:

1. To provide the students the background and a working knowledge of steps involved in planning and designing an automotive product such as a car, a truck, or a sport utility vehicle.
2. To gather data on product design issues and conduct analyses using the methods covered in this class.
3. Create class discussions on vehicle development and systems engineering applications through student project presentations.

PROJECT WORK

The objectives discussed in the previous section were accomplished by conducting the following seven projects (the percentage of the grade assigned to each project is shown in parentheses):

1. Project 1: Review of case studies on a product development of a less complex product (cyclone grinder) and an automotive product (smart car) (10% of grade) (see Chapter 23, Case Studies 4 and 5)
2. Project 2: Benchmarking, QFD, and design specifications of a future automotive product (10% of grade)
3. Project 3: Vehicle systems analyses: Requirements, interfaces, trade-offs, and verification (10% of grade)
4. Project 4: Midterm report containing a business plan and a systems management engineering plan for the proposed vehicle (25% of grade)
5. Project 5: Conceptual design of the vehicle (10% of grade)
6. Project 6: Trends in new technologies, applications, and assembly details (10% of grade)
7. Project 7: Final report on the project: vehicle brochure illustrating specifications, design, features/options, and validation plan (25% of grade).

Students were encouraged to work in a team of two or three to gather data on various issues and conduct analyses that would be common to their vehicle platform, shared systems (or components), and assembly operations. The distance learning (online) students could also work with students in the campus class.

Each team was required to submit a written report for each project on a specified date provided in the course schedule. In addition, the teams were asked to present their accomplishments in the class by making short presentations. The students made short (5–10 minutes) PowerPoint presentations on the highlights of their projects 2, 3, 5, and 6 in the class. Each student was also required to make about a 20-minute PowerPoint presentation on his or her midterm (project 4) and final (project 7) reports in the class on the report due dates.

Projects 2 through 7 were based on a target vehicle that the students were asked to select for their projects. The selection of the target vehicle was based on (1) a current automotive product that the student team would enjoy in developing its future model and (2) have access to its recent model for the entire semester to use it as the reference vehicle to study for the entire set of projects (e.g., to study its layout, packaging of occupants, hardware and storage areas, configurations of systems and their interfaces with other systems, take pictures, and make measurements). The target vehicle was assumed to be introduced in the U.S. market in 5 years. Each student working in a team was asked to select a different target vehicle that can be a variation (e.g., a different body style, type of powertrain [electric vs. internal combustion], or brand) of the selected reference vehicle.

BRIEF DESCRIPTIONS OF THE PROJECTS

Brief descriptions of the contents of each project are given here. Additional information is provided in Appendices 1 through 7.

Project 1: Introduction to product development and automotive production
- Review a case study on product development of a less complex product (Ingersoll Rand Cyclone Grinder; see Chapter 23, Case Study 4)
- Review a case study on an automotive product (smart vehicle development and assembly; see Chapter 23, Case Study 5)
- Study customer needs and engineering specifications of complex products
- Decompose the product into systems, subsystems, and components
- Understand product development process and development issues.

Project 2: Develop design specifications
- Select a target vehicle (type, size, and market segment) and make assumptions related to its organizations (the automotive company and suppliers), their resources, constraints, requirements (corporate, federal, and other), and external factors
- Conduct benchmarking using data of competitive vehicles
- Determine its customers and their characteristics
- List customer needs of the vehicle (by interviewing a few customers)

- Develop a QFD for a vehicle system (each team member was asked to select a different vehicle system and prepare a QFD for the selected system)
- Determine specifications of the vehicle.

Project 3: Vehicle systems analyses: requirements, interfaces, trade-offs, and verification (see Chapter 22)
- Develop requirements for the selected vehicle system
- List of subsystems of the selected system
- Prepare an interface diagram of the subsystems of the selected system and other major vehicle systems
- Prepare an interface matrix for the subsystems of the selected system and other major vehicle systems
- Cascade requirements of the selected system to its subsystems and develop tests required to verify requirements on the subsystems
- Provide descriptions of specific issues, considerations, and trade-offs with interfaces and observations/findings from the aforementioned exercise.

Project 4: Business plan and a systems engineering management plan for the proposed vehicle
- Description and specifications of the proposed (target) vehicle
- Competitors (makes and models) of the proposed vehicle
- Systems engineering V model-based timing plan and gateways
- Sales projections
- Costs and revenue estimation table and plots of curves of life cycle costs and revenues for the vehicle program
- Systems engineering management plan (describing tasks, timings and deliverables, disciplines/departments involved and techniques/tools to be used).

Project 5: Conceptual design of the vehicle
- Pugh diagram, concept improvements, and vehicle definition
- Sketches, drawings showing basic exterior and interior dimensions in side and plan views of the vehicle
- Vehicle configuration and preliminary packaging (locations and envelopes of major vehicle entities).
- Selected technologies and features.

Project 6: Refining the vehicle design and the vehicle assembly process
- Refining vehicle design with new technology applications (assumptions and features)
- Descriptions of new features and supporting materials (e.g., new technological developments)
- Packaging layout sketches, spaces allocated to various systems, observations, and issues
- Analyses and calculations of package parameters
- Assembly plan and assumptions: assembly process chart and plant configuration.

Project 7: Term project: final report
- Vehicle brochure (for future customers)
- Major selling points—features that will satisfy and please customers
- Product description: sketches, drawings, and features/options
- Technical superiority and technology implementations
- Major engineering accomplishments
- Analyses performed to support statements/claims
- Vehicle evaluations: verification and validation plans for selected vehicle attributes.

CONCLUDING REMARKS

Future successes in implementing new product programs will depend on those who have the knowledge and resources to integrate available and new knowledge, apply available tools, and incorporate technological changes. The chapter covered many challenges facing the systems engineering profession. Large corporations and government organizations (e.g., the National Aeronautics and Space Administration, Department of Defense, and their major suppliers) have better capabilities (e.g., management tools and documentation of the lesson learned) to manage complex programs. However, their ability to incorporate new technologies may not be substantially better than some specialized technology companies. Integration of software systems with hardware systems is increasingly important for successful development of complex products as the software design tasks consume a larger percentage of the product development budget. The usability issues of software-intensive products are also demanding greater involvement of human factors engineers in the product design process.

REFERENCES

Bhise, Vivek D. 2017. *Automotive Product Development: A Systems Engineering Implementation*. ISBN: 978-1-4987-0681-0. Boca Raton, FL: CRC Press.

Boehm, B. 2005. *The Future of Software and Systems Engineering Processes*. Los Angeles, CA: University of Southern California. http://csse.usc.edu/csse/TECHRPTS/2005/usccse2005-507/usccse2005-507.pdf (accessed July 8, 2012).

Boehm, B. 2006. Some Future Trends and Implications for Systems and Software Engineering Processes. *Systems Engineering*, 9(1): 1–19. Hoboken, NJ: Wiley Periodicals, Inc.

Friedman, T. L. 2005. *The World Is Flat: A Brief History of the Twenty-First Century*. New York: Farrar, Straus & Giroux.

Kamrani, A. K. and M. Azimi (Eds.). 2011. *Systems Engineering Tools and Methods*. Boca Raton, FL: CRC Press.

Nidiffer, K. E. 2007. Addressing the Software Engineering Challenges over the Years and into the Future. Data Analysis Center for Software, Software Engineering Institute, Carnegie Mellon University, *Software TECH News*, Vol. 10, pp. 15–21.

Appendix 1
Product Development Case Studies

OBJECTIVES

1. To understand phases, issues (customer, technical, and business), and management of a product development process
2. To understand automotive product design, assembly, and business issues by studying the available information on the Smart car.

METHOD

Part I

Read the Ingersoll Rand Cyclone Grinder development story from the following:Ingersoll Rand: Barbecues, Drag Tests, Medieval Warriors; and Slowing Down to Speed Things Up, the chapter on Cyclone grinder development story in *Liberation Management* by T. J. Peters. New York: Harper & Row, 1992.

1. Gather information about models of and parts in the Cyclone grinders by conducting an Internet search.
2. Answer the following questions:
   a. Develop a project activity and milestones chart showing all the key activities involved in the development of the Cyclone grinder (use format from *Systems Engineering and Analysis* by B. S. Blanchard and W. J. Fabrycky, 2011, p. 652).
   b. List customer needs of the grinder (as a customer/user would mention his/her words).
   c. Generate a list of functional requirements of the grinder (as an engineer would specify).
   d. Draw a product decomposition tree of the grinder showing its systems, subsystems, and components within each system.
   e. Describe key concepts/ideas that contributed to the successful design of the Cyclone grinder.

Part II

1. Read the following two articles on the Smart car:Siekman, P. 2002. The Smart Car is Looking More So. *Fortune*, April 15, 2002, pp. 310[I]–310[P]. Brooke, L. 2008. Little Car, Big Job. *Automotive Engineering (aei)*, February, 2008, pp. 76–77.
2. Also, review information on Smart car available on the following websites:
   http://www.smartcarofamerica.com/
   http://www.smartusa.com/
   http://en.wikipedia.org/wiki/Smart_(automobile)

http://www.ifm.eng.cam.ac.uk/ctm/idm/cases/smart.html
3. Answer the following questions:
   a. List customer needs of Smart car.
   b. Provide a comparison of product specifications and available features of Smart car with two other vehicles made by different vehicle manufacturers that can compete with it in the U.S. market.
   c. Describe important product design/development issues in creating the Smart car including safety-related changes.
   d. Describe key business issues in involving suppliers at the assembly location.

Appendix 2
Benchmarking, Quality Function Deployment, and Design Specifications

OBJECTIVES

1. To conduct benchmarking of competitor vehicles with the target vehicle for the project work
2. To understand development of product specifications through the application of the Quality Function Deployment (QFD) to the selected vehicle chunk (or system)
3. To develop specifications of the target vehicle.

PROCEDURE

1. Select a reference vehicle (for your target vehicle to be studied/developed for Projects #2 to #7). The reference vehicle must be preferably a recent model year vehicle (2011–2012). (Note: each student working in a team must select a target vehicle that has a different body style or brand but can still share many systems across the vehicle family.)
2. Select at least two other recent vehicles that will compete with your target vehicle in the same market segment.
3. Conduct a benchmarking exercise. Collect data on vehicle dimensions and features from Internet search and your own measurements, and take photographs of the vehicles and their chunks and systems for side-by-side comparisons. Provide a table comparing the reference vehicle with its comparators based on exterior and interior dimensions and characteristics of their corresponding systems and features.
4. Select one of the following automotive chunks (one chunk per student) of the reference vehicle for your QFD. Each student in the team must select a different chunk so that together your team will have information on several different chunks of the reference vehicle.
    a. Instrument Panel (layout, instrument cluster, center stack units [radio, climate controls, navigation system], glove box, passenger airbag, materials, etc.)
    b. Interior trim panel of the driver's door (trim panel layout, materials, attachment to the door frame, inside door release and grab handles, side glass movement mechanism, armrest, switches, map pocket, etc.)
    c. Center console (size, dimensions, storage space, cup holders, arm rest, lamps, powerpoint, coin tray, controls, etc.)
    d. Steering wheel and column (steering wheel spokes and hub design, air bag, switches mounted on the wheel, stalks, column and wheel adjustments, etc.)

e. Front suspension (control arms, springs, dampers, links, etc.)
   f. Rear end (bumper or fascia, trunk lid or liftgate, storage area, lamps, etc.)
5. Determine the selected chunk's customer needs (interview at least six customers to understand what they would like and dislike in the chunk).
6. Determine its functional specifications/requirements through discussions in your team (members from different functional areas will provide more complete information on design issues).
7. Develop relationship and correlation matrices (using the QFD symbols to convey strengths).
8. Estimate importance ratings of the customer requirements.
9. Evaluate your product chunk along with the same chunks in two other competitors' products.
10. Determine relative importance rating scores (last row).
11. Your written team report should include the following: (1) benchmarking comparison table with an additional column showing selected values (specifications) for the target vehicle and (2) QFD analysis including your completed QFD analysis charts, pictures/sketches of the selected chunk (showing parts/features), list of findings/observations, discussions on three to five most important functional specifications and other findings, and conclusions.

Appendix 3
Vehicle Systems Analyses: Requirements, Interfaces, Trade-Offs, and Verification

OBJECTIVES

1. To understand automotive systems, subsystems, interfaces between systems, and their requirements
2. To cascade system requirements into subsystem and component requirements and their verification testing issues
3. To understand coordination in system design tasks between different design and engineering activities and issues associated with trade-offs, packaging, and assembly.

METHOD

1. Select one of the following vehicle systems of your reference vehicle for this project:
    a. Climate control system
    b. Front suspension
    c. Vehicle braking system
    d. Vehicle safety system
    e. Powertrain system
2. Study the selected system in the selected vehicle and list all subsystems of the system.
3. Develop an interface diagram showing interfaces (links) between the subsystems of the selected system and other systems in the vehicle.
4. State at least three important engineering requirements on the selected system.
5. Develop an interface matrix including all of the subsystems (of the selected system) and other vehicle systems. Specify characteristics of each of the interfaces (e.g., type of interface: physical [P], spatial-packaging space [S], energy transfer [E], material flow [M], information flow [I], or none [0]) corresponding to each cell in the matrix.
6. Cascade system requirements (mentioned in item 4) to each of the subsystems. Using a tabular format, list the cascaded requirements on each subsystem, and briefly describe engineering tests that you will undertake to evaluate each of the subsystem requirements.

7. Describe at least three major trade-offs that you need to consider in designing the system to fit and work with other systems in the vehicle.
8. Prepare a report including all the above items ("1" to "7") and summarize your observations and insights gained from this project.

Appendix 4
Business Plan and Systems Engineering Management Plan for the Proposed Vehicle

OBJECTIVES
1. To provide an interim report on the progress of your project work
2. To present a business plan and systems engineering management plan for your proposed vehicle program.

CONTENTS OF THE REPORT
1. Business plan should include the following:
   a. Description and specifications of the proposed (target) vehicle, including vehicle features, options, and unique characteristics of vehicle systems
   b. Competitors (makes and models) of the proposed vehicle and comparisons of key dimensions
   c. Description of market segment
   d. Characteristics of anticipated customers
   e. Selling price and sales projections
   f. Timing plan and gateways
   g. Costs and revenue estimation table and plots of curves of life-cycle costs and revenues for the vehicle program
   h. Risks (major risks)
2. Systems engineering management plan should include the following:
   a. Customer needs
   b. Vehicle attributes and important subattributes
   c. Pugh diagram of vehicle attributes showing comparison of the proposed vehicle with reference vehicle (datum) and competitor vehicles
   d. Attribute requirements (vehicle level and system level)
   e. Systems engineering tasks, timings, and gateways
   f. Organizational needs (functional areas and team structure)
   g. Verification plan and tests
   h. Validation plan and vehicle evaluations
   i. Program status and open issues

3. Conclusions and discussions
    a. Summarize major accomplishments and findings. Describe why your vehicle with proposed characteristics will sell well.
    b. Discuss what worked well and what failed or did not get done to your satisfaction, and describe lessons learned and recommendations for future work.

Appendix 5
Conceptual Design of the Proposed Vehicle

OBJECTIVES

1. To search for additional information on trends in vehicle design, new technologies, and features to refine your vehicle definition
2. To illustrate vehicle configuration and preliminary packaging with key interior and exterior dimensions of the vehicle concept
3. To present a technology plan for selected technologies and features.

PROCEDURE

Here, you will assume that the business plan and the systems engineering plan that you submitted in Project #4 (Appendix 4) was accepted by the senior company management. Now, to kick off your concept design process, you will need to gather your team and describe to them your overall vehicle program, timings and milestones, tasks, and responsibilities of major teams, and key open issues.

In your next design team meeting, you should present an initial package drawing of the vehicle concept. This drawing will help visualize the overall vehicle size, package, and engineering issues and start the work in the design studios to create initial sketches, computer-aided design (CAD) models, and some exterior and interior surfaces. You also need to start a technology plan to define new features in the vehicle.

1. Prepare an initial package drawing containing side view and plan view (drawn to scale, either hand drawn or using a CAD application) of the vehicle with the following details:
    a. Overall vehicle envelope showing overall length, overall width, and overall height.
    b. Select a vehicle origin point and show x, y, and z axes for locations of major entities of the vehicle.
    c. Show locations of the four wheels by determining wheelbase, front and rear overhangs, and front and rear tread width.
    d. Show cowl and deck points, tire envelopes, engine envelope, firewall, back of rear seatback, vehicle floor and headliner height, gas pedal location, front and rear seating reference points, and center of the steering wheel.
    e. Show locations and envelopes of major entities such as gas tank, batteries, and drivetrain.

2. Prepare an initial technology plan using a tabular format. Your table should include all major vehicle systems as rows in its first column. The second column should describe major changes planned (one-line bullet points). The third column should describe major technological challenges. The fourth column should describe comments and key open issues such as possible modifications of existing hardware/software, make versus buy decision recommendations, possible suppliers, and the next suggested action.

To develop your technology plan, search for information on the latest advances and developments in the following areas:

a. Powertrain technologies to meet upcoming U.S. Environmental Protection Agency and the National Highway Traffic Safety Administration fuel economy and emissions requirements
b. Other fuel-saving technologies such as low-friction bearings, low rolling resistance tires, and power regenerative methods.
c. Applications of new lightweight and other recyclable automotive materials
d. Safety technologies for active and passive safety devices (e.g., driver warning systems, collision avoidance systems, and driver assistance systems)
e. Telematics devices (i.e., applications of information, communications, computers, wireless, and the global positioning system)
f. Automotive electronics (applications of microprocessors, sensors, actuators, and integrations of electronic control units [ECUs])
g. Electrical systems architecture (configuration of the electrical system)
h. Driver interface technologies (e.g., steering wheel mounted controls, touch screens, Bluetooth, programmable controls and displays, display technologies, and voice controls)
i. Vehicle lighting technologies (e.g., LED lamps, fiber optics, and smart headlamps).

Appendix 6
Vehicle Assembly Process Plan

OBJECTIVES
1. To describe vehicle assembly problems related to quality and costs
2. To present unique product design and assembly ideas or methods to improve product quality and assembly operations
3. To develop an assembly plan for your vehicle showing plant configuration and product flow through the plant.

CONTENTS OF THE REPORT
1. List at least 10 types of vehicle assembly-related problems that you need to consider to ensure that you will meet your quality and cost objectives.
2. Describe at least 10 product design or vehicle assembly ideas or methods that you will implement to improve costs and quality in building the vehicles.
3. Prepare a plan view drawing (or a sketch) showing the layout of your assembly plant floor and locations of major assembly operations/areas. Your floor plan should show sequence of assembly operations that you will need to assemble the vehicle.

Appendix 7
Term Project: Final Report

OBJECTIVES

1. Prepare a vehicle evaluation plan for the vehicle validation
2. Prepare a brochure for the vehicle
3. Conduct an engineering analysis to show superiority of the vehicle.

CONTENTS OF THE REPORT

1. Present your plan for validation of the vehicle. The plan should be presented in a tabular format as follows: (a) column 1: serial number of evaluation, (b) column 2: vehicle attribute, (c) column 3: key vehicle attribute requirements, (d) column 4: brief description of the evaluation test (what would be measured and how the measurements will be made), and (e) column 5: validation criteria (what values of the evaluation measures will constitute successful validation). The rows of the table should cover all vehicle attributes.
2. Prepare a brochure for the vehicle for prospective customers. The brochure should include the following information: (a) vehicle exterior and interior dimensions, (b) key selling points, standard and optional features/contents of the vehicle, and technical superiority-related considerations (e.g., major engineering accomplishments and comparisons with leading competitors showing why your vehicle is better than some of its key competitors), and (c) sketches and drawings to show capabilities of the vehicle.
3. Present an engineering analysis on one topic of your interest to show improvements (or capabilities of new features) incorporated in your vehicle. Some examples of areas of engineering analysis are (a) vehicle body design/construction, (b) power train capacity and fuel consumption, (c) electrical power load computations, (d) anthropometric analyses supporting interior dimensions, (e) decision analysis to support selection of vehicle characteristics (e.g., type of power train, weight reduction, and suspension selection). Your analysis report should include the following sections: Objectives, Background, Method, Results, and Conclusions and Recommendations.

Appendix 8
Calculations of Centerline and Control Limits for Control Charts

The following notation is used to define parameters used to compute centerline and control limits for different control charts.

NOTATIONS COMMON TO BOTH VARIABLES AND ATTRIBUTES CONTROL CHARTS

n = the sample size (number of items in a subgroup)
m = the number of samples (or subgroups) in the data set used to set up a control chart
i = an item number in a sample (i varies from 1 to n)
j = a subgroup number (j varies from 1 to m).

NOTATIONS FOR VARIABLES CONTROL CHARTS

x_{ij} = a measurement of ith item in jth subgroup. (x_{ij}s' are measurements of a product characteristic X [measured as a continuous variable], selected to monitor and improve the production process.)
$\bar{x}_j$ = a sample mean for the jth subgroup or sample.
$\bar{\bar{x}}$ = the grand mean [or mean of means] (over all values of j's).
R_j = a range of jth subgroup.
$\bar{R}$ = the mean range [mean of ranges (R_j)] (over all values of j's).
s_j = a standard deviation of jth subgroup.
$\bar{s}$ = the mean of standard deviations (over all values of j's).
R_{Mj} = a moving subgroup sample range for the jth artificial subgroup or sample.
$\bar{R}_M$ = the mean of moving range values (over all artificial subgroups).
$\hat{\sigma}$ = the estimate of population standard deviation of x based on sample measurements.

A_2, A_3, B_3, B_4, D_3, D_4, d_2, and c_4 are constants. Table A8.1 presents their values as functions of subgroup size n.

PROCESS CAPABILITY MEASUREMENTS

C_p = Process capability index that measures potential or inherent capability of the production process-based variable X. $C_p = \dfrac{(USL - LSL)}{(6\hat{\sigma})}$

TABLE A8.1
Values of Constants Used in Variables Control Chart Limits

Subgroup Size (n)	A2	A3	B3	B4	D3	D4	d2	c4
2	1.88	2.66	0.00	3.27	0.00	3.27	1.128	0.7979
3	1.02	1.95	0.00	2.57	0.00	2.57	1.693	0.8862
4	0.73	1.63	0.00	2.27	0.00	2.28	2.059	0.9213
5	0.58	1.43	0.00	2.09	0.00	2.11	2.326	0.9400
6	0.48	1.29	0.03	1.97	0.00	2.00	2.534	0.9515
7	0.42	1.18	0.12	1.88	0.08	1.92	2.704	0.9594
8	0.37	1.10	0.19	1.81	0.14	1.86	2.847	0.9650
9	0.34	1.03	0.24	1.76	0.18	1.82	2.970	0.9693
10	0.31	0.98	0.28	1.72	0.22	1.78	3.078	0.9727
15	0.22	0.79	0.43	1.57	0.35	1.65	3.472	0.9823
20	0.18	0.68	0.51	1.49	0.41	1.59	3.735	0.9869
30	*	0.55	0.60	1.40	*	*	4.086	0.9914
50	*	0.43	0.70	1.30	*	*	4.498	0.9949

* R-chart constants are given only up to n = 20; for larger subgroup sizes S chart should be used.

C_{pk} = Process capability index that measures actual or realized capability of the production process-based variable X. $C_{pk} = \min\left\{\dfrac{(USL - \bar{\bar{x}})}{3\hat{\sigma}}, \dfrac{(\bar{\bar{x}} - LSL)}{3\hat{\sigma}}\right\}$

USL = Upper specification limit on X
LSL = Lower specification limit on X

Note: (1) USL and LSL must be based on the customer perception of quality and product acceptability. (2) C_p and C_{pk} should be measured after the process is stabilized (i.e., process is improved by removing special or assignable causes that produce outlier points that fall outside the control limits [defined later] of the control charts).

NOTATIONS FOR ATTRIBUTES CONTROL CHARTS

n_j = number of samples (or items) in jth subgroup
P_j = the proportion of defective or nonconforming items in jth subgroup or sample (computed as a decimal value between 0.0 and 1.0)
$\bar{p}$ = the mean of the proportions of defective (or nonconforming) items (over all subgroups)
np_j = number of defectives or nonconforming items in jth subgroup or sample
$n\bar{p}$ = the mean number of defective (or nonconforming) items (over all subgroups)
u_j = number of defects (or nonconformities) in jth subgroup or sample
$\bar{u}$ = the mean number of defects (or nonconformities) in a sample (over all subgroups)
c = number of defects in a sample
$\bar{c}$ = mean number of defects in a sample (over all subgroups).

Appendix 8

The following mathematic expressions for computing upper and lower controls limits and centerlines can be obtained from Besterfield et al. (2003), NIST (2010), or Kolarik (1995).

3-SIGMA CONTROL UPPER AND LOWER LIMITS AND CENTERLINES

X-Bar and R Charts

$$\text{Upper control limit (UCL) of X-barchart} = \bar{\bar{x}} + A_2\bar{R}$$
$$\text{Lower control limit (LCL) of X-barchart} = \bar{\bar{x}} - A_2\bar{R}$$
$$\text{Centerline of X-barchart} = \bar{\bar{x}}$$
$$\text{UCL of R chart} = D_4\bar{R}$$
$$\text{LCL of R chart} = D_3\bar{R}$$
$$\text{Centerline of R chart} = \bar{R}$$
$$\hat{\sigma} = \frac{\bar{R}}{d_2}$$

X-Bar and S Charts

$$\text{UCL of X-bar chart} = \bar{\bar{x}} + A_3\bar{s}$$
$$\text{LCL of X-bar chart} = \bar{\bar{x}} - A_3\bar{s}$$
$$\text{Centerline of X-bar chart} = \bar{\bar{x}}$$
$$\text{UCL of S chart} = B_4\bar{s}$$
$$\text{LCL of S chart} = B_3\bar{s}$$
$$\text{Centerline of S chart} = \bar{s}$$
$$\hat{\sigma} = \frac{\bar{S}}{c_4}$$

X and R_M Charts

$$\text{UCL of X chart} = \bar{x} + \left(\frac{3\bar{R}_M}{d_2}\right)$$
$$\text{LCL of X chart} = \bar{x} - \left(\frac{3\bar{R}_M}{d_2}\right)$$
$$\text{Centerline of X-bar chart} = \bar{x}$$
$$\text{UCL of } R_M \text{ chart} = D_4\bar{R}_M$$
$$\text{LCL of } R_M \text{ chart} = D_3\bar{R}_M$$
$$\text{Centerline of } R_M \text{ chart} = \bar{R}_M$$

P Chart (For Constant Subgroup Size of n)

$$\text{UCL of P chart} = \bar{p} + 3\sqrt{\left[\frac{\bar{p}(1-\bar{p})}{n}\right]}$$

$$\text{LCL of P chart} = \bar{p} - 3\sqrt{\left[\frac{\bar{p}(1-\bar{p})}{n}\right]}$$

Centerline of P chart = $\bar{p}$

P CHART (FOR VARIABLE SUBGROUP SIZE OF N_j)

$$\text{UCL of P chart} = \bar{p} + 3\sqrt{\left[\frac{\bar{p}(1-\bar{p})}{n_j}\right]}$$

$$\text{LCL of P chart} = \bar{p} - 3\sqrt{\left[\frac{\bar{p}(1-\bar{p})}{n_j}\right]}$$

Centerline of P chart = $\bar{p}$

NP CHART

$$\text{UCL of NP chart} = n\bar{p} + 3\sqrt{n\bar{p}(1-\bar{p})}$$

$$\text{LCL of NP chart} = n\bar{p} - 3\sqrt{n\bar{p}(1-\bar{p})}$$

Centerline of NP chart = $n\bar{p}$

U CHART (FOR SUBGROUP SIZE OF N_j)

$$\text{UCL of U chart} = \bar{u} + 3\sqrt{\frac{\bar{u}}{n_j}}$$

$$\text{LCL of U chart} = \bar{u} - 3\sqrt{\frac{\bar{u}}{n_j}}$$

Centerline of U chart = $\bar{u}$

Note: For constant sample size, $n_j = n$ (for all j's)

C CHART (FOR CONSTANT SUBGROUP SIZE OF N)

$$\text{UCL of C chart} = \bar{c} + 3\sqrt{\bar{c}}$$

$$\text{LCL of C chart} = \bar{c} - 3\sqrt{\bar{c}}$$

Centerline of C chart = $\bar{c}$

REFERENCES

Besterfield, D. H., Besterfield-Michna, C., Besterfield, G. H., and M. Besterfield-Scare. 2003. *Total Quality Management.* Third Edition. ISBN 0-13-099306-9. Upper Saddle River, NJ: Prentice Hall.

Kolarik, W. J. 1995. *Creating Quality—Concepts, Systems, Strategies, and Tools.* New York, NY: McGraw-Hill, Inc.

National Institute of Standards and Technology (NIST). 2010. *NIST/SEMATECH e-Handbook of Statistical Methods (Engineering Statistics Handbook).* U.S. Department of Commerce. Website: http://www.itl.nist.gov/div898/handbook/ (accessed August 30, 2012).

Index

0–9

3-point direction magnitude scale, 146, 148–149
3-sigma control upper and lower limits, 331

A

Accident, 230–237
    analysis methods, 237
    causation theories, 231–233
    definition, 230
    vs. hazard, 237
Accident-based safety performance measures, 234–236
    advantages and disadvantages, 236
    currently used, 234
    incident rates, 235
Accident data analyses, 238, 390–392
    method, 237
Accident data collection, flow of, 391
Accident data reporting thresholds, 391
Accident investigation method, 391
Accident prevention costs, 181
Accident prone theory, 231
Act of God theory of accident, 231
Affinity diagram, 341
Air induction system, 110
Allocation of requirement, 114
Analytical hierarchical method, 58
AND gate, 384
    event rule, 384
Anthropometric characteristics, 357–358
Anthropometric human models, 358
Appraisal costs, 180
Arrow diagram, 348–349
Aspiration level, principle of, 55
Assembly costs, 180
Assembly engineer's recommendations to component designers, 307
Assembly evaluations, 503
Assembly line configuration, 505
Attributes control charts, 336
Attributes of product, *see* Product attributes
Automated guided vehicle (AGV), 478
Automated storage and retrieval system (ASRS), 478
Automobile—An Integrated System, 519
Automotive concept, Pugh analysis of, 277
Automotive control evaluation checklist, 359
Automotive fuel system interfaces, 108
Automotive interior evaluation, ergonomic scorecard for, 359
Automotive powertrain system, 437–450
    attributes, 440
    decomposition tree for, 440
    interfaces, 440
    steps and objectives, 437
    systems, subsystems and sub-subsystems, 438
    vehicle attributes requirements to, 444
Automotive product, 108, 183
    attributes, 83, 87–92
    development, 19, 83
    program cash flow, 183
    relating attribute structure to, 86
Avoidability, degree of, 230

B

Baldrige Quality Award, 200
Before vs after studies, 396
Behavioral human performance measurement, 218
Behavioral measures of user performance, 218
Behavioral sampling, 394
Behavioral variables, 369
Benchmarking, 267
    vs. breakthrough, 267–277
    motorcycle clutch controls, 275
    motorcycle dimensions, 269–273
    motorcycle headlamps, 274
    smart car, 4
    study of steering wheels, 456–458
Biomechanical characteristics, 212
Biomechanical human models, 358
Boeing 787 Dreamliner design, 474–476
Boeing 777 product development, problems during, 472–473
Boolean algebra, application of, 389
Boolean algebraic equations, 384
Boolean variables, 384
Boothroyd et al, assembly evaluation methods, 310, 317
Breakthrough approach, benchmarking vs., 267–277
Business plan, 294

C

CAD, *see* Computer-aided design
CAE, *see* Computer-aided engineering
CAE applications, *see* Computer-assisted engineering applications
CAM, *see* Computer-aided manufacturing
Carbon fiber fuselage, 474
Cascading quality function deployment, 287

543

Cascading requirements, 84
    from product level to component level, 34
Cause and effect (C-E) diagram, 323
C chart, 448
C-E diagram, *see* Cause and effect diagram
Centerline calculations, 331, 537
Chain of multiple events of accident, 231
Checklists., 359
    for automotive control evaluation, 359
    hazards, 381
CHESS model, *see* Comprehensive Headlamp Environment Systems Simulation model
CIT, *see* Critical incident technique Climate control system requirements
Clustering of matrix data, 115
CMM, *see* Coordinate measurement machines Commission errors
Communication methods, 143
Competitor products, 268, 277
Complex product, 4, 9
    attributes, 107
    designing, 9
    development tools, 514
    managing, 15, 26
Components, 85
    commonality, 516
    development, subsystems and, 33
    of grinder, 464
    and requirements, 77, 444, 463
    sharing of, 179
Comprehensive Headlamp Environment Systems Simulation (CHESS) model, 365
Computation methodology, problems with RPN, 382
Computer-aided design (CAD), 516
Computer-aided engineering (CAE), 45, 516
Computer-assisted engineering (CAE) applications, 516
Computer-assisted technologies, 299
    in product design, 45
Computer modeling methods, 299
Concept design, 12
Concept selection in product design, 12
Conceptual phase, product phases, 12
Configuration management plan, 168
Consensus standards, 94
Continuous improvement principle, 199
Control charts, 331–336, 537
    attributes, 336
    description, 332
    flow diagram of process improvement using, 333
    purpose, 331
    variables, 333
Control limits, calculations of, 537

Control loops, 28
Control operational errors, 215
Cooling system, engine sub-subsystem, 439
Cost-based measures of quality
Cost-benefit analysis, 175, 396, 485–512
Cost management, 175
Costs, 175–182
    due to accidents, 181
    effect of time on, 182
    manufacturing, 180
    safety, 181
    total life cycle, 182
    types, 175
CPM, *see* Critical path method
CPSC, *see* U.S. Consumer Products Safety Commission
"Creating Quality", 193
Critical incident technique (CIT), 393
Critical path method (CPM), 162
Critical to customer satisfaction (CTS)
CTS, *see* Critical to customer satisfaction
Cumulative cash flow curve, 184
Customer focused approach of SE, 24
Customer focus, quality management principles, 198
Customer needs, 6
Customer requirements, 79
    for grinder, 463
Customers, 5
Customer satisfaction, 193
Cyclone grinder development, 462
    components, 464
    customer requirements, 463
    project activities and milestone chart, 463
    risk management, 467
    success of grinder design, 468
Cylinder head, engine sub-subsystem, 438

D

DFD and DFA guidelines, 250
DMAIC process, *see* Design, Analyze, Measure, Improve, and Control process
Databases on human characteristics and capabilities, 357
Data collection methods, 141
Data gathering methods, classification of, 143
Data management plan, 48
Data presentation methods, 257, 262
Datum, 277
Decision evaluation matrix, 53
Decision making, 51–74
    product design, 41–74
    in product programs, *see* Product programs, decision making in
Decomposition process, 26

Index

Decomposition tree, 26
    for automotive powertrain system, 440
    of motorcycle into systems, 452
    of product, 26
Defective die-cast car bodies, 337
Defects in painted car body, check sheet for, 328
Defendant, 223
Descriptive research of human factors, 209
Descriptors, 127
    10-point scales with, 129, 130
Design, Measure, Analyze. Improve, and Control (DMAIC) process, 263
Designed-to-conform requirements *vs.* manufactured-to-conform requirements, 81
Design for disassembly, 249
Design for environment, 247–248
Design for manufacturing and assembly., 303–320
Design for Six Sigma (DFSS), 200
    IDOV process, 264
Design standards *vs.* performance standards, 94
Design structure matrix (DSM) approach, 116
Design synthesis, 31
Desired direction of functional specification, 281–282
Detailed design phase, product phases, 121
Detailed Engineering Design during product development, 121
Detection error, 215
DFSS, *see* Design for Six Sigma
Digital human models, *see* Manikin models
Direction magnitude scales, 145, 148–149
Discrimination error, 215
DMAIC (Define, Measure, Analyze, Improve and Control), 263–264
Domino theory of accident, 231
Door trim defects, checklist for, 327
Drivetrain sub-subsystems, 438
DSM approach, *see* Design structure matrix approach
Durability tests, 140

E

Effective performance measures, characteristics of, 218
Energy exchange model of accident, 231
Epidemiological model of accident, 231
Equivalency-based measures of user performance, 218
Ergonomically designed products, characteristics of, 211
Ergonomics, 516
    scorecard for automotive interior evaluation, 359
Error, definition of, 215

Error-free results, effective safety performance measures, 218
Evaluation of steering wheels, 458
Evaluations of product, *see* Product evaluations
Evaluative research of human factors, 209
Experimental research of human factors, 209
Experiment design, 349–354
    description, 349
    examples, 350
    multivariate, 352
    Taguchi experiments, 354
    Taguchi's product robustness and quadratic costs, 353
    Taguchi's three-step product design approach, 353
Expressed warranty, principle of, 230

F

Factual approach to decision making, quality management principles, 199
Failure modes and effects analysis (FMEA), 289–293
    example, 290
    failure mode, 289
    rating scale, 290
Fasteners, 250, 307
Fault tree analysis (FTA), 384
    Boolean algebra, application of, 384
    fault tree development rules, 387
    AND gate, 385
    OR gate, 385
    purpose, 384
Fault tree development rules, 287
Federal agencies, standard setting process in, 82, 97
Federal Motor Vehicle Safety Standard, 82, 95
Federal standards, 97
Fish diagram, *see* Cause and effect (C-E) diagram
Fixed *vs.* variable costs, 179
Flexible assembly line for laptop computers, 477
FMEA, *see* Failure modes and effects analysis
Forgetting errors, *see* Omission errors
FTA, *see* Fault tree analysis
Functional analysis, 31
    and allocation, 31
Functional interface, 100
Functional requirements, 79
    for grinder, 463

G

Gantt charts, 162
Gates, 385
Gateways or milestones, 18–20
Generic product development process, phases of, 40

Global design teams, coordination of, 515
Good requirement, characteristics of, 78
Grouping causes of headlamp misaim, 341

H

Hazard, 237
 accident *vs.*, 237
 analysis methods, 379
Hazard analysis, 238, 379
 checklists, 381
"Head event", 384
Headlamp misaim, 341
 grouping causes of, 341
 understanding causation of, 341
Histogram, 328
Horizontal *vs.* vertical standards, 94
House of quality, 280
Human anthropometric dimensions, 212
Human biomechanical characteristics, 212
Human body dimensions, measuring, 212
Human capabilities, factors affecting, 212, 214
Human errors, 214–215
"Human factored" equipment, 211
Human factors check lists, 359
Human factors engineering, 206
 approach, 206
 importance of, 211
 methods in, 220
 research studies, 209
 responsibilities in designing complex
  products, 210
Human factors engineering tools, 220
 anthropometric and biomechanical human
  models, 358
 checklist, 359
 laboratory, simulator and field studies, 366
 task analysis, 362–365
Human factors guidelines, applications of, 221
Human factors knowledge, application of, 221
Human factors methods, 220, 357–375
Human factors requirements, 81
Human factors research studies, 209
Human information processing capabilities, 213
Human information processing model, 213
Human interface, 216
Human-machine interface (HMI), 216
Human operator workload, measurement methods,
  370–373
Human performance evaluation models, 365
Human performance measurement methods, 217,
  367
 physiological, 370
 range of, 367
 secondary task, 373
 subjective assessments, 371
 types and categories of, 218, 368

Human performance models, 220, 365
Human vision models, 365
Hurwicz principle, 55

I

Identify, Design, Optimize, and Verify (IDOV)
  process, 264
Implied warranty, principle of, 239
Inadequate response error, 216
Incident rates, accident-based, 234–235
Information processing capabilities, 213
Ingersoll Rand Cyclone Grinder development,
  462–468
Ingersoll Rand pneumatic grinder, 462
Insurance costs, 175
Intake system, engine sub-subsystem, 438
Integration of tools in applications, 263
Intention, degree of, 230
Interaction matrix, 104
Interest/inflation, effect of, 182
Interface control plan, 170
Interface diagram, 104
 examples, 106
Interface matrix, 104
 clustering and sequencing of, 115
 examples, 106
 of motorcycle systems, 451
 and N-squared diagram, 104
 of powertrain system, 440
Interfaces, 99–120, 440
 automotive fuel system, 108
 automotive powertrain system, 440
 control, establishment of
 laptop computer, 106
 requirements, 102
 types, 100
 visualizing, 103
Interior trim panel, 327
Internal customers, 5
Internal failure costs, 180
International Organization for Standards (ISO)
  9000, 8, 199
Interpretational errors, 215
Interpretation error, 215
ISO 14001 Standard for managing environmental
  programs, 248
ISO 9000, *see* International Organization for
  Standards 9000
Iterative process, 35
IVIS DEMAnD model, 365

J

JEDI Model applications, 493
"Job stopper", 166
Juran's trilogy, 194

Index

K

Kano model, 196
  of quality, 195
Knowledge management, 160

L

Laplace principle, 55
Laptop computers, 75, 106
  flexible assembly line for, 477
  functions, systems, and components of, 27
  interfaces, 106
Leadership, quality management principles, 198–200
Legibility errors, 215
Liability, defined, 239
Life cycle analysis, 423–433
Life cycle cost analysis, 242, 246, 429, 512
Lower levels attribute requirements, cascading, 84
Lubrication system, transmission sub-subsystem, 439

M

Magnetic interface, 100
Maintainability, 396
Make vs. buy decisions, 177
Malcolm Baldridge award criteria, 200
Manageable levels, product divided into, 85
Management process of SE, 15, 26, 34
Mandatory vs. voluntary standards, 94
Manikin models, 358
Manual assembly time estimating methods, 312–316
Manufactured-to-conform requirements, designed-to-conform vs., 94
Manufacturing and assembly costs, 307
Manufacturing costs, 307
Manufacturing organization, processes in, 303
Manufacturing processes, 303
Market price-minus profit approach, 189
Material transfer interface, 100
Matrix data, 346
  analysis, 346
  attributes and systems in motorcycle, 453
Matrix diagram, 345
Maximax principle, 55
Maximin principle, 55
Maximum expected value principle, 53
Mechanical interface, 100
Mental effort load scale, SWAT, 371
Methods Time Measurement (MTM), 310
Milestones, 18–20
Model-based Systems Engineering, 296
Motorcycle systems, 451–453
  interface matrix of, 455

MSRP, see Manufacturer suggested retail price
MTM, see Methods Time Measurement
Multidisciplinary approach of SE, 26

N

NASA projects interface control process for, 43, 73, 169
NASA Systems Engineering Handbook, 40
National Aeronautics and Space Administration Task Load Index (NASA TLX), 371
Negligence principle, 239
Noise, vibrations, and harshness (NVH) performance, 458
Nonaccident measurement, interview and observational techniques for, 393
Nonaccident safety performance measures, 236
Nonrecurring costs, 175
Normalized preference values of products, 58
Normalized weights of products, computation of, 58

O

Objective measures, 144
  and data analysis methods, 138, 150
Observable human responses of human performance measurement methods, 143
Observation methods, 143
Occurrence, duration of, 368
Occurrence sampling technique, 394
Omission errors, 215
Operational phase, product phases, 38
Operator performance measurement methods, 217
Operator performance models, 365, 370
Operator's subjective responses of human performance measurement methods, 371
Optical interface, 100
Ordinal scale, 234
OR gate, 384
  event rules, 384
Orthogonal arrays, 353
Overhead costs, 180

P

Paired comparison-based scales, 151
Parameter design, 353–354
Pareto chart, 321
  of customer complaints, 322
Pareto principle, 321
People involvement, quality management principles, 194
Performance attribute, 91, 124
Performance requirements, 80

Performance standards, design standards *vs.*, 94
PERT, *see* Program (or project) evaluation and review technique
Physical capabilities of human, 212
Physical interface, 100
Physical mock-ups, 141, 150, 154
Physiological measurement
    of human performance, 370
Plaintiff, 239
Plan for validation, vehicle, 154
Plotting methods, 257
Power conversion, 430
    engine sub-subsystem, 438
    transmission sub-subsystem, 439
Present value, 182, 417
Prevention costs, 181
Problem-solving approaches, decision making in product programs, 52
Process, 6
Process capability measurements, 537–538
Process decision program chart (PDPC), 347
Product financial plan, 183
Product attributes, 75–98
    development, 77
    of exterior lighting system, 87
    importance of, 76
    of motorcycle requirements, 76
Product concept, Pugh analysis of, 277
Product conceptual design, 12–14
Product design, 8
    defects, 240
    description, 4
    safety engineering, *see* Safety engineering
    security considerations in, 242
Product development, 12
    case studies, 23
    early decisions importance during, 64, 73
    flow diagram, 14
    phases, 13
    processes in, 12
    risks, 68
    SE processes in verification and validations test during, 152–154
Product development process, 12
Product development tools, 267–301
    benchmarking, 268
    breakthrough, 267, 276
    QFD, *see* Quality function deployment
Product discontinuation, 182
Product evaluations, 137, 152
    communication methods tools in, 143
    issues, 138
    methods, 141
    types, 140
Product families, 11
Production phase, product phases, 14
Production process, 8
Product liability, 239
Product life cycle
    considerations of SE, 423
Product life cycle phases, 424
    methods, 423
Product manufacturing defects, 240
Product pricing approaches, 188–189
Product programs, life cycle phases of, 424
Product programs, decision making in, 53
    alternatives, outcomes, payoffs, and risks, 53
    alternatives selected by principles, 55
    maximum expected value principle, 53
    weighted total score for concepts election, 277–278
Product psychophysics, 374
Product quality measurements, 195
Product realization processes, 8
    NASA defining, 8
Product requirements, 76–82
    external factors affecting, 93
    internal factors affecting, 93
Product safety and liability, 239
    product defects, 240
    product litigations, terms and principles used in, 239
    warnings, 240
Product validation, 138, 152
Product verification, 138, 152
Program Cost Flow by Months, 183
Program (or project) evaluation and review technique (PERT), 162
Program management, 161
    complexity in, 171
    vs. project management, 159
Program planning and management, 159, 162
Program status chart, 172
Program timing charts, 162
Project financial plan, 183
Project management, 161
    challenges in, 172
    program *vs.*, 159
    software systems, 166
    timings, 162
Project planning, 159, 162
    development, 160
    steps in, 162
    tools used in, 162
Proposed vehicle
    business plan and SEMP for, 166
    conceptual design of, 531
Pugh chart, 277
Pugh diagram, 277
    for product concept selection, 458

Index

Q

QFD, *see* Quality function deployment
QMS, *see* Quality Management System
Quality
   definition of, 193
   initiatives, 198
   tools, 262, 321–356
Quality costs, 180
Quality Engineering, methods in, 321
Quality function deployment (QFD), 280
   advantages and disadvantages, 287
   cascading, 287
   description, 280
   structure, 280
Quality management, 193
   key concepts in, 194
   principles, 199–200
   tools used in, 201
Quality management system (QMS), 199
   standards, 199
Quality-related costs, 180
   types of, 175
Quality tools affinity diagram, 341
   arrow diagrams (networks), 348
   matrix data analysis, 346
   matrix diagram, 345
   PDPC, 347
   relations diagram, 338
   systematic diagram, 341
Quantifiable, effective safety performance measures, 234

R

Range charts, 333
Rating scales, 145, 151
   for detection and occurrence, 290–291
   direction magnitude and acceptance, 145–149
   for severity, 290
Ratio scale, 151
R chart, 333
Recovered error, 215
Recurring costs, 176
Recursion process, 28
Recycling and material recovery, 251–253
Reduce product development time, systematic diagram for, 343
"Red-Yellow-Green" charts, 296
Relations diagram, 338
Relationship matrix, 280
Reliability definitions of, 396
   designing for, 400
   effective safety performance measures engineer's tasks, 404

   of hybrid system, 400
   improvements, approaches for, 403
   of parallel system, 399
   requirements, 400
   of series system, 397
Requirements, 75–83
   allocation and analysis, 82
   analysis, 30
   attribute, 77
   developing, 77
   for exterior lighting system, 87–89
   factors affecting, 93
   loop, 35
   role of standards in setting, 94
   for subattributes, 87
   types, 79–81
Response variables, 350
Reversal errors, 215
Ring model of desirability, 208
Risk analysis, 69–72
   flow diagram of product failures and, 67
Risk assessment phase, 67, 69
Risk identification phase, 67, 69
Risk management plan, 65
Risk matrix, 70
Risk priority number (RPN), 70
Risks, 65–72
   measurements problems, 72
   product development, 65
   product uses, 69
Robotic assembly, 317–320
"Robust" design, 353
"Robust" products, 353
RPN, *see* Risk priority number

S

Safety
   analysis methods, 370–405
   costs, 181, 241
   countermeasures, 227
Safety analysis methodologies, 237
   accident analysis methods, 237
   accident *vs.* hazard, 237
   hazard analysis methods, 238
Safety engineering, 225–242
   3Es of, 228
   Approach, 227
   definition, 225
   historic background, 229
   in product design, 225–242
   importance and need of, 227
   methods in, 228
   safety problems, 226
Safety engineering tools, 379–405
   accident data analysis tools, 390–392

hazard identification and risk reduction tools, 379–382
reliability analyses, 396–404
safety performance monitoring, evaluation, and control, 393
systems safety analysis tools, 382
Safety performance measures, 234, 393
accident-based, *see* Accident-based safety performance measures
description, 234
nonaccident, 236
Scatter diagram, 329
standing height *vs.* sitting height, 330
S chart, 333
Schedule risk, 294
Scorecards, human factors, 359
SE, *see* Systems engineering
Secondary task performance measurement, 373
Security requirements, 242
SEMP, *see* Systems engineering management plan
Sequential error, 215
Simple product, 4
Simulation, 366
Simulator driving, evaluation in, 366, 458
Six Sigma approaches, 200, 263
Six-Sigma improvement projects, DMAIC process, 263
Six-Sigma methodologies, 200
Smart car, 468–472
Smart car product design, 472
business and supply chain issues, 472
comparison with small cars, 471
customer needs, 469
development issues, 472
"Smiley faces" chart, 220
Software interface, 100
Specifications for electric car, 480
Standards in setting requirements, 296
Standards, product design, 296
Steering wheels, 456
benchmarking and evaluation of, 456
Strict liability, principle of, 239
Subjective measurements, 143, 145
and data analysis methods, 145
paired comparison-based methods, 151
rating on scale, 145
10-point ratings data analysis, 150
Subjective workload assessment technique (SWAT), 370
Subjective workload measurement techniques, 371
Substitution errors, 215
Subsystems
of automotive exterior lighting system, 87–89
development, 38
Sustainability, 245–253
SWAT, *see* Subjective workload assessment technique

Systematic diagram, 341
Systems, 10
definitions, 9
work with other systems, 10–11
Systems engineering (SE), 23–49
advantages and disadvantages, 47
cascading requirements, 34
challenges in, 513
fundamentals, 23
importance, 46
loops in, 28
major tasks in, 30
managing, 26
management plan, 166
NASA description, 40
process., 27
in product development, 27
subsystems and components development, 10
systems engineers, role, 42
teaching, 518
"V" model, 38
Systems engineering management plan (SEMP), 166–171
Systems engineering tools applications, 87, 91
automotive powertrain system, 91
Systems engineers
future for, 517
role, 43
Systems, relating attribute structure to vehicle exterior lighting system, 86
vehicle suspension systems, 87
Systems safety analysis tools, 382
FMEA, 382
FTA, 384–390

T

Tabular formatted methods, 262
Taguchi experiments, features of, 354
Task analysis, human factors engineering tools, 362–365
Task decision gates, WBS, 165
Task deliverables, WBS, 165
Teaching systems engineering, 518
Technical process of SE, 40, 42
Technical risk, 65
Technology plan, 169
Testing, 137
involving human subjects, 138, 143
verification and validation, 137
TGRs, *see* Things gone right
TGWs, *see* Things gone wrong
Things gone right (TGRs), 195
Things gone wrong (TGWs), 195
Thurstone's Method of Paired Comparisons, 151
Time management, 172
Total quality management (TQM), 198

Index

cause and effect diagram, 323
C-E process diagram, 325
check sheet, 316
control charts, 331–336
histogram, 328–329
Pareto chart, 321–322
scatter diagram, 329–330
stratification, 330–331
Total weighted score for concept selection, 62

U

U chart, 331
Unexpectedness, degree of, 230
Unrecovered error, 215
Unsafe acts theory of accident, 233

V

Validation, 32
    loop, 27–28
    plan, 154
    tests, 154
Validity, effective safety performance measures, 234–236
Variable costs, *see* Recurring costs
Variables control charts, 333
Vehicle assembly process
    plan, 533
Verification, 32
    loop, 27–28
    tests, 153
Verification plan, 153
Visual performance prediction models, 366
"V" model, 38, 128

W

Warning, degree of, 230
WBS, *see* Work breakdown structure
Weighted total score for concept selection, 62
Work breakdown structure (WBS), 165
Workload profile (WP) method, 372
Work sampling, *see* Behavioral sampling
WP method, *see* Workload profile method

X

X-bar, 334
    chart, 335

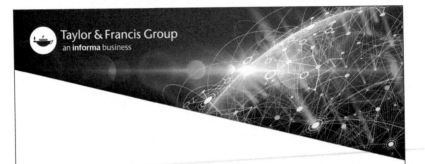